GROUND WATER MANUAL

A WATER RESOURCES TECHNICAL PUBLICATION

A guide for the investigation,
development, and management of
ground-water resources

FIRST PRINTING 1977
SECOND PRINTING 1981
THIRD PRINTING 1985
SECOND EDITION 1995

U.S. Department of the Interior
Bureau of Reclamation

As the Nation's principal conservation agency, the Department of the Interior has responsibility for most of our nationally owned public lands and natural resources. This includes fostering sound use of our land and water resources; protecting our fish, wildlife, and biological diversity; preserving the environmental and cultural values of our national parks and historical places; and providing for the enjoyment of life through outdoor recreation. The Department assesses our energy and mineral resources and works to ensure that their development is in the best interests of all our people by encouraging stewardship and citizen participation in their care. The Department also has a major responsibility for American Indian reservation communities and for people who live in island territories under U.S. Administration.

U.S. Government Printing Office

For sale by the Superintendent of Documents, U.S. Government Printing Office, Washington, DC 20402.

ISBN 978-1-78039-355-1

PREFACE

This manual has been prepared as a guide to field personnel in the more practical aspects and commonly encountered problems of ground-water investigations, development, and management.

Information is presented concerning such aspects as ground-water occurrence and movement, well-aquifer relationships, ground-water investigations, aquifer test analyses, estimating aquifer yield, data collection, and geophysical investigations. In addition, permeability tests, well design, dewatering systems, well specifications and drilling, well sterilization, pumps, and other aspects have been discussed. An extensive bibliography has also been included.

The manual has been developed over a period of years, and its many contributors have diversified technical backgrounds. Contributors include personnel from the Bureau of Reclamation Engineering and Research Center (now Technical Service Center) and field offices, other agencies, foreign governments, and many individual scientists and engineers.

Principal Bureau of Reclamation contributors include W.T. Moody, R.E. Glover, R.W. Ribbens, D. Jarvis, C.N. Zangar, H.H. Ham, W.A. Pennington, T.P. Ahrens, D. Wantland, H.R. McDonald, L.A. Johnson, A.C. Barlow, W.N. Tapp, C.R. Maierhofer, R.J. Winter, Jr., W.E. Foote, and R.D. Mohr. All references to their works are included in the bibliographies. The works of non-Bureau of Reclamation authors including C.V. Theis, M.I. Rorabaugh, W.C. Walton, C.E. Jacob, R.W. Stallman, M.S. Hantush, S.W. Lohman, F.G. Driscoll, and other scientists and engineers have also been cited.

The second edition of the Ground Water Manual has been re-organized with the objective to make the material more accessible to the occasional user. No material has been removed unless it was obviously obsolete. New material has been added to stay abreast with modern technology. Also, metric units have been added except where: (1) actual field data is reproduced, (2) units given in examples are incidental to the concept, and (3) exact dimensions are critical to the user and industry has not retooled to metric. In such cases, nominal metric units may be given in parentheses. Approximate dimensions are converted using the approximate conversion table located in the appendix.

Principal contributors to the second edition include L.V. Block, R.P. Burnett A.J. Cunningham, K.D. Didricksen, J.L. Hamilton, J.E. Lacey, P.J. Matuska, N.W. Prince, T.D. Pruitt, R.A. Rappmund, C.R. Reeves, G.D. Sanders, S.J. Shadix, R. Bianchi, W.R. Talbot, and D.E. Watt.

CONTENTS

Chapter II. Planning Ground-Water Investigations and Presentation of Results

Chapter III. Initial Operations and Aquifer Yield Estimates

Page

Chapter IV. Geophysical Investigations

Chapter V. Definitions and Theory of Saturated Ground-Water Flow and Factors Affecting Ground-Water Flow

Chapter VI. Well and Aquifer Relationships

Page

Chapter VII. Artificial Recharge, Artificial Storage and Recovery, and Subsidence

Chapter VIII. Pumping Tests to Determine Aquifer Characteristics

Chapter IX. Analysis of Discharging Well and Other Test Data

Page

Chapter X. Permeability Tests in Individual Drill Holes and Wells

Page

Page

Chapter XII. Water Well Drilling and Development

Chapter XIII. Infiltration Galleries and Horizontal Wells

Page

Chapter XIV. Dewatering Systems

Chapter XV. Water Well Pumps

Chapter XVI. Well and Pump Costs, Operation and Maintenance, and Rehabilitation

TABLES

TABLES

TABLES

FIGURES

FIGURES

FIGURES

FIGURES

FIGURES

FIGURES

FIGURES

FIGURES

FIGURES

FIGURES

GLOSSARY

Alluvial
Pertaining to or composed of alluvium or deposited by a stream or running water.

Aquiclude
A term for a saturated, but poorly permeable bed, formation, or group of formations that does not yield water freely to a well or spring.

Aquifer
A formation, group of formations, or part of a formation that contains sufficient saturated permeable material to yield economical quantities of water to wells or springs.

Aquifuge
A material or rock which contains no interconnected openings or interstices and therefore neither absorbs nor transmits fluids.

Aquitard
A term for a geologic bed, formation, group of formations, or part of a formation with relatively very low permeabilities through which virtually no water moves. Commonly referred to as a confining unit.

Coefficient
A number, constant for a given substance, used as a multiplier in measuring the change in some property of the substance under given conditions.

Collector pipe
A pipe or system of piping used to intercept and redirect surface or subsurface flows.

Confined aquifer
An aquifer bounded above and below by impermeable or distinctly lower permeability beds.

Conjunctive
Connected or joined together; serving to connect or join together.

Consolidated material
Firm coherent rock.

Drilling mud

Any substance mixed with drilling water to increase the viscosity of the water.

Exponential

Of or relating to an exponent; involving a variable or unknown quantity as an exponent.

Gravel pack

Also called a filter pack; smooth, clean, uniform, well-rounded, siliceous sand or gravel that is placed in the annulus of a well between the borehole wall and the well screen to prevent formation material from entering the screen.

Infiltration gallery

One or more horizontal screens placed adjacent to or beneath a water body in permeable alluvial materials.

Invert

The elevation of the flow line (lowest point on the inside) of a pipe.

Leakance

The rate of flow across a unit (horizontal) area of a semipervious layer into (or out of) an aquifer under one unit of head difference across this layer. The leakance equals the vertical hydraulic conductivity divided by the thickness of the semipervious layer.

Lens

A body of material that is thick in the middle and thin at the edges.

ln

Log to base e.

Perched aquifer

Unconfined ground water separated from an underlying main body of ground water by an unsaturated zone.

Perennial stream

A stream that flows year round and from the source to the mouth.

Permeable materials

A material with the property or capacity for transmitting a fluid.

Screen
Also called a well screen; a filtering device used as the intake section of a water well to keep sediment from entering a water well. Usually constructed of casing with slots cut into it, or of specially constructed, continuous slot, wire-wrapped screens.

Sediment-colloid
Extremely small solid particles, 0.0001 to 1 micron in size, which will not settle out of a solution; intermediate between a true dissolved particle and a suspended particle which will settle out of solution.

Sink
An area where ground water evaporates or is otherwise removed from the hydrologic system.

Stickup
The height of the measuring point of a well above natural ground, usually the top of the casing.

Subsurface
Underground; zone below the surface whose geologic features are interpreted on the basis of drill records and various kinds of geophysical evidence.

Subsurface drain
A drain installed to enhance subsurface drainage for the removal or control of ground water and the removal or control of soil salts.

Sump
A hole or pit which serves for the collection of fluids.

Storativity
Also called the coefficient of storage; a measure of the volume of water an aquifer releases from or takes into storage per unit surface area of the aquifer per unit change in head.

Time yield curve
A curve showing the change in yield of a well over time.

Unconfined aquifer
Also free aquifer; an aquifer having a water-table which is at atmospheric pressure at the water surface.

Unconsolidated material

Earth materials which are not firm coherent rock.

Winters Doctrine

A 1908 court decision concerning Native American reserved water rights.

Yield

The measured or estimated volume of water discharged from a well or released from an aquifer.

Chapter I

GROUND-WATER OCCURRENCE, PROPERTIES, AND CONTROLS

1-1. Introduction.—Ground-water engineering is the art and science of investigating, developing, and managing ground water for the benefit of man. The technology involves specialized fields of oil science, hydraulics, hydrology, drainage, geophysics, geology, mathematics, agronomy, metallurgy, bacteriology, and electrical, mechanical, and chemical engineering. The ever-increasing demand for water will make ground-water engineering increasingly important.

In addition to the solution of ground-water recovery problems for water supply, ground-water engineering is important in problems concerning seepage from surface reservoirs and canals, the effects of bank storage, stability of slopes, recharging of ground-water reservoirs, controlling of saltwater intrusion, dewatering of excavations, subsurface drainage, and construction, land subsidence, waste disposal, and contamination control.

Ground-water engineering involves the determination of aquifer properties and characteristics and the application of hydraulic principles to ground-water behavior for the solution of engineering problems. Determination of aquifer characteristics and the application of those data by appropriate mathematical and other methods are essential to the solution of complex problems in which ground water is a factor. The extent to which the determination of aquifer properties and characteristics must be made depends upon the complexity of the problem involved. The required investigation may range from cursory to detailed. It may entail study or consideration of all or only one or two aquifer properties and hydraulic principles. Conditions often may be so complex as to preclude the determination of finite values and the application of available theory to the solution of some problems. Such circumstances require an understanding of the hydrologic cycle developed in the latter part of the 17th century. During the 18th century, fundamentals in geology were established that provided a basis for understanding the occurrence and movement of ground water (Todd). The French engineer Henry Darcy (1803-58) studied the movement of water through sand and developed the fundamental law of ground-water flow which was largely subjective. The reliability of these principles depends upon the experience and judgment of the ground-water technical specialist.

1-2. History of Use.—The first use of ground water as a source of supply is lost in antiquity. Ancient man obtained water from springs, but hand-dug wells were widely used in the earliest of Biblical times, and the ancient Chinese are generally regarded as the inventors of drilled and cased wells (McWhorter and Sunada, 1984). For centuries (Tolman, 1957; U.S. Department of Agriculture, 1956) ground-water use was limited by developmental difficulties and by the absence of a clear understanding of its origin and occurrence.

Shallow, hand-dug wells and crude water-lifting devices marked the early exploitation of ground water. The introduction of well-drilling machinery and motor-driven pumps allowed the recovery of ground water at increased depths. Expanded knowledge of ground-water hydrology and other sciences added to man's ability to understand and use this resource.

As technology has improved, the benefits of ground-water development have become increasingly important. The use of water for domestic purposes (human and animal consumption) usually has the highest priority, followed by individual requirements and then agricultural usage (irrigation). Development of the ground-water resources of the United States has been increasing in recent years as development of surface-water sources approaches the point of full potential.

1-3. Origin.—

(a) The Hydrologic Cycle.—Precipitation, storage, runoff, and evaporation of the earth's water follow an unending sequence known as the hydrologic cycle (Meinzer, 1949; Todd, 1980; U.S. Department of Agriculture, 1956). During this cycle, the total amount of water in the atmosphere and in or on the earth remains the same; however, its form may change. Although minor quantities of magmatic water or water from other deep-seated sources may find its way to the surface, all water is assumed to be part of the hydrologic cycle.

The movement of water within the hydrologic cycle is shown on figure 1-1. Water vapor in the atmosphere is condensed into ice crystals or water droplets that fall to the earth as rain or snow. A portion evaporates and returns to the atmosphere. Another portion flows across the ground surface until it reaches a stream and then flows to the ocean. The remaining portion infiltrates directly into the ground and seeps downward. Some of this portion may be

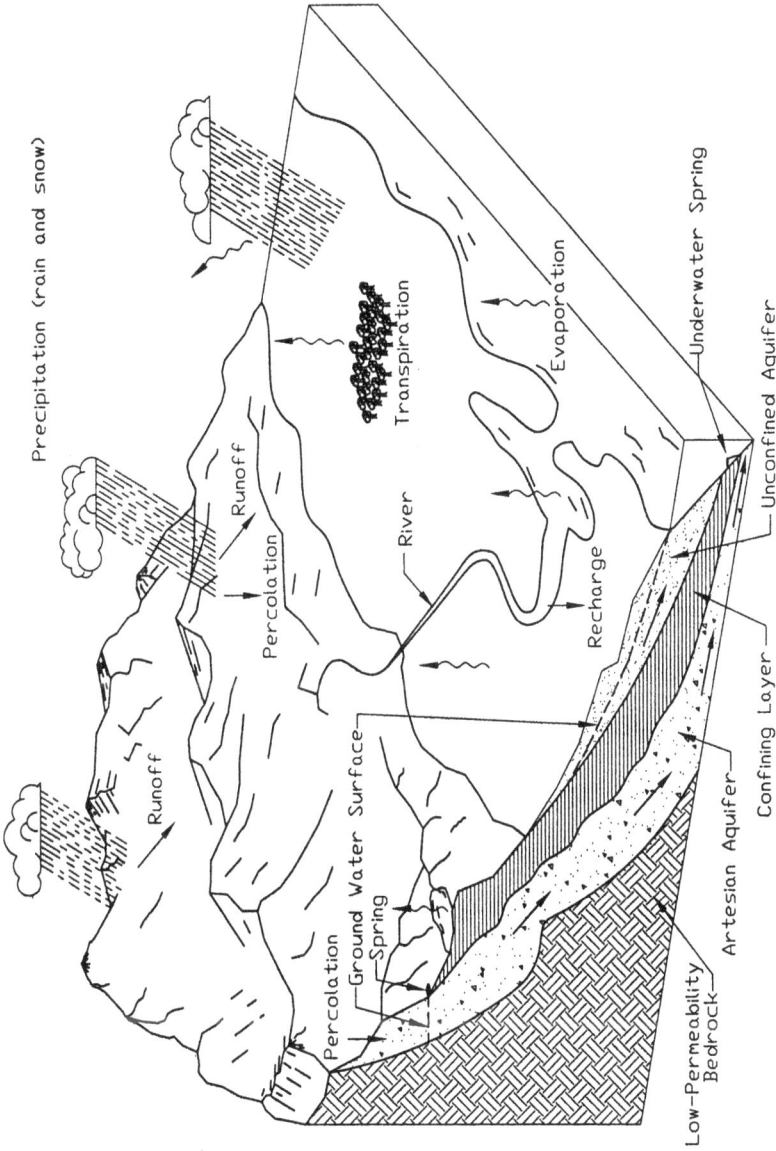

Figure 1-1.—The hydrologic cycle.

transpired by the roots of plants or moved back to the ground surface by capillarity and evaporated. The remainder seeps downward to join the ground-water body.

Ground water returns to the ground surface through springs and seepage to streams where it is subject to evaporation or is directly evaporated from the ground surface or transpired by vegetation. The water vapor rises into the atmosphere and the cycle continues.

The elements of the hydrologic cycle for any area can be quantified in an equation. For ground-water investigations, the equation can be expressed in terms of ground-water components. However, it may be necessary to evaluate the broad hydrologic picture to quantify the ground-water components. Determination of components in the ground-water equation is tedious and time consuming, and the results, at best, are only approximate. Therefore, an analysis should be made to determine that such an evaluation is necessary and justified before it is undertaken.

(b) Ground-Water Equation.—A basic ground-water equation (Meinzer, 1949; Todd, 1980), which will permit an approach to a quantitative estimate of ground-water availability, can be established for an area to account for those factors of the hydrologic cycle that directly affect flow and storage of ground water. The equation can be stated as:

$$\Delta S_{gw} = recharge-discharge \qquad\qquad 1\text{-}1$$

where ΔS_{gw} is the change in ground-water storage during the period of study. Theoretically, under natural conditions and over a long period of time, which includes both wet and dry cycles, ΔS_{gw} will be zero and inflow (recharge) will equal outflow (discharge). However, man's activities can significantly affect the equation, resulting in long-term increases or decreases in ground-water storage.

The natural recharge to the ground-water body includes deep percolation from precipitation, seepage from streams and lakes, and subsurface underflow. Artificial recharge includes deep percolation from irrigation and water spreading, seepage from canals and reservoirs, and recharge from recharge wells. The natural discharge or outflow from the ground-water body consists of seepage to streams, flow from springs, subsurface underflow, transpiration, and evaporation. Artificial discharge occurs by wells or drains. If ground-water storage in an area is less at the end of the selected period of time than at the beginning, discharge is indicated as having exceeded recharge. Conversely, recharge may exceed discharge.

(c) *Recharge to and Discharge from Aquifers.*—Recharge from natural sources includes the following:

- *Deep percolation from precipitation.*—Deep percolation of precipitation is one of the most important sources of ground-water recharge. The amount of recharge in a particular area is influenced by vegetative cover, topography, nature of soils, as well as the type, intensity, and frequency of precipitation.

- *Seepage from streams and lakes.*—Seepage from streams, lakes, and other water bodies is another important source of recharge. In humid and subhumid areas where ground-water levels may be high, the influence of seepage may be limited in extent and may be seasonal. However, in regions where the entire flow of streams may be lost to an aquifer, seepage may be of major significance.

- *Underflow from another aquifer.*—An aquifer may be recharged by underflow from a nearby, hydraulically connected aquifer. The amount of this recharge depends on the head differential, the nature of the connection, and the hydraulic properties of aquifers.

- *Artificial recharge.*—Artificial recharge to the ground water may be achieved through planned systems, or may be unforeseen or unintentional. Planned major contributions to the ground-water reservoir may be made through spreading grounds, infiltration ponds, and recharge wells. Irrigation applications, sewage effluent spreading grounds, septic tank seepage fields, and other activities have a similar, but usually unintentional effect. Seepage from reservoirs, canals, drainage ditches, ponds, and similar water impounding and conveyance structures may serve as local sources of major ground-water recharge. Recharge from such sources can completely change the ground-water regimen over a considerable area.

(d) Ground-Water Discharge.—Losses from the ground-water reservoir occur in the following four ways:

- Seepage to streams.—In certain reaches of streams and in certain seasons of the year, ground water may discharge into streams and maintain their baseflows. This condition is more prevalent in humid areas than in semiarid areas.

- Flow from springs and seeps.—Springs and seeps exist where the water table intersects the land surface or a confined aquifer outlets to the surface.

- Evaporation and transpiration.—Ground water may be lost by evaporation if the water table is near enough to the land surface to maintain flow by capillary rise. Also, plants may transpire ground water from the capillary fringe or the saturated zone.

- Artificial discharge.—Wells and drains are imposed artificial withdrawals on ground-water storage and in some areas are responsible for the major depletion.

Ground water moves in response to a hydraulic gradient in the same manner as water flowing in an open channel or pipe. However, the flow of ground water is appreciably restricted by friction with the porous medium through which it flows. This friction results in low velocities and high head losses compared to open channel or pipe flow.

1-4. Occurrence of Ground Water.—

(a) General.—Webster defines an aquifer as "a water-bearing bed or stratum of earth, gravel, or porous stone." Some strata are good aquifers, whereas others are poor. The most important requirement is that the stratum must have interconnected openings or pores through which water can move. The nature of each aquifer depends on the material of which it is composed, its origin, the relationship of the constituent grains or particles and associated surface, its exposure to a recharge source, and other factors.

Ground water occurs in almost all types of unconsolidated and consolidated section 1-2 sedimentary material and, to a lesser extent, in fractured igneous and metamorphic rocks. The potential of aquifers depends not only on lithology, but also on stratigraphy and geologic structure. In general, coarse-grained sediments, whether unconsolidated or consolidated, are the best aquifers.

(b) Sedimentary Material and Rocks.—In general, the best aquifers are the coarse-grained, saturated portions of the unconsolidated, granular sedimentary material (Hen, 1959) which covers the consolidated rocks over much of the surface of the earth. Widespread presence of unconsolidated sediments is more common at lower elevations in proximity to streams. These sediments consist of stream alluvium, glacial outwash, wind-deposited sand, alluvial fans, and similar water- or wind-deposited, coarse-grained, granular materials. In addition, some residual materials resulting from the in-place weathering of consolidated rock are good aquifers.

The coarser-grained, consolidated sedimentary rocks, such as conglomerates and sandstones, are often good aquifers, but consolidated sedimentary rocks are usually found below the granular sedimentary deposits. Their value as aquifers depends to a large extent on the degree of cementation and fracturing to which they have been subjected. Sandstones may have both primary (between grain) and secondary (fracture) permeability. In many cases, particularly where the sandstones are well indurated, secondary permeability contributes the majority of the yield. Some massive sedimentary rocks such as limestone, dolomite, and gypsum may also be good aquifers. These rocks are relatively soluble and, over the years, solution along fractures or partings may form voids which range in size from several millimeters to several hundred meters (a fraction of an inch to several hundred feet). Some of the best known and most productive aquifers are cavernous limestones.

(c) Igneous and Metamorphic Rocks.—The value of igneous and metamorphic rocks as aquifers depends greatly on the amount of stress and weathering to which they have been subjected after their initial formation. In general, the igneous rocks are very poor aquifers if they remain undisturbed. However, mechanical and other stresses cause fractures and faults in these rocks in which ground water may occur. Such openings may range from hairline cracks to voids several centimeters wide. In general, these openings disappear with depth and do not yield significant quantities of water below depths of about 300 meters (1,000 feet). Also, although initial flows from fractures may be quite high, such high yields generally decrease with time.

In coarse-grained igneous rocks, where in-place weathering has occurred, a thin permeable zone may be found in the transition zone between the unweathered rock and the thoroughly weathered. Some lavas, especially those of viscous basaltic composition, may contain good to excellent aquifers in the zones between successive flows. The scoriaceous upper and lower surfaces of flows are usually porous and permeable, and cooling fractures may be present in a zone extending into the flow from the upper and lower surfaces. Furthermore, coarse-grained sedimentary material may also be present between flows.

   (1) Unconfined Aquifers.—An unconfined aquifer (figure 1-2) does not have an overlying confining layer. It is often referred to as a free or "water table" aquifer or as being under "water-table" conditions. Water infiltrating into the ground surface percolates downward through air-filled interstices of the material above the saturated zone and joins the ground-water body. The water table, or upper surface of the saturated ground-water body, is in direct contact with the atmosphere through the open pores of the material above and is in balance with atmospheric pressure at all points. Movement of the ground water is in direct response to gravity.

   (2) Confined Aquifers.—A confined or artesian aquifer (figure 1-2) has an overlying, confining layer of lower permeability than the aquifer and has only an indirect or distant connection with the atmosphere. Water in an artesian aquifer is under pressure and when the aquifer is penetrated by a tightly cased well or piezometer, the water will rise above the bottom of the confining bed to an elevation at which it is in balance with the atmospheric pressure and that reflects the pressure in the aquifer at the point of penetration. If this elevation is greater than that of the land surface at the well, water will flow from the well. The imaginary surface, conforming to the elevations to which water will rise in wells penetrating an artesian aquifer, is known as the potentiometric or piezometric surface. The confining bed may be almost completely impermeable, or it may permit some flow. Types of confining beds include:

- Aquiclude.—A saturated but relatively impermeable material that does not yield appreciable quantities of water to wells; clay is an example.

- Aquifuge.—A relatively impermeable formation that neither contains nor transmits water; solid granite is an example.

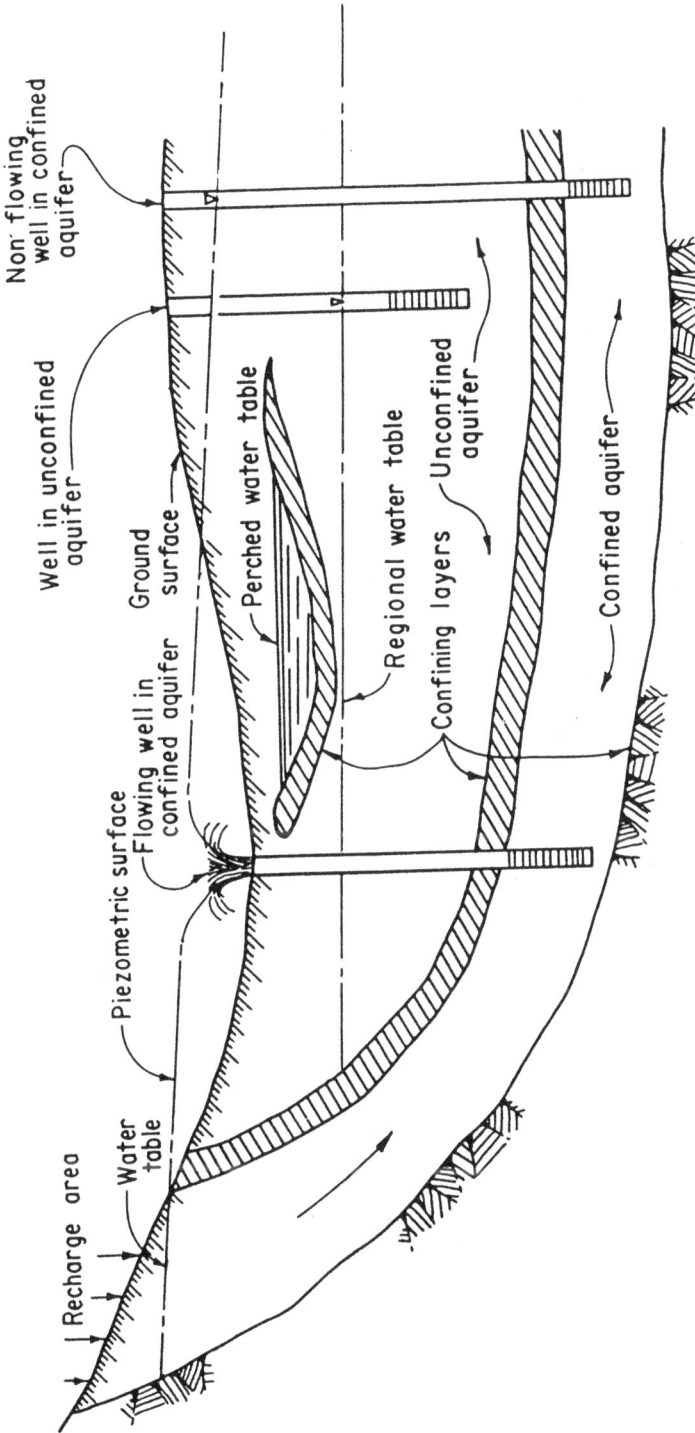

Figure 1-2.—Type of aquifers.

• Acquitard.—A saturated but poorly permeable stratum that impedes ground-water movement and does not yield water freely to wells but that may transmit appreciable water to or from adjacent aquifers and, if sufficiently thick, may serve as an important ground-water storage zone; sandy clay is an example (Todd, 1980).

(3) Perched Aquifers.—Beds of clay or silt, unfractured consolidated rock, or other material with relatively lower permeability than the surrounding materials may be present in some areas above the regional water table. Downward percolating water may be intercepted and a saturated zone of limited areal extent may be formed. This process results in a perched aquifer with a perched water table (Meinzer, 1949). An unsaturated zone is present between the bottom of the perching bed and the regional water table. A perched aquifer is a special case of an unconfined aquifer. Depending on climatic conditions or overlying land use, a perched water table may be a permanent phenomenon or may be seasonally intermittent (figure 1-2).

(d) Zones of Moisture.—Water may occur in several recognizable subsurface zones under different conditions, as shown in table 1-1, which was adapted from Meinzer (1949).

Table 1-1.—Status of water in various soil zones

Zone	Horizon	Condition of water	Condition of soil
Aeration (above water table)	Soil water	Under tension	Unsaturated
	Intermediate	Under tension	Unsaturated
	Capillary fringe	Under tension	Saturated and unsaturated
Saturation (below water table)	Unconfined ground water	Under pressure but upper surface at atmospheric pressure	Saturated
	Confined or artisan ground water	Under pressure but upper surface above atmospheric pressure	Saturated

The thickness of each zone above the zone of rock flowage varies according to the area and with time. During a period of recharge, the zone of saturation thickens at the expense of the zone of aeration. When discharge exceeds recharge, the zone of saturation thins and the zone of aeration thickens. During periods of recharge, a temporary downward migrating saturated lens may move through the zone of aeration.

The foregoing comments refer to ground water in temperate and tropical areas. However, in the colder areas of the northern and southern hemispheres, permafrost or permanently frozen ground may extend to considerable depths and influence ground-water conditions. The engineering problems associated with such conditions may be unusual and are not considered in this manual.

1-5. Ground-Water Quality.—

(a) General.—Precipitation usually contains minute amounts of silica and other minerals, and dissolved gases such as carbon dioxide, sulphur dioxide, nitrogen, and oxygen, which are present in the air and become entrained as droplets, form and fall. As a result, the pH value of most precipitation is below 7.0 (acidic) and the water is slightly corrosive. Upon reaching the earth's surface, the rainfall may pick up organic acids from humus and similar materials which increase its corrosive characteristics. While the acidic water is percolating through soil and rock, minerals may be attacked and dissolved, forming salts which are taken into solution. The relative concentrations and variety of the salts depend upon the initial chemical composition of the water; the mineralogy exposed and the weathered state of the rock and soil encountered; and the temperature, pressure, and duration of contact.

Nearly all elements may be present in ground water, and its mineral content varies from aquifer to aquifer and from place to place within an aquifer. The most commonly encountered elements and compounds are listed in table 1-2.

Table 1-2.—Chemical constituents
commonly found in ground water

Cations	Anions
Calcium, Ca	Bicarbonate, HCO_3
Magnesium, Mg	Sulphate, SO_4
Sodium, Na	Chloride, Cl
Potassium, K	Nitrate, NO_2
Iron, Fe	Fluoride, F
	Silica, SiO_2

Less commonly encountered constituents which are nevertheless important because of their known beneficial or detrimental effects on the use of water are: arsenic (As), barium (Ba), boron (B), cadmium (Cd), carbon dioxide (CO_2), copper (Cu), hydrogen sulfide (H_2S), lead (Pb), manganese (Mn), mercury (Hg), methane (CH_4), oxygen (0_2), selenium (Se), trihalomethanes (THM), various radionuclides, volatile organic compounds (VOC's), and polychlorinated biphenyls (PCB's). Many of these latter contaminants are manmade recent additions to ground water.

(b) Acceptable Limits for Chemical Constituents in Water.— Standards for ground water are currently determined by water use or aquifer classification. The Environmental Protection Agency (EPA) currently designates aquifers (450 CFR Part 149) as sole source aquifers (SSA), which carry special project review criteria for Federal actions possibly affecting designated aquifers. Ground water withdrawn for public drinking water supplies currently falls under the Safe Drinking Water Act (SDWA) (Public Law 93-523) regulations, as amended and reauthorized (1974). Current (July 1993) maximum contaminant levels (MCL's), or National Primary Drinking Water Standards, which are human health based, and Secondary Standards, governing aesthetic qualities, are shown in tables 1-3a and 1-3b. Note that the number of regulated constituents and their respective MCL's are frequently updated. Also note that standards for aquatic plant and animal life may be considerably more stringent. Obtain a current copy of the regulations (40 CFR Part 141), which includes monitoring requirements, or call the Safe Drinking Water Hotline (1-800-426-4791) for current MCL's. Noncompliance is covered in 40 CFR 142, "National primary drinking water regulations implementation."

Water discharged by municipalities, corporations and other entities identified as point sources, and certain other nonpoint sources, which may well become ground water, is regulated by the Clean Water Act (Public Law 95-217) (U.S. Geological Survey [USGS], 1977) as amended. Section 402 of the Act deals with the National Pollution Discharge Elimination System (NPDES) permitting program; Section 404 deals with dredge and fill permits; Section 319 covers nonpoint source pollution; and Section 320 concerns the National Estuary Program. Regulations promulgated under the Act, including the NPDES permits and oil spill criteria, are found in 40 CFR 109, 110, 112, 113, 114, 121, 122, 125, 129, 130, 131, and 133. The Coastal Zone Act Reauthorization Amendments (CZARA) of 1990, Section 6217, fill the gaps and complement existing nonpoint source pollution

Table 1-3a.—Safe Drinking Water Act Standards (July 1993)

Parameter	Primary MCL[1] (milligrams per liter [mg/L], unless otherwise noted)	Secondary MCL[2] (mg/L, unless otherwise noted)
Inorganics/Esthetics		
Aluminum		0.05 to 0.2
Arsenic	0.05	
Antimony	0.006	
Asbestos	7MF/L >10µm	
Barium	2	
Beryllium	0.004	
Cadmium	0.005	
Chloride		250
Chromium	0.1	
Color		15 color units
Copper	1.3 AL[3]	1.0
Corrosivity		noncorrosive
Cyanide	0.2	
Fluoride	4.0	2.0
Foaming agents		0.5
Iron		0.3
Lead	0.015 AL	
Manganese		0.05
Mercury	0.002	
Nickel	0.1	
Nitrate as N	10	
Nitrite as N	1	
Nitrate+Nitrite as N	10	
Odor		3 threshold odor Nos.
pH		6.5 - 8.5
Selenium	0.05	
Silver		0.1
Sulphate		250
Thallium	0.002	
TDS		500
Turbidity	0.5 - 1.0 NTU	
Zinc		5

[1] See 40 CFR 141 G for applicable water-supply systems.
[2] See 40 CFR 143 for applicable water-supply systems.
[3] AL = action level.

Table 1-3b.—Safe Drinking Water Act Standards (July 1993)

Parameter	MCL in mg/L (unless otherwise noted)
Organics - Pesticides, PCB's, Herbicides	
Adipates (diethylhexyl) - synthetic	0.5
Alachlor	0.002
Aldicarb	0.003
Aldicarb sulfone	0.002
Aldicarb sulfoxide	0.004
Atrazine	0.003
Carbofuran	0.04
Chlordane	0.002
2,4-D	0.07
Dalapon	0.2
Di[2-ethylhexyl]adipate - synthetic	0.4
Dibromochloropropane (DBCP)	0.0002
Dinoseb	0.007
Diquat	0.02
Endothall	0.1
Endrin	0.002
Ethylene dibromide (EDB)	0.00005
Glyphosate	0.7
Heptachlor	0.0004
Heptachlor epoxide	0.0002
Lindane	0.0002
Methoxychlor	0.04
Oxamyl (Vydate)	0.2
Pentachlorophenol	0.001
Picloram	0.5
Polychlorinated biphenyls (PCB) - synthetic	0.0005
Simazine	0.004
2,3,7,8-TCDD (Dioxin) - synthetic	0.00000003
Toxaphene	0.003
2,4,5-TP (Silvex)	0.05

Table 1-3b.—Safe Drinking Water Act Standards (July 1993) - continued

Parameter	MCL in mg/L (unless otherwise noted)
Organics - Volatile (VOC's)	
Benz(a)anthracene (PAH)	0.0001
Benzene	0.005
Benzo(a)pyrene (PAH)	0.0002
Benzo(b)fluoranthene (PAH)	0.0002
Benzo(k)fluoranthene (PAH)	0.0002
Butyl benzyl phthalate (PAE)	0.1
Carbon tetrachloride	0.005
Chrysene (PAH)	0.0002
Dibenz(a,h)anthracene (PAH)	0.0003
Dichlorobenzene o-	0.6
Dichlorobenzene m-	0.6
Dichlorobenzene p-	0.075
Dichloroethane (1,2-)	0.005
Dichloroethylene (1,1-)	0.007
Dichloroethylene (cis-1,2-)	0.07
Dichloroethylene (trans-1,2-)	0.1
Dichloromethane	0.005
Dichloropropane (1,2-)	0.005
Diethylhexyl phthalate (PAE)	0.006
Ethylbenzene	0.7
Hexachlorobenzene - synthetic	0.001
Hexachlorocyclopentadiene (HEX) - synthetic	0.05
Monochlorobenzene	0.1
Styrene	0.1
Tetrachloroethylene (PCE)	0.005
Toluene	1.0
Trichlorobenzene (1,2,4-)	0.07
Trichloroethane (1,1,1-)	0.2
Trichloroethane (1,1,2-)	0.005
Trichloroethylene (TCE)	0.005
Vinyl chloride	0.002
Xylenes	10.0
Organics - Chlorination Disinfection Byproducts (THM's)	
Bromodichloromethane	0.1
Bromoform	0.1
Chlorodibromomethane	0.1
Chloroform	0.1

Table 1-3b.—Safe Drinking Water Act Standards (July 1993) - continued

Parameter	MCL in mg/L (unless otherwise noted)
Microbiology	
Giardia lamblia	TT[1]
Legionella	TT
Standard plate count	TT
Total coliforms	[2]
Viruses	TT
Radionuclides	
Beta particle and photon activity	4 mrem/y[3]
Gross alpha particle activity	15 pCi/L[3]
Radium 226	20 pCi/L[3]
Radium 228	20 pCi/L[3]
Radon	300 pCi/L[3]
Uranium	20 µg/L[3]

[1] TT = treatment technique. Disinfection or filtration is required to deactivate or remove.

[2] ≤1 positive sample per month for systems collecting less than 40 samples per month, ≤5 percent positive for systems collecting over 40 samples per month.

[3] Proposed.

regulations for coastal States in five major source categories: (1) urban, construction, highways, airports/bridges, and septic systems; (2) agriculture; (3) forestry; (4) marinas and recreational boating; and (5) hydromodification and wetlands.

Chemical constituents in drinking and wastewater are to be determined according to standard test methods. Those methods are specified in 4 CFR 141 C for drinking water and 40 CFR 136 for wastewater.

The following chemicals have been listed for action, but no primary MCL has been set: aluminum, boron, chlorate, chlorite, manganese, molybdenum, strontium, vanadium, zinc, and zinc chloride (as of December 1994).

The following chemical and organisms are listed, but primary MCL's have not been set: bromacil, bromobenzene, bromochloroacetonitrile, chloroethane, chloromethane, chloropicrin, chlorotoluene o-, chlorotoluene p-, cyanogen chloride, DCPA (Dacthal), dibromoacetonitrile, dibromomethane, dicamba, dichloroacetaldehyde, dichloroacetonitrile, dichlorodifluoromethane, dichloroethane (1-1-), dichloropropane (1,3-) and (2,2-), dichloropropene (1,1-), dinitrotoluene (2,4-) and (2,6-), ETU, fluorotrichloromethane, hexachloroethane, isophorone, methomyl, methyl tert butyl ether, metolachlor, metribuzin, monochloroacetic acid, prometon, 2,4,5-T, tetrachloroethane (1,1,1,2-) and (1,1,2,2-), trichloroacetonitrile, trichloroethanol (2,2,2-), trichlorophenol (2,4,6-), trichloropropane (1,2,3-), trifluralin, and cryptosporidium (as of December 1994).

The determination of required water quality for irrigation purposes is a complex process. Many factors, such as soil, drainage, climate, and crop, must be considered. The U.S. Department of Agriculture Handbook No. 60 (1954) has been the standard guide on the acceptability of certain waters for irrigation and on the relationship of the chemicals in water to soils. This handbook also contains laboratory procedures for the analysis of irrigation water. More recent references include *Water Quality for Agriculture* (Ayers and Westcot, 1985) and *Irrigation Induced Water Quality Problems* (National Research Council, 1989) also published by the U.S. Department of Agriculture.

Water quality is also important because of its influence on the operating efficiency and life of equipment and materials, including pumps, well screens, and piping. Acidic water is usually corrosive, whereas alkaline water (pH >7) forms deposits more readily. Hard (alkaline) water may form deposits if it contains large amounts of sulphate, bicarbonate, and chloride radicals. Entrained gases, such as hydrogen sulfide, carbon dioxide, methane, nitrogen, and oxygen may cause corrosion and cavitation damage. Care should be taken to specify materials compatible with water quality for minimum operation and maintenance costs.

(c) Contamination and Pollution.—Contaminated or polluted water contains organisms and/or substances which make it unfit for an intended purpose. Ground water may become contaminated from traditional sources like septic tanks or their associated leach fields, garbage dumps or landfills, or improper manufacturing waste disposal activities (all of which are on the decline) or through natural processes. Other sources of pollution include improperly sealed wells, mining activities (including radioactive ores), aviation and military activities, oil field brine injection wells, and unlined or leaking industrial waste evaporation ponds. On the

increase are potential ground-water contamination activities
involving illegal dumping of regulated hazardous wastes and
illegal disposal of wastes down constructed or abandoned wells.

The horizontal and vertical distance from the source a
contaminant may migrate depends upon the contaminant, its
introductory path, the local and regional soil and geological
character and structure, and the local and regional hydrology.
Microbiological organisms and disinfectant byproducts, i.e.,
trihalomethanes (THM's), do not appear to have much subsurface
viability or mobility. Other organic and inorganic constituents
may be very persistent and highly mobile under favorable
conditions. Two newly recognized classes of chemicals, dense
nonaqueous phase liquids (DNAPL's) and light nonaqueous phase
liquids (LNAPL's) are proving to be persistent and difficult to
detect and remove once introduced into the subsurface hydrology.
Accordingly, no ground water should be assumed suitable for an
intended use without chemical analysis verification.

(d) Other Uses of Water Quality Data.—A study of the difference
and changes in the chemical content of water may be useful in
determining the source or sources of recharge, direction of flow,
and presence of boundaries (Hem, 1959; Todd, 1980). The age of
water determined by tritium content, carbon dating, and similar
analyses may be useful in estimating time in the ground, recharge
conditions, or paleohydrology. Current uses of ground-water
quality data include determining proposed inject water
compatibility, contaminant plume vector, THM fate and transport,
and differentiating native ground water from inject water for water
banking accounting.

1-6. Ground- and Surface-Water Relationships.—

(a) Humid Area Relationships.—Ground water in humid areas
maintains the baseflow of streams by seepage into stream
channels. However, the headwater reaches of some streams may
be above the water table, and therefore are dry during seasons of
low precipitation. In such reaches, seepage from the streambed
may charge an underlying aquifer. Consequently, some reaches of
a stream may be replenished by ground water and others may lose
water to the ground-water reservoir.

(b) Arid Area Relationships.—In many arid drainage basins, the
perennial master streams receive seepage from the ground-water
reservoir; whereas other streams may be above the water table and

streamflow occurs only during periods of high surface runoff. Where the water table is below the streambed, practically all the streamflow may be lost by seepage to the ground-water reservoir. Beneath many streambeds, considerable underflow may be present in the channel fill although the channel is dry.

In semiarid to arid areas, where irrigation is usually practiced, water losses from canals and deep percolation from irrigation applications frequently alter natural ground-water conditions. Such alterations include water-table rise and waterlogging and salination of soils. Artificial drainage by open or buried pipe drains, wells, or other means is often required to lower the water table, maintain a salt balance, and permit the continued production of crops.

(c) Artificial Ground-Water Recharge.—In recent years, much interest has developed in recharging ground-water reservoirs with excess surface water (Rima et al., 1971; Signor et al., 1970). The purposes for artificial recharge include: (1) ground-water (well field) management, (2) reduction of land subsidence, (3) renovation of wastewater, (4) improvement of ground-water quality, (5) storage of stream water during periods of high or excessive flow, (6) reduction of floodflows, (7) well yield increase, (8) decrease the size of the areas needed for water-supply systems, (9) reduction of saltwater intrusion or leakage of mineralized water, (10) increase streamflow, (11) store fresh water derived from rain and snowmelt, and (12) secondary recovery of oil (Pettyjohn, 1981). In addition, pollutants such as oil field brines and toxic and radioactive industrial wastes are often disposed of by storing them in deep, isolated, nonpotable aquifers; however, the injection wells are generally referred to as disposal wells rather than artificial recharge wells.

Artificial recharge can be accomplished by surface spreading or by injection well. The choice of a particular method is governed by the local topographic, geologic, and soil conditions; the quality of the water to be recharged, water use, land value, water quality, and climate (Todd, 1980). In general, recharge water must be potable to prevent potential bacterial or chemical contamination of the aquifer. Although recharge generally uses excess surface water, increasing use is being made of tertiary-treated sewage effluent. This usage is particularly the case where sewage effluent would otherwise be transported to the ocean for final disposal.

(1) Surface Spreading.—Surface spreading facilities can be constructed by excavating or installing low dams or berms. Initial construction costs for surface spreading facilities are generally less expensive than for wells, but maintenance costs can be high. In addition, land costs can be excessive in developed areas. High land costs can sometimes be overcome by installation of facilities in the riverbed where flooding potential precludes other development. Inflatable rubber dams can be used to retain flood or other flows. These dams can be deflated to permit passage of initial debris-laden floodwaters and then reinflated to retain the later flow.

If water used for spreading contains fines, frequent scraping of the sides and bottom of the spreading grounds may be required to maintain permeability. Also, deposition of iron or other materials may reduce infiltration. Composition of the proposed infiltration water, as well as hydrogeologic conditions at the site, should be carefully evaluated prior to initial design of surface spreading facilities.

(2) Injection Wells.—Design and construction of injection wells is generally more complicated than design of production wells because of clogging potential. Injected water must be clear and free of fines. Also, compatibility of injected water with ambient water must be evaluated to ensure that adverse chemical reactions which could cause clogging of the well screen or filter pack will not occur.

In cases where injection and production occur in alternating sequence, "bulbs" of injection water may be created, with little mixing with the aquifer water. This procedure can permit temporary storage of fresh water even in saline aquifers.

(d) Ground-Water Reservoirs.—Suitable surface water reservoir sites are becoming scarce. Consequently, interest has increased in the underground storage of water. While underground reservoirs are not as obvious or as readily delineated as surface reservoirs, they offer a possible alternative in many areas where conventional storage would be costly or otherwise undesirable. As is true of all alternative solutions, each type of reservoir offers advantages and disadvantages. To assist in the evaluation of the alternatives, table 1-4 lists the major advantages of each type of reservoir.

Table 1-4.—Advantage of surface versus subsurface reservoirs

Subsurface reservoirs	Surface reservoirs
Many large capacity sites available	Few new sites available
Slight to no evaporation loss	High evaporation loss even in humid climate
Require little land area	Require large land area
Slight to no danger of catastrophic structure failure	Ever-present danger of catastrophic failure
Uniform water temperature	Fluctuating water temperature
High biological purity	Easily contaminated
Safe from immediate radioactive fallout	Easily contaminated by radioactive material
Reservoir serves as conveyance system - canals or pipeline across lands of others unnecessary	Water must be conveyed
Water must be pumped	Water may be available by gravity flow
Storage and conveyance use only	Multiple use
Water may be mineralized	Water generally of relatively low mineral content
Minor flood control value	Maximum flood control value
Limited flow at any point	Large flows
Power head usually not available	Power head available
Difficult and costly to investigate, evaluate, and manage	Relatively easy to evaluate, investigate, and manage
Recharge opportunity usually depends on surplus surface flows	Recharge depends on annual precipitation
Recharge water may require expensive treatment	No treatment required
Continuous, expensive treatment of recharge areas or wells	Little treatment required

1-7. Ground-Water Rights.—

(a) General.—In the United States, doctrines of law and statutes relating to the ownership and use of water (Thomas, 1953) are the responsibility of the courts and legislative bodies of the States. No

Federal statutes exist under which a water right can be acquired; that is, a right granted by law to use or take possession of water in a natural source and put it to a beneficial use. However, Indian water rights granted by treaty may take precedence over State water laws. In the past, both surface- and ground-water rights for Reclamation projects were obtained in conformance with the laws of the States in which the project was located. This procedure is still followed except where Indian water rights pre-empting State water rights are involved.

Two entirely different systems for acquiring water rights are followed in the contiguous 48 States. These systems are the doctrine of riparian rights, recognized in the 31 predominantly Eastern States, and the doctrine of prior appropriation, recognized in the 17 Western States (figure 1-3). Indian water rights generally are determined based on the Winters Doctrine, which is summarized below.

(b) *Doctrine of Riparian Rights.*—The doctrine of riparian rights is based on the common law of England and stems from ownership of land contiguous to a natural water source such as a stream or lake. For ground water, ownership of land overlying an aquifer is sufficient to establish a ground-water right. This doctrine is often referred to as the *English rule of unlimited use.*

(c) *Doctrine of Prior Appropriation.*—Under this doctrine, owner-ship of water is vested in the State (i.e., the common property of the people). An appropriator who is first in time to beneficially use a certain water source has a prior right to its use. However, water for domestic use usually is not subject to the need for appropriation.

(d) *Prescriptive Rights.*—In some States where the doctrine of prior appropriation is followed, a prescriptive water right can be acquired by taking and putting to beneficial use, for some number of consecutive years, water to which other landowners or prior appropriators have rights.

(e) *Indian Water Rights.*—Indian water rights established for reservations have been determined to be based on potential need rather than present use, and in many cases have not been quantified. The Winters Doctrine (*Winters* v. *United States*, 207 US 564) established in 1908 that the Federal Government's reservation of land for the Indians implicitly carried with it a reservation of water needed to make the land "adequate and

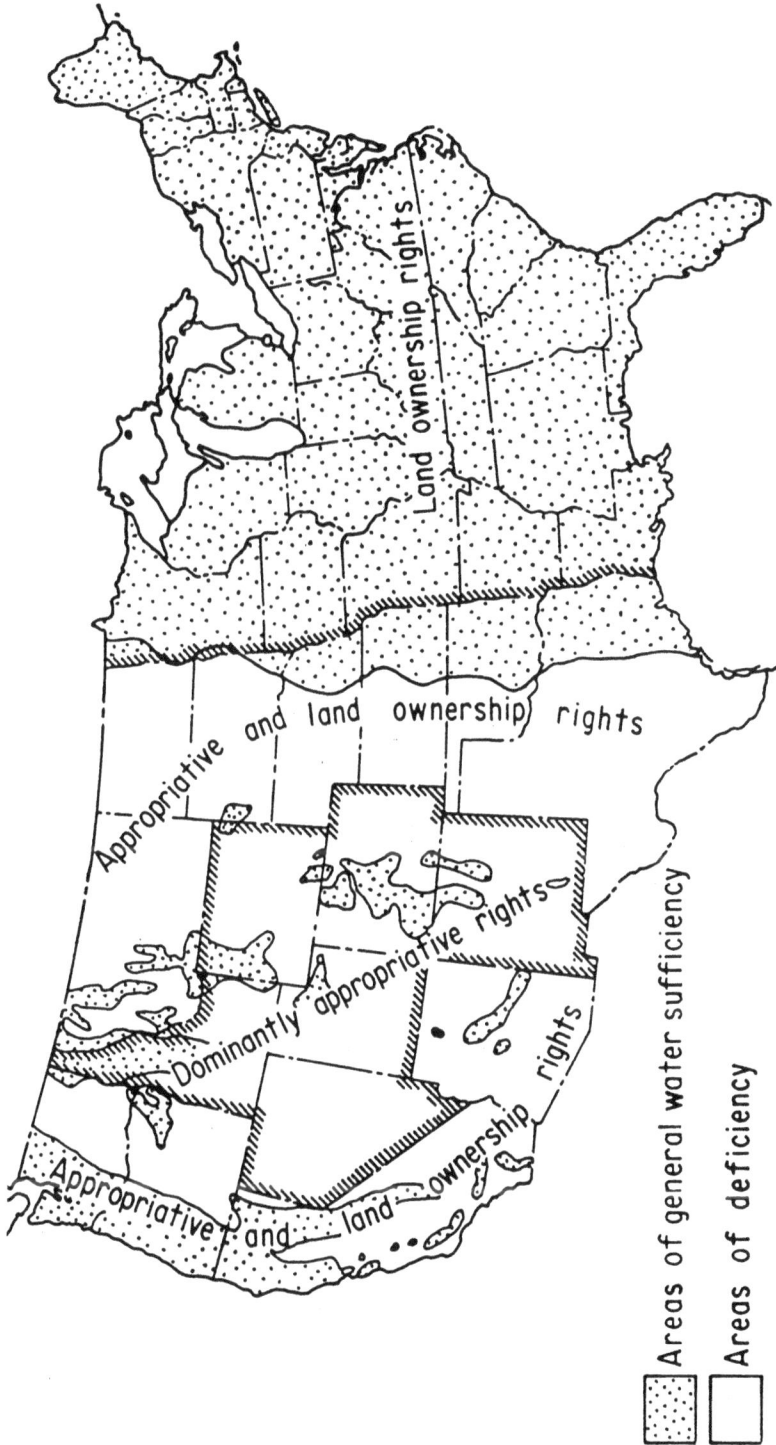

Figure 1-3.—Water rights doctrines by States and water supply sufficiency of the contiguous United States.

valuable" for the inhabitants. Also, it established that the rights of
Indians to such waters could not be diminished by the application
of State law (Price and Weatherford, 1976). This decision
recognized a power in the Federal Government to reserve and
exempt water from appropriation under State law, and
implicitlyreserved from appropriation under State law an amount
of water sufficient for irrigation purposes, its appropriation in this
case relating back to the treaty date (Nelson, 1977). The quantity
of water is measured by the amount necessary to fulfill the
purposes of the reservation.

The Winters Doctrine has been clarified and expanded over the
years in numerous court cases. It has generally been held that the
Winters Doctrine applies to ground water as well as surface water
(Nelson, 1977).

(f) *Ground-Water Regulations.*—In addition to those providing for
water rights, other statutes and rules relating to the administration and control of ground water have been established in some
States to protect the public interest and to provide for orderly
development of this resource. Some of the more common
regulations provide for licensing and bonding of well drillers,
obtaining permits to drill new wells or to rehabilitate existing
wells, filing of geologic logs of new wells, and following
construction practices that ensure against contamination. Also,
some States have regulations restricting the subsurface disposal of
pollutants, such as brines and industrial wastes that might
contaminate the public ground-water supplies.

(g) *Conjunctive Use of Surface and Ground Water.*—Conjunctive
use is any scheme that capitalizes on the flexibility and efficiency
that can be gained through integrated management of surface- and
ground-water supplies. Conjunctive use involves the coordinated
and planned operation of surface-water and ground-water
resources to meet water demands. It particularly applies to ground
water in alluvium, which may either deplete or recharge the
adjacent stream. Particularly, in the Western United States,
streams may go dry during the summer months while considerable
water still flows in the alluvium. In some States, holders of
surface water rights can alluvial ground water at such times.

On a larger scale, conjunctive use involves river basin planning,
reservoir storage and operation, and ground-water recharge.
Ideally, from a water-supply standpoint, surface-water reservoirs
would be operated to maximize ground-water recharge. However,

other factors generally are included in reservoir operation which may restrict or make infeasible such operation. Also, less than full conjunctive use may be subject to legal restrictions. Planning for conjunctive use requires evaluation of physical (geologic, topographic, and hydrologic) conditions, legal aspects, and public acceptance.

1-8. Application of Ground-Water Engineering.—

(a) Water Supply.—The major application of ground-water engineering has been, and probably always will be, the provision of a water supply by means of wells and infiltration galleries. Facilities range from isolated individual small wells yielding a few liters per minute for domestic and stock purposes to well fields consisting of a number of irrigation, municipal, or industrial water-supply wells with individual discharges in excess of 20,000 liters per minute. The small individual well seldom presents a problem if it is designed according to good engineering practice. The larger installations, particularly those with numerous wells, require evaluation of the aquifer characteristics, estimates of well spacing, drawdowns, quality of water, and possibly recharge-discharge relationships. Wells must be designed and pumps selected for economical, long, and trouble-free operation within the capabilities of the aquifer, with the consideration of any possible corrosion and encrustation problems which may be present.

Proposed development may be further complicated by restrictions imposed by overlying or underlying saline aquifers, salt-water intrusion, influences on the discharge of adjacent surface water streams, and land subsidence.

Some aquifers have little measurable recharge or discharge but contain large quantities of water in storage which have accumulated over long periods. Estimates can be made of the desirability of mining the water and the probable economic life of such aquifers under various degrees of development.

(b) Ground-Water Reservoirs and Artificial Recharge.—The storage of surface waters in underground reservoirs and the recharge of depleted ground-water reservoirs are other aspects of ground-water engineering of growing importance and interest. Recharge wells, basins, channels, and waste disposal facilities present special problems of aquifer plugging caused by chemical, biological, and physical factors, and of contamination of overlying or adjacent potable aquifers.

The maintenance of minimum streamflows by supplementing surface water with pumped ground water during low flow periods and recharging the ground-water reservoir during high runoff is also of growing interest.

(c) Drainage.—Drainage may involve the lowering of ground-water levels beneath irrigated lands to permit crop growth, the lowering of water levels or prevention of boils in limited areas to permit excavation and construction activities in the dry, reduction of pressures to maintain stability of slopes, and the reduction of pressures and exit velocities to ensure stability of dams and similar structures incident to reservoir and dam construction. Applications of ground-water hydraulics and engineering are involved in all such problems.

(d) Contamination Problems.—In recent years, ground-water engineering has become increasingly important in investigation, evaluation, and mitigation of subsurface contamination. Ground-water modeling is usually a major aspect of predicting direction and rate of contamination potential and mitigation. Situations involving contamination are generally much more complex than those involving water supply or drainage because of complex chemical and biological interactions. In addition, flow of contaminants may not coincide with direction and rate of ground-water flow. The ground-water engineer involved in these projects must work closely with chemists, biochemists, and geologists, in evaluating conditions.

1-9. Bibliography.—

American Society of Civil Engineers, 1961, "Ground Water Basin Management," *ASCE Manual of Engineering Practice No. 40.*

Ayers, R.S., and D.W. Westcot, 1985, Water Quality for Agriculture: Food and Agriculture Organization of the United Nations, FAO Irrigation and Drainage Paper 29, Rome, Italy.

Hem, J.D., 1959, "Study and Interpretation of the Chemical Characteristics of Natural Water," U.S. Geological Survey Water-Supply Paper 1473.

Meinzer, O.E. (editor), 1949, "Physics of the Earth—IX, Hydrology," Dover Publications, New York, New York.

National Research Council (U.S.), 1989, "Irrigation-Induced Water Quality Problems," Committee on Irrigation-Induced Water Quality Problems, National Academy of Sciences, National Academy Press, Washington, DC.

Nelson, Michael C., 1977, "The Winters Doctrine: Seventy Years of Application of Reserved Water Rights to Indian Reservations," University of Arizona, Office of Arid Lands Studies, Tucson, Arizona.

Price, Monroe E., and Gary D. Weatherford, 1976, "Indian Water Rights in Theory and Practice: Navajo Experience in the Colorado River Basin," In: Law and Contemporary Problems, vol. 40, No. 1, Quarterly Journal, Duke University School, pp. 97-131.

Pyne, R. David G., 1995, "Groundwater Recharge and Wells: A Guide to Aquifer Storage and Recovery."

Rima, D.R., E.B. Chase, and B.M. Myers, 1971, Subsurface Disposal by Means of Wells—A Selected Annotated Bibliography," U.S. Geological Survey Water-Supply Paper 2020.

Signor, D.C., D.J. Growitz, and W. Kam, 1990, "Annotated Bibliography on Artificial Recharge of Ground Water, 1955-67," U.S. Geological Survey Water-Supply Paper.

Thomas, H.E., 1953, "Ground Water Law," Transcript of Lecture Presented at Ground Water Short Course, U.S. Geological Survey and Bureau of Reclamation, Fort Collins, Colorado.

Todd, D.K., 1980, "Ground Water Hydrology," John Wiley & Sons, New York, New York.

Tolman, C.F., 1957, "Ground Water," McGraw-Hill, New York, New York.

U.S. Department of Agriculture, 1954, "Diagnosis and Improvement of Saline and Alkaline Soils," Handbook No. 60.

U.S. Department of Agriculture, 1956, "Water" Agricultural Year Book for 1955.

U.S. Geological Survey, 1985, "Study and Interpretation of the
    Chemical Characteristics of Natural Water, 3d edition, Water-
    Supply Paper 2254, p. 263.

U.S. Government, 1993, Safe Drinking Water Act: 42 USC 300f
    et seq.

U.S. Government, 1977, Clean Water Act Amendments.

Chapter II

PLANNING GROUND-WATER INVESTIGATIONS AND PRESENTATION OF RESULTS

2-1. Introduction.—In ground-water investigations, each study is unique in the problems presented and the solutions available. Guidelines are available, but no single step-by-step approach will be very successful over the range of investigations encountered by ground-water professionals. Most ground-water investigations proceed in four stages:

- Planning
- Data collection and field work
- Data analysis
- Report preparation

Planning a ground-water investigation or project requires a thorough appreciation of the purpose, the scope of the work required, the areal extent and geologic complexity of the area involved, and the limitations imposed by available financing and allotted time. Ground-water hydrology is a dynamic and inexact science. The accuracy and reliability of acquired data usually increase with the time available for observation and interpretation, and much of the success and value of such an investigation depends on the imagination, experience, and judgment of the ground-water technical specialists involved. Ground-water investigations generally are costly because of the time factor and the need for extensive subsurface and data collection.

Some typical purposes of a ground-water investigation include:

- Locating a small domestic or stock water well

- Designing a large well field to furnish irrigation, industrial, or municipal water

- Lowering the water table where drainage is required

- Locating and designing ground-water recharge facilities

- Estimating the safety and economic aspects of water loss and effect on adjacent lands of seepage from a reservoir or canal

- Estimating the average annual volume of water recoverable and the storage space available in a ground-water reservoir

- Dewatering an excavation for construction purposes

- Planning conjunctive surface- and ground-water uses

- Investigating the nature and extent of contaminated or poor quality ground water to identify the source (natural or man-made) of the ground-water degradation

- Defining the hydraulic properties of wetland complexes

- Design water treatment facility

Each purpose may present unique problems and require different concepts, data, approaches, funding, and time. The location of a single small well may require only a cursory reconnaissance of an area and an examination of a few existing wells, all of which may be accomplished in a day or two. Investigations leading to dewatering of an excavation of limited size may require one or more test wells and a pumping test, which usually can be completed in several weeks to several months. In other instances, where conditions are complex and cover a large area, the work may entail many months or even years of study, investigations, and modeling. Layout of sizeable well fields for any purpose may require a comprehensive ground-water inventory to determine the relationships between climate, long-time, ground-water fluctuations, ground- and surface-water interaction, spatial variation in aquifer characteristics, recharge and discharge, contaminant distribution, and other similar factors.

Ground-water data based on short-term investigations may be more indicative than substantive. When reliable quantitative information is required, provision should be made for refinement of data by continued observation and data collection.

In the planning of an investigation, a review of previous work provides a basis for planning additional work, and reconnaissance field surveys provide the information needed to determine field conditions, obstacles, limits, and possible alternative methods for completing any additional work contemplated.

When the required field work has been tentatively determined, the minimum number and type of field personnel, cooperative arrangements with other offices, necessary equipment, and the time and fund requirements can be estimated. Adjustments can then be made to conform to the requirements of the overall project. The program and plan should be kept flexible, allowing for curtailment or expansion as determined from information acquired as the investigation progresses.

Upon completion of the field investigations and data collection, a final review of the data should be conducted and a written summary of the field investigations should be prepared. The final report, which presents the results of the investigation, should contain a compilation of the data, the results of the analysis of the data, and the supporting maps, figures, and tables.

2-2. Ground-Water Modeling.—The following brief introduction to the basic concepts of ground-water modeling, which focuses primarily on deterministic numerical ground-water models, is a reprint with slight revision from Mercer and Faust (1986).[1]

Numerical models have been extensively used for ground-water analysis since the mid-1960's, yet confusion and misunderstanding over their application still exists. As a result, some hydrologists have become disillusioned and have overreacted, concluding that models are worthless. At the other extreme are those who have been willing to accept any model results, regardless of whether or not they make hydrologic sense.

(a) Modeling Approaches.—Simulation of a ground-water system refers to the construction and operation of a model whose behavior assumes the appearance of the actual aquifer behavior. The model can be physical (for example, a laboratory sand tank or Hele-Shaw model), electrical analog, or mathematical. Other model divisions may be found in Karplus (1976) and Thomas (1973). A mathematical model is simply a set of equations which, subject to certain assumptions, describes the physical processes active in the aquifer. Although the model itself obviously lacks the detailed reality of the real ground-water system, the behavior of a valid model approximates that of the aquifer. Mathematical models may be deter-

---

[1] The book *Ground-Water Modeling* can be purchased from the National Ground Water Association, 6375 Riverside Drive, Dublin, Ohio, 43017. The material presented here is reprinted with minor revision from "Chapter 1 — Ground-Water Modeling: An Overview" with the authors' permission.

ministic, statistical, or some combination of the two. This section
is restricted to deterministic models (i.e., those that define cause
and effect relationships based on an understanding of the physical
system).

The procedure for developing a deterministic, mathematical
model of any physical system can be generalized as shown on
figure 2-1. The first step is to understand the physical behavior of
the system. Cause-effect relationships are determined and a
conceptual model of how the system operates is formulated. For
ground-water flow, these relationships are generally well known
and are expressed using concepts such as hydraulic gradient to
indicate flow direction. For the movement of hazardous wastes,
these relationships, especially those involving physical-chemical
behavior, are only partially understood.

The next step is to translate the physics into mathematical terms
(i.e., make appropriate simplifying assumptions and develop the
governing equations). This process constitutes the mathematical
model. The mathematical model for ground-water flow consists of
a partial differential equation together with appropriate boundary
and initial conditions that express conservation of mass and that
describe continuous variables (for example, hydraulic head) over
the region of interest. In addition, the mathematical model entails

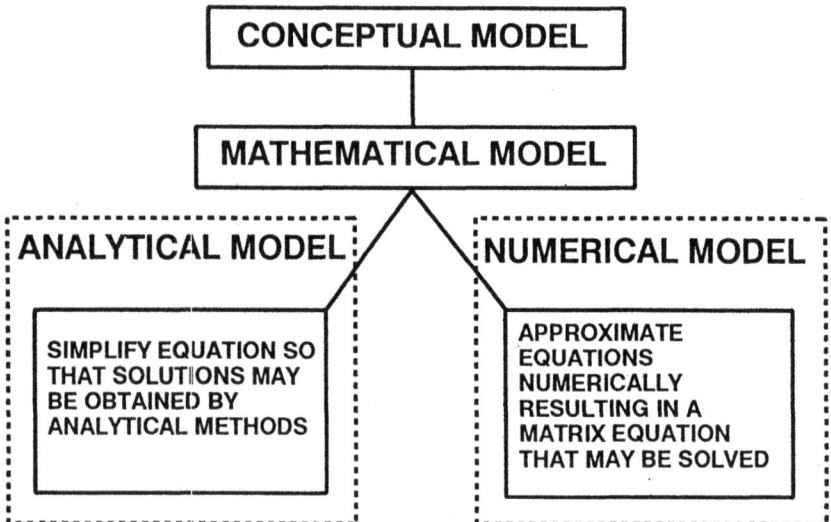

```
                  ┌─────────────────────────┐
                  │   CONCEPTUAL MODEL      │
                  └────────────┬────────────┘
                               │
                  ┌────────────┴────────────┐
                  │  MATHEMATICAL MODEL     │
                  └────────────┬────────────┘

    ┌- - - - - - - - - - - ┐ / \ ┌- - - - - - - - - - - - ┐
    : ANALYTICAL MODEL :   /   \  : NUMERICAL MODEL :
    :   ┌──────────────┐   :      :   ┌────────────────┐  :
    :   │ SIMPLIFY     │   :      :   │ APPROXIMATE    │  :
    :   │ EQUATION SO  │   :      :   │ EQUATIONS      │  :
    :   │ THAT         │   :      :   │ NUMERICALLY    │  :
    :   │ SOLUTIONS MAY│   :      :   │ RESULTING IN A │  :
    :   │ BE OBTAINED  │   :      :   │ MATRIX EQUATION│  :
    :   │ BY ANALYTICAL│   :      :   │ THAT MAY BE    │  :
    :   │ METHODS      │   :      :   │ SOLVED         │  :
    :   └──────────────┘   :      :   └────────────────┘  :
    └- - - - - - - - - - - ┘      └- - - - - - - - - - - - ┘
```

Figure 2-1.—Logic diagram for developing a mathematical model.

various phenomenological "laws" describing the rate processes active in the aquifer. An example is Darcy's law for fluid flow through porous media; this law is generally used to express conservation of momentum. Finally, various assumptions may be invoked such as those of one- or two-dimensional flow and artesian or water-table conditions.

For solute (e.g., hazardous wastes) and heat transport, additional partial differential equations with appropriate boundary and initial conditions are required to express conservation of mass for the chemical species considered and conservation of energy, respectively. Examples of corresponding phenomenological relationships are Fick's law for chemical diffusion and Fourier's law for heat conduction.

Once the mathematical model is formulated, the next step is to obtain a solution using one of two general approaches. The ground-water flow equation can be simplified further (e.g., assuming radial flow and infinite aquifer extent, to form a subset of the general equation that is amenable to analytical solution). The equations and solutions of this subset are referred to as *analytical models*. The familiar Theis-type curve represents the solution of one such analytical model.

Alternatively, for problems where the simplified analytical models no longer describe the physics of the situation, the partial differential equations can be approximated numerically (e.g., with finite-difference techniques or with the finite-element method). In so doing, one replaces continuous variables with discrete variables that are defined as grid blocks (or nodes). Thus, the continuous differential equation, defining hydraulic head everywhere in an aquifer, is replaced by a finite number of algebraic equations that define hydraulic head at specific points. This system of algebraic equations is generally solved using matrix techniques. This approach constitutes a *numerical model* and, generally, a computer program to solve the equations on a digital computer.

Probably the most frequent application of ground-water models is that of history matching and prediction of site-specific aquifer behavior. Of the various types of models discussed, the numerical model offers the most general tool for simulating aquifer behavior. Physical models usually offer the most intuitive insight into aquifer behavior but are limited in application (once constructed) and have the difficulty of scaling results to field level.

Electric analog models can be applied to field problems but are usually site specific and expensive to construct. Deterministic mathematical models (both analytical and numerical) retain a good measure of physical insight while permitting a larger class of problems to be considered with the same model. Analytical methods, such as type curve analysis, are relatively easy to use. Numerical models, although more difficult to apply, are not limited by many of the simplifying assumptions necessary for the analytical methods. Finally, purely statistical methods are useful in classifying data and describing poorly understood systems but generally offer little physical insight.

Each type of model has advantages and disadvantages. Consequently, no single approach should be considered superior to others for all applications. The selection of a particular approach should be based on the specific aquifer problem addressed. Whichever approach is taken, the final step in modeling a ground-water flow system is to translate the mathematical results back to their physical meanings. In addition, these results must be interpreted in terms of both their agreement with reality and their effectiveness in answering the hydrologic questions that motivated the model study.

(b) Types of Ground-Water Models.—Four general types of ground-water models are listed on figure 2-2. The problem of water supply is normally described by one equation, usually in terms of hydraulic head. The resulting model providing a solution for this equation is referred to as *a ground-water flow model.* If the problem involves water quality, then an additional equation(s) in the ground-water flow equation must be solved for concentration(s) of the chemical species. Such a model is referred to as a *solute transport model.* Problems involving heat also require an equation in addition to the ground-water flow equation, similar to the solute transport equation, but in terms of temperature. This model is referred to as a *heat transport model.* Finally, a *deformation model* combines a ground-water flow model with a set of equations that describe aquifer deformation.

Ground-water flow models have been most extensively used for such problems as regional aquifer studies, ground-water basin analysis, and near-well performance. More recently, solute transport models have been used to aid in understanding and predicting the effects of problems involving hazardous wastes. Some of the applications include: seawater intrusion, underground storage of radioactive wastes, movement of leachate from sanitary

Model Types			
GROUND-WATER FLOW	**SOLUTE TRANSPORT**	**HEAT TRANSPORT**	**DEFORMATION**
WATER SUPPLY	SEA-WATER INTRUSION	GEOTHERMAL	LAND SUBSIDENCE
REGIONAL AQUIFER ANALYSES	LAND FILLS	THERMAL STORAGE	
NEAR-WELL PERFORMANCE	WASTE INJECTION	HEAT PUMP	
GROUND WATER/ SURFACE WATER INTERACTION	RADIOACTIVE WASTE STORAGE	THERMAL POLLUTION	
	HOLDING PONDS		
DEWATERING OPERATIONS	GROUND WATER POLLUTION		

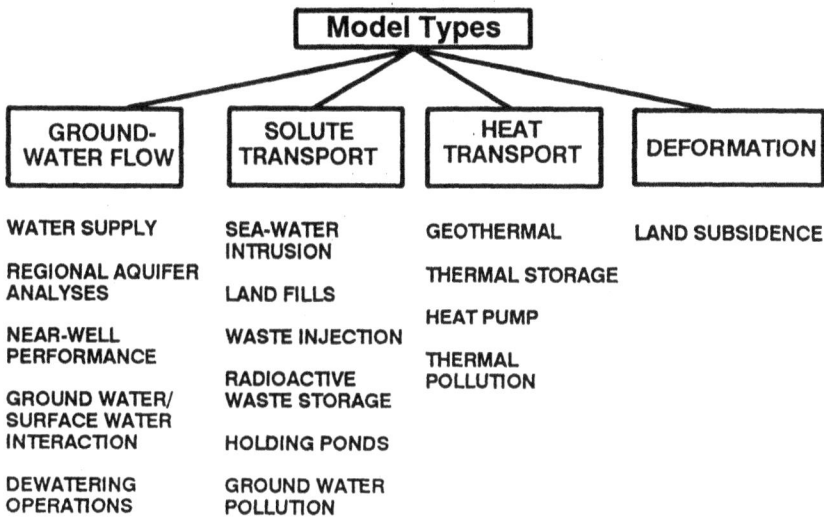

Figure 2-2.—Types of ground-water models and typical applications.

landfills, ground-water contamination from holding ponds, and waste injection through deep wells. Heat transport models have been applied to problems concerning geothermal energy, heat storage in aquifers, and thermal problems associated with high-level radioactive waste storage. Deformation models have been used to examine field problems where fluid withdrawal has decreased pressures and caused consolidation. This compaction of sediments results in subsidence at the land surface.

This classification of ground-water models is by no means complete. All of the above models can be further subdivided into those describing porous media and those describing fractured media. Ground-water models can be combined with statistical techniques in an effort to characterize uncertainty in model parameters. These models can also be used to estimate aquifer parameters. In addition, other models deal with multifluid flow (e.g., oil and water) and multiphase flow (e.g., unsaturated zone problems). Some resource management models combine flow models and linear programs, which are used to optimize certain decision parameters, like pumping rates. Other models combine some or all of the models on figure 2-2 (e.g., a thermal loading problem may require that a heat transport model be combined with a deformation model). The type of model used will obviously depend on the application. For further information on the various models and their availability, the interested reader is referred to Bachmat et al. (1978) and Appel and Bredehoeft (1976).

A numerical model is most appropriate for general problems involving aquifers having irregular boundaries, heterogeneities, or highly variable pumping and recharge rates. The remaining sections are therefore generally concerned with numerical ground-water models giving the most emphasis to ground-water flow models and the least emphasis to deformation models.

(c) Model Use.--Because the number of ground-water models available today is large, when beginning a study, the first question that may come to mind is, "Which one should I use?" Actually, the first question should be, "Do I need a numerical model study for this problem?" The answers to both of these questions can be determined by first considering the following: (1) What are the study objectives? (2) How much is known about the aquifer system (i.e., what data are available)? (3) Does the study include plans to obtain additional data?

The study objectives may be such that a numerical model is unnecessary. Or, if necessary, objectives may require only a very simple model. Additionally, lack of data may not justify a sophisticated model; however, if a field study is in its initial stages, the ideal approach is to integrate the data collection and analysis with a model. Once it is decided that a model is necessary, the one used will, in part, depend on the study objectives (e.g., if the drawdowns near a well are of interest, then a regional model, where the local effects are lost because of the large spacing between nodes, should not be used). Instead, perhaps a radial flow model with small grid spacing would be sufficient.

The application of a ground-water model to an aquifer involves several areas of effort. These areas are shown on figure 2-3 and include: data collection, data preparation for the model, history matching, and predictive simulation. These tasks should not be considered separate steps of a chronological procedure; rather, they should be considered as a feedback approach. The model is best used not only as a predictive tool, but also as an aid in conceptualizing the aquifer behavior. For example, a model used in the early stages of a field study can help in determining which and how much data should be collected.

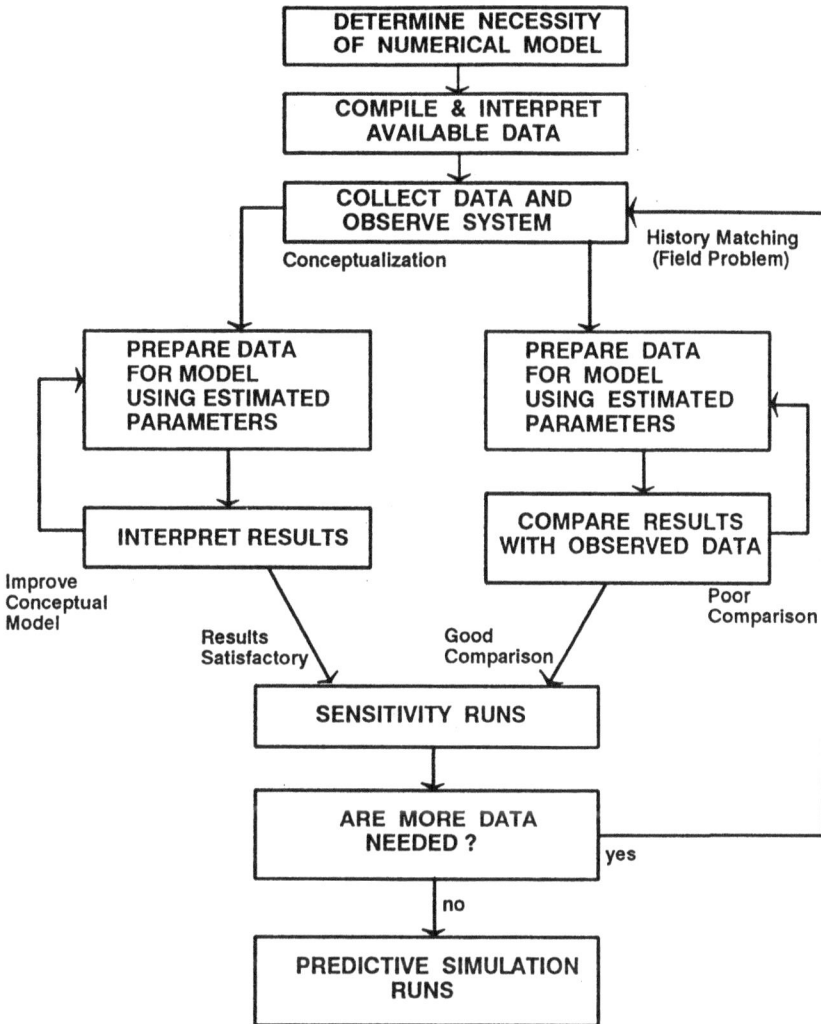

Figure 2-3.—Diagram showing model use.

Data preparation for the ground-water model first involves determining the boundaries of the region to be modeled. The boundaries may be physical (impermeable or no flow, recharge or specified flux, and constant head) or merely convenient (small subregion of a large aquifer). Once the boundaries of the aquifer are determined, the region must be discretized (i.e., subdivided into a grid). Depending on the numerical procedure used, the grid may have rectangular or irregular polygonal subdivisions. Figure 2-4 shows a hypothetical aquifer with a well field development. Figures 2-4a, 4b, and 4c show typical two-dimensional grid patterns for both the finite-difference and finite-element methods.

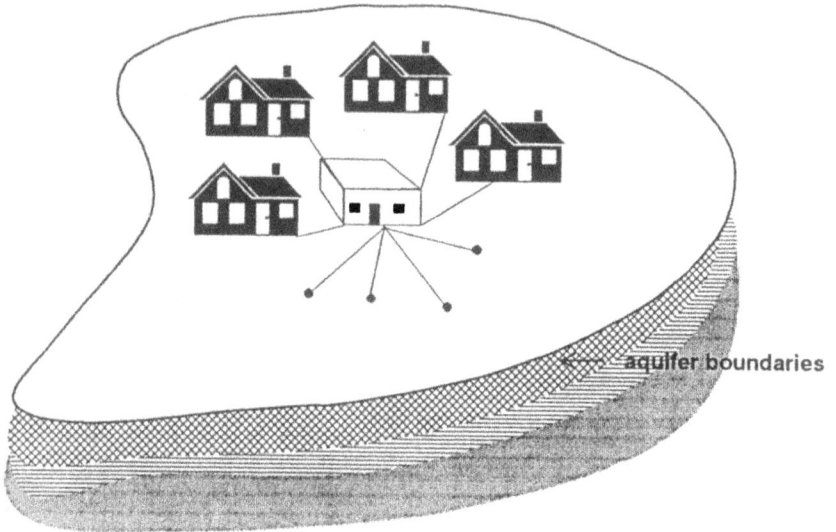

Figure 2-4a.—Map view of aquifer showing well field and boundaries.

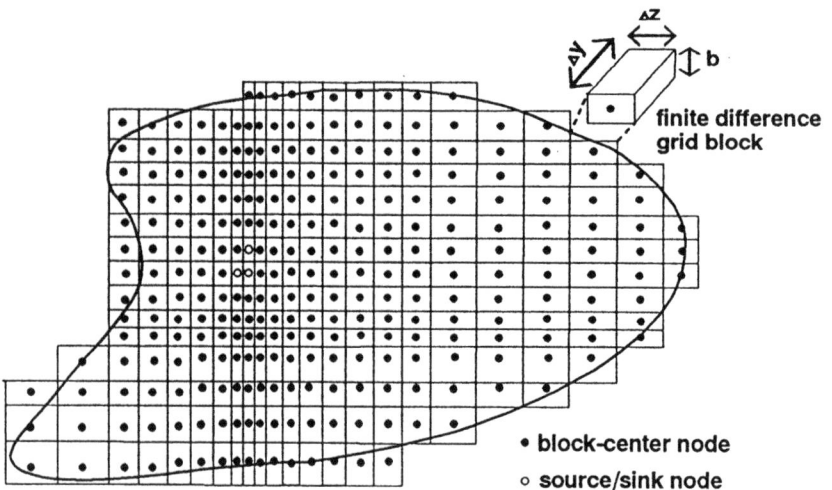

Figure 2-4b.—Finite-difference grid for aquifer model, where Δx is the spacing in the x direction, Δy is the spacing in the y direction, and b is the aquifer thickness.

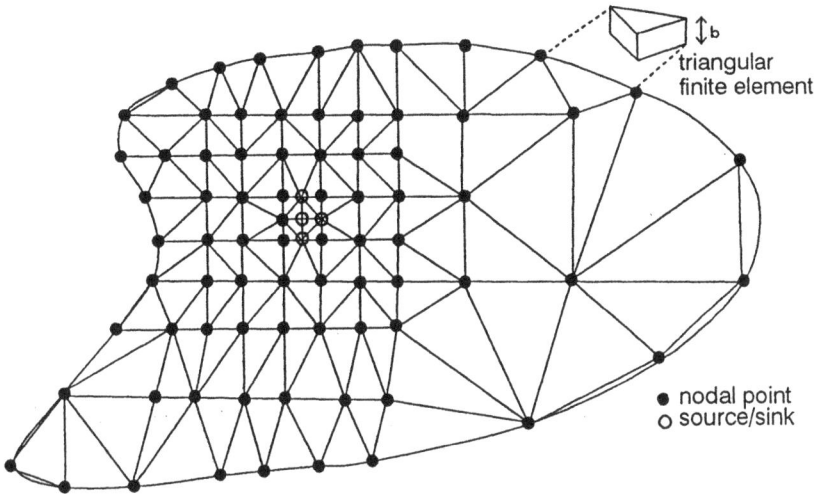

Figure 2-4c.—Finite-element configuration for aquifer model,
where *b* is the aquifer thickness.

Once the grid is designed, aquifer parameters and initial data
must be specified for the grid. For descriptive purposes, the
following discussion refers to the finite-difference method, using a
rectangular grid. Required program input data include aquifer
properties for each grid block, such as storage coefficients and
transmissivities (see table 2-1). For solute transport (i.e., programs
used for tracking hazardous wastes) and heat transport, additional
data are required, such as hydrodynamic dispersion properties and
thermal conductivity, respectively. Computed results generally
consist of hydraulic heads at each of the grid blocks throughout the
aquifer. These spatial distributions of hydraulic head are
determined at each of a sequence of time levels covering the period
of interest. For transport problems, computed results might also
include concentrations and temperatures at each of the grid blocks.

Initial estimates of aquifer parameters constitute the first step in
a trial-and-error procedure known as history matching. The
matching procedure (often referred to as model calibration) is used
to refine initial estimates of aquifer properties and to determine
boundaries (i.e., the areal and vertical extent of the aquifer) and
the flow conditions at the boundaries (boundary conditions);
aquifer tests generally provide the initial estimates for storage
coefficients and transmissivities. For certain ground-water
problems, steady-state (or equilibrium) heads must also be

determined and used as initial or beginning conditions. Simulated
wells in the aquifer grid system are then allowed to pump at the
observed rates, and computed (simulated) drawdowns are
compared with observed drawdowns.

Assuming that the model is correct, comparison between these
two indicates the accuracy of the initial estimates of input data.
Some of the input data may require modification until observed
and calculated data compare sufficiently well. In the past, this
procedure has been done by trial and error; more recently, the
amount of work in matching has been reduced by using parameter
estimation methods that modify initial estimates of input data in a
more objective fashion.

No hard and fast rules exist to indicate when a satisfactory
match is obtained. The number of "runs" required to produce a
satisfactory match depends on the objectives of the analysis, the
complexity of the flow system and length of observed history, as
well as the patience of the hydrologist. Once completed, the model
can be used to predict the future behavior of the aquifer. Of
course, confidence in any predictive results must be based on (1) a
thorough understanding of model limitations, (2) the accuracy of
the match with observed historical behavior, and (3) knowledge of
data reliability and aquifer characteristics.

The main purpose of prediction is estimation of aquifer
performance under a variety of development schemes. Although
the aquifer can be developed only once at considerable expense, a
model can be run many times at low expense over a short period of
time. Observation of model performance under differing
development schemes then aids in selecting an optimum set of
operating conditions for using the ground-water resource. More
specifically, ground-water modeling allows estimates of:
(1) recharge (both natural and induced) caused by leakage from
confining beds, (2) effects of boundaries and boundary conditions,
(3) effects of well locations and spacing, and (4) effects of various
withdrawal (or injection) rates.

Other purposes for prediction include estimating the rates of
movement of hazardous wastes from sanitary landfills and other
containment areas. Models are used to predict the encroachment
rate of saltwater in coastal regions caused by fresh-water
withdrawal. They are also used to help determine what, if any,
remedial action is best to take in a contamination situation.

Table 2-1.—Data requirements to be considered for a
predictive model (after Moore, 1979)

I Physical Framework	II Stresses on System
A. Ground-Water Flow	*A. Ground-Water Flow*

I Physical Framework

A. Ground-Water Flow

1. Hydrogeologic map showing areal extent, boundaries, and boundary conditions of all aquifers.
2. Topographic map showing surface-water bodies.
3. Water-table, bedrock-configuration, and saturated - thickness maps.
4. Transmissivity map showing aquifer and boundaries.
5. Transmissivity and specific storage map of confining bed.
6. Map showing variation in storage coefficient of aquifer.
7. Relation of saturated thickness to transmissivity.
8. Relation of stream and aquifer (hydraulic connection).

B. Solute Transport (in addition to above)

9. Estimates of the parameters that comprise hydrodynamic dispersion.
10. Effective porosity distribution.
11. Background information on natural concentration distribution (water quality) in aquifer.
12. Estimates of fluid density variations and relationship of density to concentration.
13. Hydraulic head distributions (used to determine ground-water velocities).
14. Boundary conditions for concentrations.

C. Heat Transport (in addition to above)

15. Estimates of thermal conductivities and specific heats of rock and water.
16. Background information on natural temperature distribution in aquifer, including heat flow measurements.
17. Estimates of fluid density variations and relationships of density and viscosity to temperature.
18. Boundary conditions for temperature.

II Stresses on System

A. Ground-Water Flow

1. Type and extent of recharge areas (irrigated areas, recharge basins, recharge wells, etc.).
2. Surface-water diversions.
3. Ground-water pumpage (distributed in time and space).
4. Streamflow (distributed in time and space).
5. Precipitation.

B. Solute Transport (in addition to above)

6. Areal and temporal distribution of water quality in aquifer.
7. Streamflow quality (distribution in time and space).
8. Sources and strengths of pollution.

C. Heat Transport (in addition to above)

9. Areal and temporal distribution of temperature in aquifer.
10. Strengths of heat sources.

III Other Factors

A. Ground-Water Flow and Transport

1. Economic information of water supply.
2. Legal and administrative rules.
3. Environmental factors.
4. Planned changes in water and land use.

Finally, heat transport models are used to help predict the behavior of geothermal reservoirs and aquifers used for thermal storage.

In addition to these site-specific applications, models are also used to examine general problems. Hypothetical (but typical) aquifer problems may be designed to study various types of flow behavior, such as ground-water and surface-water interactions or flow around a deep radioactive waste repository. The feasibility of certain proposed mechanisms for observed behavior can be tested. Parameters may be changed to learn what effect they may have on the overall process. This process is sometimes referred to as a *sensitivity analysis* because results from these runs will indicate what parameters the computed hydraulic heads are most sensitive to. Sensitivity analysis is also useful for site-specific applications to indicate what additional data need to be determined and areas where additional data are needed.

(d) Model Misuse.—Models can be misused in a variety of ways (Prickett, 1979). Three common and related misuses are: overkill, inappropriate prediction, and misinterpretation. The temptation to apply the most sophisticated computational tool to a problem is difficult to resist. A question often arises regarding under what circumstances should simulation be three-dimensional as opposed to two- or even one-dimensional. Inclusion of flow in the third (nearly vertical) direction is often recommended only if aquifer thickness is "large" in relation to areal extent or if pronounced heterogeneity exists in the vertical direction (e.g., high stratification). Another type of overkill involves using grid sizes that are finer (smaller) than necessary considering available information about aquifer properties; this error results in additional work and expense.

In some applications, complex models are used too early in the study. For example, one generally should not begin the study of a hazardous waste problem with a solute transport model. Rather, the first step is to be sure the ground-water hydrology (velocity in particular) can be characterized satisfactorily; therefore, one begins by modeling ground-water flow alone. Once this modeling is done to satisfaction, then solute transport can be included. One must assess the complexities of the problem, the quantity of data that are available, and the objectives of the analysis, and then determine the best approach for the particular situation. A

general rule might be to start with the simplest model and a
coarse aquifer description and refine the model and data until the
desired estimation of aquifer performance is obtained.

One must always be aware that the history match portion of the
simulation occurred under a given set of field conditions, and that
these conditions are subject to change during the prediction portion
(e.g., during the history match portion, the aquifer may be confined
but may also be on the verge of becoming desaturated). Using a
confined model for prediction will give erroneous results because
the saturated thickness and storage coefficient will be incorrect.
Because ground-water models deal with the subsurface, unknown
factors always exist that could affect results. In general, one
should not predict more than about twice the period used for
matching, and only then, under similar pumping schemes.

Perhaps the worst possible misuse of a model is blind faith in
model results. Calculations that contradict normal hydrologic
intuition almost always are the result of some data entry mistake,
a "bug" in the computer program, or misapplication of the model to
a problem for which it was not designed. Proper application of a
ground-water model requires an understanding of the specific
aquifer. Without this conceptual understanding, the whole
exercise may become a meaningless waste of time and money.

(e) Limitations and Sources of Error in Modeling.—To avoid
model misuse, the limitations and possible sources of error in
numerical models must be known and understood. All numerical
models are based on a set of simplifying assumptions which limit
their use for certain problems. To avoid applying an otherwise
valid model to an inappropriate field situation, it is not only
important to understand the field behavior but also to understand
all of the assumptions that form the basis of the model. An areal
(two-dimensional) model, for example, should be applied with care
to a three-dimensional problem involving a series of aquifers,
hydrologically connected by confining beds, because the model
results may not be indicative of the field's behavior. Errors of this
type are considered conceptual errors.

In addition to these limitations, several potential sources of error
exist in the numerical model results. First, replacement of the
model differential equations by a set of algebraic equations
introduces truncation error (i.e., the exact solution of the algebraic
equations differs from the solution of the original differential

equations). Second, the exact solution of the algebraic equations is not obtained because of *round-off* error, as a result of the finite accuracy of computer calculations. Finally, and perhaps most importantly, aquifer description data (for example, transmissivities, storage coefficients, and the distribution of heads within the aquifer) are seldom known accurately or completely, thus producing *data error.*

The level of truncation error in computed results may be estimated by repeating runs or portions of runs with smaller space or time increments. Significant sensitivity of computed results to changes in these increment sizes indicates a significant level of truncation error and the corresponding need for smaller spatial or time increments. Compared to the other error sources, round-off error is generally negligible.

Error caused by erroneous aquifer description data is difficult to assess because the true aquifer description is never known. An adage used to describe the error associated with these data is, "Garbage in, garbage out." A combination of core analysis, aquifer tests, and geological studies often give valuable insight into the nature of transmissivity, storage coefficients, and aquifer geometry. However, much of this information may be very local in extent and should be regarded carefully when used in a model of a large area. As discussed, the final parameters that characterize the aquifer are usually determined by obtaining the best agreement between calculated and observed aquifer behavior during some historical period.

(f) Summary.—Numerical ground-water models are an important tool for the ground-water specialist. They can be used to simulate the behavior of complex aquifers including the effects of irregular boundaries, heterogeneity, and different processes such as ground-water flow, solute transport, and heat transport. The use of numerical models involves data collection, data preparation, history matching, and prediction. The process of constructing a model for an aquifer study forces one to develop a conceptual understanding of how the aquifer behaves. Models, therefore, can be used in all phases of the aquifer study, including conceptualization and data collection, as well as prediction. To be most effective, the specialist must have a thorough understanding of the specific aquifer studied, must be familiar with alternative modeling techniques, and must realize the limitations and sources of error in models. Upon meeting these criteria, a successful model study will

not only improve one's understanding of the particular hydrologic system but should also provide appropriate prediction and analysis of the problem under study.

2-3. Planning.—

(a) Purpose and Scope.—In the planning stage, the sequence of activities preceding the start of field investigations, the purpose, scope, and requirements of the project should be clearly defined and documented. During the planning stage, the organizational structure for the management of all stages of the project will need to be established. The planner(s) of the project should identify what data are currently available on the project and what additional data are required for the completion of the project. If the project will require additional field work, a field reconnaissance of the project area should be made. To complete the planning stage, a plan of study (POS) that defines all of the tasks that will need to be completed to produce the final product should be written. If a ground-water model is to be used in the analysis, the modeler should be heavily involved in the planning of data collection. Otherwise, the data collected may not fit the modeling needs.

As a minimum, the following information is needed: (1) required field work and additional data collection, (2) ground-water modeling needs, (3) the size and location of the study area, (4) the scheduled start, duration, and due date of the project, (5) level of study (preliminary examination, reconnaissance, feasibility, construction), and (6) any special problems or requirements associated with the project (permits, accessibility, hazardous conditions, design etc.).

After review of the existing data, a determination of the additional data requirements should be made. In the data collection and field work stage, additional work that may be required, depending on the scope of the investigation, includes:

- Preparing new or supplemental planimetric, topographic, and geologic maps of suitable scale

- Geologic field mapping to obtain, clarify, or add information on structure, stratigraphy, and lithology

- Inventories of wells and similar facilities

- Initiating ground-water level measurement and sampling programs

- Locating test sites and the selection of existing wells or the design, preparation of specifications, and construction of new wells for pump tests, together with the location, design, construction, and sampling of exploration holes, observation wells, and piezometers

- Establishing gauging stations on streams and springs

- Determining the location and measuring point elevations on wells and springs

- The logging of all new drill holes and wells

- The selection and location of geophysical surveys

- The selection of type of borehole logs, electrical resistivity, gamma ray, etc.

- Providing for mechanical analysis of drill hole samples

- Providing for chemical and bacterial analysis of water samples

- Completing required National Environmental Policy Act (NEPA) compliance documents

A POS should be developed that presents the results of the planning stage activities discussed above and defines the remainder of the tasks to be completed in order to produce the final product. The size and content of the report will be dictated by the purpose and scope of the investigation. The descriptions of the tasks to be completed (for all data collection, compilation, analysis, and report preparation) should be as detailed as possible, including equipment, staff, schedule, and reporting requirements. Flowcharts showing work elements, sequence, and duration should be developed, especially in complex projects, to help identify and prevent scheduling problems.

A field reconnaissance survey of the area should be made, especially if the project will require additional field work. The survey could be conducted any time prior to the preparation of the POS. The field survey will provide information needed to

determine field conditions, obstacles, limits, and possible
alternative methods for completing any additional work
contemplated.

Consideration should be given to the accessibility of the area by
personnel and equipment. Such aspects of the area as location,
extent, topography, transportation facilities, land ownership
patterns, cultural development, climate, permits, and potential for
exposure to hazardous materials or activities should be determined
prior to planning of field investigations. In many cases, an
archeological review is required before work may begin.

The organizational structure for all stages of the project will
need to be established early in the planning stage. A project team
is generally created to manage the project. The team leader
acquires required authorizations and funding. The size and
composition of the project team will vary considerably depending
on the scope of the project. For a large project, the team members
may represent a variety of Federal, State, local, and private
agencies and groups. On the other hand, for a very small project,
one individual may serve as the project team, team leader, and
team member.

(b) *Field Investigations.*—A successful field investigation does not
just happen; it must be carefully planned to achieve the desired
results. Choosing the most effective methods, outlining the tasks,
mobilizing personnel and equipment, acquiring materials,
establishing criteria for reporting results, and arranging for
reviews are all part of the planning process. Field investigations
are controlled by data requirements, available personnel and
equipment, funding, and accessibility. In most cases, some
compromise must be made between the ideal and the affordable.

To find this compromise, the planner must have a very clear
understanding of what data are truly required to answer the
question being investigated.

Data needs vary according to the level of study and the method
of analysis to be used. At the reconnaissance or appraisal level,
field work may be limited to observations of local geology and
topography and gathering data from existing wells or springs. At
the other end of the spectrum are extensive drilling and logging,
hydraulic conductivity testing, water quality testing, and other

activities that will promote understanding of the problem at hand. If a model is to be used, the type of data needed may differ from those needed for more basic analytic procedures.

Equipment needs and availability must be considered. Choices often must be made between methods which have different strengths and weaknesses. For instance, hollow stem augers may obtain better samples than a rotary system but cannot be used to install a large diameter test well. The planner must decide which option is more critical or whether both are warranted. Personnel are generally more flexible than equipment, but their abilities may still influence the approach to the study.

Funding almost always imposes restraints on the amount and type of data that can be obtained. The planner must determine the most efficient uses of available funds, and if available funds are inadequate, must seek more funding or consider terminating the project. Time restraints are similar to, and often related to, funding restraints.

2-4. Data Collection and Field Work.—

(a) Identify Existing Data.—In estimating funding, time, and staff requirements, a review of previous reports on an area is essential. The U.S. Geological Survey (USGS) reports on geology and hydrology often provide valuable information. Other pertinent data may be found in the records of the National Weather Service and in U.S. Department of Agriculture reports. A search of engineering and geological bibliographies may provide additional references. Many State engineers, State geological surveys, water resource centers, State colleges and universities, and similar agencies have records of wells and other subsurface investigations which may include location, logs, yields, and methods of construction. The references obtained should be abstracted, analyzed, and summarized; then one can determine additional data required, methods of acquisition, and the time, manpower, and funds necessary to accomplish the work.

(b) Subsurface Investigations.—Information on the stratigraphy, structure, and hydraulic characteristics of the subsurface materials, and water-table and piezometric surface levels and fluctuations are important. Information can be obtained from logs of wells previously drilled in the area, samples of material from

wells, well pump tests, and records of levels of the water-table or
piezometric surface. Some of this information may be available
from local well drillers, but care must be exercised in using it.

Surface geophysical surveys and borehole geophysical logs
combined with test drilling may provide valuable information on
subsurface conditions including approximate depth to water and
bedrock. Prior to undertaking geophysical investigations, an
experienced geophysicist should be consulted regarding the
probable value of geophysics and the best procedures to use in
solving a particular problem. Federal and State agencies and oil
and mining companies are sources of geophysical data. Chapter 4
of this manual gives an overview of common geophysical methods
and their typical applications in ground-water investigations.

(c) Water Quality Data.—The chemical and bacterial qualities of
water may be items of necessary information in ground-water
investigations. For water intended for human consumption, the
bacterial and chemical qualities of the water must be known to
determine its suitability and also to furnish a guide for the type
and intensity of treatment required to make it potable. The
chemical quality must also be known for industrial and irrigation
water supplies because the presence of selected chemical
constituents may not only make water unfit for consumption by
either humans or livestock, but also unsuitable for industrial or
irrigation use.

Chemical analyses are also helpful in preparing well and pump
designs and specifications for permanent facilities where corrosive
or encrusting waters are or may be present. In addition, chemical
analyses can often be used to determine the source of the water or
its contaminants. State or local health agencies may have records
of bacterial and chemical analyses of ground water within their
area of responsibility. The USGS Water-Supply Paper 2254 (Hem,
1989) contains an excellent discussion on the interpretation of the
chemical characteristics of natural water.

Finally, the quality of water usually must meet Federal, State,
and local water quality standards if it is to be used for aquifer
recharge or discharged to a surface-water body.

(d) Climatic Data.—In major ground-water investigations,
records of precipitation, temperatures, wind movement,
evaporation, and humidity may be essential or useful supplemental

data. The source of such records in the United States is the National Weather Service. In ground-water studies, climatic data are used principally for estimating the seasonal variations and amounts of precipitation which may be available for ground-water recharge. This precipitation availability estimate must be determined for any complete estimate of ground-water availability. However, in many studies of limited extent, such detail is not necessary or justifiable. If the determination is needed, the detailed methods can be found in any complete text on hydrology (American Society of Civil Engineers [ASCE], 1952; Criddle, 1958; Freeze and Cherry, 1979; Fetter, 1988; Hamon, 1961; Skeat, 1969; Kazmann, 1965; Linsley et al., 1949; Linsley et al., 1958; Lowry and Johnson, 1942; Rouse, 1949; Todd, 1980; Wisler, 1959).

(e) Streamflow and Runoff.—Surface-water data may be essential in solving the ground-water equation because seepage to or from streams is a major element of discharge or recharge of ground water. Records of water use, runoff distribution, reservoir capacities, return flows, and stream section gains or losses may be available on the area under investigation. The best records on streamflow are those obtained from continuously recording gauges, but some information can be obtained from staff gauges and rating curves if the gauges have been read frequently. If the study is sufficiently critical, the installation of continuous recorders may be justified. The Water Resources Division of the USGS and State and local water resource agencies are sources of streamflow data.

(f) Soil and Vegetative Cover.—Soil maps and reports are readily available for most areas of the United States and are very useful in estimating recharge rates. Soil maps and reports supply information on soil characteristics and surface gradients which influence runoff and infiltration. Vegetative cover maps serve multiple purposes. They may show areas of phreatophytes where the ground water is close to the surface and may indicate the density and type of vegetation which intercepts precipitation, retards runoff, and transpires moisture. Both soil and vegetative cover maps can usually be obtained from the U.S. Department of Agriculture, State colleges and universities, or other Federal and State agencies interested in forestry, grazing, and agriculture. Where maps are not available, field observations and notes may be adequate for interpretative purposes.

2-5. Data Analysis.—

(a) Maps and Diagrams.—Analysis and evaluation of subsurface data for a ground-water study are readily performed using maps, cross sections, fence diagrams, and other similar illustrations. The size, scale, and symbols used for illustrations during the investigation stage are largely a matter of convenience and ease of use. Many drawings are maintained in an incomplete stage, and new data are added as they become available until the work is practically completed. However, consistent with Reclamation practice, the size, scale, and symbols used in final illustrations intended for inclusion in final reports should conform to a set standard such as Appendix A, Drafting Standards (Bureau of Reclamation, 1972) and the *Engineering Geology Office Manual* (Bureau of Reclamation, 1988). Whenever feasible, the scales of such illustrations showing related or interconnected information should be uniform to permit ready comparison and interpretation through overlays and other similar means.

The number and types of illustrations may vary depending upon the scope and intensity of the work and the complexity of the area. It may be advisable to construct a geographic information system (GIS) base as a tool for understanding and manipulating the data that are recorded on various maps and diagrams. The term GIS encompasses the concepts of both automated mapping and data base management and uses computer graphics to show the spatial relationships of information contained therein. A GIS can be very useful for data manipulation. The process allows system and user to ask logical questions to extract meaningful information from a GIS data base. However, the data base may require a considerable upfront effort to construct; sufficient data may be available to make it worth the effort, and simply knowing the mechanics does not guarantee a useful tool. Like most analytical tools, a GIS requires considerable ground-water-related experience and judgment, in addition to computer skills, to be of much value. The information presented in the following section summarizes the maps more commonly used in ground-water studies and interpretations.

(1) Topographic Maps.—Although topographic maps may not be necessary for all ground-water studies, appreciation and understanding of topography are useful if not essential. For some reconnaissance studies, either a good planimetric map or aerial photographs may be used in the field study instead of a

topographic map. However, for more detailed studies, good topographic maps are a necessity. Topographic maps supply information on surface gradients and drainage patterns and are used as the basis for construction of cross sections and maps showing geology, depth to water, surface and water-table gradients, contributing and recharge areas, and related features and phenomena. Depending upon the type of terrain and the detail required, scales of satisfactory topographic maps range from 1/2 inch to the mile (1:126,700) to 4 inches to the mile (1:15,800). At times, maps with a scale of 1 inch to 400 feet (1:4,800) may be desirable for the detailed study of local phenomena within larger areas of interest. Desirable contour intervals range from 1 foot in areas of low relief or for large-scale detailed maps to 25 to 50 feet for rugged areas or small scale maps.

The USGS is the primary source of topographic maps, but other Federal agencies, including the U.S. Department of Agriculture and the U.S. Army Corps of Engineers, as well as various State agencies, are also sources of suitable maps. If a satisfactory map is not available, one may need to be prepared.

   (2) *Aerial Photographs.*—Aerial photographs must serve as a substitute for topographic maps in many areas. Photographs are available either as contact prints or enlargements at scales ranging from 1:20,000 to 1:4,000. Where the photographs have been taken with sufficient overlap, they may be used with a stereoscope to obtain a three-dimensional view of the terrain. Also, mosaics compiled from numerous individual pictures covering large areas are frequently available.

   The U.S. Department of Agriculture Soil Conservation Service and the USGS are major sources of aerial photographs and mosaics. These agencies, as well as Bureau of Reclamation, will usually have access to other sources of photographs. In addition to conventional black and white and color photography, side-looking radar, infrared photography, thermal scanner imagery, and other remote sensing techniques are often very useful.

   (3) *Geologic Maps and Sections.*—Geologic maps and sections (figure 2-5 is an example), especially when accompanied by adequate reports, are useful in most ground-water investigations and are essential where complex stratigraphy and structures are involved. Analyses of reports and maps give information on recharge areas, possible aquifers, water-level conditions, structural

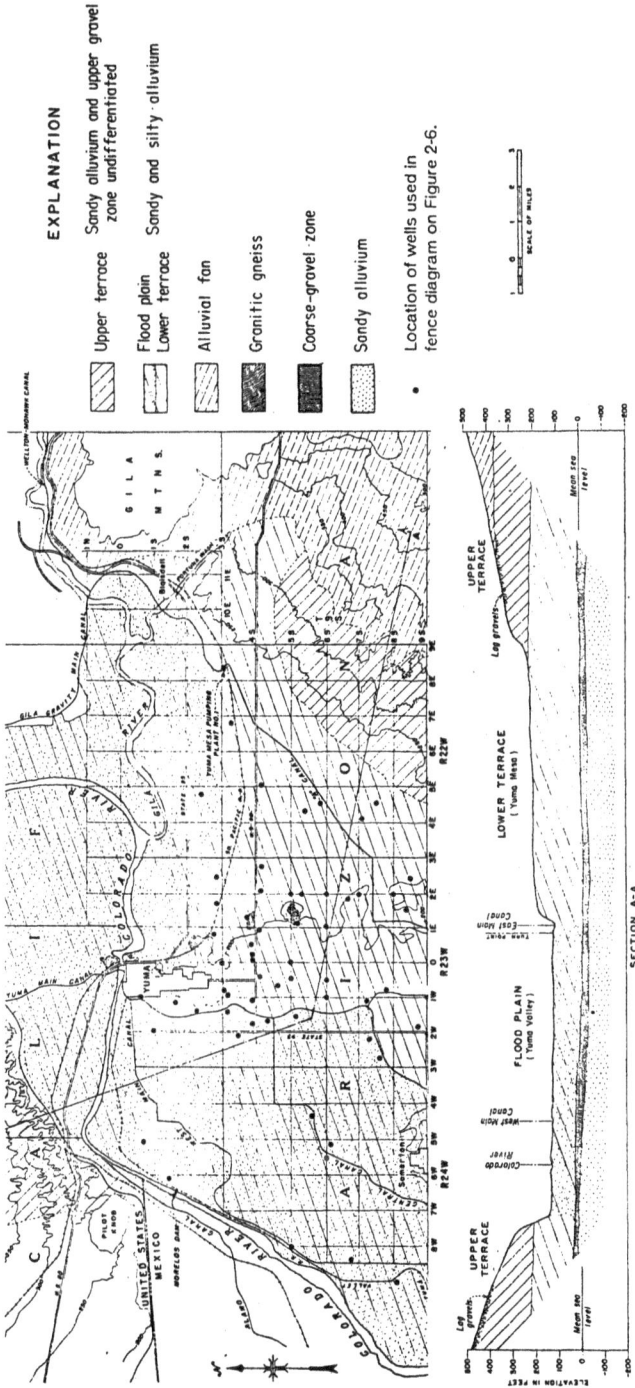

Figure 2-5.—Generalized geologic map and cross section of a
portion of the Yuma, Arizona, area.

and stratigraphic control of water movement, and related factors. The USGS and State geological agencies are primary sources of these materials.

Universities and colleges, geological societies, oil and mining companies, and other similar organizations also have data which may be obtainable. In areas for which no geologic reports or maps exist, a reconnaissance geologic investigation may be necessary as a minimum alternative.

(4) *Water-Table Contour Maps.*—A water-table contour map is the most commonly constructed and most useful map for studies of unconfined ground water. It is a topographic map of the water table, and the contour lines are usually lines of equal elevation (figure 2-6).

The map is constructed using water-level elevation in observation wells, stream and lake surfaces, and spring discharge points for controls.

(5) *Piezometric Surface Maps.*—A piezometric surface map is similar to a water-table contour map, except that it is based on the piezometric potential developed in piezometer or tightly sealed wells which penetrate a single confined aquifer (figure 2-7).

(6) *Depth-to-Water-Table Maps.*—Depth-to-water-table maps are of particular interest when considering drainage and dewatering problems (see figure 2-8).

They are most easily prepared by overlaying a water-table contour map on a surface topographic map. The points at which the contours intersect are a whole number of feet apart in elevation and are the control points for drawing a contour map of depth to water. They can also be prepared by calculating the depth to water from the ground surface and placing this depth figure on a map at the location of the observation well. Contours are then drawn connecting these points.

Care should be exercised in the preparation, use, and evaluation of ground-water level and depth maps. Initially, it should be remembered that only a limited number of spaced control points (observation wells, etc.) can normally be used and that ground-water conditions between the points may deviate widely from the expected. Furthermore, unless the control point facilities are constructed to reflect a specific condition, a composite condition

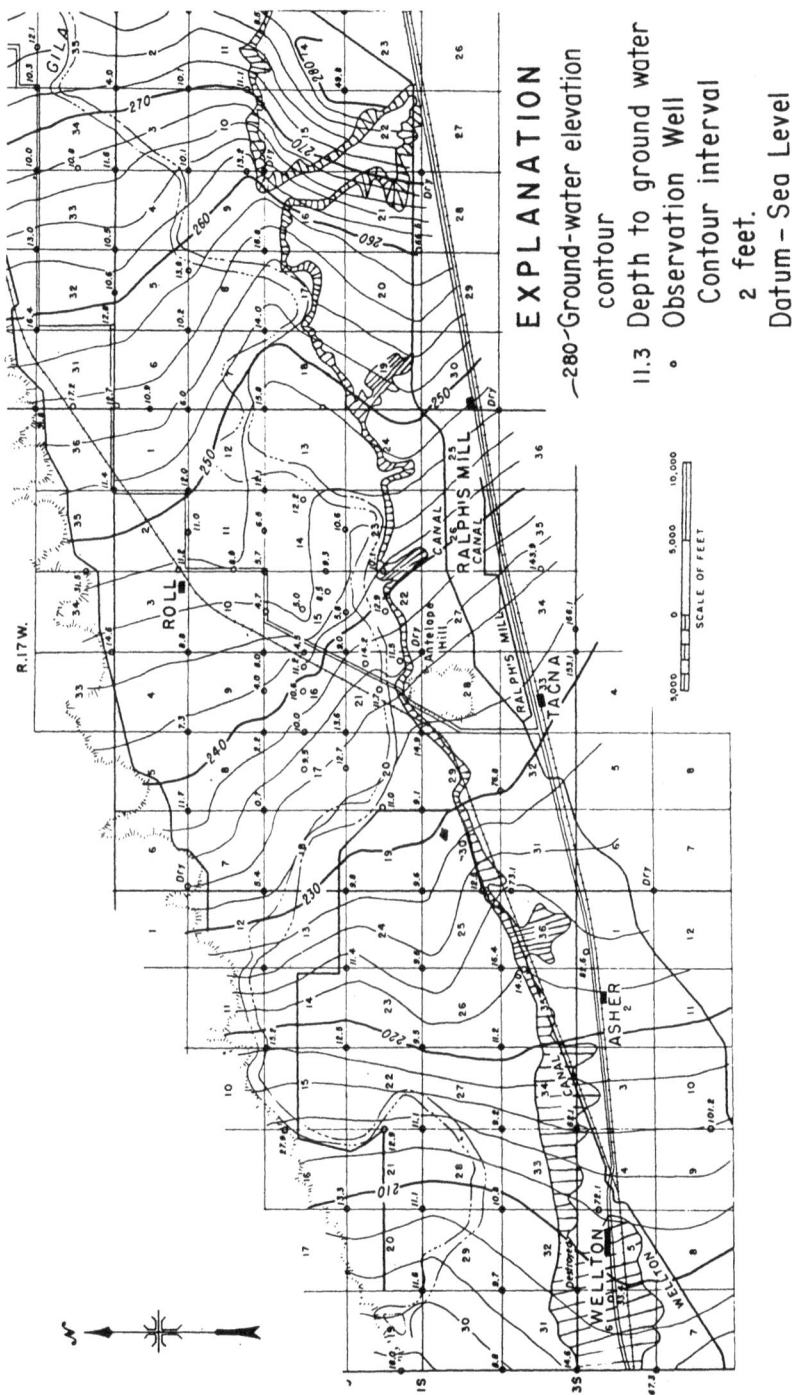

Figure 2-6.—Water-table contour map.

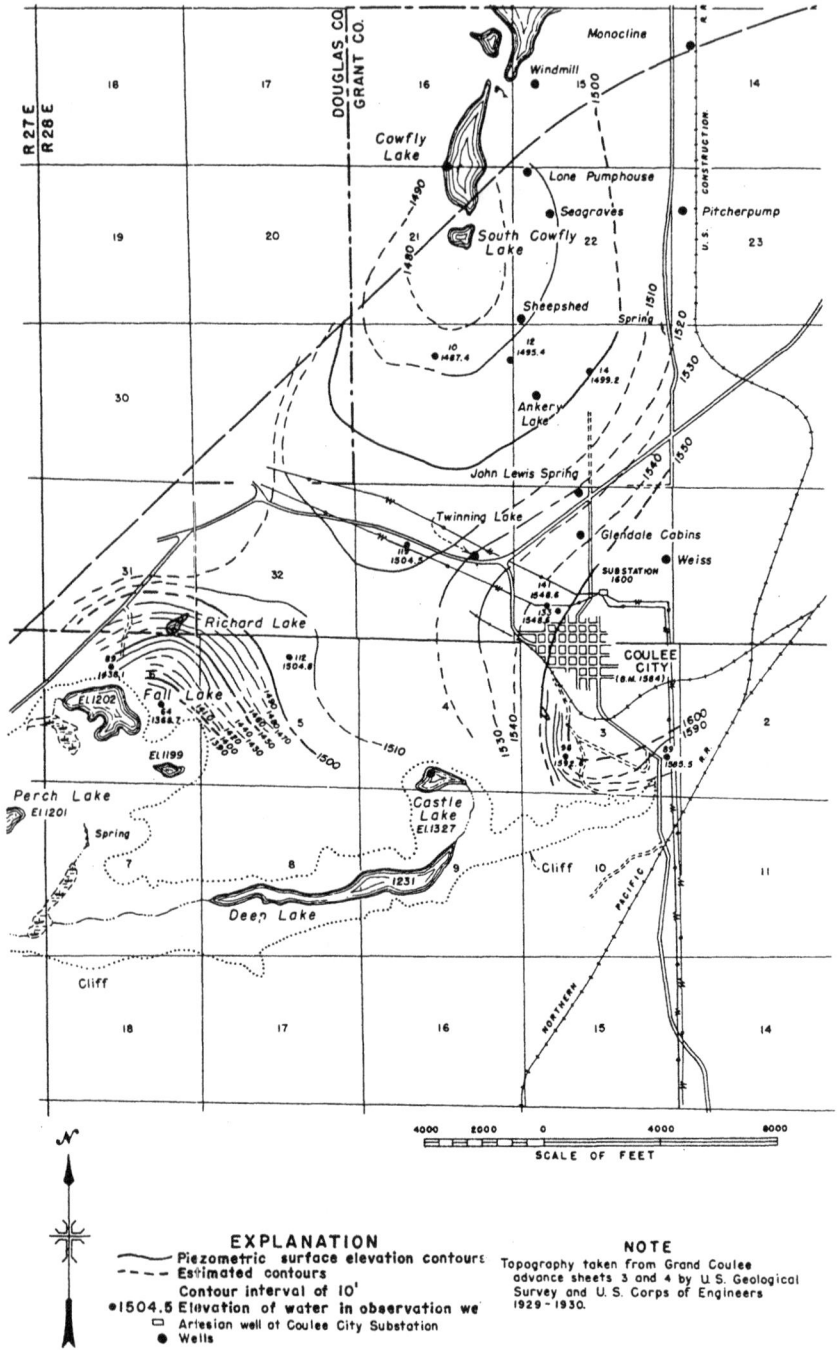

Figure 2-7.—Piezometric surface contour map.

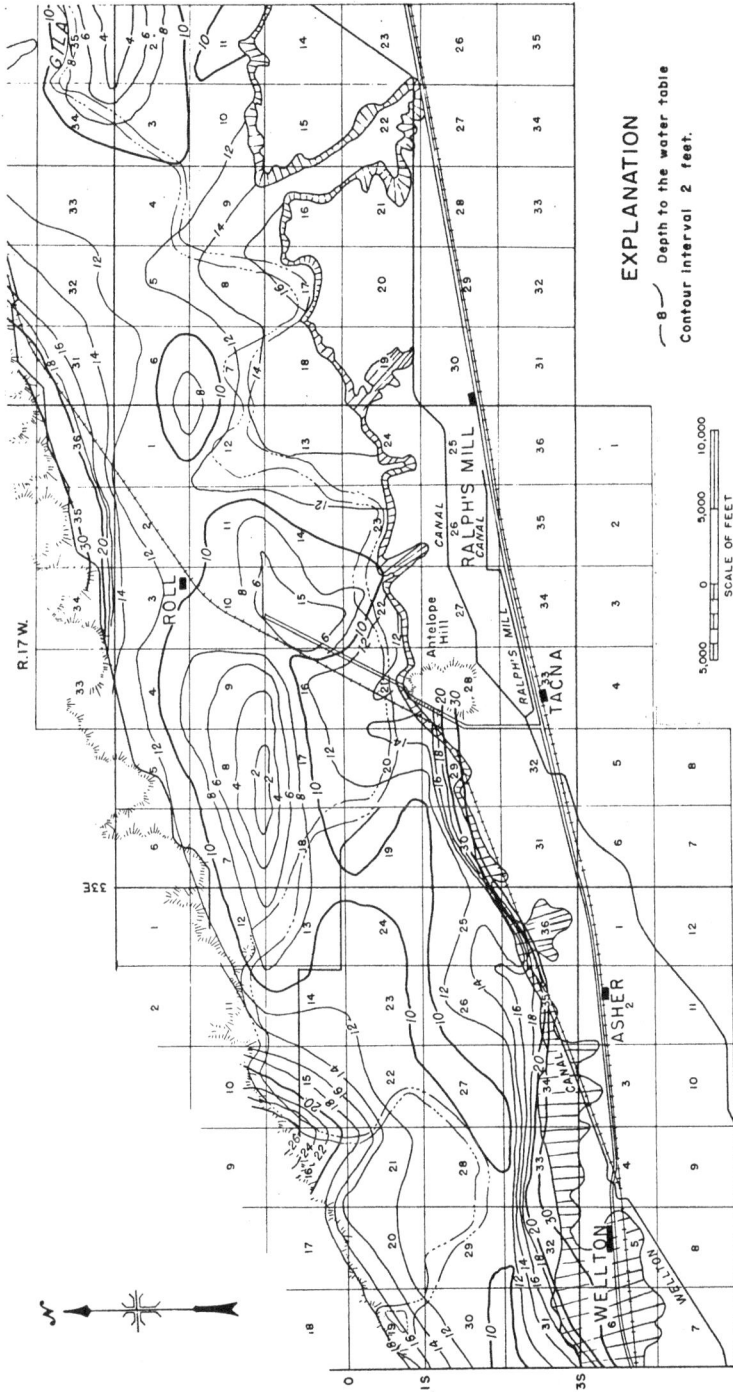

Figure 2-8.—Depth-to-water-table contour map.

such as a combined water-table and piezometric level may be reflected. This condition could yield erroneous and misleading data.

(7) Profiles or Cross Sections.—Vertical geologic and hydrogeologic profiles drawn through lines of wells or drill holes depict information on subsurface conditions by spatially relating surface features and subsurface conditions (see figure 2-9).

At each location, the geologic log of the hole is plotted vertically to show the top and bottom of each stratum that can be identified, and adjacent holes are compared to show continuity of strata. Unconfined water-table or piezometric surface levels can also be plotted at each well location for one reading or for a series of readings taken over a period of time. This plot will show the relative location of the free water-table or piezometric surface and its fluctuation during the period of the readings. Professional judgment is used to augment the available data. Cross sections should be referenced to a map for convenience in location. The horizontal scale of the section should conform to that of the map, but the vertical scale generally will need to be larger than the horizontal scale to make the drawing understandable. The vertical scale should be large enough so the smallest significant feature can be easily identified. This scale size may require a broken scale to show a thin stratum in a relatively deep geologic log.

(8) Isopach Map.—The isopach map is a thickness drawing shown as contours. It may show the thickness of saturated materials of a free aquifer or the thickness of an artesian aquifer between the upper and lower confining beds. A similar map may be drawn to show the thickness of a confining bed. Construction of maps of this type, of course, depends upon the availability of the logs of holes and wells that fully penetrate the beds of interest.

(9) Structure Contour Maps.—Structure contour maps are drawn to show the upper surface of a particular stratum or formation. These maps are primarily useful in conjunction with stratigraphy in interpreting structural features, such as faults and folds, which may control ground-water movement beneath an area.

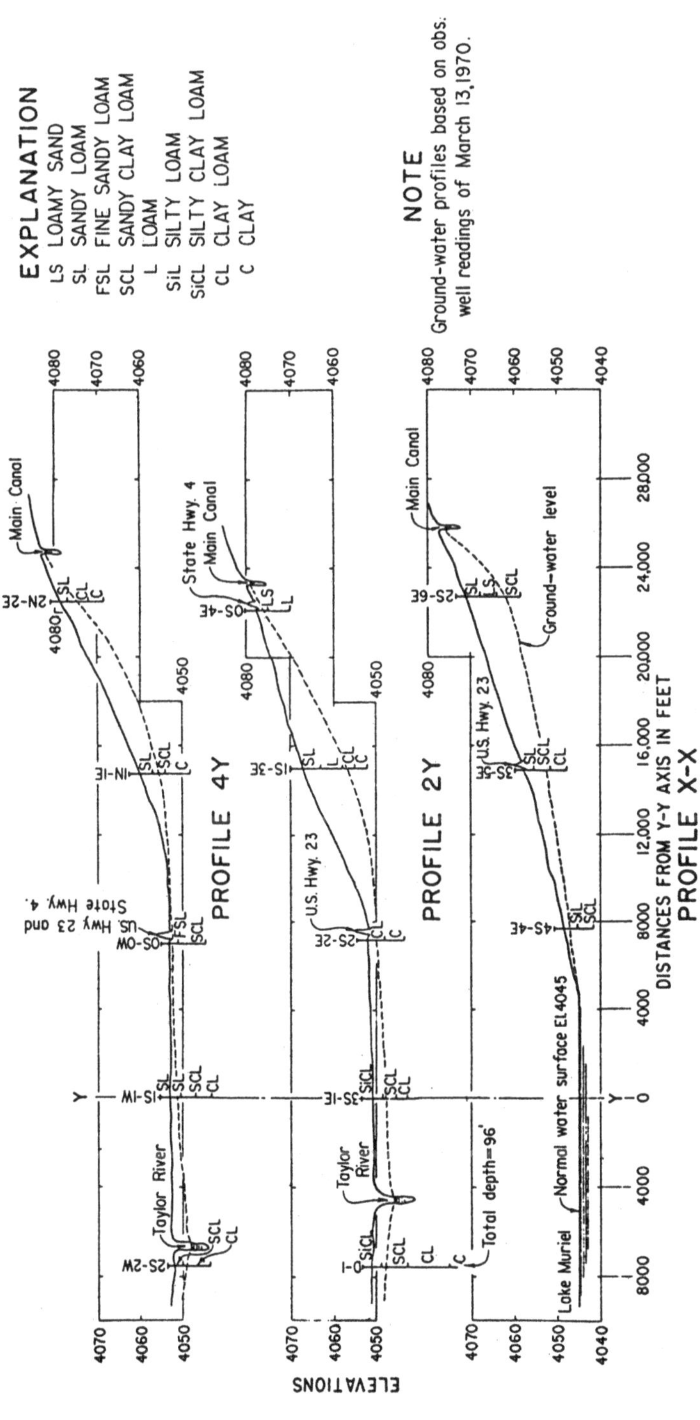

Figure 2-9.—Hydrogeologic profiles.

(10) Fence Diagrams.—Fence diagrams are three-dimensional cross sections that are helpful in presenting an areal picture of geologic and ground-water conditions. As with the sections, they are based on the logs of the holes, measurements of ground-water levels, and topography (figure 2-10).

(11) Hydrographs.—Hydrographs of individual observation wells and piezometers are essential in depicting ground-water fluctuations, trends, and other time-related factors. Hydrography is plotted on cross-section paper with water elevations as the ordinate and time as the abscissa (figure 2-11). Plotting the geologic log at the left margin usually enhances the value of the hydrograph.

(b) Ground-Water Map Interpretation.—The basic principle of ground-water flow holds that water moves from a higher level or potential toward the lower. The contours on ground-water elevation contour maps are those of equal potential and the direction of movement is at right angles to the contours. This movement is true whether the contours are of an unconfined water surface or of a piezometric surface. In an unconfined free aquifer, the contours often tend to parallel the land surface contours. In many instances, however, little apparent relationship exists between surface and subsurface flow.

Ground-water mounds can result from downward seepage of surface water or upward leakage from deeper artesian aquifers in areas of local recharge. In an ideal aquifer, gradients from the center of a recharge mound will decrease radially and at a declining rate. An impermeable boundary or change in transmissivity will affect this pattern and may provide clues in determining such changes.

Analysis of conditions revealed by ground-water contours is in accordance with Darcy's law, $Q=KiA$, which is discussed in section 5-1. Accordingly, the spacing of contours (the gradient) depends on the flow rate and on the aquifer thickness and permeability. If continuity of the flow rate is assumed, the spacing depends only on aquifer thickness and permeability. Thus, areal changes in contour spacing may be indicative of changes in aquifer conditions. However, in view of the heterogeneity of most aquifers, changes in gradients must be carefully interpreted with consideration of all possible combinations of factors.

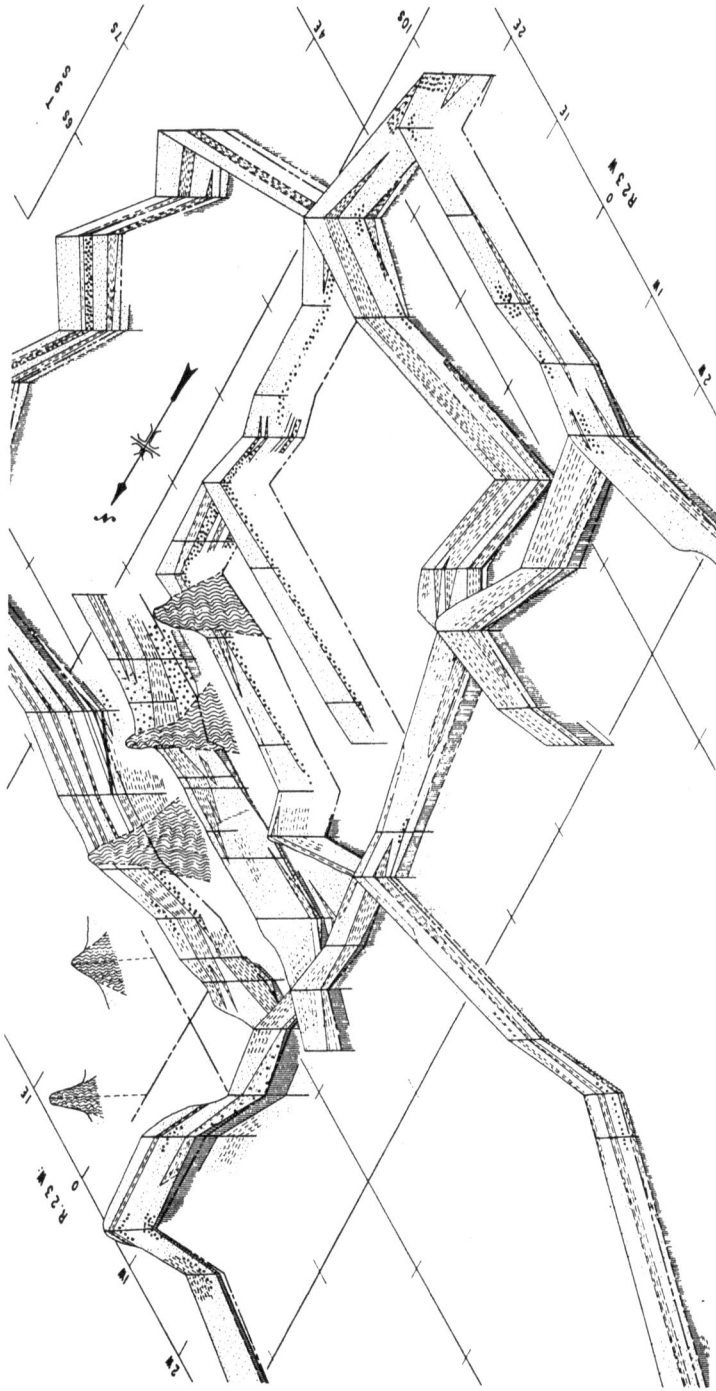

Figure 2-10.—Isometric fence diagram showing subsurface stratigraphy of a Yuma, Arizona, area (sheet 1 of 2).

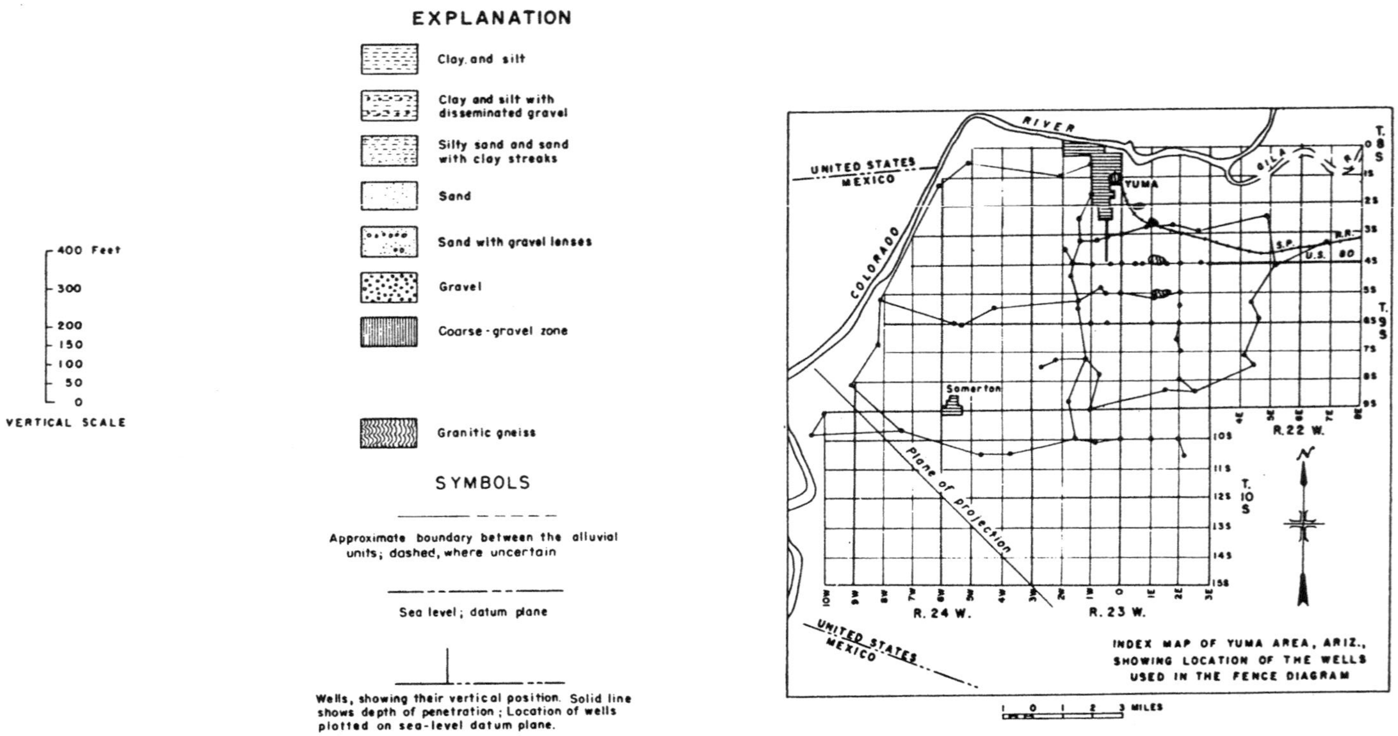

Figure 2-10.—Isometric fence diagram showing subsurface stratigraphy of a Yuma, Arizona, area (sheet 2 of 2).

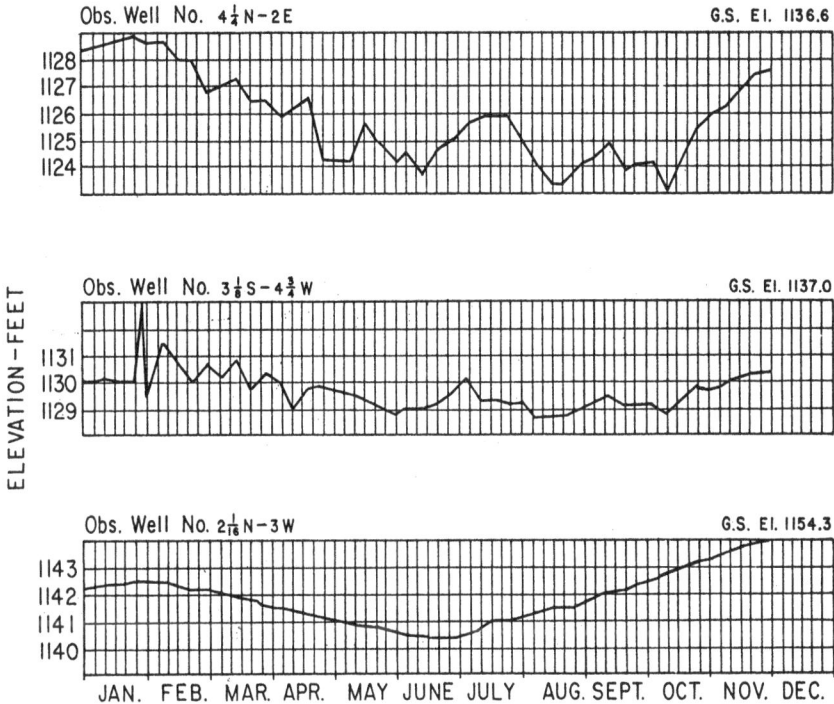

Figure 2-11.—Hydrographs of observation wells.

Pumping from a relatively small area of an extensive aquifer may cause little change in static water level over the unpumped area although the water level in the pumped portion continues to lower rapidly. This occurrence is the result of the pumpage exceeding the ability of the aquifer to transmit water to the pumped area, a condition that can be recognized by contours of the water levels within the aquifer.

An overlay of two ground-water contour maps made from measurements taken at different times permits an estimate of the change in ground-water storage which has occurred in the interval between the two series of measurements if the storativity (section 5-4) is known. Similarly, the same volume of change multiplied by the porosity gives an estimate of the change in gross storage. The latter is useful only in the event of a rising water table which saturates a volume that previously was relatively dry. The volume of water required to saturate the material can be estimated in this manner. To obtain the volume of water released from storage when the water table lowers, the storativity factor must be applied because the entire pore space will not be evacuated.

If the permeability and cross-sectional area (or transmissivity and width) of the aquifer are known and the gradient is available from a contour map, an estimate may be obtained of the rate of flow by applying Darcy's law.

Because aquifers act both as reservoirs and conduits, periodic estimates of the change in storage during the year may permit an estimate of the annual recharge. Similar estimates for a number of years may give an estimate of the average annual recharge.

The accuracy of the foregoing estimates depends upon the uniformity of the aquifer and the overall applicability of the aquifer characteristics as determined from pumping or other tests. Although the theory is simple, the heterogeneity of most aquifers necessitates caution and requires considerable judgment in the application of resultant data.

At some point the question should be asked, do the data base and analyses meet the objectives of the investigation? The data collection and analysis task is almost always an iterative process in which the results of early investigations are used to guide the direction and magnitude of ongoing investigations. The investigator must focus on stated objectives of the project and work toward these objectives, usually in an evolving plan of study. Only the very simplest investigations can be economically planned and carried out in a single stage. This process must incorporate a balance between economy and accuracy. We can never learn all there is to know about a ground-water unit, so experience and judgment must determine when the data base is sufficient to meet the objectives of the investigation.

Each discrete ground-water activity and product is treated in some part of this manual. The user should use this information as a guide—but not a constraint—as the investigation moves toward the objectives.

2-6. Report Preparation.—A ground-water report may range from less than a page long, containing a statement of the problem and a conclusion or recommendation, or both, to a voluminous work of many pages containing text, numerous maps, charts, graphs, and tables. The importance and complexity of the task, and the time and funds expended, generally determine the length and content of the report.

The author of the report should exercise judgment in determining the type of information that is necessary and the amount of detail required. The main body of the report may contain some or all of the following data in greater or lesser detail as required to clearly state the problems, conclusions, and recommendations. An outline of a typical report could be as follows:

A. Problem or purpose of study

B. Location and size of the area of interest

C. Cultural features of the area:
  1. Public utilities.—Electric power availability, location of existing lines, number of phases, and power rate schedule.
  2. Natural gas facilities.—Location, capacity, and rate schedule.
  3. Water supplies.—Domestic, municipal, industrial, irrigation, and stock. Sources, capacities, quality of raw and treated water, reliability of sources, and rate schedules.
  4. Sewage disposal.—Location and capacity of treatment plant, type of treatment, method of disposal of effluent, quality of effluent, and method of disposal of residue.
  5. Transportation.—Highways, roads, railroads, and shipping points.
  6. Settlement.—History if pertinent, location and size of towns, land ownership, and present and contemplated use.
  7. Cover and crops.—Vegetative cover, natural types and densities, crops, and crop acreage.
  8. Irrigation.—Extent, practices, and trends.

D. Climatic summary:
  1. Amount, rates, distribution, seasonal occurrence, and type of precipitation.
  2. Temperature extremes, monthly means, length of growing season.
  3. Wind directions, velocity, and seasonal occurrence.

4. Humidity.

5. Evapotranspiration.

E. Surface hydrology:

1. Natural surface drainage, channel characteristics, runoff volumes and characteristics, flood potential, location of gauging stations, and the losing and gaining reaches of channels.

2. Surface-water bodies including natural lakes, swamps, reservoirs, etc., with their location, size, capacity, and fluctuations in water levels.

3. Present and proposed canals and drains: location, size, length, capacity, lining, losing and gaining reaches, and physical condition.

4. Quality of surface water: chemical, bacteriological, seasonal fluctuations, and trends in quality.

F. Geology and geomorphology:

1. Summary of the physiography.

2. Elevations and relief.

3. Surface gradients.

4. Summary of the regional geology.

5. Stratigraphy and lithology.

6. Geologic structure.

7. Summary of the more important hydrogeologic provinces.

8. Unstable formations from standpoint of well (drilling, construction, and design).

9. Areas of possible subsidence.

10. Earthquake danger or potential.

G. Ground-water hydrology:

1. Location, depth, thickness, lithology, areal extent, and type of aquifer or aquifers present.

2. Water-table and piezometric surface gradients, direction of flow, recharge and discharge areas, areas of artesian pressure, contributing areas.

3. Seasonal and annual fluctuations in ground-water levels, extremes, and long-time trends.

4. Present ground-water development, including: number of wells, locations, depths, screen diameters, settings, lengths, and types; casing diameter, type, and weights; yields; drawdowns; pumping lifts; and annual pumpage.

5. Well history, average life, experience with encrustation, corrosion, sand pumping, collapse, and surface caving.

6. Transmissivity and storativity of aquifers.

7. Quality of ground water, chemical, bacterial, and trends; corrosivity of water and soils.
8. Suitability of aquifers for proposed development or use.

H. Analytical methods used:
1. Techniques used.
2. Discussion of assumptions and measured values.
3. Modeling.
  a. Description of models.
  b. Format of results.

I. Proposed ground-water development program:
1. Number, location, and spacing of proposed production wells.
2. Probable capacitor, pump lifts, and horsepower requirement of proposed wells.
3. Proposed well design.
4. Recharge possibilities; location, type, and design of facilities required; sources, volume, and quality of recharge water; probable operation and maintenance problems.

J. Factors and facilities for ground-water development:
1. Number of drilling contractors in area; number, type, and capacities of rigs.
2. State and local laws and regulations governing ground-water rights, drilling permits, design and construction of wells, and licensing of drillers; name and location of State or local offices administering such rules and regulations.
3. Water well supply dealers, pipe dealers, chemical supply houses, well logging and geophysical survey companies, laboratories capable of making mechanical analyses of samples, chemical and bacterial water analyses and soil tests, and sources of materials such as gravel for packs.

K. Maps of the study and adjacent area are usually of uniform scale except for the location map. The following is a list of maps which frequently appear in reports:

  • General location map.—shows location at study area within a larger area which includes known features such as counties, cities, and towns. An inset usually shows location within the State.
  • Planimetric.—Shows county and land office subdivision lines, existing location of wells, towns, highways, railroads, public utilities, etc. May be used as a base for other maps.

- Topographic.—May be used as base for other maps.
- Geologic.—Usually shows surficial geology with symbols indicating structural features, cross-section lines, geophysical surveys, etc. (figure 2-5).
- Ground-water and piezometric surface maps showing water surface elevation contours at minimum and maximum periods; the location and elevation notations at each point of measurement; and contributing, recharge, and discharge areas and flowing well areas, if present. Several maps may be used to show changes with time or season (figure 2-6).
- Isobathic or depth of water.—Similar to No. 5 above, but shows depth to water by contours (figure 2-8).
- Isopachic of aquifer or aquifers.—Similar to No. 6, but shows thickness of aquifers by contours.
- Surface-water map showing natural surface drainage, surface- water bodies, existing and proposed dams, canals and drains, and stream gauging stations.
- Land ownership.—Farm unit boundaries.
- Land use and vegetative cover.
- Quality of water.—Chemical and bacteriological.
- Aquifer characteristics.—Variations in transmissivity and storativity values by contours or by areas containing values within a given range.
- Isohyetal or Theisen polygons for precipitation showing location of weather stations.
- Well field and service area, ground-water facilities, and plans.

L. Cross sections, fence diagrams, and hydrographs:
   1. Geologic.—Includes control exploration hole designations.
   2. Hydrologic.—Could be several cross sections showing seasonal variations, including measuring point well locations (figure 2-9).

M. Graphs, charts, and tables:
   1. Temperature range and growing season.
   2. Average annual monthly precipitation.
   3. Annual precipitation, minimum, mean, and extreme.
   4. Cumulative precipitation.
   5. Stream and lake hydrographs—baseflows.
   6. Ground-water observation well hydrographs.
   7. Quality of water, both areal and seasonal.
   8. Projections of water use, population, power, etc.
   9. Ground- and surface-water use.
   10. Mechanical analyses of aquifer samples.
   11. Chemical and bacterial analyses of water samples.

12. Pump test measurements and analyses.
13. Well logs.—Drillers, geologist, resistivity, etc.
14. Geophysical surveys.
15. Evapotranspiration.—Records or estimates.

N. Drawings:
  1. Well and infiltration gallery designs.
  2. Test site layouts.
  3. Special equipment designs.

No report will include all of the items listed here, but preparers of reports can use this outline as a checklist to produce complete and useful reports. The outline may also be used in the planning stage to provide insight into what activities will take place.

2-7. Bibliography.—

American Society of Civil Engineers (ASCE), 1952, *Hydrology Handbook, ASCE Manual of Engineering Practice No. 28,* New York.

Appel, C.A. and J.D. Bredehoeft, 1976, "Status of Groundwater Modeling in the U.S. Geological Survey," U.S. Geological Survey Circular 737, 9 pp.

Bachmat, Y., B. Andrews, D. Holtz, and S. Sebastian, 1978, "Utilization of Numerical Groundwater Models for Water Resource Management," Environmental Protection Agency Report EPA-600/8-78-012, 178 pp.

Bureau of Reclamation, April 1988, *Engineering Geology Office Manual,* Denver, Colorado.

Bureau of Reclamation, December 1972, "Reclamation Instructions - Series 130, Design - Part 134, Drawings - Appendix A, Drafting Standards," Washington, DC.

Butler, S.S., 1957, *Engineering Hydrology,* Prentice-Hall, Englewood Cliffs, New Jersey.

Coats, K.H., November 1969, "Use and Misuse of Reservoir Simulation Models," *J. Pet. Tech.,* pp. 1391-1398.

Criddle, W.D., January 1958, "Methods of Computing Consumptive Use of Water," Proceedings of the ASCE, Journal of Irrigation and Drainage Division, vol. 84, No. IR1, Paper 1507.

Davis, S.N. and R.J.M. DeWiest, 1966, *Hydrogeology*, John Wiley & Sons, New York.

Hamon, W.R., May 1961, "Estimating Potential Evapotrans-piration," Proceedings of the ASCE, Journal of the Hydraulics Division, vol. 87, No. HY3, pp. 107-120.

Hem, J.D., 1989, "Study and Interpretation of the Chemical Characteristics of Natural Water," 3d edition, U.S. Geological Survey Water-Supply Paper 2254.

Karplus, W.J., 1976, "The Future of Mathematical Models of Water Resource Systems," In: *System Simulation in Water Resources* (G.C. Vansteenkiste, editor), North-Holland Publishing Co., pp. 11-18.

Kazmann, R.G., 1965, *Modern Hydrology*, Harper and Row, New York.

Linsley, R K., Jr., M.A. Kohler, and J.L.H. Paulhus, 1949, *Applied Hydrology*, McGraw-Hill, New York.

Linsley, R.K., Jr., M.A. Kohler, and J.L.H. Paulhus, 1958, *Hydrology for Engineers*, McGraw-Hill, New York.

Lowry, R.L., Jr. and A.F. Johnson, 1942, "Consumptive Use of Water for Agriculture," Transactions of the ASCE, vol. 107, Paper 2158, pp. 1243-1266.

Mercer, J.W. and C.R. Faust, 1986, *Ground-Water Modeling*, 2d edition, National Water Well Association, Dublin, Ohio.

Moore, J.E., 1979, "Contribution of Ground-Water Modeling to Planning," Journal of Hydrology, vol. 43 (Oct.), pp. 121-128.

Prickett, T.A., 1979, "Ground-Water Computer Models State of the Art," *Ground Water*, vol. 17, No. 2, pp. 167-173.

Rouse, H. (editor), 1949, "Engineering Hydraulics," Proceedings of the Hydraulics Conference, University of Iowa, Iowa City, Iowa.

Skeat, W.0. (editor), 1969, *Manual of British Water Engineering Practice*, 4th edition, vol. 11, Engineering Practice, W. Heller and Sons, Cambridge.

Thomas, R.G., 1973, "Groundwater Models," Irrigation and Drainage Paper 21, Food and Agriculture Organization of the United Nations, Rome, 192 pp.

Van Poollen, H.K., H.C. Bixel, and J.R. Jargon, July 1969, "Reservoir Modeling - 1: What it Is, what it Does," Oil and Gas Journal, pp. 158-160.

Wisler, C.O., and E.R. Brater, 1959, *Hydrology*, 2d edition, John Wiley & Sons, Inc., New York.

Chapter III

INITIAL OPERATIONS AND AQUIFER YIELD ESTIMATES

3-1. Introduction.—Proper management of a resource such as ground water requires knowledge of the magnitude, distribution, depletion, and replenishment, if any, of the resource. Without such ·an assessment, the effects of past development and predictions of the influences of future development cannot be adequately determined. Budgets and inventories provide the means of assessment of ground-water resources and involve such factors as storage, recharge, and discharge. Because of the interrelationship of surface and ground water, comprehensive, quantitative budgets and inventories must consider both modes of water occurrence.

The ground-water basin boundaries, both vertical and horizontal, and aquifer dimensions and characteristics must be determined before ground-water storage capacity can be estimated. Aquifer characteristics are rarely uniform over large areas, so the variations must be quantified and delineated using subsurface geologic and hydrologic conditions, well capacities, and similar differentiating factors.

Ground water is a dynamic resource with constantly changing water levels caused by natural or artificial influences. Interpretation is facilitated by careful analysis of water level fluctuations as related to these influences. Such interpretation may need to be supplemented by pumping tests located on the basis of the initial studies to clarify localized variations in aquifer characteristics and boundaries.

Such data are essential to aquifer analysis, including electric analog or digital computer analysis of aquifer response to development. For assessments of long-time aquifer yield and performance, evaluations are usually based on an average annual basis and maximum high and low water conditions. The basic results of a ground-water inventory are the determination of the total water in storage and the annual change. Further studies involve the response of the system to various schemes of development, possibilities of induced additional recharge caused by development, artificial recharge, desirability and probable life of a water mining operation, design of wells, conjunctive use of surface and ground water, and possibilities of subsidence.

The techniques of a ground-water evaluation are relatively
subjective, and the degree of accuracy and the reliability of the
initial result are often questionable. Many evaluation studies
involve a continued reassessment and refinement of the estimates
as more data on actual response of the aquifer to development
become available.

As previously mentioned, a ground-water study involves consideration not only of ground water but of surface water. The boundaries of a ground-water reservoir may or may not coincide with
those of an overlying surface-water basin. If they do, the study
may be simplified. Many investigations may require the setting of
arbitrary boundaries.

Most ground-water inventory methods were revised for application in semiarid and arid zones, generally in response to obvious
overdevelopment. No wholly standardized investigation procedure
exists because conditions and needs vary. Methods will vary
depending upon areal development; complexity of the geology;
climate; availability of existing data; time, funds, equipment, and
manpower available to obtain data; and similar factors. The
following sections summarize many of the factors involved in
ground-water inventory and methods of procedure.

3-2. Hydrologic Budgets.—The hydrologic budget is a
quantitative evaluation of the total water gained or lost from a
basin or part of a basin during a specific period of time. It
considers all water, whether surface or ground water, entering,
leaving, or stored within the area of study. The hydrologic budget
is summarized in the following equation:

$$\Delta S = P - E \pm R \pm U \qquad\qquad 3\text{-}1$$

where:

ΔS = changes in storage in channel and reservoirs, in ground-
water storage, and in soil moisture
P = precipitation on the area of study
E = evapotranspiration from the area
R = net surface-water inflow or outflow
U = net ground-water outflow and inflow

The components of ΔS are:

ΔS_s, changes in surface storage which may be available in the form of reservoir or lake capacity curves

ΔS_c, changes in stream channel storage, which are of minor importance in a long-time budget (usually ignored in the budget analysis)

ΔS_m, changes in soil moisture which are also of minor importance, hence ignored

ΔS_g, changes in ground-water storage

which can be estimated from contour maps of the water table or piezometric surface and the storativity of the aquifer at the beginning, during, and end of the study period or periods.

Precipitation consists of all the rain and snow falling on the area. Records are usually available from weather stations in or adjacent to the area of study. Methods of analysis of the various factors are discussed in several references (Butler, 1957; DeWiest, 1966; Skeat, 1969; Kazmann, 1965; Wisler and Brater, 1959).

Estimating the long-term evapotranspiration is an intricate process which has been studied and discussed by numerous authors. Many methods of analysis have been developed, all of which are approximate. American Society of Civil Engineers (ASCE) - Methods and Reports on Engineering Practice - No. 70, *Evapotranspiration and Irrigation Water Requirements*, 1990, provides and indepth discussion of these parameters. Generally, it is advisable to seek the assistance of someone familiar with the subject of evapotranspiration when estimating quantities for a water budget.

Streamflow, R, consists of surface runoff of precipitation within the area, R_s; surface inflow to the area, R_i; water pumped from aquifers and exported from the basin, R_p; and ground-water seepage to streams, R_g. The value of R_i can be estimated from stream gauging records. A number of methods for estimating R_g are discussed in the literature, particularly for individual basins (ASCE, 1952; Butler, 1957; DeWiest, 1966; Skeat, 1969; Kazmann, 1965; Linsley et al., 1949; Linsley et al., 1958; Rouse, 1950; Wisler and Brater, 1959).

Ground-water flow components are U_t, the underflow from, and U_o, the underflow to adjacent basins. Flow can be estimated by determining the width of the flow path from knowledge of the aquifer dimensions, the gradient from water level contour maps, and the transmissivity from results of pumping tests or other sources. These factors can be applied to the solution of Darcy's equation (chapter II) to determine total underflow.

3-3. Ground-Water Inventories.—The ground-water components of the hydrologic cycle used in estimating a ground-water budget are summarized in the equation:

$$G - D = \Delta S_g \qquad\qquad 3\text{-}2$$

where:

G = recharge to the aquifer
D = discharge from the aquifer
ΔS_g = change in storage in the aquifer

The components of G may include: deep percolation from precipitation; seepage from surface-water bodies; ground-water underflow from adjacent areas; artificial recharge including deep percolation from irrigation, sewage disposal facilities, and recharge wells; and leakage through confining beds.

Components of D may include evapotranspiration; seepage to surface-water bodies; ground-water underflow to adjacent areas; discharge of springs; artificial discharge including drainage systems, wells, and infiltration galleries; and discharge through confining beds.

Adequate records to permit an accurate and reliable appraisal of all the factors involved are seldom available.

Changes in ground-water storage are reflected by fluctuations in the ground-water levels. Because most assessments are based on long-term averages with the beginning and end of the study period occurring at about the same season of the year, changes in soil moisture can usually be ignored.

Estimates on the portion of rainfall which may enter an aquifer are generally made by an analysis of the fluctuations of ground-water level as the result of a specific isolated storm or on the basis of a long-time correlation between water level hydrographs and precipitation records. The deep percolation from an individual storm is influenced by the intensity and duration of the precipitation and the deficiency of soil moisture at the beginning of the storm. The long-period correlation will therefore usually give a more nearly effective average. In using such analyses, it must be recognized that during the recharge, discharge from the aquifer continues, and the net change in the water table represents the difference between the recharge and discharge. Where artificial recharge, specifically for recharge of aquifers, is practiced, adequate records of rate and volume of inflow usually are available. Recharge from irrigation may be estimated from the difference between the consumptive use and water deliveries less the surface waste.

Where surface- and ground-water basins do not coincide, underflow may be a major factor in an inventory. Estimates of such flow should be based on dimensions of the aquifer, gradient, and transmissivity as described earlier.

Rough estimates of ground-water withdrawals for irrigation during a given period can be made by several methods including:

- Survey of well owners to determine rates and duration of pumping of all sizable wells in the area

- Survey of landowners and agricultural agencies to determine total acreages of common crops and normal water application for such crops

- Survey of utility companies to determine installed ratings of all sizable pumps and the power usage of these pumps

These data can be combined with other data, such as areawide depths to ground-water and pump efficiencies, to arrive at reasonably reliable rough estimates of withdrawals. Similar information for wells serving municipal and industrial uses may be available from owners, utility companies, and local or State regulatory agencies.

Evapotranspiration can be estimated by use of Blaney-Criddle (Criddle, 1958), Lowry-Johnson (Lowry and Johnson, 1942), or similar equations if data on crop types and acreages and vegetative cover maps are available.

Estimates of seepage to or from streams can be made based on the difference between surface inflow and surface outflow plus evapotranspiration. Based on ground-water contour maps, reaches of the stream where seepage predominates can be segregated and inflow-outflow measurements of each section can be made. Similar estimates of ground-water seepage for lakes or reservoirs are sometimes possible on the basis of inflow-outflow data corrected for evaporation plus or minus the change in storage.

Surface discharge of springs often can be measured. They are seldom uniform, however, so periodic measurements taken when the water-table elevations are measured are recommended to obtain a value of average discharge or to correlate the discharge with ground-water elevation.

As discussed previously, ground-water inventory studies are often subjective and influenced by local condition, availability of data, time and funds, and climatic variations.

3-4. Perennial Yield Estimates of Aquifers.—An estimate is often desired of the probable perennial yield of an aquifer. A number of methods that have been derived for such estimates in arid areas are summarized by Todd (1959, 1980).

O.E. Meinzer (1932) wrote: "The most urgent problems in ground-water hydrology at the present time are those relating to the rate at which rock formations will supply water to wells in specified areas—not during a day, a month, or a year, but perennially" (McWhorter and Sunada, 1984).

The concept that is presently used is "safe yield," which is defined as the amount of naturally occurring ground water that can be economically and legally withdrawn from an aquifer on a sustained basis without impairing the native ground-water quality or creating an undesirable effect such as environmental damage (Fetter, 1988). In this case, recharge must equal discharge.

Because safe yield is based on a long-term average recharge, pumping and recharge may be out of balance for any given year. Water levels will usually rise in wet years and fall in drought years, reflecting this imbalance.

3-5. Initial Operations.—Before initiating field work, the field force should be familiar with the requirements and procedures outlined in chapters II and VIII of this manual and should review the basic data available for the area involved. A preliminary reconnaissance survey should then be made of the area, paying particular attention to those subareas for which published reports lack data or indicate the existence of problems such as ground-water overdevelopment, waterlogging or poor drainage, and saline or alkaline soils. In addition, observations should be made of:

- Geomorphological features which might influence the occurrence of ground water

- Surface elevations and gradients

- Soil and rock textures

- Stream pattern, gradients, and bed characteristics springs, seeps, and marshy areas

- Vegetation types and densities; distribution, density, and type of water wells

- Land-use patterns, size of farm units, and land ownership

- Present water use and the relationships of these features to the general geology

During the survey, information on the location and capabilities of laboratory facilities, drilling contractors, well service companies, and similar organizations whose services might be required should be obtained and, if possible, initial contacts should be made. The survey should permit a delineation of those subareas requiring additional geologic, topographic, or other mapping, and the outline of an initial program for the work to be done.

On large projects, much of the required information on political, social, and economic factors, such as utilities, land use, and ownership, will have previously been prepared within the Bureau of Reclamation or by other agencies. Similarly, these offices may have data on climate and surface-water hydrology. When the data are not readily available and the ground-water hydrologist is unfamiliar with surface-water hydrology, the type of information that may be required should be carefully determined and knowledgeable advice should be sought on obtaining these data.

3-6. Records of Wells, Springs, Seeps, and Marshes.—One of the early activities usually undertaken in a ground-water investigation is an inventory of existing ground-water facilities and collection of well logs. In many States, the State Engineer's Office, the Water Resources Board, or similar organizations will usually have files of well records giving location, depth, formation, logs, casing and screen used, static water level, pumping water level, yield, date drilled, the driller, and similar data. Copies of the records should be obtained, and the data for each well should be entered on a form similar to the form shown on figure 3-1. Each well should be numbered, and a map showing its location should be prepared.

The records in the State offices are often incomplete or questionable, so each well should be checked in the field. If not included in the State records, the ground surface elevation at the well should be determined either by leveling or, if a suitable topographical map is available, by observation and interpolation. In the case of especially useful wells, the owner should be contacted for additional data and permission to measure the depth of water in the well and, if required, to make a pumping test. Also, the driller (if available) can often furnish data.

Wells which are not on the State records may be found in the field, and such wells should be inventoried. Also, data on new wells drilled during the study should be obtained from the driller and owner.

Wells are usually tied to the state plane coordinate system or the township-range system shown on figure 3-2 used by the U.S. Geological Survey (USGS) for location and mapping purposes.

One of the first steps normally taken during a ground-water investigation is the measurement of water levels in wells. Because ground-water levels are dynamic, measurements should be made in all wells in as short a period as possible. Figure 3-3 is a typical form used in making such a survey. On completion of the first complete ground-water elevation measuring program, a ground-water contour map should be prepared, preferably using a topographic map as a basis. A study of the map will permit recognition of those points where control is poor or lacking and will serve as a guide in locating observation wells that must be drilled. The contour intervals on the topographic map control the precision of interpretation. Larger contour intervals are subject to greater inaccuracy. In addition, the map can be used to indicate possible ground- and surface-water interrelationships. It also indicates locations for observation wells near streams, canals, lakes

WELL RECORD

Region_____ Date_____ 19____ Recorded by_____
Project   Well No.
or Unit _____ or Name_____
Source of data_____
_____

1. Location: State_____ County_____
   Map_____
   _____ ¼ _____ ¼ Sec._____ T_____ N/S R_____ E/W
2. Owner:_____ Address_____
   Tenant_____ Address_____
   Driller_____ Address_____
3. Topography_____
4. Elevation_____ m(ft)
5. Type: Dug, drilled, driven, bored, jetted
   date__ _____ 19____
6. Depth: Reported_____ m(ft)
            Measured_____ m(ft)
7. Casing: Diam._____ mm(in),to
   _____ mm(in), type_____
   Depth_____ m(ft), Finish_____
8. Chief aquifer_____ From_____ m(ft)to_____ m(ft)
   Others_____
9. Water level_____ m(ft)rept./meas._____ 19__ below_____
   _____ which is _____ m(ft)below surface
10. Pump:Type_____ Capacity_____ l/min(gal/min)
    Power:Kind_____ Horsepower_____
11. Yield:Flow_____ l/min(gal/min),Pump_____ l/min(gal/min)
    Drawdown_____ m(ft)after_____ hours pumping_____ l/min
12. Use: Dom.,Stock, PS, R.R., Ind., Irr., Obs._____
    Adequacy, permanence_____
13. Quality_____ Temp_____ ºF
    Taste, ordor, color_____ Sample yes/no
    Unfit for_____
14. Remarks_____
    _____
    _____

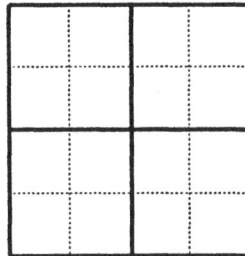

Figure 3-1.—Typical well record form.

R. 29 W.

6	5	4	3	2	1
7	8	9	10	11	12
18	17	16	15	14	13
19	20	21	22	23	24
30	29	28	27	26	25
31	32	33	34	35	36

T.
7
S.

Well No. 7 - 29 - 12 aad

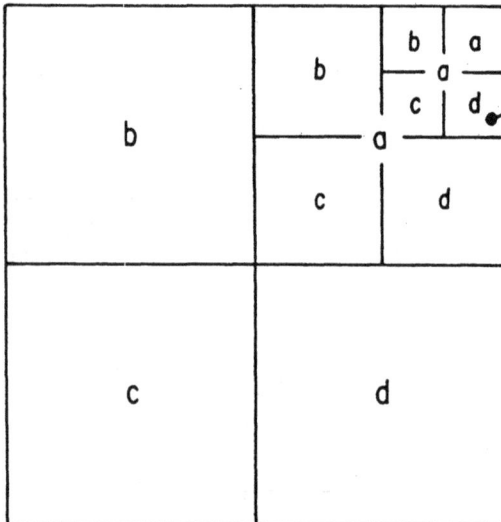

SECTION 12

Figure 3-2.—USGS township-range numbering system.

WATER LEVEL MEASUREMENTS (Field)

Measured by

Location of project

DATE	HOUR	WELL	TAPE READING AT-		DEPTH TO WATER	REMARKS
			MEAS.POINT	WATER LV.		

Figure 3-3.—Typical water level record form.

reservoirs, and other bodies of water, as well as locations for staff gauges (Bureau of Reclamation, 1967). Staff gauge locations are also plotted on the map, and a form similar to the one on figure 3-4 can be completed.

The presence of springs, seeps, and marshes is usually the result of the water table intersecting the ground surface or of leakage from an artesian aquifer. Accordingly, the location, discharge, and water level elevation of such features may be significant. During the initial well survey, all springs, seeps, and marshes should be visited, given a distinctive symbol and number, plotted on the map, and recorded on a form similar to the one on figure 3-4. The approximate water level elevation should be determined, a gauge or weir should be installed, and an effort should be made to determine the geologic and hydrologic conditions giving rise to the feature. Flow may vary diurnally, seasonally, or in some other regular or irregular pattern. If these variations are significant, they should be measured and recorded for use in analyzing the ground-water system.

Remote sensing can often detect the discharge of seeps and springs because the wet areas exhibit more vegetation growth during dry weather than adjoining areas. On summer nights, the wet areas are cooler than the surrounding ground, and on winter nights, the wet areas are warmer. The contrast can be detected by airborne thermal sensing and infrared film that senses the 0.9- to 1.0-μm wavelength and that detects subtle differences in the ground surface temperature (Zimmie and Riggs, 1981).

3-7. Initiation and Frequency of Ground-Water Level Measurements.—Water level measurements may be made of the water table or the piezometric surface under numerous conditions. Continuous measurements such as those supplied by a continuous water stage recorder or of periodic measurements with a time interval extending from less than 1 minute (for a pump test) to 6 months may be required. The frequency of measurement should be adjusted to the circumstances. In some instances, only a few measurements are possible or expedient to make, but in other instances, frequent measurements over a long period of time may be required. The possibility of error in interpretation decreases as the frequency of measurement and length of record increase. In ground-water inventory and drainage investigations, water level observations may continue for many years. Initially, measurements are made often until the annual regimen is established. The frequency may then be reduced to about four a year with the

STAFF GAGE or WEIR RECORD (field)

Date:_____

Region:_____Recorded by:_____

Project or Unit:_____Staff Gage or Weir No._____

1. Location:State_____County_____

   Map_____

   _____¼_____¼ Sec_____T_____N/S R_____E/W

2. Owner of property:_____
   Address:_____

   Tenant of property:_____
   Address:_____

   Installed by:_____

3. Gage name and type:_____
   (Staff gage, cipoletti weir, etc.)
4. Feature measured, name and type:_____
   (Stream, lake, marsh, spring, etc.)
5. Elevation, zero index on gage or weir crest:_____m (ft)

6. Measured water surface:_____m (ft)

7. Elevation measured water surface:_____m (ft)

8. Discharge if determined:_____m^3/s (ft^3/s)

9. Date of measurement:_____19_____

10. Measured by:_____

11. Remarks:_____
   _____
   _____

Figure 3-4.—Staff gauge or weir record form.

exception of a few carefully selected observation wells. These wells may be read 6 to 24 times a year or equipped with continuous recorders.

Where water storage structures or new irrigated areas are contemplated, it is advisable to install a number of observation wells at least 2 years prior to construction and to take measurements monthly or bimonthly to determine pre-existing groundwater conditions. The program should be continued after construction and full operation to permit comparison of pre-existing and post-facility conditions. Such data may be invaluable in the event of claims or suits for damage.

3-8. Water Level Measuring Devices.—Measurements may be made with a number of different devices and procedures (figure 3-5). The most common device for measuring static water levels is probably the chalked steel tape, which has a weight attached on the lower end. The weight keeps the tape taut and aids in lowering it into the well. The tape is chalked with carpenter's chalk, ordinary blackboard chalk, or dry soil which changes shade upon becoming wet. The line of the color change denotes the length of tape immersed in water. Subtracting this length from the reading at the measuring point gives the depth to water. Cascading water in a well may mask the mark of the true water level on the tape; however, this condition usually occurs only in a well that is being pumped and may require the use of another measurement method. In small-diameter wells, the volume of the weight may cause the water level to rise in the pipe, and the measurement may be inaccurate.

Electric sounders are becoming more widely used to measure the depth to water in wells. A number of commercial models are available. Many sounders use brass or other metal indicators clamped around a conductor wire at 1.5-meter (5-foot) intervals to indicate the depth to water when the meter indicates contact. The spacing of these indicators should be checked periodically with a surveyor's tape to ensure accurate and reliable readings.

Some electric sounders use a single-wire line and probe and rely on grounding to the casing to complete the circuit; others use a two-wire line and double contacts on the electrode. Most sounders are powered with flashlight batteries, and the closing of the circuit by immersion in water is registered on a milliammeter. Experience has shown the two-wire circuits with a battery are by far the most satisfactory. Electric sounders are generally more

Chalked line

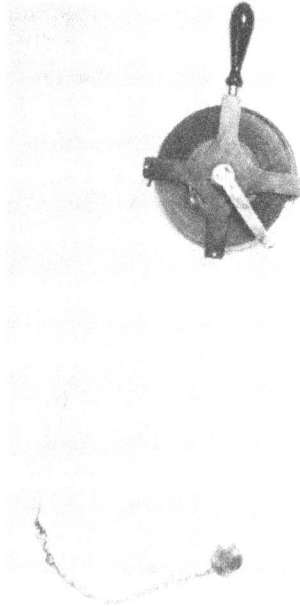

Popper.

Electric sounder.

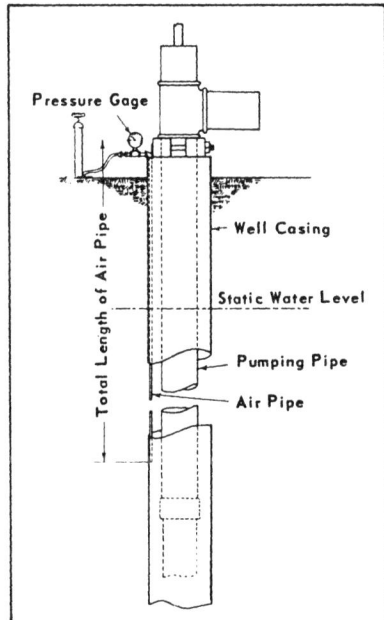

Air line.

Figure 3-5.—Devices for measuring depth to water in well.

suitable than other devices for measuring the depth to water in wells that are being pumped because they generally do not require removal from the well for each reading. However, when oil is present on the water, water is cascading into the well, or the well has a turbulent water surface, measuring with an electric sounder may be difficult. Oil not only insulates the contacts of the probe, but also results in an erroneous reading if it is considerably thick. Some instances may require insertion of a small pipe in the well between the column pipe and the casing from the ground surface to about 0.6 meter (2 feet) above the top of the pump bowls. This pipe should be plugged at the bottom with a cork or similar seal which is blown out after the pipe is set. Measurements with the electric sounder can then be made in the smaller pipe where the disturbances are eliminated or dampened, the true water level is measured, and the insulating oil is absent. Figure 3-6 illustrates a convenient arrangement for direct measurement of drawdown during pumping tests. A marker on the sounder wire is referred to a value on the tape, and the same marker is used as a reference to determine drawdown by changes on the tape when contact with the water is made. A new marker is used each time the water level drops a 1.5-meter (5-foot) increment.

A simple and reliable method for measuring the depth to water in observation holes between 40 and 150 millimeters (1 1/2 and 6 inches) in diameter is a steel tape with a popper (figure 3-5). The popper is a metal cylinder 25 to 40 millimeters (1 to 1 1/2 inches) in diameter and 50 to 75 millimeters (2 to 3 inches) long with a concave undersurface and is fastened to the end of a steel tape. The popper is raised a short distance and then dropped to hit the water surface, where it makes a distinct "pop." By adjusting the length of tape, the point at which the popper just hits the surface is rapidly determined. Poppers generally are not satisfactory for measuring pumping wells because of the operating noise and lack of clearance.

Electronic data loggers offer significant advantages over most other water measurement devices and are highly suitable for aquifer testing and long-term monitoring. Data loggers can be set to record data at intervals of 0.1 second at the beginning of a test and logarithmically increase the interval according to a present pattern. The data are automatically stored in electronic memory and can be transferred directly to a printer or to a personal computer. The water level is sensed by a pressure transducer located at some point below the lowest water level expected during

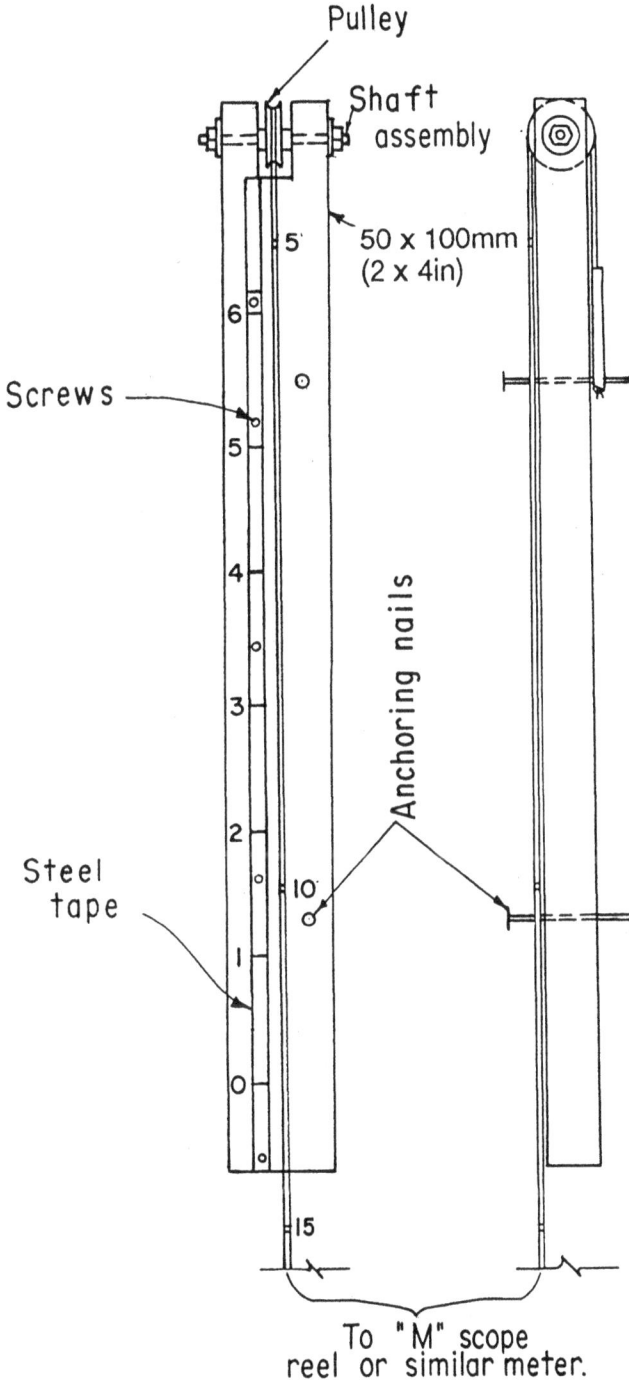

Figure 3-6.—Direct drawdown measuring board.

the test. Therefore, readings are not influenced by noise, and cascading water, or oil on top of the water column can be compensated for in the initial setup. The disadvantages are initial investment in the instruments and the technical knowledge required to operate the equipment. Also, because of the required setup time at each measurement point, data loggers generally are not suitable for a reconnaissance or survey of water levels in a well field or in widely scattered wells.

Permanent pump installations should always be equipped with an access hole for probe insertion or an air line and gauge, or preferably both, to measure drawdown during pumping. An air line is accurate only to about 0.15 meter (0.5 foot) unless calibrated against a tape for various drawdowns, but is sufficiently accurate for checking well performance.

Artesian wells with piezometric heads above the ground surface are conveniently measured by capping the well with a cap that has been drilled, tapped, and fitted with a plug which is removed for the insertion of a pressure gauge or mercury manometer stem. The static level is determined from the gauge or manometer reading after the pressure has stabilized. Figure 3-7 is a drawing of a mercury gauge, designed by S.W. Lohman of the USGS, that is particularly suited to field use, especially when running a recovery test after constant head tests of artesian aquifers (section 9-9). A recording pressure gauge may be used for continuous records.

Continuous records of a well can be obtained by mechanical devices, electrical devices, or a pressure transducer connected to an electronic data logger. Data loggers can be programmed to transmit data directly to a computer for processing.

3-9. Records of Water Level Measurements.—Accurate permanent records should be kept of all water level measurements and should include:

- Identification of the well by number and location

- Location and elevation of reference point

- Elevation of ground surface

- Date of measurement

① One - 6mm (¼in) stainless steel stop cock.
② One - 1.2m (4ft) length of 16mm (⅝in) i.d. rubber hose.
③ One - 50mm (2in) dia. ink bottle.
④ One - 3 holed No. 8 rubber stop.
⑤ One - 19mm (¾in) hose coupling.
⑥ One - 1200mm (48in) length of 2mm i.d. glass tubing.
⑦ One - 1125mm (45in) length of stainless steel strip with graduations which give readings in m (ft) of water.
⑧ One - 100mm (4in) length of 6mm (¼in) o.d. stainless steel tubing with fittings.
⑨ One - 8mm (⁵/₁₆in) stainless steel stop cock.
⑩ One - 100mm (4in) length of 8mm (⁵/₁₆in) o.d. stainless steel or plastic tubing with fitting.
Assorted lumber (marine plywood)
Assorted 3mm (⅛in) bolts with nuts.

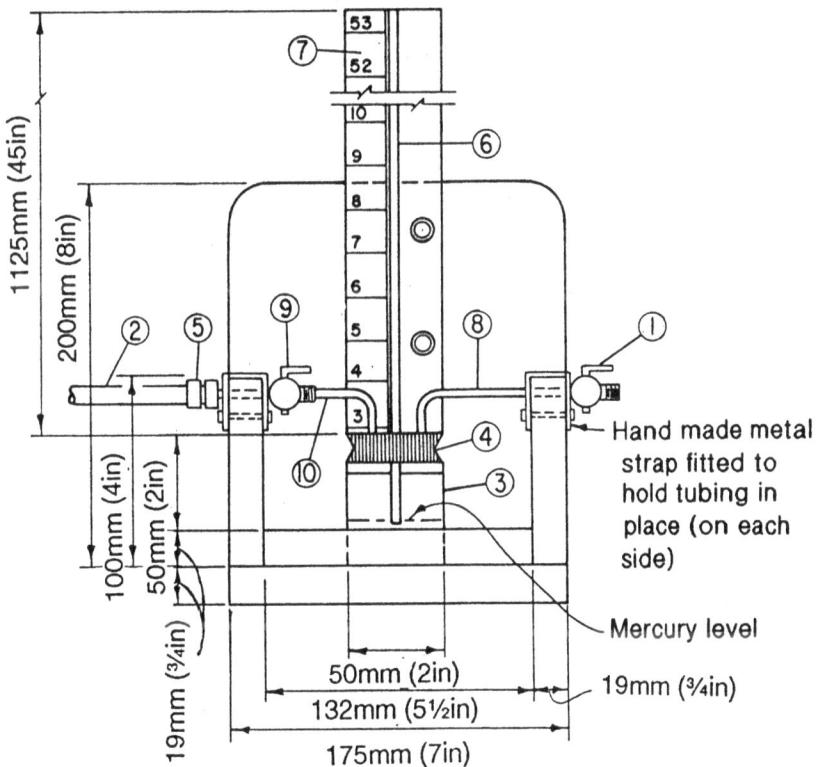

(After S. W. Lohman)

Figure 3-7.—Mercury manometer for measuring artesian heads.

- Measured depth to water or to the bottom of the hole, if dry

- Computed elevation of the water table or piezometric surface

- For piezometers, the aquifer or other zone represented by the reading

- A note whether the well was being pumped when measured, was pumped recently, or whether a nearby well was pumping during the measurement

Water level records are typically displayed as well hydrographs, which can be easily converted to water level contour maps, water level change maps, water level profiles, depth-to-water maps, and piezometric surface maps. When preparing piezometric surface maps, it is important to use only those wells that represent the subject aquifer and to avoid wells that have contact with more than one aquifer.

Most water level fluctuations are caused by: (1) changes in ground-water storage, (2) atmospheric pressure in contact with the water surface in the well, (3) deformation of the aquifer, (4) disturbances within the well, and (5) chemical or thermal changes in and near the well. Minor fluctuations can be caused by earthquakes, trains, earth-moving machinery, explosions, and other sources of temporary stresses in the aquifer (Davis and DeWiest, 1966).

3-10. Exploration Holes, Observation Well, Piezometer, and Monitoring Well Installation.—Areas may be encountered containing wells for which logs are not available, where well construction features preclude measurement of water levels, or where the wells have not been drilled. Exploratory drilling is often necessary in such areas. Drilling should be tailored to the needs of the investigation and gaps in available data. In many instances, holes drilled for stratigraphic or other data can be converted for use as observation wells or piezometers.

- Stratigraphic holes are drilled primarily for the purpose of determining the nature, depth, and thickness of the geologic formations.

- Pilot holes are usually drilled to obtain data on which to base the design of wells.

- Observation wells are usually constructed for the purpose of measuring water levels where subsurface conditions are relatively simple.

- Monitoring wells are usually constructed to obtain water samples of the aquifer.

- Piezometers are a special type of observation well so finished as to permit the measurement of the water level in a particular stratum or zone.

After exploratory or similar holes are completed, a permanent record should be made of each. This record should include all as-built drawings of the facility, showing the elevation of the point from which measurements of the depth to water in the hole will be made; the elevation of the average ground surface in the vicinity of the well; the depth of hole; the length, size, and type of casing; location of seals and packers; and the location of the screen or perforations.

The record should also show subsurface geologic conditions, water level data, the location of the hole with respect to landlines or whatever land subdivision system is used in the area, and the identification number of the hole (section 3-6).

A monitoring well should be constructed, if available, using the simplest, narrowest diameter pipe which will permit development, accommodate the sampling equipment; and minimize the need to purge large volumes of potentially contaminated water (Environmental Protection Agency [EPA], 1987). Three well volumes are usually required for development of a monitoring well.

3-11. Installation of Exploratory Holes, Observation Wells, Piezometers, and Monitoring Wells.—Many methods and combinations of methods can be used to drill exploratory holes and wells. Classified according to method of installation, the most common holes are dug, drilled, bored, driven, vibrated, or jetted. Briefly, these methods are described as follows:

- Dug holes are usually restricted to shallow depths where information is not needed for more than 1 or 2 meters (3 or 6 feet) below the water table. This type hole is rarely used in the United States.

- Drilled holes may be put down by any of the well drilling methods in common use, but the type of rig and tools used and the diameter of hole drilled will depend upon the materials to be drilled and the data to be obtained. In general, a 100-millimeter (4-inch) hole is about the smallest that is satisfactory in unconsolidated materials, and a 75-millimeter (3-inch) hole is the smallest satisfactory size in consolidated rock. The hole is cased if the material will not stand for the period of time required for use. Drilled holes can be put down to great depths (hundreds of meters) and through any material. The quality of samples obtained depends largely upon the type of drill rig used.

In drilling *any type of well*, regardless of the drilling method used, the driller should keep an accurate log or record of the well. This information is invaluable if additional wells are to be drilled nearby or if the well requires any repair or reconditioning.

The most important function performed by the drill crew is to keep the log and mark the cores accurately; but, unfortunately, in an effort to get the core and still make progress, this phase of the operation can occasionally slip by the drill crew, and then much of their work is wasted (Acker, 1974).

In some situations (hazardous materials), the drilling rig and tools should be steam cleaned to minimize the potential for cross-contamination between formations or successive borings (EPA, 1987).

- Sonic or rotary-vibratory drilling may be used as an alternative to direct or reverse rotary drilling. In sonic drilling, the drill head is vibrated at 50 to 120 cycles per second in addition to the rotary motion. The vibration frequency is adjusted to produce resonance in the drill line. Resonance is a function of the length of the drill line. This process enhances drilling speed in most geologic materials, and it can be accomplished without drilling fluids.

- Bored or augered holes may be drilled manually or by machine-driven augers. This type of hole can only be used in unconsolidated fine- to medium-grained material. The depth limitation for hand-augered holes is about 12 meters (40 feet), whereas machine-driven augers may penetrate to more than 100 meters (330 feet). When holes are bored in unstable material below the water table, caving may prevent further progress. Samples obtained by augering may range from nonexistent in saturated coarse-grained materials to disturbed but representative samples of fine-grained materials. Hollow stem augers can often retrieve relatively undisturbed samples.

- Driven or vibrated holes are advanced by driving a pipe, usually equipped with a well point, into the material. Neither samples nor a log can be obtained, and this method is suitable only for measuring water levels. Installation is restricted to shallow depths in fine- and medium-grained unconsolidated materials.

- Jetted holes are similar to driven holes except that the pipe is put down by hydraulic jetting and often can be installed to greater depths. Badly mixed washed samples and a rough log may be obtained when holes are jetted.

Where conditions are uniform, it may be satisfactory to install observation holes on a grid with holes spaced at uniform intervals. Where conditions are not uniform, wells should be located to conform to the local variations in conditions.

The magnitude and type of the study will also affect the spacing and location of holes. In a reconnaissance study to obtain general information on an area, a wide spacing is satisfactory; for a detailed study, the spacing must be reduced to provide the necessary detail.

For ground-water inventory or development studies, the holes should be deep enough to penetrate at least 3 meters (10 feet) below the lowest water table of record or to the top of an artesian aquifer. If information on thickness of an aquifer is required, one or more holes should be drilled through the aquifer. An indication of the required hole depth can usually be obtained from an inventory of existing well records. Separate wells or piezometers may be required when two or more aquifers are involved.

For protection against damage, holes completed for observation or piezometric measurements should be located, if possible, in a fence row or adjacent to a permanent structure.

The practice of installing observation wells on a step-by-step or stage basis is recommended both from a technical and an economic standpoint.

Casing installed in observation wells should be designed to the purpose of the facility and means of obtaining data. All casing must be sealed when the first aquitard is encountered to prevent cross contamination. Generally, if water levels are to be measured by a wetted tape or electric probe, a 19- to 30-millimeter- (3/4- to 1-1/4-inch) diameter steel or plastic pipe is suitable. However, if a standard water stage recorder is to be used or water samples are taken from the facility, a minimum 100-millimeter (4-inch) casing may be required. Suitable perforations should be made opposite the saturated zone to ensure reliable readings.

Piezometer installations (figure 3-8), rather than simple observation wells, are essential to a clear understanding of ground-water conditions where subsurface conditions are complex. The presence of a confined zone or several zones each with a different water level requires use of piezometers to confine and separate each level. Observation of pumping test influence may especially require the use of piezometers, even in apparently homogeneous aquifers. Installation of piezometers, especially in slowly permeable materials, may require strict design considerations to minimize time lag and other similar problems.

Each piezometer should consist of three essential components:

- A watertight standpipe of the smallest possible diameter consistent with the method of reading, attached to the tip and extending to the surface.

- A tip consisting of a well screen, porous tube, or other similar feature, and in fine-grained materials, a surrounding zone of filter sand.

- A seal consisting of cement grout, bentonite slurry, or other similar slowly permeable material placed between the standpipe and the hole to isolate the zone.

Where several piezometers are required at a given location, it may be possible, as a cost-saving feature, to install them in a single hole, as shown on figure 3-8.

In addition to the described standpipe-type piezometer, several commercially available instruments are operated by hydraulic or pneumatic pressure, or by an electric signal. Such instruments may be especially valuable for unusual subsurface or monitoring conditions, such as in very slowly permeable materials.

Drill 3mm (⅛ in)
hole in std. cap

150mm (6 in) Hole

Clay backfill

Sand-portland
cement grout

Centering guides

2.5 m (8ft)

1.2 m (4 ft)

19mm (¾ in) steel pipe

300mm (12 in) Fine sand
{passing #50 screen}
or 150 mm (6 in) tamped
plastic bentonite

50 x 1200mm (2 x 48 in)
Wellpoint with 20 - 40
slot screen

1.2 m (4 ft)

Saturated clean sand
pack, with 100%
passing N. 10, 100%
retained on No. 16

2.5m (8ft)

Clay backfill

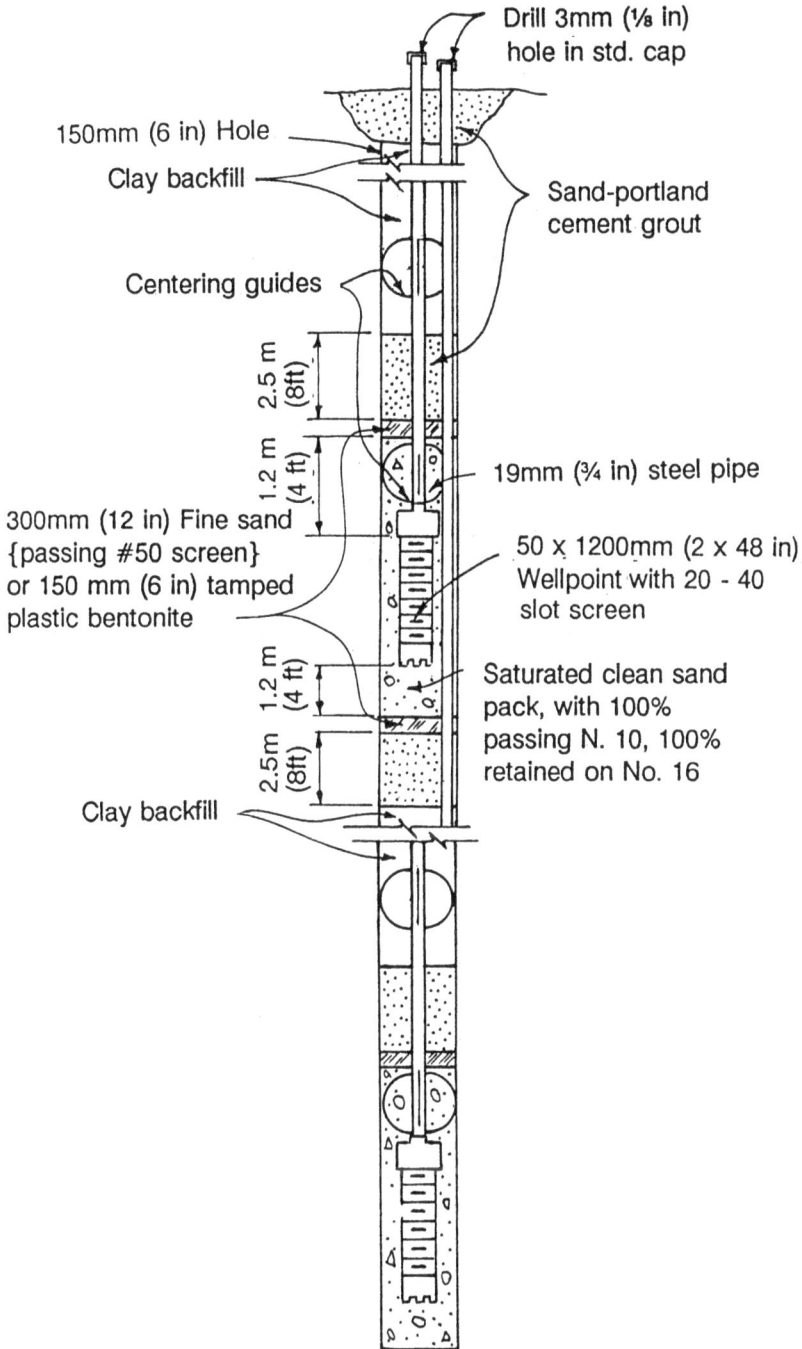

Figure 3-8.—Typical dual piezometer installation.

The casing or pipes in an observation well or piezometer usually extend above the ground surface at least 0.3 meter (1 foot) unless pit installation is necessary. The top of the casing or each pipe should be fitted with a screwcap or locking cap containing a small hole to permit adjustment of air pressure in the pipe in response to water level fluctuations or barometric changes. Where artesian flow conditions are present, a tight-fitting cap which has been drilled and tapped for a pressure gauge or mercury manometer should be used. If climatic conditions require protection against freezing, a suitable shelter equipped with heating facilities or replacement of the water in the upper portion of the piezometer by a nonfreezing fluid may be necessary.

Facilities should be protected against standing surface water and leakage alongside the casing by proper grading and placement of grout or clay seals at the surface.

Observation wells or piezometers that must be located in the open where damage by livestock or farm machinery may occur should be adequately identified and protected.

When drilling a monitoring well, the selection of the drilling technique should depend on the geology of the site, the expected depth of the well, and the suitability of the drilling equipment for the contaminants of interest. Monitoring in the vadose zone is attractive because it should provide an element of early detection capability of contamination. Soil gas sampling techniques have been commercially developed and are useful in monitoring underground storage tanks.

Monitoring wells should be developed to provide water free of suspended solids for sampling, and the additional time and money spent for well development will expedite sample filtration and result in samples that are more representative of the water chemistry in the formation.

3-12. Sampling and Logging of Exploration and Observation Holes.—The Bureau of Reclamation's *Earth Manual* (1985) describes methods and equipment for drilling and sampling which are applicable to ground-water investigations. The Bureau of Reclamation's *Engineering Geology Field Manual* offers terminology and descriptors of the physical properties of rock and soil. Undisturbed samples generally are not required for ground-water investigations. However, representative samples which preserve grain size and gradation relationships of granular materials are often required, especially for design criteria. So far as possible, the drilling method and equipment should be capable of yielding the necessary samples. Sampling applicable to well drilling is also described in chapter XII.

Each hole should be carefully logged with regard to depth and material as the samples are obtained. For field logging of unconsolidated materials, the Unified Soil Classification symbols and nomenclature described in the *Earth Manual* should be used. About a liter of representative samples should be saved of each of the primarily sandy or coarser materials. Samples need not be taken of clayey or predominantly silty materials unless unusual conditions are found or data are needed, but such materials should be described and accurately located in the logs. If gravel larger than 25 millimeters (1 inch) in diameter is obtained in the samples, it may be removed, and the size range and approximate percentage of the sample it represents should be noted.

(a) Undisturbed Samples of Unconsolidated Material.— Undisturbed samples taken by drive sampling or coring should be described as homogeneous, layered, stratified, etc. When layers consist of different materials such as clay and fine sand, the nature, thickness, and color of the layers should be recorded and the coarse fraction should be separated, if possible, and mechanically analyzed.

Samples of coarse, granular materials of a more homogenous nature should be described on the basis of visual examination according to the Unified Soil Classification given in the *Earth Manual* (1985) and mechanically analyzed.

(b) Disturbed Samples of Unconsolidated Material.—Disturbed samples, such as those obtained with a cable tool, rotary, or reverse circulation rig, usually represent a mixture of the materials in the interval sampled. The sample should be examined carefully for larger cohesive fragments which may indicate the nature of the material in place. Any material adhering to the bit, auger, or bail should be scraped off and included with the sample unless it is obvious that the material was scraped off the hole wall while being withdrawn. Samples other than those obtained with a direct-circulation rotary rig should not be washed prior to being sent to the laboratory. Samples taken from the ditch when using a direct-circulation rotary rig and clay-based drilling fluid should be placed in a 20-liter (5-gallon) container filled with water, stirred vigorously, and permitted to settle for at least 20 minutes. The

muddy water then should be decanted, and the material from the
bottom of the container should be taken for a sample. The total
volume of cuttings representing each drilled interval should be
mixed and quartered until a 2-liter (2-quart) volume of
representative material remains.

Geologic samples (formation samples) are often collected at the
surface; however, because of the lag time for cuttings to come to
the surface and the amount of mixing the cuttings may undergo as
they come up the borehole, the only way to truly know what the
subsurface materials look like is to stop drilling and collect a
sample (EPA, 1987).

(c) *Mechanical Analyses of Samples.*—Samples should be washed
on a No. 200 sieve and the percentage of minus 200 material
should be determined. A hydrometer analysis of the minus 200
size normally is not necessary. The plus 200 sizes should be
screened through a nest of 3/8 and No. 4, 8, 16, 30, 50, and
100 sieves. Forms 7-1451 and 7-141 illustrated on figures 3-9 and
3-10, respectively, should be prepared for each sample. The
washed and sieved samples less the minus 200 sieve sizes should
be recombined for visual study.

(d) *Visual Examination of Samples.*—The washed samples should
be examined with a binocular microscope or hand lens and
adequately described, including grain size and roundness,
mineralogy, and other characteristics.

(e) *Drill Core Samples of Consolidated Rock.*—Cores should be
identified regarding the type of rock, color, cementation, fractures,
and other similar characteristics. Sandstone and conglomerate
cores, if readily friable, should be crumbled and mechanically
analyzed. In many instances, the field logs can be refined after a
mechanical analysis and a visual study of the samples.

**3-13. Water Samples from Boreholes, Wells, and Surface
Sources.**—The type of investigation and purpose of the study
determine, to a large degree, the need for water samples, sampling
locations, and the frequency of collection. Ground-water quality
may vary from hole to hole in the same aquifer and sometimes
with depth in a relatively homogeneous aquifer.

When drilling uncased holes with augers and cable tools, a representative water sample can usually be obtained from the first

LABORATORY SAMPLE NO. <u>1</u>

FEATURE <u>WELL NO.1</u> AREA <u>UPPER BENCH</u> EXC. NO.___ DEPTH <u>150'</u> TO <u>155'</u>

SAMPLE PREPARATION								
PREPARED BY <u>A.X. BOBB</u> % MOIST + NO. 4 __ WET WT. TOTAL SAMPLE ___								
DATE <u>SEPTEMBER 16,19</u> % MOIST − NO. 4 __ DRY WT. TOTAL SAMPLE <u>111.9</u>								
SIEVE SIZE	5"	3"	1−1/2"	3/4"	3/8"	NO.4	TOTAL WT. PASSING NO. 4	
WT. PAN+RETAINED MAT.					422.9			
WT. PAN					311.0			
WET WT. RETAINED							_____ WET	
DRY WT. RETAINED					1.1	4.0	106.8 DRY	
DRY WT. PASSING					110.8	106.8		
% OF TOTAL PASSING					99.0	95.4 = W%		

SIEVE AND HYDROMETER ANALYSIS

DISH NO. <u>1</u> DRY WT. OF SAMPLE(W)=<u>106.8</u> gms.   FACTOR $(F) = \dfrac{W\%}{W} = \dfrac{95.4}{106.8} = 0.893$

DRY WT. SAMPLE (SIEVED) <u>106.8</u>

SIEVING TIME <u>15 MINUTES</u> DATE <u>SEPTEMBER 16,19</u>

SIEVE NO.	WEIGHT RETAINED	WEIGHT PASSING	% WEIGHT PASSING =	% OF TOTAL PASSING	PARTICLE DIA (mm)	REMARKS
8	3.3	103.5		92.4	2.380	
16	13.9	89.6	F X WEIGHT PASSING = % OF TOTAL PASSING	80.0	1.190	
30	47.0	42.6		38.0	0.590	
50	35.4	7.2		6.4	0.297	
100	4.8	2.4		2.1	0.149	
200	0.6	1.8		1.6	0.074	
PAN	1.8					
TOTAL	106.8					

TESTED AND COMPUTED BY <u>R.E. SMITH</u> CHECKED BY <u>L.R. JONES</u> DATE <u>SEPTEMBER 16,19</u>

HYDROMETER ANALYSIS

HYDROMETER NO. _____ DISPERSING AGENT _____

STARTING TIME _____ DATE _____ AMOUNT _____ ml

TIME	TEMP C°	HYD READ	HYD CORR	READ CORR	F X CORRECT READ = % OF TOTAL PASSING	% OF TOTAL PASSING	PARTICLE DIA.(mm)	REMARKS
.5 MIN*							0.050	
1 MIN							0.037	
4 MIN							0.019	
19 MIN							0.009	
60 MIN							0.005	
HR. 15 MIN*							0.002	
25 HR. 45 MIN*							0.001	

TESTED AND COMPUTED BY _____ CHECKED BY _____ DATE ____

*NOT REQUIRED FOR STANDARD TEST.

Figure 3-9.—Typical mechanical analysis form.

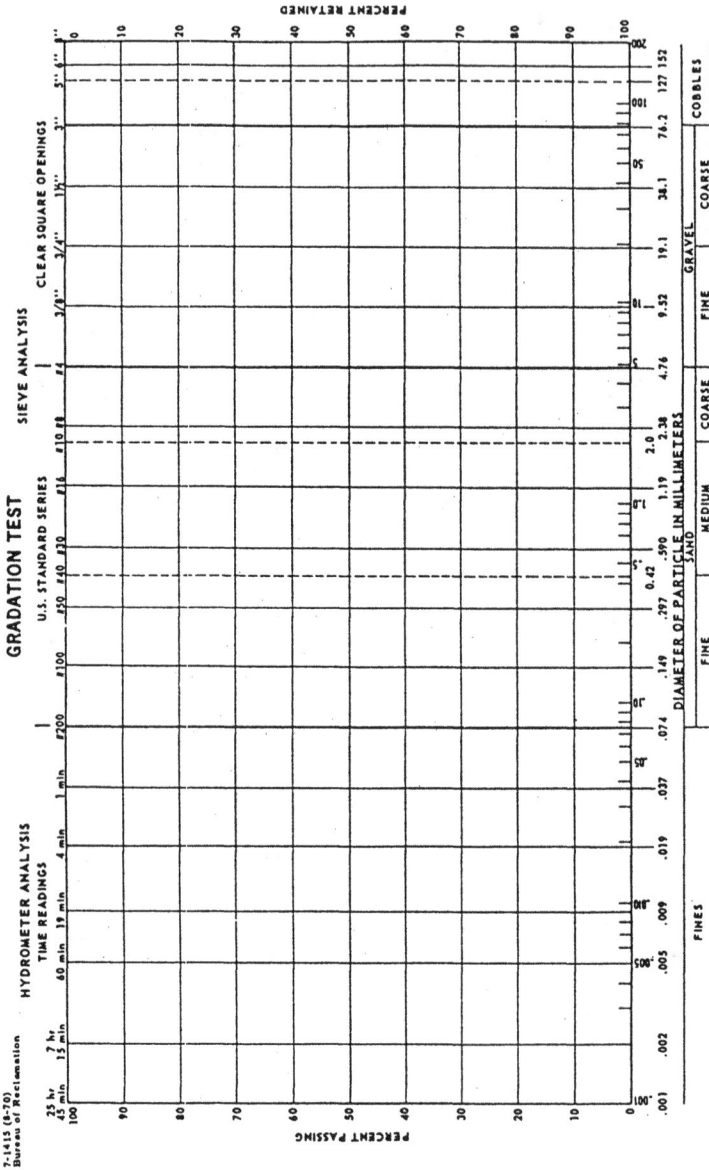

Figure 3-10.—Typical gradation test form.

water encountered by bailing the hole dry, permitting the water level to rise, and then bailing a sample from the hole. However, representative samples from levels deeper in the formation or from deeper aquifers cannot be obtained unless the hole is cased.

In rotary drilled holes, samples from individual aquifers or specific depths cannot readily be obtained except by drill stem tests or other similar procedures. On completion of the hole, flushing out the drilling fluid and pumping or bailing of the hole for a sufficient time will permit obtaining a fairly representative composite water sample. In addition, an electrical log of an uncased hole may sometimes be interpreted to give some idea of the relative quality of the water in different aquifers and at different depths (Pryor, 1956).

In consolidated rock, where casing is not usually used, a composite water sample can be taken. When water samples are required from specific depths or aquifers, a pump equipped with inflatable packers above and below the intake screen may sometimes be a practical solution to the problem (figure 3-11).

Prior to taking a water sample, sufficient water should be bailed or pumped from the hole to ensure a representative sample. An amount equal to twice that stored in the hole is normally adequate.

Before purging or sampling a well, it is important to measure and record the water level in the well. These measurements are needed to estimate the amount of water to be pumped from the well prior to sample collection.

Samples from existing wells are usually taken at the discharge. The well should be pumped for a sufficient length of time to ensure a representative sample. Temperature may be taken by inserting the thermometer in the stream as it discharges from the pipe.

One-liter (1-quart) water samples are usually adequate for most chemical analyses, but if pesticides and similar contaminants must be determined, several liters may be required.

Polyethylene sample bottles are the most satisfactory for Reclamation purposes. The new bottles should be thoroughly rinsed, filled with water, allowed to stand for about a week, then emptied, rinsed once or twice with tap water, and lastly rinsed with distilled water. After drying, they should be capped and not opened until used. A similar treatment is recommended before a

Gland, water tight
coupling.

Stainless steel
clamps.

Inflatable rubber
packer.

12 Times diameter of
hole.

Perforated section
of drop pipe.

Stainless steel
clamps.

Gland.

Coupling

Power line.

Pump

75 mm (3 in.) diameter or
larger submersible pump.

Intake screen.

NOTE
Since water circulation
to cool motor will be at
a minimum, pump should
never be operated in
excess of 4 or 5
minutes at a time,
with 10 minute
cooling period between
each pumping period.

Motor

Water line

Welded brace

12 Times diameter
of holes

Perforated adaptor pipe
25 mm (1 in.) smaller i.d. than
pump welded or other-
wise attached to bottom
Plate of motor
stainless steel
clamps.

Perforation
Seal

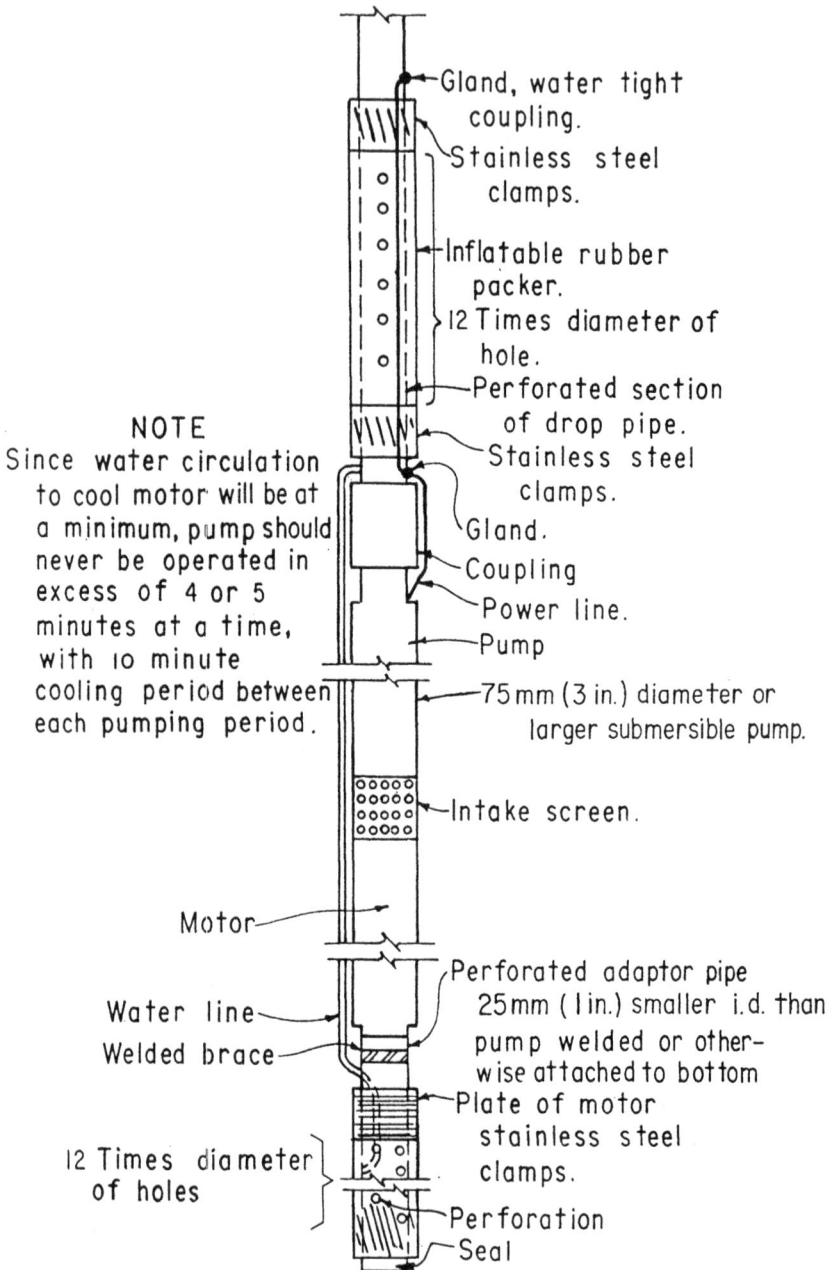

Figure 3-11.—Packer-equipped pump for selective sampling
of water from wells.

bottle is reused. Bottles which are discolored or contain visible deposits not readily removed by rinsing should not be reused. The polyethylene bottles should have relatively long screw caps with a positive seal lip on the bottle. They should be filled to the rim of the seal so that little or no air is contained in the bottle. Such bottles are not subject to breakage by shock or freezing, nor are they likely to lose fluid because of changes in atmospheric pressure. However, precautions against freezing should be taken because freezing and subthawing may change the character of the water. Samples should be transported to the laboratory and analyzed as soon as possible.

Each sample bottle should be tagged or otherwise identified, and the following applicable sample data should be recorded:

- Well or hole number and location

- Depth of well

- Source (aquifer or formation) of the water

- Method of collection and time since pumping or bailing started

- Depth or interval from which sample was taken

- Water temperature

- Date and time of collection

- Appearance at time of collection (i.e., clear, milky, colorless, etc.)

- Initials or name of collector

- Type of analysis required (i.e., comprehensive or key constituents only)

- Field analyses made, if any

Samples for bacterial analyses are usually taken in 0.1- to 0.2-liter (4- to 6-ounce) sterile glass bottles provided by a health agency or other laboratory (Rainwater and Thatcher, 1960). The caps should not be removed until a sample is to be taken. When taking a sample, care should be exercised not to touch the inside of the cap or bottle with the fingers, nor should water be permitted to flow over the hands or fingers and into the bottle or inside of the cap.

The samples should be kept cool or refrigerated, if possible, during transport to the laboratory. No more than 48 hours should elapse between taking a sample and its delivery to the laboratory.

If water samples collected need to be analyzed for dissolved inorganic chemical constituents (e.g., metals, alkalinity, and anionic species), the water sample should be filtered in the field (EPA, 1987). After proper preservation, most samples can be held for the EPA recommended maximum holding time (EPA, 1987).

Any surface-water samples needed should be taken with the same procedures outlined above for ground-water samples and should be treated in the same manner.

Additional treatment of samples in the field may be required for more specialized studies, special purposes, and conditions. A USGS paper outlines many of these studies or treatments (Rainwater and Thatcher, 1960).

The frequency of water sampling and the type of analyses to be made usually cannot be predetermined but should be developed on the basis of experience and needs. Samples for chemical analysis should be taken on completion of any borehole or well, and a comprehensive analysis should be made and printed (figure 3-12). The first known sample taken from an existing well should always be given a comprehensive analysis. Subsequent samples might be taken at low and high water stages, seasonally, or annually. Analyses may be comprehensive or for key constituents only, depending on conditions and requirements.

This procedure is also common for bacterial analyses. Samples are usually checked for pathogenic organisms or for indication of sewage or similar contamination. In some instances, however, the examination may require determination of the presence of sulphate-reducing or similar nonpathogenic but corrosion-fostering or other economically deleterious organisms.

Lab No.	Field identification	K x 10⁶ at 25° C	pH	% Na	TDS	B	Ca	Mg	Na	K	CO₃	HCO₃	SO₄	Cl	NO₃
To convert me/l to p/m, multiply by							20.0	12.2	23.0	39.1	30.0	61.0	48.0	35.5	62.0

REPORT OF WATER ANALYSES
Design and Construction Division
Engineering Laboratories Branch—Chemical Laboratory

Date shipped _____ By _____   Sheet ___ of ___
Date rec'd _____ Analyst _____   Date ___

Parts per million, p/m

Milligram equivalents per liter, me/l

Figure 3-12.—Typical chemical analyses form.

Cost-effective water quality sampling is difficult in ground-water systems because proven field procedures have not been extensively documented and because of the time, manpower, and cost of most water quality monitoring equipment (EPA, 1987).

3-14. Bibliography.—

Acker, W.L., III, 1974, "Basic Procedures for Soil Sampling and Core Drilling," Acker Drill Company, Inc.

American Society of Civil Engineers, 1990, *Evapotranspiration and Irrigation Water Requirements*, American Society of Civil Engineers - Methods and Reports on Engineering Practice, No. 70, M.E. Jensen, R.D. Burman, and R.G. Allen (editors).

Bureau of Reclamation, 1974, *Earth Manual*, 2d edition (1st reprinting, 1980; 2d reprinting, 1985).

_____, 1967, *Water Measurement Manual*, 2d edition.

Butler, S.S., 1957, *Engineering Hydrology*, Prentice-Hall, Englewood Cliffs, New Jersey.

Criddle, W.D., January 1958, "Methods of Computing Consumptive Use of Water," Proceedings of the American Society of Civil Engineers, *Journal of Irrigation and Drainage Division*, vol. 84, No. IR1, Paper 1507.

Davis, S.N., and R.J. DeWiest, 1966, *Hydrogeology*, John Wiley & Sons.

DeWiest, R.J.M., 1966, *Geohydrology*, John Wiley & Sons, New York.

Environmental Protection Agency, 1987, *Handbook-Groundwater*, EPA 625/6-87/016.

Fetter, C.W., 1988, *Applied Hydrogeology*, 2d edition, Charles E. Merrill Publishing Company.

Hamon, W.R., May 1961, "Estimating Potential Evapotransporation, "Proceedings of the American Society of Civil Engineers, *Journal of the Hydraulics Division*, vol. 87, No. HY3, pp. 107-120.

Kazmann, R.G., 1965, *Modern Hydrology*, Harper and Row, New York.

Linsley, R.K., Jr., M.A. Kohler, and J.L.H. Paulhus, 1949, *Applied Hydrology*, McGraw-Hill, New York.

_____, 1958, Hydrology for Engineers, McGraw-Hill, New York.

Lowry, R.L., Jr., and A.F. Johnson, 1942, *Consumptive Use of Water for Agriculture*, Transactions of the American Society of Civil Engineers, vol. 107, paper 2158, pp. 1243-1266.

McWhorter, D.B., and D.K. Sunda, "Ground-Water Hydrology and Hydraulic," 1948, (1984 reprint) (1977 1st edition), Water Resources Publication, Ft. Collins, Colorado.

Meinzer, O.E., 1932, "Outline of Methods for Estimating Ground-Water Supplies," U.S. Geological Survey Water-Supply Paper 638-C.

Pryor, W.A., 1956, "Quality of Water Estimated from Electric Resistivity Logs," Illinois State Geological Survey Circular 215, Urbana.

Rainwater, F.H., and L.L. Thatcher, 1960, "Methods for Collection and Analysis of Water Samples," U.S. Geological Survey Water-Supply Paper 1454.

Rouse H., 1950, "Engineering Hydraulics," Proceedings of the Hydraulics Conference, University of Iowa, Iowa City.

Schict, R.J., and W.C. Walton, 1961, "Hydrologic Budgets for Three Small Water Sheds in Illinois," Illinois State Water Survey Report of Investigation No. 46, Urbana.

Skeat, W.O., 1969, *Manual of British Water Engineering Practice*, 4th edition, vol. II, Engineering Practice, W. Heffer and Sons, Cambridge.

Thornthwaite, C.W., 1948, "An Approach to a Rational Classification of Climate," *Geographical Review*, vol. 38, pp. 55-74.

Todd, D.K., 1959, *Ground Water Hydrology*, John Wiley & Sons, New York.

_____, 1980, *Groundwater Hydrology*, 2d edition, John Wiley & Sons, New York.

Williams, C.C., and S.W. Lohman, February 1947, "Methods Used in Estimating Groundwater Supply in the Wichita, Kansas, Wellfield Area," *Transactions of the American Geophysical Union*, vol. 28, No. 1, pp. 120-131.

Wisler, C.O., and E.F. Brater, 1959, *Hydrology*, 2d edition, John Wiley & Sons, New York.

Zimmie, T.F. and C.O. Riggs, 1979, "Permeability and Groundwater Contaminant Transport," Symposium sponsored by the American Society for Testing and Materials, ASTM Special Technical Publication 746, p. 82, Philadelphia, Pennsylvania.

Chapter IV

GEOPHYSICAL INVESTIGATIONS

4-1. Introduction.—Geophysical investigations involve determining physical properties of subsurface materials by analyzing measurements made on or above the ground surface or in a borehole. Geophysical measurements detect subsurface variations in such physical properties as elasticity (bulk and shear moduli), electrical resistivity, density, magnetic susceptibility, and radioactivity. Geophysical methods can be used to obtain information of interest in ground-water studies, such as the configuration of bedrock, depth to the water table, geometry of aquifers and aquacludes, relative salinity of ground water, porosities of geologic layers, and locations of fracture zones and faults. Geophysical surveys may reduce drilling requirements and overall costs of a ground-water investigation.

Geophysical techniques can be broadly classified into two groups: surface methods and borehole methods. Surface methods are performed by making measurements with instruments that are placed on the ground surface, carried above the ground, or in a few cases towed behind an airplane. Borehole methods involve making measurements with a tool that is lowered into a borehole. Depending on the type of borehole survey, the measurements are either taken with the tool stationary or with the tool continuously moving in the borehole.

Many types of surface and borehole geophysical methods exist depending on the physical property measured and the measurement technique used. The most appropriate technique for a particular problem depends on several factors, including the objective, or targeting, of the survey, the size of the survey area, the required depth of investigation, and the degree of detail desired. Examples of survey targets and geophysical techniques that may potentially be used for each target are given in table 4-1.

The most common geophysical techniques used in ground-water studies are described in the remainder of the chapter. This chapter is not intended to provide a comprehensive description of all available methods, but rather to provide an overview of common geophysical methods and their typical applications in ground-water investigations. Many references are available for more detailed information on geophysical investigations. Reclamation's *Engineering Geology Field Manual* (1988) contains descriptions of

Table 4-1.—Examples of geologic/hydrologic targets and
geophysical methods that may be applicable for each target

Survey target	Geophysical methods	
	Surface methods	Borehole methods
Bedrock configuration	Seismic refraction or reflection, electrical resistivity, EM[1], magnetics, gravity	
Stratigraphy	Seismic refraction or reflection, electrical resistivity, EM	Sonic, electrical, or radiation logging; seismic tomography
Regional fault patterns	Gravity, magnetics	
Local fracture zones/ faults	Seismic reflection, electrical resistivity, EM, SP[2]	Sonic logging, borehole imaging, seismic tomography
Seepage/ground-water flow	SP	Temperature logging, flowmeters
Top of water table	Seismic refraction or reflection, electrical resistivity, EM	
Porosity of geologic materials		Sonic, electrical, or radiation logging
Density of geologic materials	Gravity	Radiation logging
Clay content/ mapping aquifers and aquacludes	Electrical resistivity, EM	Electrical or radiation logging
Relative salinity of ground water	Electrical resistivity, EM	Electrical logging

[1] EM = electromagnetic
[2] SP = self-potential

geophysical techniques used for engineering applications, many of
which are applicable to ground-water studies. Griffiths and King
(1981) and Labo (1986) provide basic descriptions of geophysical
methods, whereas Telford et al. (1976), Hallenburg (1984), and
Paillet et al. (1990) give thorough technical descriptions.

Descriptions of geophysical techniques with emphasis on ground-water problems are found in Keys and MacCary (1971), Ward (1990), Haeni (1986a and 1986b), Fetter (1988), Driscoll (1986), Freeze and Cherry (1979), Todd (1980), and Wright State University (1989). Applications of geophysical techniques to ground-water problems can also be found in the annual Proceedings of the Symposium on the Application of Geophysics to Engineering and Environmental Problems (SAGEEP), sponsored by the Environmental and Engineering Geophysical Society.

4-2. Surface Geophysical Methods.—

(a) Seismic Methods.—Surface seismic methods are based on the generation and recording of seismic waves (mechanical waves) traveling through subsurface materials. The seismic energy is normally generated with either a large sledge hammer, a mechanical vibrator, or an explosive source. As the seismic waves radiate from the point source, they are *refracted* along interfaces between materials having different physical properties, and they are *reflected* from the interfaces. These seismic interfaces may correspond to geologic contacts, such as the soil/bedrock interface, or to other physical changes such as the top of the water table. The seismic waves are recorded by geophones placed on the ground surface, usually equally spaced in a straight line. The seismic refraction method uses the seismic energy that is refracted from the seismic interfaces, and the seismic reflection method uses the energy that is reflected from the interfaces.

(1) Seismic Refraction Method.—When the seismic wave velocity increases across the interface between one material and the underlying material, a refracted seismic wave traveling along the interface is produced. Seismic velocity depends on the material density and elastic (bulk and shear) moduli. In general, seismic velocity is lowest for unconsolidated materials and increases with the degree of consolidation or cementation. For one type of seismic wave, the compressional or P wave, seismic velocity also increases with increasing degree of saturation. In the seismic refraction method, seismic waves refracted along interfaces are analyzed to determine the depths to the interfaces and the seismic velocities of the subsurface materials. The final result of a seismic refraction survey is usually a cross section showing the configuration of the seismic interfaces and the seismic velocities within each layer. Contour maps of the depths to the interfaces may be produced, if refraction data are collected along several closely spaced lines.

The seismic refraction method is useful for mapping the depth to bedrock and the depth to the water table. It may also be used to determine the configuration of any geologic unit that is sufficiently thick and has sufficient seismic velocity contrast with overlying and/or underlying materials. The seismic refraction method has the limitation that seismic velocity must increase with depth for the interface depths to be computed correctly. Therefore, this method is not recommended for areas where a relatively thick, low-velocity layer may exist at depth.

(2) *Seismic Reflection Method.*—A reflected seismic wave is produced when either a seismic velocity contrast or a material density contrast exists across an interface. In the seismic reflection method, these reflected waves are recorded and processed to produce a "time section" showing the reflections from all of the seismic interfaces encountered. This time section looks like a cross section; but is plotted as a function of recorded time rather than depth. The time section can be directly examined to determine the configuration of the water table and the top of bedrock. The configuration and continuity of stratigraphic layers and the presence of faults can also be determined. If sufficient velocity information can be obtained either from the seismic reflection data or from other geophysical data, the time section may be converted to a depth cross section. The seismic reflection method does not have the limitation of the refraction method of increasing velocity with depth and therefore is useful in areas where low-velocity layers are present at depth. Also, the reflection method may provide better resolution of the stratigraphy than the refraction method, especially if lateral discontinuities such as pinch-outs or faulting are present. The reflection method has the disadvantage compared to the refraction method of more complex data acquisition and processing procedures.

(b) *Electrical Methods.*—

(1) *Electrical Resistivity Methods.*—These methods involve sending an electric current of known intensity into the ground through a pair of electrodes and measuring the resulting electric potential (voltage) between another pair of electrodes. Apparent resistivities of subsurface materials are then computed. The electrical resistivity of a material is a measure of its resistance to electric current flow. Almost all of the electric current passing through rock or clay-free soil is carried by ions in the pore fluid. Hence, the electrical resistivity of such a material is determined largely by its porosity, permeability, degree of saturation, and the

salinity of the pore fluid. An increase in any of these properties decreases the resistivity of the soil or rock. The degree of compaction and the grain size distribution indirectly affect the resistivity by changing the porosity and permeability. Increasing temperature decreases the resistivity by lowering the viscosity of the pore fluid and thereby increasing the mobility of the ions. The presence of clay minerals greatly reduces the resistivity of a material because of high electrical conductivity along the surfaces of clay particles. Changes in resistivity within a survey area may be caused by any combination of the above factors. Hence, to determine which factor is affecting the resistivity in a particular area, other geological or geophysical information is required.

Lateral or vertical variations of electrical resistivity within a given survey area can provide a useful indication of relative changes of subsurface soil, rock, or ground-water properties. These methods are used for aquifer and aquiclude delineations, salinity studies, bedrock mapping, and identification of faults or fracture zones. Two types of resistivity surveys are commonly performed: Vertical electrical soundings (VES) and electrical profiling surveys.

- VES.—This type of survey measures apparent resistivity values at one location. The electrodes are moved farther and farther apart, and as a result, the electric current penetrates progressively deeper into the subsurface. A plot of apparent resistivity versus electrode separation is constructed. These data are then modeled to obtain a one-dimensional, layered resistivity-depth model. This method is used to investigate variations of resistivity with depth at fixed locations and is used to help constrain the results of the electrical profiling survey discussed below.

- Electrical Profiling.—The electrical profiling survey measures apparent resistivity values along the length of a survey line. By using different electrode spacings, apparent resistivity profiles representing different depth ranges are obtained. The resistivities are normally presented in a pseudosection format, with resistivity profiles representing larger depth ranges plotted progressively "deeper" in the pseudosection. The resistivity values in the pseudosection are contoured so that trends and anomalies can be easily observed. A pseudosection is simply a representation of multiple profiles of apparent resistivity values and is not a cross section. A cross section of material resistivities can be constructed by modeling the apparent resistivity data in the pseudosection. Results from at least one VES are often used to constrain the modeling process. The profiling survey method is used to investigate lateral changes in resistivity and has poorer vertical resolution than VES.

(2) Self-Potential (SP) Method.—This method involves measuring electric potentials that exist naturally within the subsurface. Electric potentials of interest in ground-water studies are created by the flow of fluid through rock or soil. Applications of the SP method include: mapping seepage flow paths associated with dams, dikes, and other containment structures; studies of regional ground-water movement; and delineation of flow patterns associated with such features as wells, faults, drainage structures, and sinkholes.

SP readings are normally made by measuring the electric potential (voltage) between a stationary base electrode and a portable measuring electrode. The portable electrode is carried along the survey line and voltage measurements are made at predetermined intervals. Sometimes a permanent electrode array is installed for repeat measurements. In this case, the electrodes are buried a few meters in the ground and wires are run to the surface. One of the electrodes is arbitrarily chosen as the base electrode. Electric potentials are measured between this base electrode and each other electrode in the array. SP measurements can also be made in rivers or lakes by towing a pair of electrodes along the water surface. For offshore surveys, the first derivative of the SP profile is obtained. The gradient SP signal recorded by the offshore measurement system is numerically integrated to give a "total-field" SP profile equivalent to that measured on land.

Much interpretation of SP data is performed qualitatively by examining profiles of field data for anomalies and by comparing profiles collected at different times. Geometric modeling techniques are also used to estimate the depth and lateral extent of the source of the SP signals. Analytical modeling techniques may also be used to estimate flow rates, but these methods are complex and require values of material resistivities, hydraulic conductivities, and cross-coupling coefficients.

(c) Electromagnetic (EM) Methods.—EM methods provide a means of measuring the electrical resistivity of subsurface materials without directly sending an electric current into the

ground. A (primary) magnetic field is either actively induced or passively existing in the subsurface, causing an electric current to flow. This subsurface electric current induces a secondary magnetic field that is measured at or above the ground surface. Variations in the secondary magnetic field are analyzed to determine the variation of electrical resistivity in the subsurface. EM methods are applied to the same types of targets as resistivity methods (section 4-2(b)(1)). Some advantages of EM methods over resistivity methods are discussed below.

Many types of electromagnetic methods exist. The most common EM techniques for ground-water applications can be grouped into three broad categories: slingram techniques, the time-domain electromagnetic method, and passive electromagnetic methods. Geophysical investigation using EM methods is a complicated and evolving discipline, and the summaries given here are not meant to be comprehensive overviews of all EM techniques. More detailed discussions of electromagnetic methods can be found in the references cited in section 4-1.

(1) Slingram Techniques.—The conventional slingram technique, also known as the horizontal loop electromagnetic (HLEM) method, consists of prospecting with two electrical coils. Electric current flowing in one coil (the transmitter coil) induces the primary magnetic field in the subsurface. The vertical or horizontal component (depending on loop orientation) of the total magnetic field (the primary field and the secondary field induced by subsurface electric currents) is measured by the second coil (the receiver coil) and recorded. The two coils are normally carried above the ground surface at a fixed separation and orientation and data profiles are constructed. Contour maps may be made if several closely spaced profiles are acquired. Specialized types of slingram systems, known as ground conductivity meters, have been designed to record bulk ground electrical conductivity (the inverse of electrical resistivity). These newer instruments are an order of magnitude more sensitive than conventional slingram systems but have a shallower exploration depth. The resulting magnetic field, conductivity profiles, or contour maps constructed from slingram surveys are analyzed for anomalies that may be related to subsurface features of interest. These methods are particularly useful for locating steeply dipping faults or fracture zones, measuring bulk ground conductivity (for mapping the extent of electrically conductive contaminants or for salinity studies), and

locating buried metal (such as contaminant drums). These EM
techniques give better lateral resolution than electrical resistivity
profiling methods.

*(2) Time-domain [Transient] Electromagnetic (TDEM)
Method.*— This method is most often used to investigate variations
of ground resistivity with depth. The survey configuration used for
these types of investigations is known as the central loop sounding
mode. A square transmitting loop is laid on the ground surface,
and a smaller receiver coil is placed in the center of the
transmitting loop. A magnetic field is induced in the subsurface by
an electric current in the transmitting loop. The transmitter is
then turned off and electrical eddy currents are instantly
generated in the subsurface, near the transmitting loop, to try to
maintain the total magnetic field at the value that existed just
before the transmitter was turned off. The eddy currents induce a
secondary magnetic field that is recorded by the receiving coil. The
eddy currents diffuse to greater depths with time and, as a result,
the secondary magnetic field decays. The manner in which the
secondary magnetic field decays is related to the variation of
material resistivity with depth. The variation of the secondary
magnetic field with time is modeled to obtain a one-dimensional,
layered resistivity-depth model. The result is similar to that
obtained from a VES. However, lateral resolution is better with
the TDEM method than with VES. Data acquisition is faster with
the TDEM method than with VES, but the TDEM field equipment
is more complicated and expensive than the VES equipment.
Applications of the TDEM method include: delineating aquifers
and aquicludes, mapping salt-water intrusions into fresh-water
aquifers, and determining depths to the water table and bedrock.

(3) Passive Electromagnetic Methods.—In these types of
methods, the primary magnetic field is not actively induced but
rather constantly exists in the subsurface. Passive electromagnetic
techniques include the audio frequency magnetic field (AFMAG)
and very low frequency (VLF) methods. In the AFMAG method,
the main sources of the primary magnetic field are lightning
strikes from worldwide thunderstorms. VLF methods employ the
electromagnetic energy from distant radio transmitters. These
methods are not as widely used in ground-water studies as other
EM techniques, but can be used to map overburden thickness,
locate steeply dipping fracture zones, and map concealed
boundaries between geologic units having different resistivities.

(d) Magnetic Method.—In a magnetic survey, the strength of the Earth's magnetic field is measured either by a ground or airborne survey. Data may be collected along a few lines for constructing profiles or along numerous closely spaced survey lines for constructing a contour map. Analysis of the profiles or contour map indicates qualitatively the relative depths to bedrock and the presence of structural features such as dikes, sills, or faults. The advantage of the magnetic method is that it can be performed quickly and relatively inexpensively. The magnetic method is good for broadly outlining a ground-water basin. Also, magnetic surveys are often performed to help interpret the results of electrical surveys because buried metal, which affects the results of electrical surveys, can be detected with a magnetic survey.

(e) Gravity Method.—The strength of the Earth's gravitational field is measured in a gravity survey. As for magnetic surveys, gravity data may be collected along a few lines for constructing profiles or along numerous closely spaced survey lines for constructing a contour map. Accurate surveying information is required for gravity work because gravity readings are affected by latitude and elevation, and therefore the observed gravity values must be corrected for variations in these parameters. Gravity readings must also be corrected for time variations caused by Earth tides and instrument drift. Sometimes, corrections for topographical effects and regional gravity variations are also applied. After the above corrections have been applied to the gravity measurements, the reduced gravity profiles or contour map are analyzed. The remaining gravity anomalies are caused by variations in the density of subsurface materials.

Inspection of the reduced gravity data may indicate qualitative information about the configuration of bedrock and the presence of other features such as faults or intrusive bodies. Modeling may be done to construct a subsurface density model that is consistent with the reduced gravity data. Because different density models may yield similar gravity profiles, other geologic or geophysical information may be needed to construct an accurate density model.

4-3. Borehole Geophysical Methods.—Numerous types of borehole geophysical methods exist. However, most borehole methods used for ground-water applications are logging methods. Geophysical borehole logging consists of measurement of various physical properties of geologic materials surrounding a borehole. A geophysical log is obtained by making measurements with an instrument lowered into a borehole and recording the data with a device located on the ground surface. Interpretation of geophysical logs may furnish qualitative information and sometimes quantitative information about the characteristics of subsurface materials.

(a) Seismic Methods.—Two seismic borehole methods are commonly used in ground-water studies: sonic logging and borehole imaging. Both methods involve emitting acoustic energy from a transmitter on a logging tool and recording the refracted or reflected energy on one or more receivers located on the same tool. In the sonic logging method, seismic waves refracted along the borehole-formation interface are analyzed and seismic travel times through the geologic formations are computed. In the borehole imaging method, seismic energy reflected from the borehole-formation interface is used to produce an image of the borehole wall. A third borehole seismic method that is used less frequently than the former techniques is crosshole seismic tomography. In this method, seismic energy is transmitted from one borehole to another, and an image of the intervening geologic materials is constructed from the transmitted energy.

(1) Sonic Logging.—Modern sonic logging tools have at least one transmitter and two or more receivers. Part of the seismic energy emitted by the transmitter is refracted along the borehole-formation interface. The difference in travel time (transit time) of the refracted seismic wave between the receivers is measured and recorded. Often, the seismic waveforms are also recorded for further analysis. The transmitter is repeatedly fired as the sonic tool is raised in the borehole and transit times (and seismic waveforms) are recorded. In this way, a curve of seismic transit time as a function of depth in the borehole is obtained. From the seismic transit times and recorded waveforms, geologic contacts and fracture zones can be identified, lithologies can be inferred, and formation porosities can be estimated.

(2) Borehole Imaging.—Borehole imaging tools (often referred to by the popular trade name "televiewer") contain a transducer that acts as both the transmitter and receiver, a direction sensor, and a motor. The motor rapidly rotates the transducer and direction sensor about the vertical axis, and at the same time, the transducer emits ultrasonic acoustic energy. The energy is reflected from the borehole wall and the reflected energy is converted to electrical impulses by the transducer and transmitted to the recording device at the surface. Continuous, 360-degree images of the borehole wall are constructed from the reflected

energy. The image intensity is proportional to the amplitude of the reflected acoustic energy, which in turn, is related to the physical condition of the borehole wall. Borehole imaging is particularly useful for detecting open fractures and cavities, which produce low-amplitude reflections. The location, orientation, aperture, and filling of such features can be determined with this technique.

(3) Crosshole Seismic Tomography.—Crosshole seismic tomography involves creating an image of geologic materials between two boreholes by sending seismic energy from one borehole to the other. A transmitter is lowered into one borehole and the transmitted seismic energy is recorded by several receivers located in the second borehole. The positions of the transmitter and receivers are varied so that the seismic energy is transmitted between the two boreholes over a large depth range and at many different angles. The arrival times of the transmitted seismic energy are used to construct an image of seismic velocity of the geologic materials between the two boreholes. In addition, the amplitudes of the transmitted signals may be used to construct an image of the apparent attenuation of the geologic materials. Attenuation is a measure of the amount of energy loss of the seismic signal and is related to such factors as material type, degree of compaction or cementation, porosity, saturation, and fracturing. Crosshole seismic tomography may be used to image solution cavities, fracture zones, and geologic contacts.

(b) Electrical Logging Methods.—Electrical logging methods involve measuring natural potentials existing in the subsurface or measuring the electrical resistance of subsurface materials to an electric current emitted or induced by the logging system. Electrical borehole logs provide information about lithology, clay content, porosity, saturation, and pore-fluid salinity of subsurface materials. In addition, correlation of electrical logs from different boreholes can provide information about the continuity of geologic layers. Several types of electrical logs are used for site investigations. Rarely is only one type of electrical borehole log acquired; rather, a few types of electrical logs are usually collected and interpreted jointly. The most common electrical logging methods can be classified into the three categories given below.

(1) Self-Potential Logging.—SP logging consists of measuring naturally occurring electric potentials (voltages) between an electrode at the ground surface and an electrode in a borehole. The borehole electrode is continuously moved and SP measurements are repeatedly made to construct an SP log.

Variations on the SP log indicate changes in electric potential between the borehole fluid and the fluids in the subsurface geologic layers. Readings opposite shales are relatively constant and form the baseline. The SP curve deflects to the left or right opposite permeable layers depending on the relative salinities of the drilling mud and the fluids in the surrounding geologic materials. The relationship of SP measurements to geologic material type and pore fluid type is shown on the following graph.

(2) *Resistivity Logging.*—In conventional resistivity logging, an electric current is forced to flow between two electrodes, and the resulting electric potential (voltage) is measured between two other electrodes. Resistivities of subsurface materials can be computed from the voltage measurements. Several variations of conventional resistivity logging are used. These resistivity logging variations differ in the arrangement and spacing of the current and measurement electrodes. In addition, less conventional logging tools are sometimes used that contain more than four electrodes. The effects of the variations in the number and arrangement of electrodes include differences in the depth of penetration into the materials surrounding the borehole and different degrees of vertical resolution. An idealized resistivity log is shown on figure 4-1.

(3) *Induction Logging.*—The borehole induction tool uses electric coils to create magnetic fields that in turn induce electric currents in the materials surrounding the borehole. These induced ground currents create magnetic fields that induce voltages in receiver coils. The intensity of the ground currents, and therefore the voltages induced in the receiver coils, is proportional to the conductivity (reciprocal of resistivity) of the geologic materials. Hence, this tool measures resistivities of subsurface materials, similar to resistivity logging tools. However, the induction logging tool can be used in dry boreholes, in boreholes containing nonconducting fluids, and in polyvinyl chloride-cased boreholes, whereas resistivity tools cannot.

(c) *Nuclear Radiation Logging Methods.*—Nuclear radiation logging consists of measuring radiation emitted by geologic materials near a borehole. Depending on the specific type of log acquired, the radiation is either naturally occurring or is induced by a radioactive source in the logging tool. Radiation logging methods are used to determine bulk density, clay content, porosity,

Self potential	Descriptive log	Apparent resistivity
−60 −20	Casing	
	Dry sand	
	Sand with fresh water	
	Clay	
−10 millivolts	Sand with fresh water	
	Clay	
	Clayey sand with fresh water	
Clay baseline	Clay	
	Sand with brackish water	
	Clay	
−40 millivolts	Sand with salt water	
	Clay	

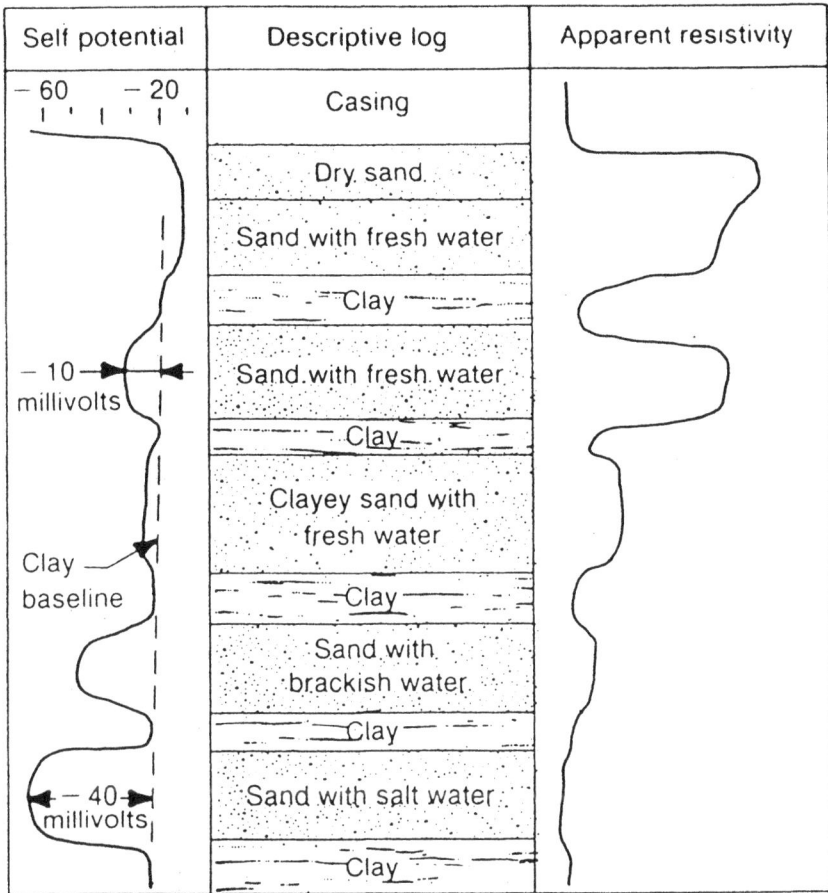

Figure 4-1.—Idealized SP curve and resistivity curve showing electric-log responses corresponding to alternating sand and clay strata; sands are saturated with fresh water, brackish water, and salt water. Note relative deflections of SP curve opposite freshwater and saltwater sands. In general, the resistivity and SP curves move in opposite directions if the drilling fluid is less saline, or fresher, than water in the formation. This is not the case for the upper sand layer, where the water in the formation has less total dissolved solids, and therefore, is less active chemically than the drilling fluid (Driscoll, 1986).

and water content of subsurface materials. These methods can also aid in determination of lithology and continuity of geologic layers between boreholes.

(1) Natural Gamma Logging.—This logging method measures natural gamma radiation emitted by radioactive elements in geologic materials surrounding the borehole. Natural gamma logs generally reflect the presence of shale and clay because the radioactive elements tend to concentrate in clay minerals.

(2) Gamma-Gamma Density Logging.—The gamma-gamma density logging tool contains a radioactive source that bombards the materials surrounding the borehole with gamma radiation. Electrons in the atoms of the surrounding geologic materials scatter the gamma rays so that some of them do not reach the detector in the logging tool. The amount of scattering is proportional to the electron density, and therefore the bulk density, of the surrounding geologic materials. Hence, the greater the bulk density of the geologic materials, the lower the gamma ray count at the detector. The gamma-gamma density log is primarily used to determine the bulk density of subsurface materials. If grain density and pore fluid density are known, the porosity can be calculated.

(3) Neutron Logging.—The neutron logging tool contains a radioactive source that emits high-energy neutron radiation. When the neutrons collide with hydrogen atoms, the neutrons lose energy. These low-energy neutrons are captured by nuclei of certain elements. When this happens, gamma rays are emitted. The detector on the neutron logging tool either measures the low-energy neutrons or the gamma rays. In either case, the measurements are related to the amount of hydrogen in the geologic materials. In water-bearing rocks, the amount of hydrogen is a direct measure of the water content. If the rocks are saturated, the porosity can also be calculated.

(d) Other Logging Methods.—

(1) Temperature Logging.—In temperature logging, temperatures are measured with a thermal resistor as the logging tool is raised or lowered in a borehole. The measured temperature is that of the borehole fluid, which may or may not be representative of the temperature in the surrounding materials.

Nevertheless, temperature logging is useful for identifying the movement of water into and out of aquifers, locating recharge water or waste discharge, and identifying fracture zones.

(2) Flowmeters.—Flowmeters can be used to measure rates of fluid flow in a borehole. Measurements can be made under natural (static) conditions or during periods of induced flow (pumping or injection). This technique can be used to determine relative fluid flow into or out of different aquifers. Flowmeters may also provide some information on the location of fracture zones.

(3) Caliper Logs.—Caliper logs show the variation in the diameter of uncased drill holes. The logs are made by running a self-actuated caliper through the hole. A recorder at the surface shows the relationship between the diameter of the hole and the hole depth. Such logs are useful in interpreting electric logs in which the apparent resistivity is influenced by hole diameter variations, and in estimating the volume of cement required for grouting in a casing or the volume of gravel which may be required for a pack. They also may show the nature of the subsurface materials because a drill hole is usually washed out to a larger diameter when poorly consolidated and noncohesive materials are penetrated by the hole. The caliper log is sometimes useful in well rehabilitation work because it will show where uncased holes have raveled or caved and where casing in cased holes has been damaged or otherwise undergone deterioration.

4-4. Bibliography.—

Bureau of Reclamation, 1988, *Engineering Geology Field and Office Manual*, (two volumes), Denver, Colorado.

Driscoll, F.W., 1986, *Groundwater and Wells*, Johnson Division, St. Paul, Minnesota.

Fetter, C.W., 1988, *Applied Hydrogeology*, 2d edition, Charles E. Merrill Publishing Company.

Freeze, R.A. and J.A. Cherry, 1979, *Groundwater*, Prentice-Hall, Inc., Englewood Cliffs, New Jersey.

Griffiths, D.H. and R.F. King, 1981, *Applied Geophysics for Geologists and Engineers, The Elements of Geophysical Prospecting*, Pergamon Press, New York, New York.

Haeni, F.P., 1986a, "Application of Seismic-Refraction Techniques to Hydrologic Studies," U.S. Geological Survey Water-Resources Investigation Report 84-746.

Haeni, F.P., 1986b, "Application of Seismic Refraction Methods in Groundwater Modeling Studies in New England," *Geophysics*, 51, pp. 236-249.

Hallenburg, J.K., 1984, *Geophysical Logging for Mineral and Engineering Applications*, PennWell Publishing Co., Tulsa, Oklahoma.

Keys, W.S. and L.M. MacCary, 1971, "Application of Borehole Geophysics to Water-Resources Investigations," *Techniques of Water-Resources Investigations of the U.S. Geological Survey*, book 2, chapter E1.

Labo, J., 1986, *A Practical Introduction to Borehole Geophysics*, Society of Exploration Geophysicists, Tulsa, Oklahoma.

Paillet, F., C. Barton, S. Luthi, F. Rambow, and J. Zemanek, 1990, *Borehole Imaging*, The Society of Professional Well Log Analysts Inc., Houston, Texas.

Telford, W.M., L.P. Geldart, R.E. Sheriff, and D.A. Keys, 1976, *Applied Geophysics*, Cambridge University Press, New York, New York, pp. 860.

Todd, K.D., 1980, *Groundwater Hydrology*, 2d edition, John Wiley & Sons.

Ward, S.H., (editor), 1990, "Geotechnical and Environmental Geophysics," Investigations in Geophysics Series, No. 5, vol. 1 - *Review and Tutorial*, and vol. 2 - *Environmental and Groundwater*, published by the Society of Exploration Geophysicists, Tulsa, Oklahoma.

Wright State University, 1989, *Groundwater Practice 1, A Workbook in Applied Hydrogeology*, IRIS Program, Department of Geological Sciences.

Chapter V

DEFINITIONS AND THEORY OF SATURATED GROUND-WATER FLOW AND FACTORS AFFECTING GROUND-WATER FLOW

5-1. Structural Geology and Stratigraphy.—Geologic factors such as stratigraphy, structure, and lithology constitute the skeleton or framework which controls the occurrence and movement of water and must be considered in the analysis and solution of ground-water problems. A proper understanding of the depositional environment and subsequent geologic activity also enhances the ground-water setting characterization. A clear understanding of the geomorphology that went into forming the deposits can provide advance knowledge before any other investigations take place and substantial savings in time and money.

Flow through various geologic features may be quite different under unsaturated conditions than under saturated conditions. This chapter deals only with saturated conditions.

5-2. Darcy's Law.—The foundation of ground-water hydraulics is Darcy's law (Muskat, 1946; Theis, 1935), which states that the flow rate through a porous medium is proportional to the head loss and inversely proportional to the length of the flow path (see figure 5-1). The law is applicable where flow is laminar, without turbulence. Formulas expressing the law are given in a number of forms, the most common of which are presented below and are derivations of $Q = AV$:

$$V = Ki \qquad\qquad 5\text{-}1$$

$$Q = KiA \qquad\qquad 5\text{-}2$$

$$Q = KA\,\frac{h_1 - h_2}{L} \qquad\qquad 5\text{-}3$$

$$K = \frac{Q}{iA} = \frac{\frac{L^3}{t}}{L^2} = \frac{L}{t} \qquad\qquad 5\text{-}4$$

where:

V	= velocity, L/t
K	= permeability or hydraulic conductivity of the porous medium, L/t
i	= hydraulic gradient h_1 - h_2/L, dimensionless
A	= area normal to the direction of flow, L^2
Q	= rate of flow L^3/t
h_1 and h_2	= the water level or potential at two points on a line parallel to the direction of flow
L	= length of flow path between h_1 and h_2
t	= time

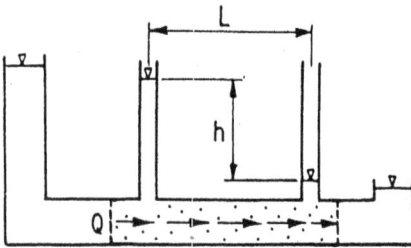

Darcy's determination of rate of flow through a porous medium:
$$Q = KA\frac{h}{L}$$

Where　Q – Volume of flow per unit time, $\frac{L^3}{t}$
K – Permeability or hydraulic conductivity of a porous medium, $\frac{L}{t}$
h – Head loss in distance L
L – Length of flow path L
A – Cross sectional area of a porous medium normal to flow, L^2

Figure 5-1.—Illustration of Darcy's law.

5-3. Hydraulic Conductivity.—Rearrangement of Darcy's equation leads to $K = V/i = Q/iA$, where K is a proportionality constant commonly known as the hydraulic conductivity (Lohman et al., 1972) and has the dimensions $L^3/t/L^2$ which reduces to L/t or velocity.

Numerous expressions, some based on Q/t, others on L/t, and with a variety of consistent and inconsistent units and i values, have been used for expressing K. In Bureau of Reclamation practice, K is usually expressed for water as L/t under a unit gradient. In laboratory work, L is usually expressed in centimeters and t in seconds; however, in field determinations from pumping tests, L/t is usually expressed in meters/day, ft^3/ft^2/day, or gal/ft^2/day. Factors for conversions between the most commonly used units are given in table 5-1.

The value of K varies for different fluids depending upon their density and viscosity as follows:

$$K = \frac{k\gamma}{\mu} \qquad\qquad 5\text{-}5$$

where γ is the specific weight and μ is the viscosity of the fluid. In this formula, k is the intrinsic permeability of the medium, i.e.:

$$V = Ki = \frac{ki\gamma}{\mu} \qquad\qquad 5\text{-}6$$

The terms "coefficient of permeability" or "permeability" are sometimes used as synonyms for hydraulic conductivity. However, to avoid confusion with intrinsic permeability, the term hydraulic conductivity is preferred.

In ground-water engineering, this refinement is seldom required. In laboratory determinations of K using water, the results are usually expressed as the value obtained at a water temperature of 16 or 20 °C (60 or 68 °F). Laboratory results neglecting the slight change in weight with temperatures can be compared to field determinations with the expression:

$$K_L = \frac{V_F}{V_L}K_F \qquad\qquad 5\text{-}7$$

Table 5-1.—Conversion factors for various units of hydraulic conductivity

(1)	(2)	(3)	(4)	(5)	(6)	(7)	(8)	(9)	(10)	(11)
ft³/ft²/yr	ft³/ft²/day	ft³/ft²/hr	ft³/ft²/min	ft³/ft²/sec	in³/in²/day	in³/in²/hr	gal/ft²/day	m³/m²/day	cm³/cm²/hr	Darcy cm³/s-cm² (atm/cm)
1	2.74×10^{-3}	1.141×10^{-4}	1.903×10^{-6}	3.171×10^{-8}	3.287×10^{-2}	1.37×10^{-3}	2.049×10^{-2}	8.35×10^{-4}	3.479×10^{-3}	1.133×10^{-3}
365	1	4.167×10^{-2}	6.945×10^{-4}	1.157×10^{-5}	12	5.0×10^{-1}	7.4805	3.05×10^{-1}	1.270	4.115×10^{-1}
8,760	24	1	1.667×10^{-2}	2.778×10^{-4}	288	12	179.5	7.32	30.48	9.872
525,600	1,440	60	1	1.667×10^{-2}	17,280	720.0	10,772	438.9	1,829	591.7
31,536,000	86,400	3,600	60	1	1,036,800	43,200	646,315	26,335	109,723	35,549
30.42	8.333×10^{-2}	3.472×10^{-3}	5.787×10^{-5}	9.645×10^{-7}	1	4.166×10^{-2}	6.234×10^{-1}	2.54×10^{-2}	1.058×10^{-1}	3.435×10^{-2}
730	2.0	8.334×10^{-2}	1.389×10^{-3}	2.315×10^{-5}	24	1	14.96	6.1×10^{-1}	2.540	8.217×10^{-1}
48.78	1.337×10^{-1}	5.569×10^{-3}	9.282×10^{-5}	1.547×10^{-6}	1.604	6.682×10^{-2}	1	4.07×10^{-2}	1.697×10^{-1}	5.494×10^{-2}
1,198	3.28	1.368×10^{-1}	2.27×10^{-3}	3.78×10^{-5}	39.38	1.64	24.54	1	4.167	1.35
287.4	7.874×10^{-1}	3.281×10^{-2}	5.469×10^{-4}	9.114×10^{-6}	9.449	3.939×10^{-1}	5.890	0.24	1	3.246×10^{-1}
886.96	2.43	10.13×10^{-2}	16.88×10^{-4}	28.13×10^{-6}	29.20	1.217	18.2	7.41×10^{-1}	3.08	1

—Example (1)

All factors computed for 68 °F with viscosity of 1.0050 centipoises.

Examples:

(1) The permeability of a soil has been determined to be 15 gal/ft²/day. What is this in in³/in²/hr? Find value of 1 in column 8 and move horizontally to value for in³/in²/hr in column 7. Multiply value in column 7 (0.0668) by 15 = 1.002 in³/in²/hr.

(2) The permeability of a soil has been determined to be 4,000 ft³/ft²/yr. What is this in gal/ft²/day? Find value of 1 in column 1 and move horizontally to value for gal/ft²/day in column 8. Multiply value in column 8 (0.02049) by 4,000 = 82.0 gal/ft²/day.

where K_L is the standard or laboratory determination, K_F is the field determination, V_F is the kinematic viscosity of water at field temperature, and V_L is the kinematic viscosity at laboratory temperature. Because ground-water temperatures at depths to 60 meters (200 feet) from the surface seldom vary more than about 1 °C (2 °F) from the average annual temperature of the area in which they occur, the above conversion for K is seldom necessary because the determined values will be used in the area where the test was made. Table 5-2 gives the variation of properties of pure water with temperature.

Laboratory determinations of hydraulic conductivity are only representative of a specific sample of aquifer material, whereas field determinations by pumping tests are usually an average value representing an integration of all the permeability variations in all directions and from place to place in an aquifer. For this reason, laboratory determinations, even if made using undisturbed samples, are not as representative as those found in actual field aquifer tests and may be misleading.

5-4. Transmissivity.—Because the term *hydraulic conductivity* fails to describe adequately the flow characteristics of an aquifer, Theis (1935) introduced the term *transmissivity* (Meinzer, 1949), $T = KM$, which is equal to the average permeability times the saturated thickness of the aquifer, to clarify this deficiency (figure 5-2). Transmissivity has dimensions of $L^3/t/L$ or L^2/t because it represents flow through a vertical strip of aquifer one unit wide. Where K may be considered as the hydraulic conductivity of a unit cross-sectional area of the aquifer, T may be considered as the hydraulic conductivity of a unit width of the full thickness of the aquifer.

5-5. Storativity.—The terms specific yield, effective porosity, coefficient of storage, and storativity (Meinzer, 1949) have often been used interchangeably to express the storage capacity of an aquifer. However, some authors have limited the use of specific yield to unconfined aquifers and coefficient of storage to confined aquifers. Because the influence is essentially the same in either case and S is the commonly used symbol to express the value regardless of the nature of the aquifer, the term *storativity* will be used herein to designate both concepts.

Table 5-2.—Variation of properties of pure water with temperature

Temp. (°C)	Temp. (°F)	Density at 1 atmosphere, [1] (gm·cm⁻³)	Dynamic viscosity, in centipoises [2] (10^{-2} dyne-sec cm⁻²)	Kinematic viscosity, in centistokes [3] (10^{-2} cm² sec⁻¹)	Surface tension against air, [4] (dyne cm⁻¹)	Vapor pressure, [5] (mm Hg)
5	41.0	0.999965	1.5188	1.5189	74.92	6.543
6	42.8	0.999941	1.4726	1.4727	74.78	7.013
7	44.6	0.999902	1.4288	1.4289	74.64	7.513
8	46.4	0.999849	1.3872	1.3874	74.50	8.045
9	48.2	0.999781	1.3476	1.3479	74.36	8.609
10	50.0	0.999700	1.3097	1.3101	74.22	9.209
11	51.8	0.999605	1.2735	1.2740	74.07	9.844
12	53.6	0.999498	1.2390	1.2396	73.93	10.518
13	55.4	0.999377	1.2061	1.2069	73.78	11.231
14	57.2	0.999244	1.1748	1.1757	73.64	11.987
15	59.0	0.999099	1.1447	1.1457	73.49	12.788
16	60.8	0.998943	1.1156	1.1168	73.34	13.634
17	62.6	0.998774	1.0875	1.0889	73.19	14.530
18	64.4	0.998595	1.0603	1.0618	73.05	15.477
19	66.2	0.998405	1.0340	1.0357	72.90	16.477
20	68.0	0.998203	1.0087	1.0105	72.75	17.535
21	69.8	0.997992	0.9843	0.9863	72.59	18.650
22	71.6	0.997770	0.9608	0.9629	72.44	19.827
23	73.4	0.997538	0.9380	0.9403	72.28	21.068
24	75.2	0.997296	0.9161	0.9186	72.13	22.377
25	77.0	0.997044	0.8949	0.8976	71.97	23.756
26	78.8	0.996783	0.8746	0.8774	71.82	25.209
27	80.6	0.996512	0.8551	0.8581	71.66	26.739
28	82.4	0.996232	0.8363	0.8395	71.50	28.349
29	84.2	0.995944	0.8181	0.8214	71.35	30.043
30	86.0	0.995646	0.8004	0.8039	71.18	31.824
31	87.8	0.995340	0.7834	0.7871	[6]71.02	33.695
32	89.6	0.995025	0.7670	0.7708	[6]70.86	35.663
33	91.4	0.994702	0.7511	0.7551	[6]70.70	37.729
34	93.2	0.994371	0.7357	0.7399	[6]70.53	39.898
35	95.0	0.99403	0.7208	0.7251	70.38	42.175
36	96.8	0.99368	0.7064	0.7109	[6]70.21	44.563
37	98.6	0.99333	0.6925	0.6971	[6]70.05	47.067
38	100.4	0.99296	0.6791	0.6839	[6]69.88	49.692

[1] *Handbook of Chemistry and Physics*, 4th edition, 1965-66: Cleveland, Chemical Rubber Publishing Company, table F-4, computed from the relative values.
[2] International critical tables of numerical data, physics, chemistry, and technology: National Academy of Sciences, vol. 5, p. 10.
[3] Dynamic viscosity divided by density.
[4] International critical tables of numerical data, physics, chemistry, and technology: National Academy of Sciences, vol. 4, p. 447.
[5] *Handbook of Chemistry and Physics*, 46th edition, 1965-66: Cleveland, Chemical Rubber Publishing Company, table D-94.
[6] Interpolated.

Piezometers 1−unit
of length apart

Unit hydraulic gradient,
1-unit of length drop in
1-unit of length of flow
distance.

Confining Material

Confining Material

Aquifer

Aquifer

Flow

M

Confining Material

Confining Material

Opening B, 1−unit
of length wide
and aquifer
height M.

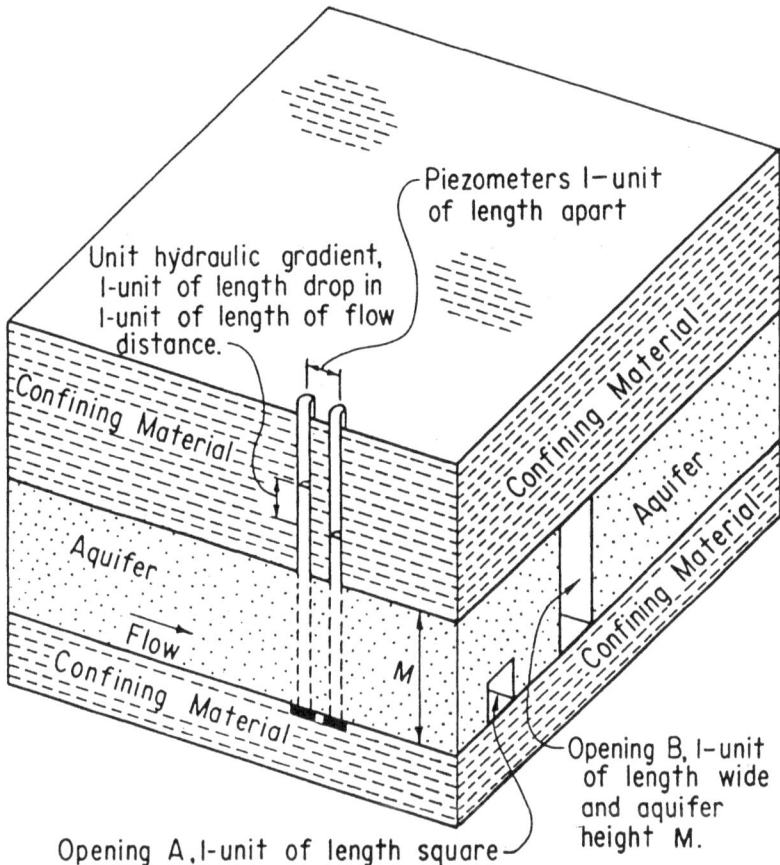

Opening A, 1-unit of length square

L − 1 Unit of length, t − Unit time,
h − Potential difference − L

$A = L^2$; $Q = \dfrac{L^3}{t}$; $i = \dfrac{h}{L} = 1$, Non-dimensional. v − velocity $\dfrac{L}{t}$

$Q = KiA = \dfrac{L^3}{t} = KiL^2$

$K = \dfrac{\frac{L^3}{t}}{L^2 i} = \dfrac{\frac{L}{t}}{i} = \dfrac{L}{t}$

M − Thickness of aquifer − Multiples of L.

$T = KM = \dfrac{\frac{L^3}{t}}{L} = \dfrac{L^2}{t}$

(After U.S. Geological Survey)

Figure 5-2.—Illustration of hydraulic conductivity and transmissivity.

Storativity is defined as the volume of water released from or taken into storage per unit surface area of the aquifer per unit change in the component of hydraulic head normal to that surface. In a vertical column with a horizontal cross section of one square unit extending through an aquifer (figure 5-3), the storativity equals the volume of water released from or gained by the aquifer when the piezometric surface or water table declines or rises one unit. Storativity is expressed as the ratio:

$$S = \frac{V'}{V} \qquad\qquad 5\text{-}8$$

where V' equals the volume of water released and V is the volume of material drained in an unconfined aquifer or the volume defined by the change in piezometric head for an artesian aquifer.

Because $V'/V = L^3/L^3$, S is dimensionless.

S must be considered in equations for unsteady (transient) flow. Release of water from the aquifer in response to a change in head generally is assumed to be instantaneous. In many cases, however, initial release is relatively rapid but decreases in rate with time. In other cases, because of the fine-grained nature of the aquifer, drainage may be so slow that the response to change in head is similar to that of a leaky aquifer for a fairly prolonged period of time (section 5-17).

In an unconfined aquifer, S is a function of the size and number of interconnected voids and represents the actual volume of water drained from the aquifer by lowering of the water table. The S value ranges from as low as 1 percent to over 40 percent, but is usually in the range of 10 to 30 percent. The less uniform, finer grained, and more dense a material is, the smaller the S value.

In a confined aquifer, where the cone of depression is not drawn below the bottom of the upper confining layer, no actual aquifer drainage occurs. Water released is caused by: (1) the small expansion of the water resulting from the reduction in pressure and (2) water being forced out of the aquifer by compaction of the aquifer skeleton because of this reduction in pressure. The value of S in an artesian aquifer may be independent of void content of the aquifer material and ranges from 1/1,000 of 1 percent to 1/10 of 1 percent (0.00001 to 0.001).

Piezometric surface

Unit cross-sectional area

Unit decline of
piezometric surface

Confining   Material

Aquifer prism of
height M

M

Aquifer

Confining Material

A. CONFINED AQUIFER

Unit cross-sectional area

Water table

Unit decline of
water table

Aquifer prism of
height M

M

Aquifer

Confining   Material

B. UNCONFINED AQUIFER

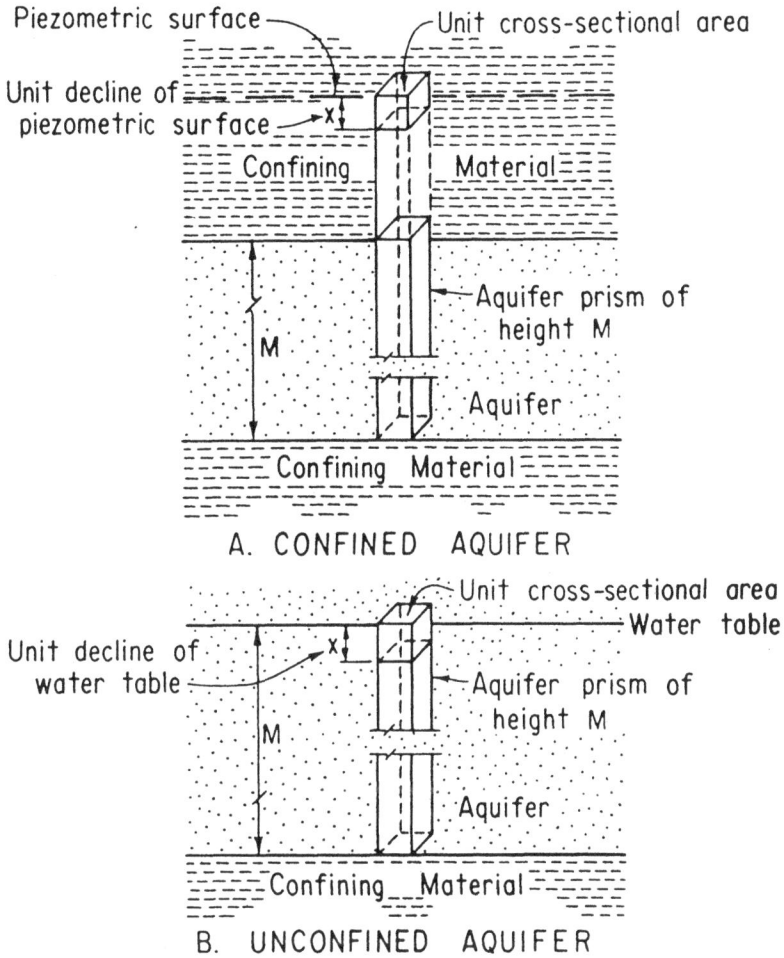

Storativity, $S = \dfrac{V'}{V}$, is non-dimensional

L - Unit length, V - The volume involved in a unit drop or
increase in piezometric surface or water-table
elevation within a prism of 1-unit^2 cross section(L^3).
V' - The volume of water released by a change in
piezometric surface or drained or recharged by a
change in water-table elevation within the prism(L^3).

(After U.S. Geological Survey)

Figure 5-3.—Illustration of storativity.

5-6. Specific Retention.—If a unit volume of dry porous material is saturated and then permitted to drain by gravity, the volume of water released is less than that required for saturation (Meinzer, 1949). The volume of water retained in the material is held by capillary action and molecular forces against the pull of gravity. The ratio of the volume of the retained water to the volume of the material is the *specific retention*, a dimensionless value expressed as a percentage. The value of specific retention ranges from less than 1 percent to nearly 100 percent. It increases with a decrease in grain size and pore size of the material.

The specific retention may be expressed as follows:

$$R_s = \frac{V_{wr}}{V} 100 \qquad\qquad 5\text{-}9$$

where:

R_s = specific retention
V_{wr} = volume of water retained, L^3
V = volume of material, L^3

5-7. Porosity.—Porosity (Meinzer, 1949) is a dimensionless value that expresses the ratio of the volume of pores to the total volume of a porous material and is usually expressed as a percentage:

$$P = \frac{V_p}{V} 100 \qquad\qquad 5\text{-}10$$

where:

P = porosity
V_p = volume of pore space, L^3
V = volume of material, L^3

Porosity ranges from less than 1 percent to as much as 80 percent in some recently deposited clays, but in most granular materials it falls between about 5 and 40 percent. In unconfined aquifers, the porosity is equal to the specific retention plus the storativity (specific yield). Porosity must be considered in recharge and storage analyses because the volume of water recharged into a dry aquifer will be greater than the recoverable volume. Some rocks may have considerable porosity represented by vugs or holes which are not interconnected. From the standpoint of ground-water flow, such rocks have zero effective porosity.

Primary porosity is attributable to the soil or rock matrix and secondary porosity is attributable to such phenomena as secondary solution or structurally controlled regional fracturing (Freeze and Cherry, 1979).

5-8. Velocity.—The discharge velocity of a porous medium (Meinzer, 1949), V, is defined as the volume of water that flows per unit time across a unit cross-sectional area normal to the direction of flow. However, in an ideal porous medium, only the void space, which is equal to the porosity, is available for flow. Hence, the actual average interstitial or seepage velocity, V_s, is the discharge velocity divided by the porosity, or V/p, and is expressed as simple velocity, L/t. Analogously, the velocity under a unit gradient is equal to the hydraulic conductivity divided by the porosity, or K/p. Another concept takes into account that water in pore spaces small enough to hold water against the force of gravity; that is, the specific retention, is probably stagnant and that flow occurs only through the area represented by the specific yield, or:

$$V_s = \frac{V}{S} \qquad \qquad 5\text{-}11$$

In this case, V_s will be somewhat higher than when the entire pore volume is considered. However, this concept should not be used to estimate the precise rate, distance, or time required for a given molecule of water to move from one place to another.

A direct method to determine a bulk ground-water velocity over a significant flow path distance is to introduce a tracer at one point in the flow field and observe its arrival at other points (Freeze and Cherry, 1979).

5-9. Hydraulic Diffusivity.—The ratio of transmissivity to storativity in transient flow conditions can be expressed in the formula:

$$\alpha = \frac{T}{S}, \qquad \qquad 5\text{-}12$$

where α is hydraulic diffusivity and has the dimensions L^2/t. In an ideal aquifer, the time of response at a distant location to an imposed stress, such as a discharging well, is inversely proportional to the diffusivity (Glover, 1960).

Transmissivity (or hydraulic conductivity times aquifer thickness) is the overriding factor in well yield in terms of yield versus drawdown or specific capacity. Under transient conditions, storativity is also a controlling factor.

Figures 5-4 and 5-5 give generalized conversion scales for units of transmissivity and hydraulic conductivity and illustrate general relationships involving potential well yield and potential and hydraulic conductivity of common aquifer materials.

5-10. Steady One-Directional Flow.—Steady flow of water in a confined aquifer of uniform thickness behaves in accordance with Darcy's law (i.e., the head decreases linearly in the direction of flow) (Weeks, 1964). For steady-state flow to occur, the magnitude and direction of the flow velocity are constant with time at any point in the flow field. However, in an unconfined aquifer, the water table is also a flow line. The shape of the water table determines the flow distribution, but conversely, the flow distribution determines the shape of the water table. Accordingly, a general analytical solution of the flow is therefore not possible.

In an attempt to simplify analysis of undirectional flow Dupuit (Muskat, 1946) made the following assumptions: (1) the velocity of flow is proportional to the tangent of the hydraulic gradient rather than the sine as determined by Darcy; and (2) the flow is horizontal and uniform everywhere in a vertical section. Dupuit derived the following equation:

$$Q = \frac{K(h_2^2 - h_1^2)}{2L} \qquad\qquad 5\text{-}13$$

where:

Q = flow of water per unit time, L^3/t, per unit width normal to the direction of flow

h_1 = saturated thickness of the aquifer at one point in the line of flow

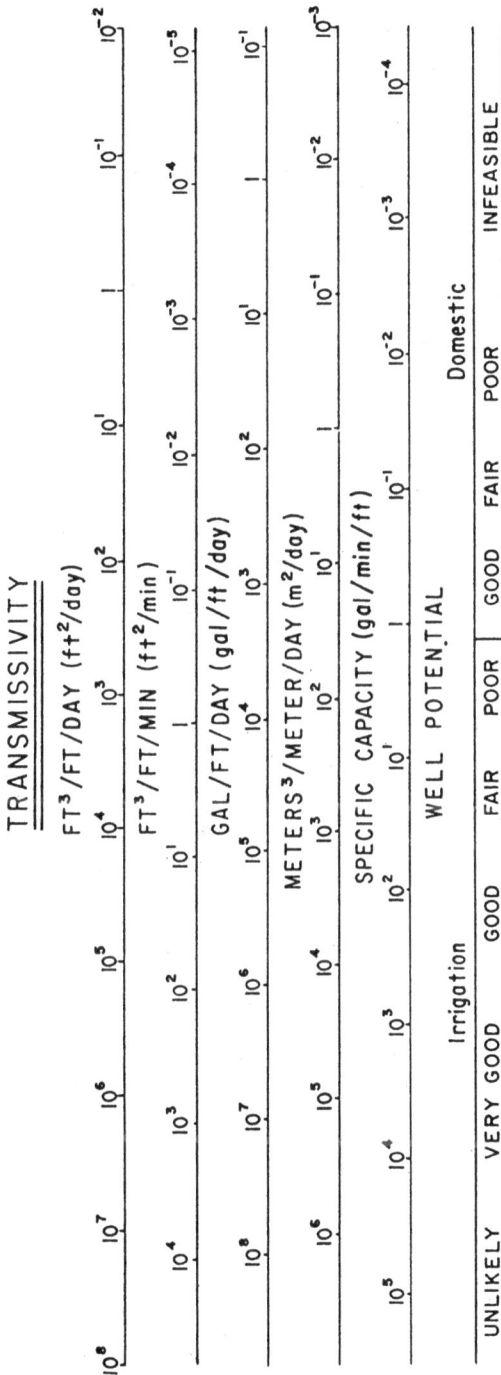

Figure 5-4.—Comparison of transmissivity, specific capacity, and well potential.

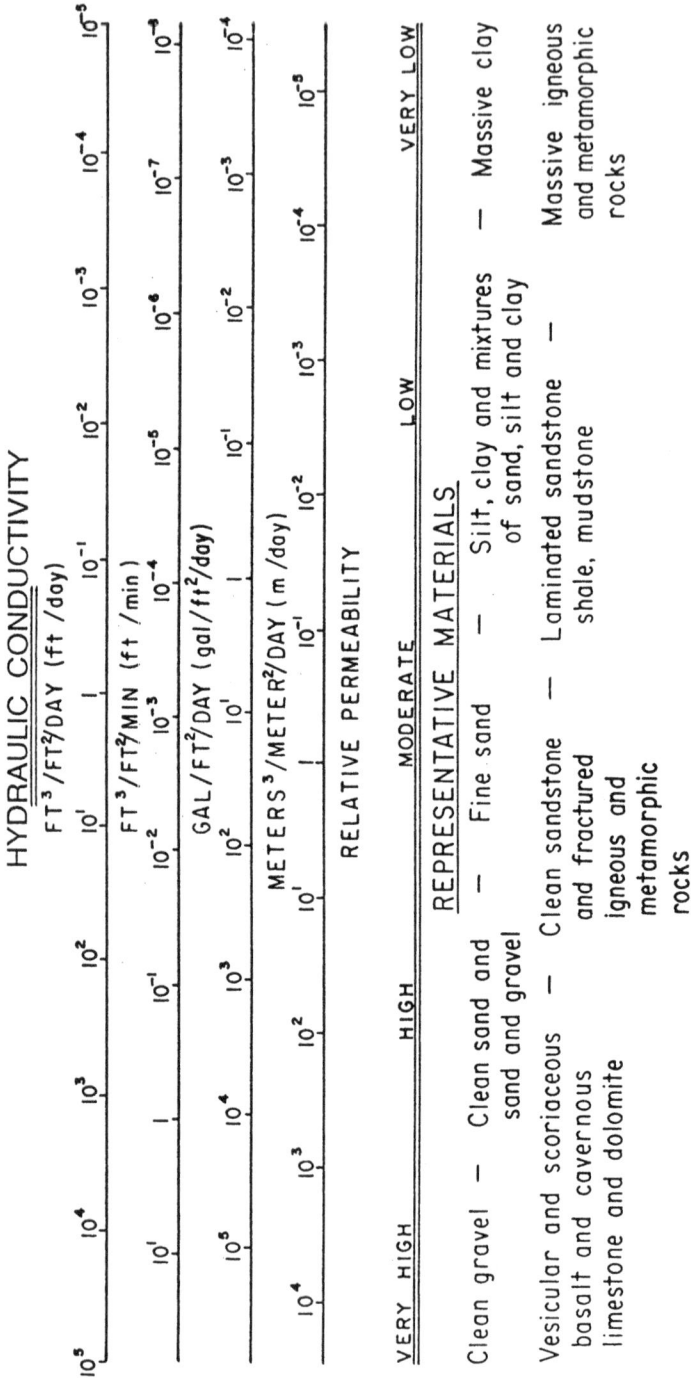

Figure 5-5.—Comparison of hydraulic conductivity and representative aquifer materials.

h_2 = saturated thickness of the aquifer at a second point in the line of flow from h_1

L = distance between the points parallel to the direction of flow, L

However, because of the assumptions of horizontal flow (no vertical component), the computed (Dupuit) water table deviates more and more from the actual water table in the direction of flow. Nevertheless, despite the simplifying assumptions, the equation closely approximates the water-table position where the sine and tangent of the slope of the water table are approximately equal. The equation is applicable under such conditions to determine Q and K but should be used with caution near a point or zone of discharge where the drawdown curve may be accentuated (Forcheimer, 1930).

5-11. Steady Radial Flow.—Early in this century, Theim (Muskat, 1946) and Forchheimer (1930) independently derived equations for steady radial flow (Todd, 1980) to a fully penetrating well with 100-percent penetration and open hole, using Darcy's law and Dupuit's assumptions. The equations, known today as the steady state, Theim, Dupuit-Forchheimer, or Theim-Forchheimer equations, can be used to determine the coefficient of permeability of an aquifer from measurements made during a pumping test using a fully penetrating well with 100-percent open hole and two or more observation wells. The equation for a confined aquifer is:

$$K = \frac{Q ln\left(\dfrac{r_2}{r_1}\right)}{2\pi M(s_1 - s_2)} \qquad \text{5-14}$$

and for an unconfined aquifer:

$$K = \frac{Q ln\left(\dfrac{r_2}{r_1}\right)}{\pi(h_2^2 - h_1^2)} \qquad \text{5-15}$$

where:

K = hydraulic conductivity, L/t
Q = discharge of the well, L^3/t
r_1 and r_2 = horizontal distances from the center of the dis-
                   charging well to the centers of observation wells
                   located on a line passing through the center of the
                   discharging well; distance increases with the value
                   of the subscript, L
M   = thickness of the aquifer, L
s_1 and s_2 = drawdown in observation wells r_1 and r_2,
                   respectively, L
h_1 and h_2 = saturated thickness of the aquifer at r_1 and r_2,
                   respectively, L
l_n = natural log = common log 2.303

The steady-state equation for an unconfined aquifer ignores vertical components of flow and curvature of the equipotential lines but recognizes decrease in aquifer thickness in the direction of the well. An additional assumption is that the well has infinitesimal diameter compared to a fixed radius of influence. Despite the simplifying assumptions, the equations give relatively reliable determinations of Q and K or T from measurements made of pumping tests of adequate duration with fully penetrating wells in confined aquifers and in unconfined aquifers where the drawdowns in observation wells do not exceed $0.25\ M$. Piezometers should be used in confined aquifers instead of observation wells to ensure reliable drawdown data. The steady-state equations do not take into consideration time and the release of water from storage. All water is assumed to originate beyond the radius of influence.

5-12. Transient One-Directional Flow.—Moody and Ribbens (1965) modified and applied equations derived by Glover (1960) for unsteady one-directional flow using much the same basic principles that are involved in the equation for unsteady radial flow (section 5-13). The assumptions on which the equations are based are as follows:

• The aquifer is homogeneous and isotropic.

• Hydraulic conductivity, K, and storativity, S, are constant with time.

• The aquifer is of infinite horizontal extent.

- The aquifer is confined or, if unconfined, the saturated thickness is large compared to the drawdown.

- Water is withdrawn at a constant rate from a fully penetrating vertical plane sink which is oriented parallel to the source of water. The drawdown is constant at a given time for all points along the sink.

The equation for transient one-directional flow is as follows:

$$s = \frac{Q}{2K}\left[\sqrt{\frac{4Kt}{\pi MS}}\ \exp\left(-\frac{r^2S}{4Tt}\right) - \frac{r}{M}\left(1 - erf\sqrt{\frac{r^2S}{4Tt}}\right)\right] \qquad \text{5-16}$$

where:

s = drawdown at any point a distance r from the vertical plane sink on a line normal to the sink, L
Q = the rate of withdrawal of water per linear foot from the sink, L^3/t
K = hydraulic conductivity of the aquifer, L/t
t = time since pumping or discharge per linear foot from the sink began, t
M = saturated thickness of the aquifer, L
S = storativity (dimensionless)
r = distance from the plane sink on a line normal to it, L
T = transmissivity of the aquifer, L^2/t
exp = the exponential function $(\exp(x) = e^x)$
erf = the error or probability function

Tables of the exponential function e^x are found in Applied Mathematics Series 14 and the error or probability function is found in Applied Mathematics Series 41 of the National Bureau of Standards.

Moody and Ribbens (1965) also give a function, equation 5-17, with which to compute the additional drawdown caused by convergence of parallel flow to a horizontal line sink (i.e., a drain or ditch instead of a fully penetrating plane sink):

$$s = \frac{Q}{2K}\left\{\frac{r}{M} - \frac{2}{\pi} \ln\left[\exp\left(\frac{\pi r}{2M}\right) - \exp\left(-\frac{\pi r}{2M}\right)\right]\right\} \qquad 5\text{-}17$$

In equation 5-17, M = the thickness of the aquifer and the other symbols are as in equation 5-16. If two line sinks exist in the neighborhood of each other, the drawdowns are additive.

5-13. Transient Radial Flow.—The limitations and errors in the steady-state well equations resulting from the simplifying assumptions made in their derivations were recognized by early investigators. Theis (1935) perceived the analogy between the flow of heat and flow of water and adapted the equation for the flow of heat in a conducting solid to the flow of water to a well in a confined aquifer. Jacob (1940) derived an identical equation from purely hydraulic considerations. The Theis, or nonequilibrium equation, which takes into consideration both time and storativity, has the form:

$$s = \frac{Q}{4\pi T} \int_u^\infty \frac{e^{-u}}{u} du \qquad 5\text{-}18$$

where:

s = the drawdown at any point r on the cone of depression, L
Q = uniform discharge of a well per unit time, L^3/t
T = KM, the transmissivity, by definition the hydraulic conductivity times the thickness of the aquifer, L^2/t
r = the distance from the center of the discharging well to the point of measurement of s, L
S = the storativity or coefficient of storage (dimensionless)
t = the time since discharge of the well began, t
u = unit of the well function $r^2 S/4Tt$ (dimensionless)

The assumptions on which the nonequilibrium equation is based are:

• The aquifer is homogeneous, isotropic, of uniform thickness, and of infinite areal extent.

• The discharging well is of infinitesimal diameter, completely penetrates, and is open to the aquifer.

- Discharge of water from storage is instantaneous with the reduction in pressure caused by drawdown.

- Flow to the well is radial and horizontal.

The exponential integral of u is frequently expressed as $W(u)$, the well function of u, and the equation can then be rewritten as:

$$s = \frac{QW(u)}{4\pi T} \qquad \text{5-19}$$

The above assumptions are rarely all present in actual conditions. Also, equation 5-19 is theoretically applicable only to confined aquifers. However, the error in the analysis of unconfined aquifers is minor provided the drawdown at the point of observation does not exceed 25 percent of the aquifer thickness. Use of the nonequilibrium equation permits analysis of aquifer conditions and predictions of aquifer behavior that change with time and involve storage. Rapid solution of the nonequilibrium equation by computers makes possible many of the modern modeling techniques used in ground-water analyses.

5-14. Anisotropy.—Most aquifers are anisotropic (i.e., flow conditions vary with direction [Hantush, 1966; Weeks, 1964]). In granular material, the particle shape and orientation, and the process and sequence of deposition, usually result in vertical permeability being less than horizontal permeability. The primary cause of anisotropy on a small scale is the orientation of clay minerals in sedimentary rocks and unconsolidated sediments. In nongranular rocks, the size, shape, orientation, and spacing of fractures and other voids may result in anisotropy. Regardless of the nature of the anisotropy, the effects on yield and drawdown by distortion of the distribution of flow are similar. Where anisotropy is the result of difference in vertical and horizontal permeability, the effect is to distort the distribution of drawdown in unconfined aquifers. The distortion is related to the distance to the observation well, the thickness of the aquifer, the ratio of the horizontal and vertical permeabilities, and the degree of aquifer penetration by the pumping well. Hantush (1966) and Weeks (1964) derived theoretical methods of determining horizontal and vertical permeability ratios and values from the analysis of pump test data.

Where anisotropy is the result of vertical being less than horizontal permeability, and the ratio K_h/K_v is relatively small, the flow distortion effect is small in flow to a well with a 100-percent open hole in a confined aquifer or in an unconfined aquifer with small drawdown. However, where the ratio of K_h/K_v and the vertical component of flow are large, such as in a partially penetrating well in a confined aquifer or in an unconfined aquifer with large drawdown, the decrease in yield or increase in drawdown compared to the ideal aquifer may be significant (see figure 5-6). The bottom of such a well may be considered to be the bottom of the aquifer.

5-15. Boundaries.—The nonequilibrium (transient) equations presented in this chapter are based on saturated flow through a homogeneous and isotropic aquifer of infinite areal extent. However, all aquifers have boundaries that modify flow conditions.

The confining beds of an artesian aquifer and the water table and the lower confining bed of an unconfined aquifer represent a type of boundary which limits transmissivity. However, as used here, boundaries are usually those limiting the horizontal extent of aquifers. These boundaries may be negative (impermeable), or positive (recharge), or both.

An impermeable boundary exhibits a significant reduction in transmissivity. Examples would be where the permeable alluvial fill of a valley abuts the buried valley sides that consist of impermeable granite, where a permeable sandstone is faulted against an impermeable shale, or where large unfractured volcanic boundaries have intruded an aquifer. An impermeable boundary influences a discharging well by retarding or stopping the expansion of the cone of depression, which results in increased drawdown between the well and the boundary and subsequent removal of water from storage in this area of increased drawdown. In the discharging well, the rate of drawdown is increased, the specific capacity is decreased, and the slope of the cone of depression not only decreases in the direction of the boundary but increases on the opposite side of the pumping well (see section 9-11).

A recharge boundary exhibits a significant increase in transmissivity (e.g., where a permeable material is in direct connection with a surface body of water or where a permeable material is faulted against a more permeable material). A recharge boundary also influences a discharging well by retarding

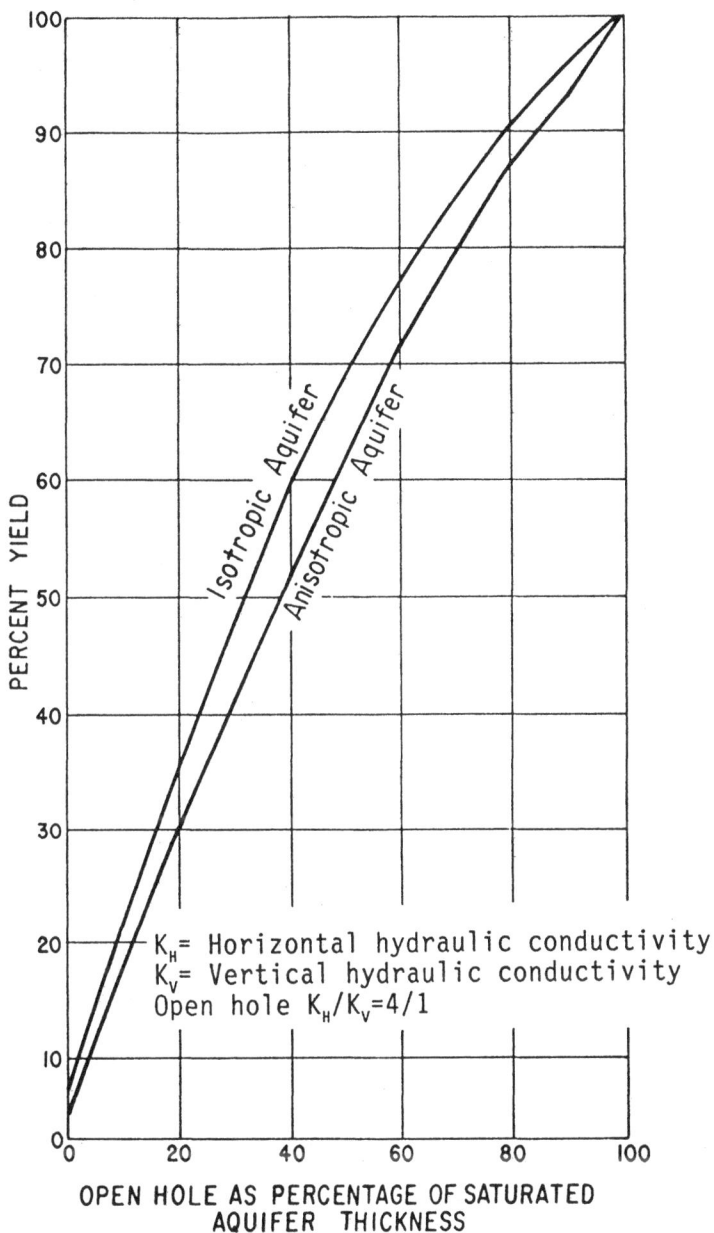

Figure 5-6.—Relationship of yield to open hole in isotropic and
anisotropic confined aquifers.

or stopping the expansion of the cone of depression. However, as the boundary provides a source of recharge to replace the normal flow from outside the boundary, the drawdown stabilizes between the well and the boundary, and removal of water from storage is limited. In the discharging well the rate of drawdown is lessened, the specific capacity is increased, and the slope of the cone of depression is not only increased in the direction of the boundary but is decreased on the opposite side of the pumping well (see section 9-11).

Boundaries are of concern when predicting the influence and probable yield and drawdown of wells. If an aquifer test is not run long enough for the area of influence to intercept a boundary, a substantial error may be made in estimating well performance. Also, a well may draw as much as 90 percent of its discharge from a stream that is hydraulically connected to the aquifer (see section 9-16).

Boundary effects can sometimes be anticipated on the basis of known geological conditions, or the conditions may be hidden and revealed only by analyses of a pumping test. In some instances, two or more boundaries may influence well performance to the extent that reliable determinations of aquifer characteristics and boundary locations are precluded. Methods of analyses of boundary effects from pumping tests are discussed in sections 9-11 and 9-12.

5-16. Leaky Aquifers.—An aquifer that receives a significant inflow from adjacent beds is called a leaky aquifer, although in reality the aquitard is leaky (Freeze and Cherry, 1979). The aquitard may be considered a type of boundary. When a well discharges from such an aquifer, the reduction in head may promote increased flow through confining beds to the aquifer or reduction in flow from the aquifer. If in a leaky confining layer in an aquifer system has enough storage, then part of the flow during the initial time period will come from storage in the confining layer (Fetter, 1980). When the area of influence has expanded sufficiently so that the amount of increased seepage into or reduced seepage out of the aquifer equals the pumping rate, the discharge drawdown, relationship, and flow pattern about the well stabilize (see section 9-8). A similar response is reflected when an unconfined aquifer overlying a confined aquifer is pumped. The lowering of the water table in the cone of depression reduces the pressure and increases upflow from the artesian aquifer. The influences on aquifer tests of boundary and leaky aquifer conditions may appear similar but usually can be differentiated (see subsection 9-5b).

5-17. Delayed Drainage.—An unconfined aquifer may consist in whole or in part of fine-grained material from which drainage is relatively slow (Boulton, 1963). A delayed water-table response can be observed when water-level drawdown in piezometers or observation wells adjacent to the pumping well in an unconfined aquifer tend to decline at a slower rate than that predicted by the Theis solution (Freeze and Cherry, 1979).

The delayed response is related to the vertical components of flow that are induced in the flow system and usually is a function of the radius, r, and time, t, (Freeze and Cherry, 1979). Pumping tests conducted under such conditions may yield unusual S-shaped plots of log t versus log s, which may be attributed to leaky aquifer influence. Furthermore, in extreme cases, an uneconomically long test might be required to differentiate between the two situations. In the final analysis, judgment based on the knowledge of subsurface and other conditions may be the principal basis for interpretation.

5-18. Recharge and Discharge Areas.—Recharge areas are those within which water enters an aquifer. The location, size, and features of the area within which recharge occurs to an aquifer are pertinent to many ground-water problems. In some unconfined aquifers, recharge occurs over the entire aquifer area; in others, it may be limited by the presence of natural or artificial impermeable materials overlying parts of the area or to aquifers connected with a body of surface water. In confined aquifers, the recharge area is limited to a large extent by the exposure of the aquifer at the surface or to its subsurface connection with another aquifer or a body of surface water.

Contributing areas may be considerably larger than recharge areas. Water may enter a recharge area from adjacent and surrounding terrain. The entire area from which water is tributary to a recharge area is the contributing area.

Discharge areas are of similar complexity and variation. Primary avenues of natural discharge include evapotranspiration,

spring flow, seepage to streams, and leakage to other aquifers.
The determination and delineation of recharge and discharge areas
are sometimes a complex problem.

**5-19. The Radius of Influence and the Cone of
Depression**.—The equilibrium equations assume creation of a
fixed radius of influence of a well and further assume that all
water pumped by the well enters the cone of depression from
beyond the radius of influence (Ferris et al., 1962; Meinzer, 1949).
This assumption can be equated to a well discharging from an
aquifer underlying a circular island and which is in hydraulic
connection with the surrounding sea. However, the nonequilibrium
equation assumes that all water comes from storage within the
radius of influence and that this radius increases with time. In an
ideal aquifer of infinite areal extent, the radius of influence and
the drawdown theoretically increase at a constantly diminishing
rate as long as the well is pumped. Under field conditions,
however, the rate of change becomes so slow after a sufficiently
long period of pumping that it is difficult to measure (see
figure 5-7). In many aquifers, the area of influence intercepts the
natural aquifer discharge or encounters recharge in sufficient
quantity to balance well discharge, and true stabilization occurs.

5-20. Well Interference.—If two or more wells discharging
from the same aquifer are close enough to each other so that their
respective areas of influence overlap, each well interferes with the
other, and the chord joining the two points of intersection of the
areas of influence (Bentall, 1963; Meinzer, 1949) then becomes a
divide across which no flow occurs. This phenomenon is called well
interference and, as a consequence, the rate of drawdown of each
well is accelerated.

A recharging well has a similar but opposite effect on a
discharging well. The chord joining the two points of intersection
of the areas of influence then becomes a line of no drawdown.
Flow across it is only in one direction and is equal to the flow
originating in the recharging well. Consequently, the rate of
drawdown is retarded. Section 9-11 contains a further discussion
of this concept.

5-21. Principle of Superposition.—If the transmissivity and
storativity of an ideal aquifer and the yield and duration of
discharge or recharge of two or more wells are known, the
combined drawdown or buildup at any point within their
interfering area of influence may be estimated by adding

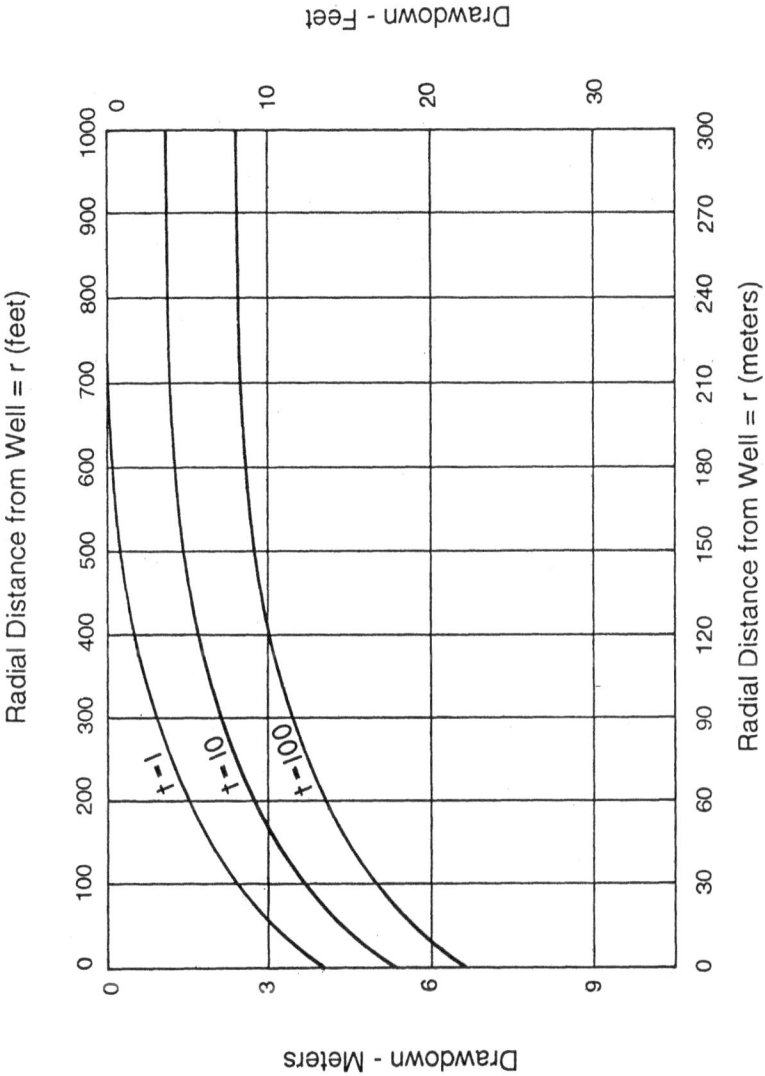

Figure 5-7.—Time-drawdown relationship with all factors constant except time, t.

algebraically the component of drawdown of each well. This principle is illustrated on figures 9-17 and 9-18 in section 9-11, which show the impressed heads of a real and image well and the resultant actual drawdown. The effects would be identical for two real wells. The principle of superposition is used to determine desirable spacing of wells in well fields, effects of recharging wells, and in the evaluation boundary conditions. Superposition is further discussed in section 9-12. The calculation of drawdown is greatly simplified through the use of computer programs.

5-22. Bibliography.—

American Geophysical Union (Transactions), 1946, "Radial Flow in a Leaky Artesian Aquifer," vol. 27, No. II, pp. 198-208.

Bentall, Ray (compiler), 1963, "Shortcuts and Special Problems in Aquifer Tests," U.S. Geological Survey Water-Supply Paper 1545-C.

Boulton, N.S., 1963, "Analysis of Data from Non-Equilibrium Pumping Tests Allowing for Delayed Yield from Storage," Proceedings of the Institution of Civil Engineers (London), paper No. 6693, vol. 26, pp. 469-478.

Ferris, J.G., D.B. Knowles, R.H. Brown, and R.W. Stallman, 1962, "Theory of Aquifer Tests," U.S. Geological Survey Water-Supply Paper 1536-E, pp. 69-174.

Fetter, C.W., Jr., 1980, *Applied Hydrogeology*, Merrill Publishing Company, Columbus, Ohio, pp. 488.

Forchheimer, Philipp, 1930, *Hydraulik*, 3d edition, B.G. Teubner Verlagsgesellschaft, Berlin.

Freeze, R. Allan and John C. Cherry, 1979, *Groundwater*, Prentice-Hall, Inc., New Jersey, pp. 604.

Glover, R.E., March 1960, "Studies of Water Movement," Bureau of Reclamation, Technical Memorandum 657.

Hantush, M.S., January 15, 1966, "Analysis of Data from Pumping Tests in Anisotropic Aquifers," *Journal of Geophysical Research*, vol. 71, No. 2, pp. 21-425.

Jacob, C.E., 1940, "On the Flow of Water in an Elastic Artesian Aquifer," *Transactions of the American Geophysical Union*, vol. 2, part II, pp. 74-586.

_____, April 1946, "Radial Flow in a Leaky Artesian Aquifer," *Transactions of the American Geophysical Union*, vol. 27, No. II, pp. 98-208.

Lohman, S.W. et al., 1972, "Definitions of Selected Ground Water Terms—Revisions and Conceptual Refinements," U.S. Geological Survey Water-Supply Paper 1988.

Meinzer, O.E. (editor), 1949, "The Physics of the Earth—IX, Hydrology," Dover Publications, New York.

Moody, W.T. and R.W. Ribbens, December 1965, "Water— Tehama-Colusa Canal Reach No. 3, Sacramento Canals Unit, Central Valley Project," Memorandum to Chief, Canals Branch, Bureau of Reclamation, Office of Chief Engineer, Denver, Colorado.

Muskat, Morris, 1946, "The Flow of Homogeneous Fluids Through Porous Media," J.W. Edwards, Ann Arbor, Michigan.

Theis, C.V., 1935, "The Relation Between the Lowering of the Piezometric Surface and the Rate and Duration of Discharge of a Well Using Water Storage," *Transactions, American Geophysical Union*, vol. 16, pp. 519-524.

Todd, D.K., 1959, *Water Hydrology*, John Wiley & Sons, New York.

_____, 1980, *Water Hydrology*, 2d edition, John Wiley & Sons, New York.

Weeks, E.P., 1964, "Field Methods for Determining Vertical Permeability and Aquifer Anisotropy," U.S. Geological Survey Professional Paper 501-D, pp. 193-198.

Chapter VI

WELL AND AQUIFER RELATIONSHIPS

6-1. Aquifer and Well Hydraulics.—Aquifer characteristics exert primary control over well performance in terms of yield versus drawdown. Accordingly, determination of the effects of well geometry on the flow and head distribution in aquifers and on the yield and drawdown of wells has been the goal of most research on well hydraulics. Methods of mathematical analysis have been developed for both steady-state and transient-state conditions. Steady-state analyses are performed according to Darcy's law and Dupuit's assumptions of horizontal radial flow and fixed radii of influence. All mathematical solutions assume ideal artesian and unconfined aquifers, which are isotropic, homogeneous, of uniform thickness, and infinite areal extent. The conclusions are generally adequate for estimating the performance of wells in artesian aquifers and in unconfined aquifers where the drawdown is a small percentage of the aquifer thickness and the discharging well is fully penetrating. Corrections for partial penetration of the discharging well, large drawdowns in unconfined aquifers, and anisotropy have been derived, but adequate data for application of the corrections are often not readily available. Much research has also been done on analogs and other models of various types, but too often, the geometry of the test apparatus has not duplicated field conditions.

However, the nature of the well is also an important factor in well performance. Experience has shown that well design features and construction practices have measurable effects on well performance and operating life and on the economic use of the aquifer. Despite this relationship, the engineering and scientific aspects of well hydraulics have received little attention because pumpage of a desired amount of water has been the principal criterion of a successful well. The few laboratory analog analyses which have been made of these relationships have seldom duplicated field conditions. As a result of this and other factors, water well design has commonly been based on experience, observations, and judgment of the designer and driller.

In this discussion, "yield" is defined as the potential production capacity of the well, which is controlled by aquifer characteristics and well design and construction. "Discharge," on the other hand, is defined as the actual production of the well, which depends on the pump and discharge pipe characteristics as well as the above parameters.

6-2. Flow to Wells.—Theoretically, on initiation of discharge from a well, the water level or head in the well is lowered relative to the undisturbed condition of the potentiometric surface or water table outside the well (Jacob, 1947; Meinzer, 1949; Muskat, 1946). The water in the aquifer surrounding the well responds by flowing radially to the lower level in the well.

In an artesian aquifer, except for a slight delay caused by inertia, the actual distribution of flow to the well conforms relatively close to the theoretical distribution shortly after pumping is started. However, in an unconfined aquifer, the materials in the cone of depression must drain and establish progressively the surface configuration of the cone. Hence, the actual distribution of flow may not conform to the theoretical. Figure 6-1 illustrates schematically the successive stages of development of flow distribution under such conditions by means of equipotential lines and flow lines around a well.

In an ideal artesian aquifer, assume a 100-percent open hole well of radius r_w surrounded by two concentric cylinders of radius r_1 and r_2 with heights equal to the thickness, M, of the aquifer. The surface area, A, of the well is $2\pi r_w M$, and of the cylinders, $2\pi r_1 M$ and $2\pi r_2 M$, respectively (see figure 6-2). Under steady-state conditions, the same quantity of water per unit time must flow through each cylinder and ultimately into the well. According to Darcy's law, $Q = KiA$ or $V = Ki$. If Q and K are constant, the gradient i and the velocity must increase in value as A decreases. Hence, if h_o is the effective head (potentiometric surface) at the radius of influence, r_o, and h_2 and h_1 are the heads at radii r_2 and r_1 about the well, the velocity and gradient must increase in the direction of the well. The result is a funnel-shaped area of lowering of the potentiometric surface centered about the well.

Figure 6-3 illustrates schematically the distribution of flow in an artesian aquifer by means of a network of flow lines and equipotential lines lying in a vertical section passing through the axis of a fully penetrating well which has a 100-percent open hole. If the piezometric surface is not drawn down below the bottom of the upper confining bed, the flow lines to the well remain parallel and horizontal and the equipotential lines remain parallel and

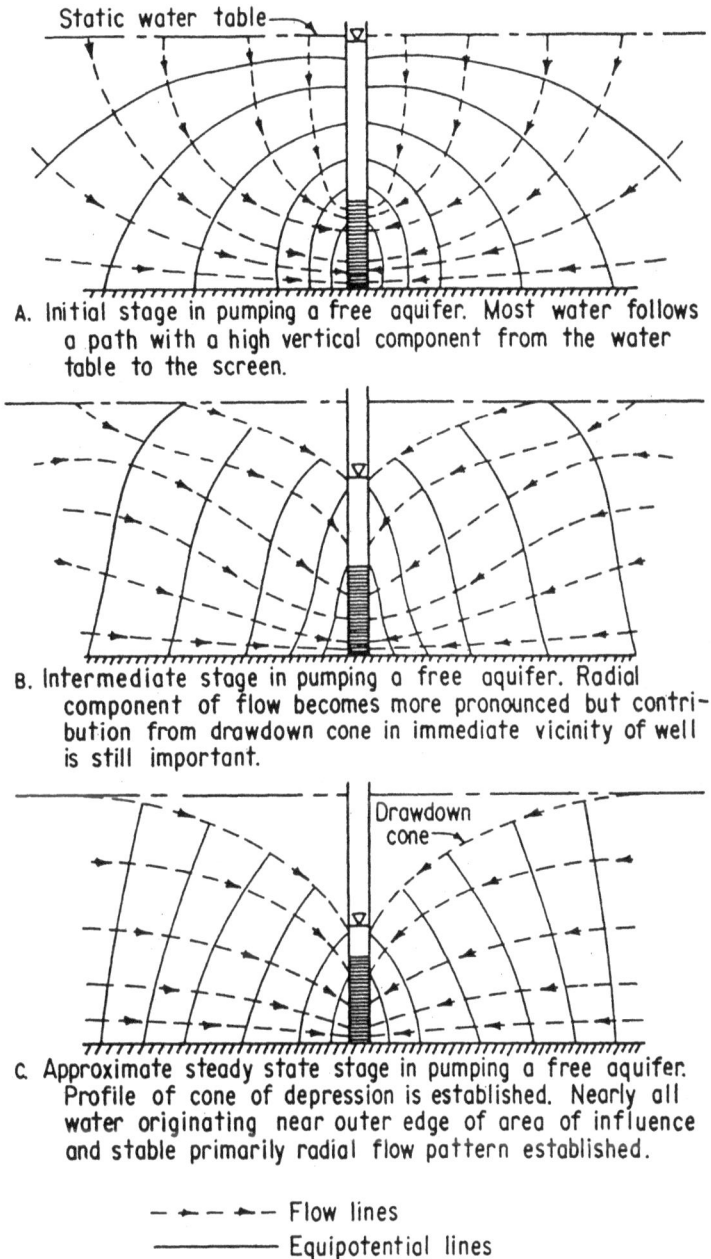

A. Initial stage in pumping a free aquifer. Most water follows a path with a high vertical component from the water table to the screen.

B. Intermediate stage in pumping a free aquifer. Radial component of flow becomes more pronounced but contribution from drawdown cone in immediate vicinity of well is still important.

C. Approximate steady state stage in pumping a free aquifer. Profile of cone of depression is established. Nearly all water originating near outer edge of area of influence and stable primarily radial flow pattern established.

— ←— ←— Flow lines

————— Equipotential lines

Figure 6-1.—Development of flow distribution about a discharging well in an unconfined aquifer—a fully penetrating and 33-percent open hole.

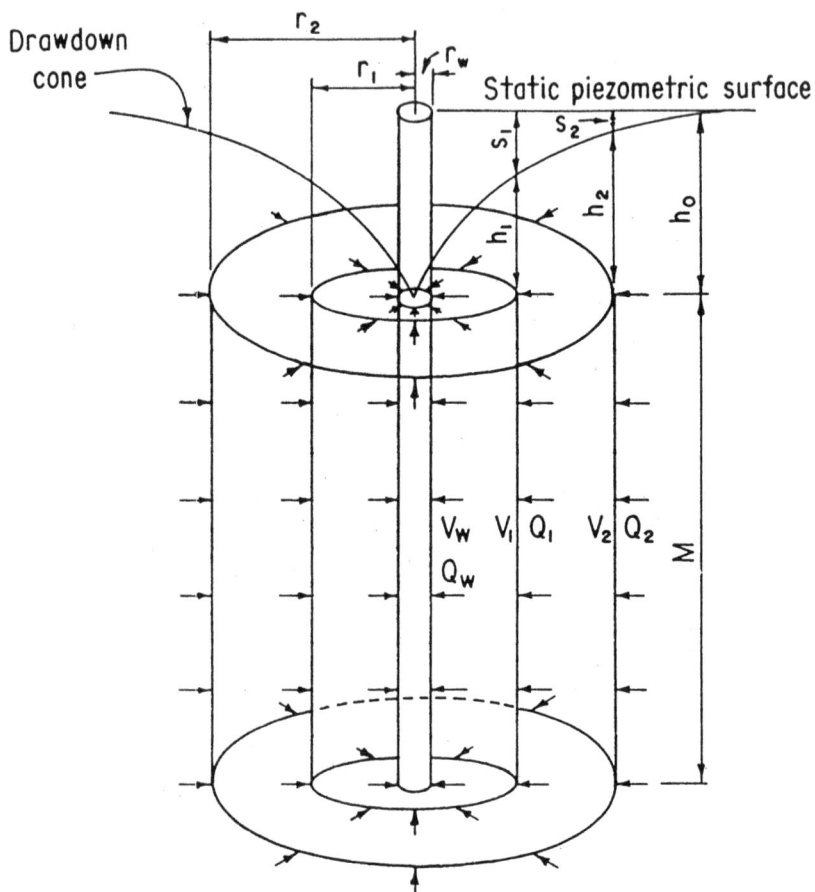

M = Thickness of aquifer

h_o = Undisturbed artesian head

h_1, h_2 = Artesian head at r_1 and r_2 respectively when well is discharging.

s_1, s_2 = Drawdown at r_1 and r_2 respectively when well is discharging

$Q_w = Q_1 = Q_2$

$A_w = 2\pi r_w M$

$A_1 = 2\pi r_1 M$

$A_2 = 2\pi r_2 M$

$V_w = \dfrac{Q_w}{A_w}$

$V_1 = \dfrac{Q_1}{A_1}$

$V_2 = \dfrac{Q_2}{A_2}$

Figure 6-2.—Flow distribution to a discharging well in an artesian aquifer—a fully penetrating and 100-percent open hole.

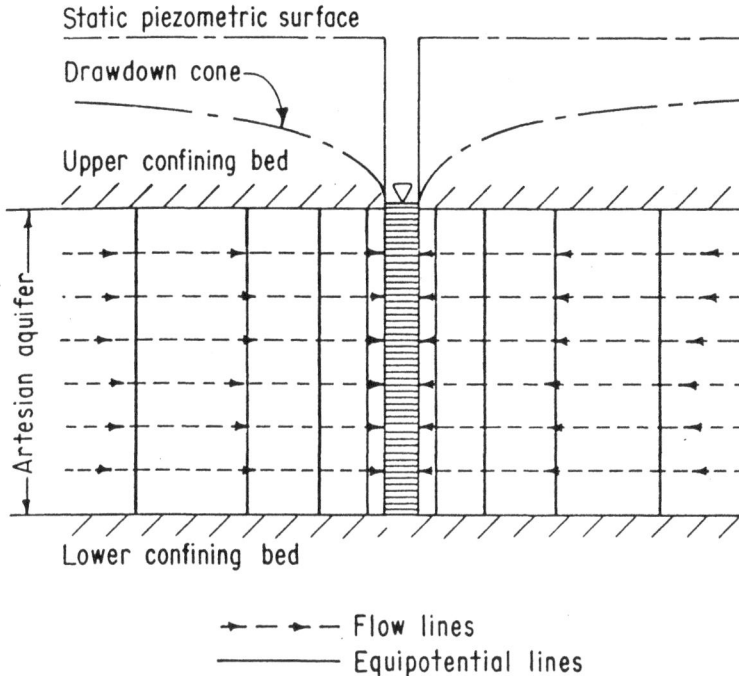

Figure 6-3.—Distribution of flow to a discharging well in an artesian aquifer—a fully penetrating and 100-percent open hole.

vertical. If the drawdown falls below the bottom of the upper confining bed, a mixed condition of artesian and free aquifer flow results, which is difficult to assess.

Figure 6-4 illustrates a well that penetrates through the upper confining bed but not into the artesian aquifer. The flow lines and equipotential lines develop hemispherically about the well radius, and a strong vertical component of flow is established out to a distance about equal to the thickness of the aquifer. At a distance about 1.5 times the thickness of the ideal aquifer, the flow lines and equipotential lines assume a relationship similar to that of the fully penetrating 100-percent open hole well.

Figure 6-5 illustrates a well penetrating into and open to about 50 percent of the thickness of an artesian aquifer. The vertical component of flow is less than in the nonpenetrating well, but the equipotential lines are still strongly curved, although not as great as on figure 6-4. The transition to strictly horizontal flow at a

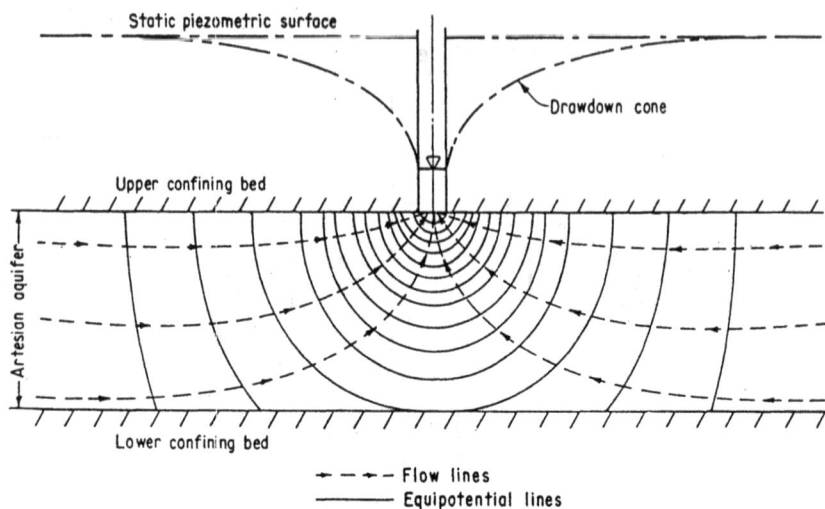

Figure 6-4.—Distribution of flow to a discharging well—just penetrating to the top of an artesian aquifer.

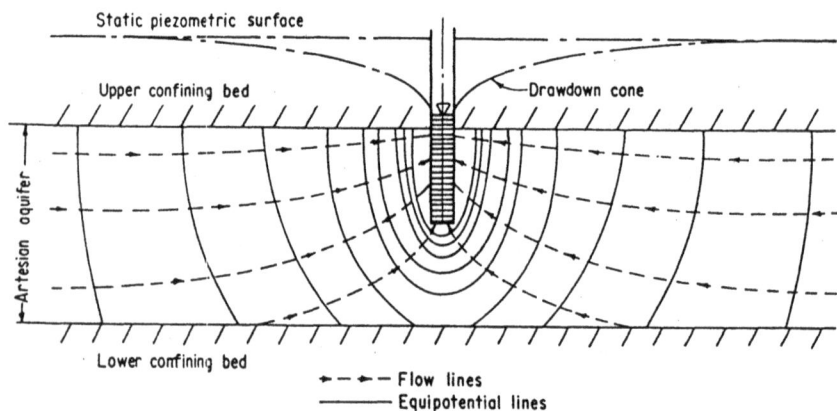

Figure 6-5.—Distribution of flow to a well in an artesian aquifer—a 50-percent penetrating and open hole.

distance from the well of about 1.5 times the thickness of the aquifers is also apparent. If the open hole was in the lower rather than the upper half of the aquifer, the pattern of flow and equipotential lines would be similar to figure 6-5 if it was inverted.

Vertical convergent flow in wells of less than 100-percent penetration and open hole results in increasing drawdown with a decreasing percentage of open hole.

Figure 6-6 illustrates schematically the distribution of flow and equipotential lines around a fully penetrating well in an unconfined aquifer with the drawdown in the well about one-half the thickness of the aquifer. Drainage of material above the drawdown cone results in a decline in aquifer thickness, M, and a similar decline in transmissivity. Also, the drawdown in an unconfined aquifer accentuates the vertical components of flow. Thus, the drawdown is accentuated by, and in turn, accentuates the decline in aquifer thickness, decline in transmissivity, and vertical component of flow.

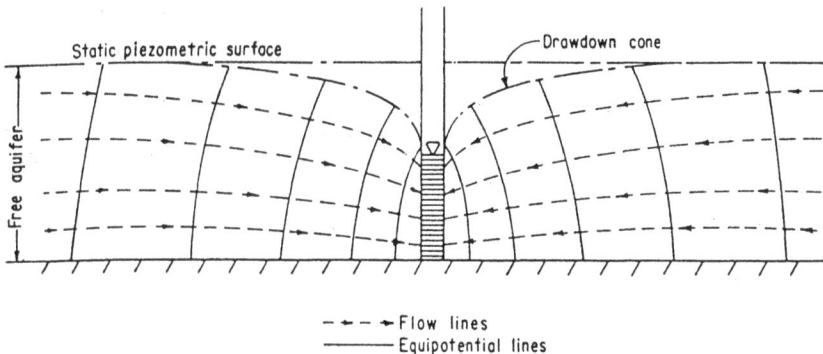

Figure 6-6.—Distribution of flow to a discharging well in an unconfined aquifer—a fully penetrating and 50-percent open hole.

Reduction in the percentage of open hole in an unconfined aquifer has an effect on the drawdown cone and flow lines similar to those in an artesian aquifer. However, the effect is further accentuated because of dewatering of the aquifer in the direction of the well.

Wells often must be drilled in aquifers of large, but unknown, thickness where the cost of full penetration would be prohibitive. Under such circumstances, the usual practice is to compromise on theoretical aspects and drill only to the depth that will furnish the desired supply of water at an acceptable lift.

Kozeny (Muskat, 1946) derived an equation for estimating the yield of partially open holes in an ideal artesian aquifer. Figure 6-7 is a graph of this equation for parameters usually encountered or used in well design. The plot shows approximate values which may be used for estimating purposes if the aquifer is fairly uniform and homogeneous and its thickness and characteristics are known. Jacob (Bentall, 1963) also derived a method for use in determining aquifer characteristics to correct observed drawdowns resulting from partial penetration to those of the ideal condition if the aquifer thickness was known. Both methods, however, are influenced by other, often unknown, factors such as aquifer anisotropy and boundaries.

6-3. Yield and Drawdown Relationships.—The following two steady-state equations (Ferris et al., 1962; Meinzer, 1949; Muskat, 1946) can be expressed in approximately the same form as the unsteady state:

The equation for the unsteady state is:

$$s_1 - s_2 = \frac{Ql_n\left|\frac{r_2}{r_1}\right|}{2\pi KM} \quad and \quad (h_2)^2 - (h_1)^2 = \frac{Ql_n\left|\frac{r_2}{r_1}\right|}{\pi K} \qquad \text{6-1}$$

The equation in the steady state is:

$$s = \frac{QW(u)}{4T\pi} \qquad \text{6-2}$$

This relationship facilitates recognition of the following:

- The drawdown at any point in the cone of depression is proportional to Q (see figure 6-8).

- At a given yield, the drawdown at any point on the cone of depression is inversely proportional to $ln\ r$ in all equations, and in the unsteady-state equation, the drawdown is also inversely proportional to storativity and proportional to $log_{10}\ t$ (time) (see figure 6-9). At a given Q, drawdown decreases with increased values of transmissivity (see figure 6-10).

The relationships are applicable for any point in the cone of depression for either fully or partially open hole discharging wells, although the actual form of the cone of depression will be distorted from the ideal by a partially open hole, anisotropy, or boundaries.

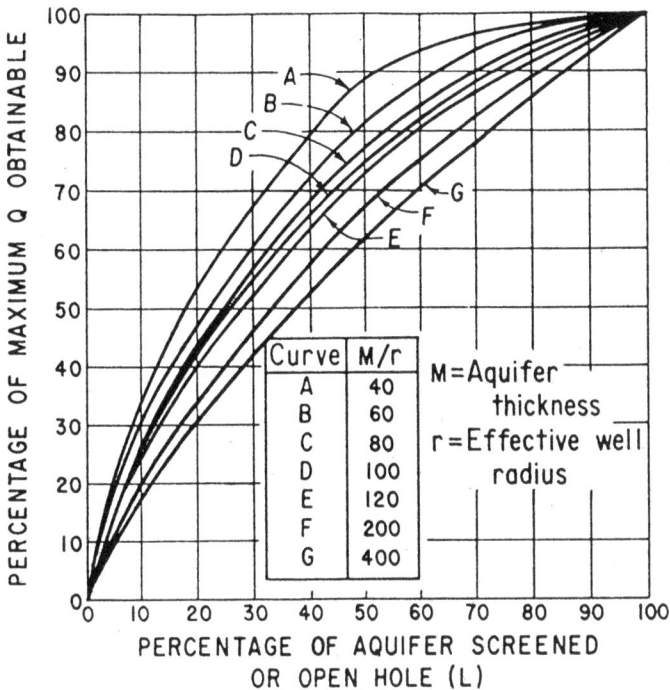

$$\frac{Q_p}{Q} = L\left[1 + 7\left(\frac{r}{2ML}\right)^{1/2} \cos \frac{\pi L}{2}\right]$$

Q_p - Yield of a partially open hole.

Q - Yield of a fully open hole.

L - Length of open hole as a fraction of aquifer thickness

r - Well radius

M - Aquifer thickness

Figure 6-7.—Graph of Kozeny's equation for relative yield and percentage of open hole in an ideal artesian aquifer.

When time is infinite, the transient-state equation becomes the same as the steady-state equation. Measurements made simultaneously with the drawdowns in the discharging well and one observation well have been used to compute transmissivity or hydraulic conductivity using the equilibrium equation. Theoretically, this computation is possible; but practically, it is not recommended because the water level inside a well generally is lower than outside because of well losses. The result is generally a computed transmissivity or hydraulic conductivity that is less than the true value.

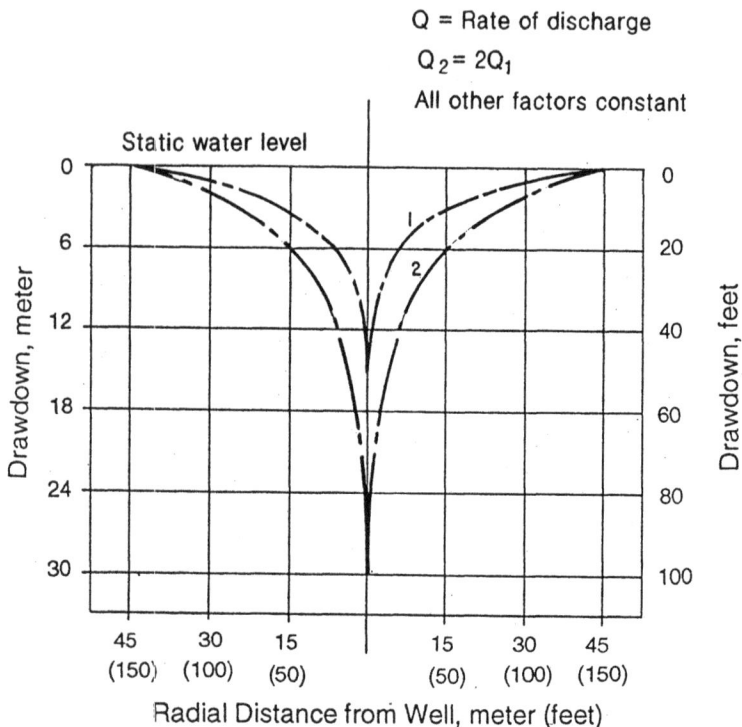

Q = Rate of discharge

$Q_2 = 2Q_1$

All other factors constant

Figure 6-8.—Influence of rate of discharge on drawdown in a well.

The previous discussion has been primarily concerned with the effect of partial and full penetration on distribution of flow to wells. From a theoretical standpoint, but of equal importance, is the effect on well performance. The steepening of the flow lines resulting from a partially open hole in an aquifer results in increased drawdown for the same yield, in other words, a decrease in specific capacity of the well. A well in an artesian aquifer should be open through the entire thickness of the aquifer. A well in an unconfined aquifer should be open in the lower one-third to one-half of the aquifer. However, this specification is not economically feasible in very deep and thick aquifers. In such aquifers, the usual practice is to penetrate a sufficient thickness of the aquifer to ensure the required discharge at an acceptable pumping lift.

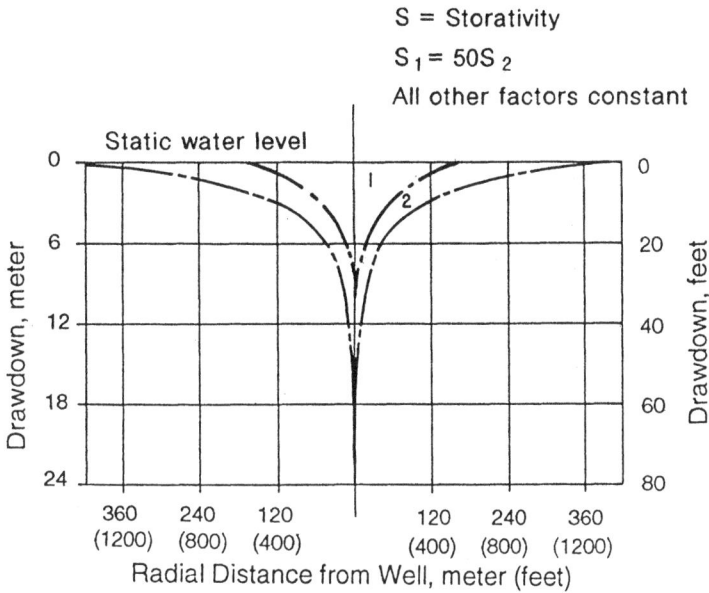

Figure 6-9.—Influence of storativity on drawdown in a well.

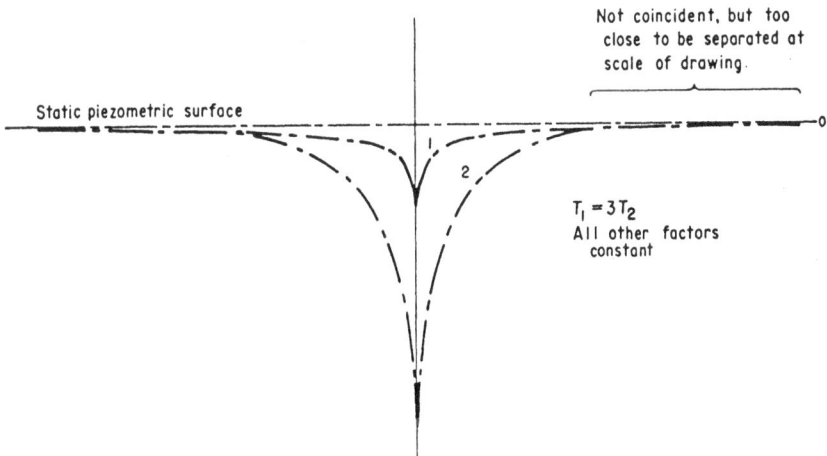

Figure 6-10.—Influence of transmissivity on drawdown in a well.

In a correctly designed well in an artesian aquifer where turbulent flow is minimal and in which the drawdown is not below the bottom of the upper confining bed, a nearly linear relationship exists between yield and drawdown. Hence, specific capacity remains relatively constant regardless of yield. The yield is approximately proportional to the drawdown in a 100-percent efficient artesian well with 100-percent open hole as long as the drawdown does not lower the hydrostatic head below the bottom of the upper confining bed. Conditions differ, however, in an unconfined aquifer in which the saturated thickness decreases with an increase in drawdown and vertical components of flow prevail. A number of equations relating yield and drawdown in an unconfined aquifer are available, but all of these equations are rough approximations. Figures 6-11 and 6-12 are inexact, composite curves which may serve only as useful guides to yield-drawdown relationships. Available equations and observations show a nearly linear relationship for yield and drawdown in an unconfined aquifer for drawdown as much as 50 percent of the saturated thickness and acceptable ratios approaching 65 percent. Beyond this limit, the specific capacity begins to fall rapidly. For these reasons, most screened wells are screened in the lower one-third or one-half of the aquifer. Figures 6-11 and 6-12 and some of the equations show maximum yield with 100-percent drawdown. This phenomenon is the result of extending steady-state equations beyond the limits established by the fundamental assumptions that define steady-state conditions. Obviously, the yield reaches a point of diminishing returns as the inlet area decreases, ultimately approaching zero. Also, in practice, the well discharge in unconfined aquifers decreases with increasing drawdown, primarily because of decreased efficiency of the pump and increased head losses.

6-4. Well Diameter and Yield.—A common misconception holds that well yield is proportional to well diameter and that doubling the diameter will increase the yield proportionately, other things being equal. Well diameter, in this case, refers to the diameter of the hole that penetrates the aquifer.

The fallacy in this reasoning can be shown by assuming all factors (other than the well diameter) constant in the equilibrium equations and rearranging the equations.

The resultant equation is:

$$Q = \frac{C}{\ln\left(\dfrac{r_e}{r_v}\right)} \qquad\qquad 6\text{-}3$$

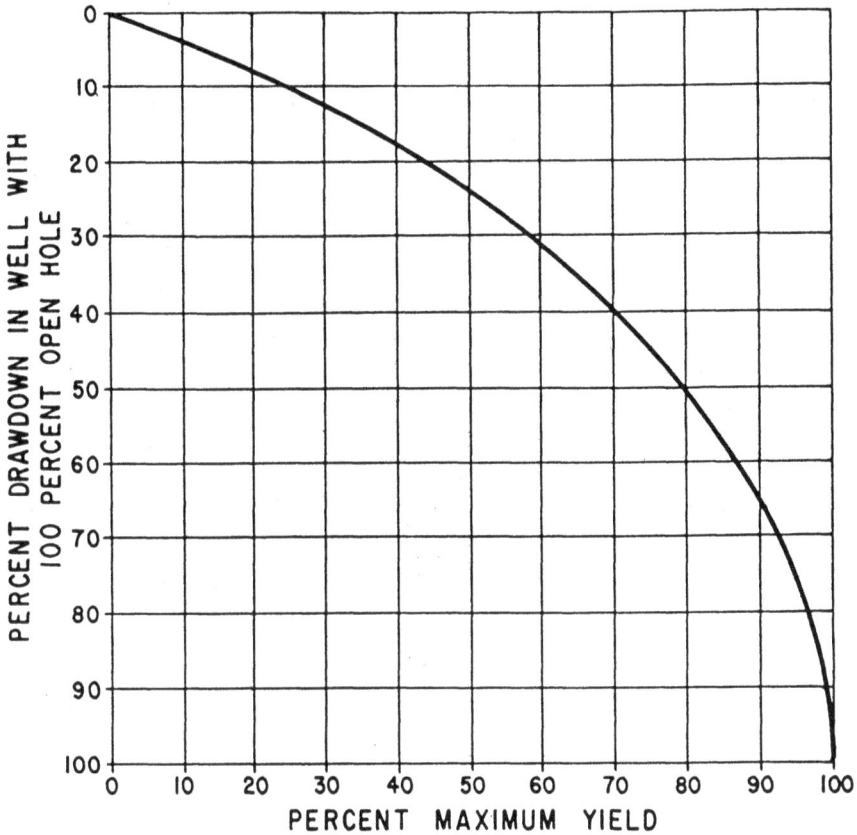

Values are approximate mean values of those
obtained from steady state, Kozeny's and other
equations, none of which is strictly accurate.

Figure 6-11.—Comparison of yield with drawdown in a
100-percent open hole in an ideal aquifer.

For an artesian aquifer: $c = 2\pi KM(S_2 - S_1)$

And for an unconfined aquifer: $C = K[(h_2)^2 - (h_1)^2]$

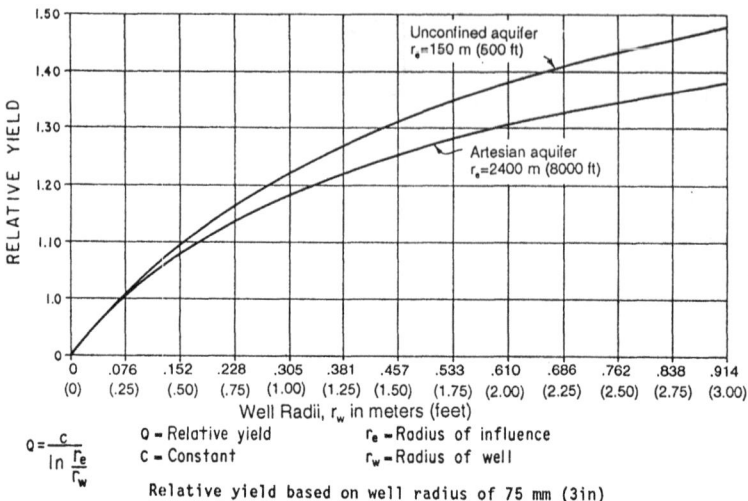

Figure 6-12.—Comparison of well radius and relative yield of a well.

Where:

Q = potential yield
c = constant
r_e = radius of influence
r_w = radius of well
K, M, s_2, s_1, h_2, and h_1 are as previously defined

By analysis, the yield can be shown to be proportional to the reciprocal of $ln\ r_e/r_w$.

Figure 6-12 illustrates the increase in yield caused by increasing the radius of a well, assuming the effective radius of influence is 2,400 meters (8,000 feet) in an artesian aquifer and 150 meters (500 feet) in an unconfined aquifer. Depending on the initial radius, the increase in yield caused by doubling the radius ranges from 8 to 13 percent for an artesian aquifer and from 10 to 17 percent for an unconfined aquifer. To theoretically double the yield within these parameters would require increasing the diameter of the well in the artesian aquifer about 90 times and in the unconfined aquifer about 45 times. This analysis assumes laminar flow and steady-state conditions. Experience has indicated, however, that doubling the diameter of wells will, in some instances, result in an increase in yield of as much as

25 percent. The difference is probably caused by a reduction in turbulence in the aquifer and in the well, increased entrance area, and other factors which reduce well losses.

Zangar and Jarvis (Zangar, 1953), on the basis of electric analog studies of recharge wells, determined that one effect of screen diameter in a partially open hole is to reduce the effective diameter of a well to that represented by a hole of the same length but having a surface area equivalent to the open area of the screen. Restated, a 300-millimeter (12-inch) screen with 25 percent open area would have an effective well diameter of 75 millimeters (3 inches) in the well and aquifer relationship. It is not known whether this relationship applies to a pumping well of any percentage of open hole because tests have not been made regarding this feature at this time.

Little uniformity appears to have been used in the past concerning the selection of screen diameters. In an attempt to rationalize the selection of screen diameters, many items such as head losses in pipes of various diameters, well and aquifer relationships, probable effects of screen length, slot size, and patterns have been evaluated for many wells of various discharges, diameters, and efficiencies. The results, which are imprecise and strictly empirical, are given in table 10-9 in section 10-4(b). They are presented not as a rigid requirement but as a suggested tentative standard for efficient well design considering initial construction and operation and maintenance costs.

6-5. Well Penetration and Yield.—The total depth of a well has little relationship to well yield except in respect to aquifer depth. As noted previously, the important factor is the percentage of saturated thickness of the aquifer penetrated by and open to the well (i.e., the percentage of open hole). In an artesian aquifer, the specific capacity of a well will vary with the percentage of open hole and is maximum when the entire thickness of the aquifer is penetrated by a screen or open hole (see figure 6-3). A similar relationship exists in an unconfined aquifer (see figures 6-13 and 6-14), but because of the thinning of the aquifer in the direction of the well as the well is pumped, the increase in specific capacity is limited depending on the amount of drawdown experienced and the desirability of limiting it to 60 to 65 percent of the aquifer thickness. But in either case, if sufficient additional aquifer thickness is available, the least expensive way to increase specific

GROUND WATER MANUAL

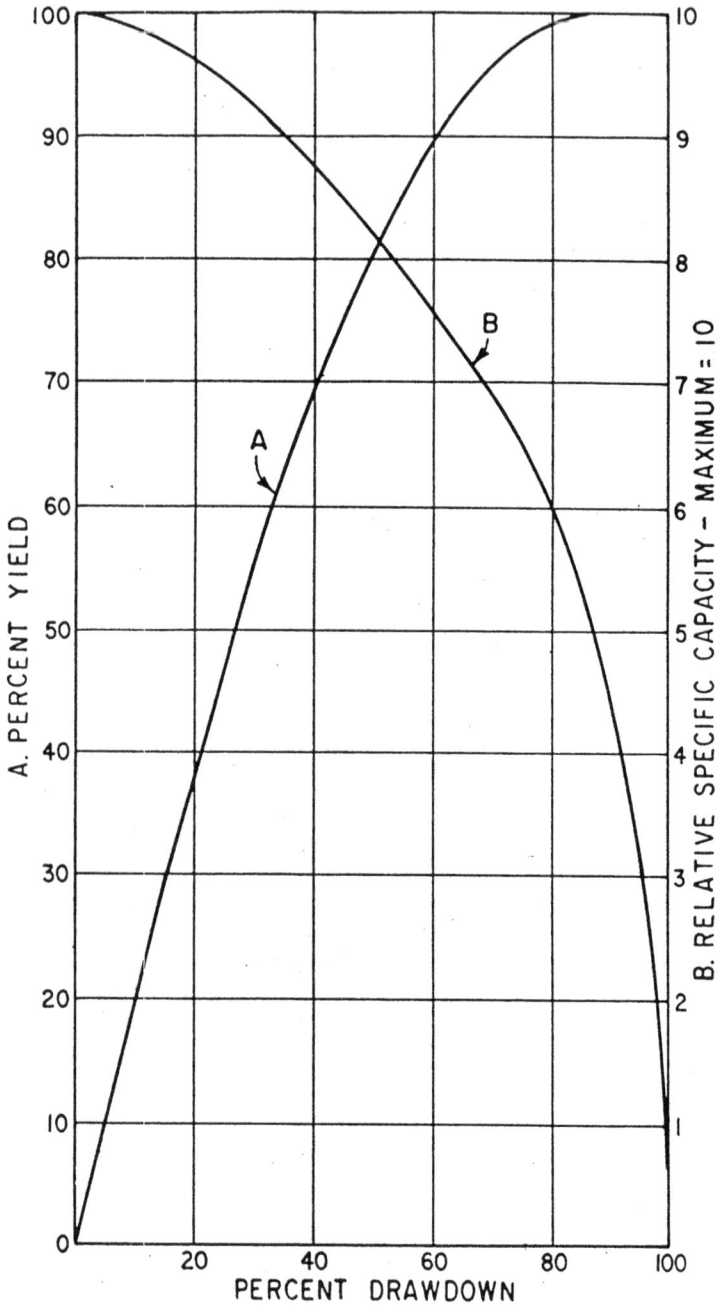

Figure 6-13.—Theoretical relationship between yield and drawdown in an unconfined aquifer with 100-percent penetration.

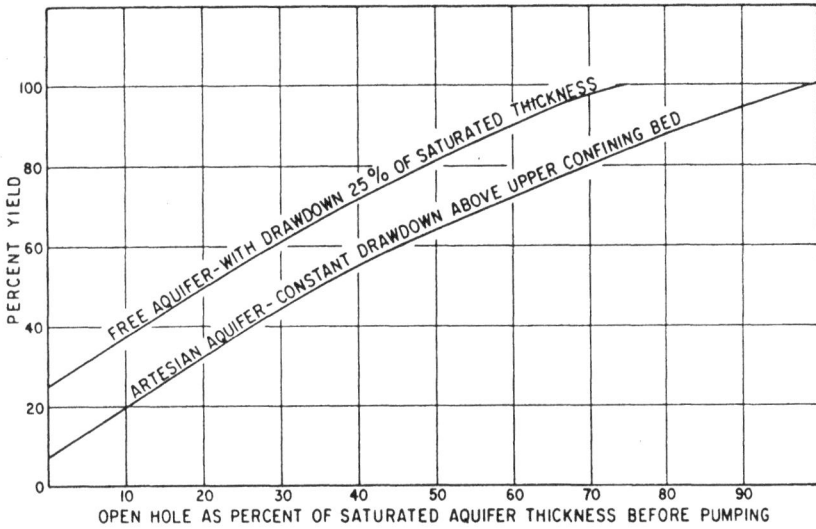

Figure 6-14.—Relationship between yield and drawdown in an
unconfined aquifer with 100-percent penetration.

capacity and yield is usually to increase either the depth of the
well, the percentage of open hole, or both, if possible (Bentall,
1963).

6-6. Entrance Velocity.—A generally acceptable principle of
well design holds that the average entrance velocity, based on the
percentage of open area of the screen and the desired yield, Q/A,
should be 0.03 meter (0.1 foot) per second or less. The hydraulic
theory behind this criteria holds that at such low velocities, flow is
entirely laminar; thus, turbulence will not contribute to well loss.
However, the average entrance velocity concept may be misleading.
Soliman (1965) and Li (1954) analyzed flow to a well, and they
showed that the entrance velocity in the upper 10 percent of a
screen was about 70 times that of the lower 10 percent in an ideal
aquifer. In every screened well, part of the entrance area is
blocked by the screen and aquifer material or gravel pack.
Depending upon the slot size and size and gradation of the grains
in the aquifer, as much as 78 percent of the open area may be lost,
although 50 percent is a more generally accepted and practical
estimate. Furthermore, individual zones in an aquifer may have
different permeabilities, and the volume of water delivered to the
screen face, other factors being equal, is a function of the
permeability of the adjacent aquifer. However, despite these many
unknown and indeterminate variations from the average entrance

velocity, the concept and practice has proved to be worthwhile in maintaining well efficiency and life. Where conditions and economics permit, the lowering of the entrance velocity to less than 0.03 meter (0.1 foot) per second, consistent with other criteria, would be advantageous.

6-7. Percentage of Open Area of Screen.—One of the major factors controlling head loss through a screen is the percentage of open area. Factors which control the percentage of open area include basic design, materials, fabrication processes, and strength requirements. In a screen of a given length and diameter, the head loss decreases rapidly with an increase in percentage of open area up to about 15 percent, less rapidly up to about 25 percent, and relatively slowly between about 25 and 60 percent. Beyond an open area of about 60 percent, practically no increase in efficiency is obtained. For practical purposes, a percentage of open area of about 15 percent is acceptable and easily obtained with many commercial screens, although not with perforated casing. If at least 15 percent of open area cannot be obtained, other criteria may have to be modified to maintain the maximum 0.03 meter (0.1 foot) per second entrance velocity. This velocity may be obtained by increasing the diameter or length of the screen or perforated casing used and, hence, the total open area for any pattern of perforations.

Peterson et al. (1953), and Vaadia and Scott (1958) experimentally determined a relationship among the length, head loss, and diameter; percentage of open area; and the type of slot of a screen. Their findings, however, have limited usefulness under field conditions.

The percentage of open area in slotted pipe ranges from about 1.0 percent for 0.50-millimeter (0.020-inch) slots to about 12 percent for 6.25-millimeter (0.250-inch) slots in slotting patterns which do not seriously weaken the pipe. Punched or slotted screens have open areas between 4 and 18 percent, depending upon pattern and size of slots. Open areas of louvered screen range from about 3 percent for 0.50-millimeter (0.020-inch) slots to about 33 percent for 5.0-millimeter (0.200-inch) slots. The open area of cage-type wire wound screens ranges from about 2 percent for 0.15-millimeter (0.006-inch) slots to as much as 62 percent for 3.75-millimeter (0.150-inch) slots.

6-8. Screen Slot Sizes and Patterns.—Uniform axial flow of water in a pipe is characterized by a stagnant zone at the wall of the pipe, and velocity and turbulence increase toward the center of the pipe. Conditions in a screen are different, however, because a screen performs as a header or collector in which each perforation operates as a radially directed jet.

As a result of this jet inflow, the stagnant layer is absent or only partially present, and velocity of the axial flow is not uniform but increases from the bottom of the screen to the top. This distribution of flow has not been studied thoroughly, but with screens of the same diameter and equal percentage of open area, the one with the smaller and more numerous slots can be assumed to have lower velocity of flow through the slot and the smaller head loss. Parallel-sided slots such as are found in many saw- or machine-perforated casings appear to be the least efficient type of orifice. In addition, they are more subject to clogging by sand grains. The thinner the edge, the more efficient the slot; thus, the V-shaped slot found in most wire-wound screens and on some perforated pipe has a small hydraulic advantage in addition to its self-cleaning properties. Furthermore, a sharp, clean, smooth slot edge not only contributes to better hydraulic efficiency, but also may reduce the rate of corrosion and encrustation.

The convergence of flow lines and continual acceleration of the water in radial flow to a well have been discussed previously (section 6-2). The spacing of screen slots can add measurably to the convergence. With two screens having equal open area, the one with large, widely spaced slots will require more convergence of flow lines and greater acceleration of the streamflow through the aquifer to the well, with a consequent greater loss of head than would be experienced with the one having small, closely spaced slots (see figure 6-15).

Use of very fine slots results in a small percentage of open area even in wire-wound screens. In addition, the small cross section of the slots results in high friction losses and may tend to promote encrustation. Slot sizes of 0.15 to 0.50 millimeter (0.006 to 0.020 inch) generally should be limited to use in small, low-capacity wells. A gravel pack is advisable in large-capacity wells if a slot size smaller than 0.75 millimeter (0.030 inch) would be required to stabilize the base material.

6-9. Gravel Packs.—The theory of gravel packs holds that surrounding the screen with a more permeable material than the aquifer increases the effective diameter of the well. Because the theoretical permeabilities of packs may be from 10 to 1,000 times

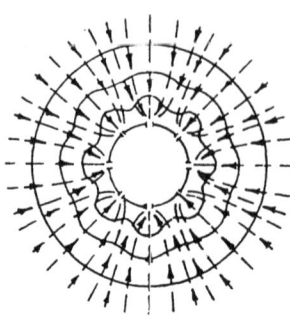

PLAN

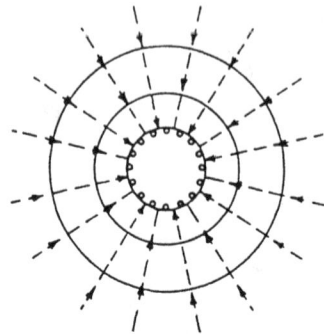

PLAN

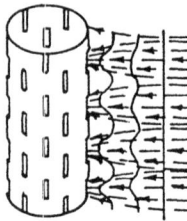

ELEVATION

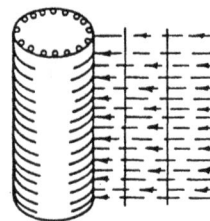

ELEVATION

FIGURE A
Flow to a slotted casing
showing convergence of flow
lines and distortion of
equipotential lines.

FIGURE B
Flow to a continuous slot
screen showing less
convergence of flow lines
and distortion of equipotential
lines.

--------- Flow Lines
--------- Equipotential Lines

Figure 6-15.—Distribution of flow to well screens.

as large as that of the aquifer, this theory is true. The high
permeabilities may also effectively increase the well screen opening
percentage to some degree. In gravity flow pipe drains, a properly
designed gravel pack theoretically has the effect of converting the
pipe from one with limited openings to one that is completely
permeable. Because gravity flow drains have much lower entrance
velocities, the theory probably does not hold equally for wells but
logically, some benefit would exist. Doubling the effective diameter

of the well in conjunction with other benefits of a gravel pack may increase the well yield by as much as 25 percent (Bureau of Reclamation, 1986). If the aquifer were sufficiently thick, a similar increase possibly could be obtained much more inexpensively by increasing the depth of the well. Gravel or sand packs are useful to minimize sand pumping in unconsolidated or semiconsolidated materials where fines or fine sand are present.

Where gravel packing is required, the grain size of the pack should be uniform and as large as possible commensurate with stabilization of the aquifer and ease of installation. The screen slot size should permit only a small percentage of the pack material to pass, and the pack should consist of firm, well-rounded grains. All these factors together with a low entrance velocity contribute to better hydraulic efficiency (see section 11-11). Thin gravel or sand packs may also be used in consolidated materials where sloughing of sides of the well bore is a possibility. The gradation should be sufficiently coarse that no more than 5 percent will pass through the screen slots. Proper development of the well will remove most of these fines before the permanent pump is installed.

Gravel pack design is described more completely in section 11-11. The thickness of the gravel pack is limited to a practical dimension that allows both proper placement and the effective removal of the rigid (impervious characteristics of the wall) cake by the water jetting well development method. The practical gravel pack thickness ranges from a minimum of 75 millimeters (3 inches) to a maximum of about 150 millimeters (6 inches) (Bureau of Reclamation, 1986).

6-10. Well Efficiency.—Well efficiency is a function of the loss of head resulting from flow through the screen and pack and axially in the well to the pump. Thus, in a 100-percent efficient well, all drawdown results from head losses in the aquifer and would be unrelated to the presence or design of the well. A reasonably accurate method is available for estimating the efficiency of a fully penetrating artesian well with 100-percent open holes if values of r_w, are known, and T and S of the aquifer are also known. In such a well, values of the measurable drawdown, S_m and t, can be determined from a pumping test. By inserting the values of T, S, r_w, and time (t) in the nonequilibrium equation, the theoretical drawdown, s_c, can be computed. The efficiency then would be the computed drawdown, s_c, divided by the measured drawdown, S_m, times 100, or

$$E = \frac{S_c}{S_m}$$ 6-4

If, however, the well does not have 100-percent open hole, the effects of partial penetration and anisotropy are difficult or even impossible to determine, but may have a major influence on the computed efficiency.

The problem is even greater for a well in an unconfined aquifer. Not only is the result influenced by the value of r_w, but drawdown negates the effectiveness of 100-percent open hole even if used. Furthermore, anisotropy may have an adverse effect regardless of the percent of open hole, and the effects may be compounded if the well does not fully penetrate the aquifer.

Therefore, any attempt to accurately determine well efficiency appears to be futile unless conditions are ideal. The most practical procedure is to apply the theoretical and empirical factors accepted as being good design practice in conjunction with adequate well development and disregard theoretical well efficiency as a significant factor.

A step test analysis and determination of the apparent efficiency by the methods of Jacob (1947) and Rorabaugh (1953) may be useful in comparing variations in apparent efficiency of an individual well with time as an aid in recognizing deterioration and possible need of rehabilitation (see subsection 5-14(a)).

6-11. Bibliography.—

Bentall, Ray (compiler), 1963, "Methods of Determining Permeability, Transmissibility, and Drawdown," U.S. Geological Survey Water-Supply Paper 1536-I.

Bureau of Reclamation, June 1986, Gravel Pack Thickness for Ground Water Wells, Report No. 1.

Ferris, J.G., D.B. Knowles, R.H. Brown, and R.W. Stallman, 1962, "Theory of Aquifer Tests," U.S. Geological Survey Water-Supply Paper 1536-E, pp. 69-174.

Jacob, C.E., 1946, "Drawdown Test to Determine the Effective Radius of an Artesian Well," *Transactions of the ASCE*, vol. 112, pp. 1047-1070.

_____, April 1946, "Radial Flow in a Leaky Artesian Aquifer," *Transactions of the American Geophysical Union*, vol. 27, No. 11, pp. 198-208.

Li, W.H., December 1954, "Interaction Between Well and Aquifer," Proceedings of the ASCE, vol. 80, Separate No. 578.

Meinzer, O.E. (editor), 1949, *The Physics of the Earth-IX, Hydrology*, Dover Publications, New York.

Muskat, Morris, 1946, *The Flow of Homogeneous Fluids Through Porous Media*, J.W. Edwards, Ann Arbor, Michigan.

Petersen, J.S., C. Rohwer, and M.L. Albertson, December 1953, "Effect of Well Screens on Flow into Wells," *Proceedings of the ASCE*, vol. 79, Separate No. 365.

Rorabaugh, M.I., December 1953, "Graphical and Theoretical Analysis of Step-Drawdown Test of Artesian Well," *Proceedings of the ASCE*, vol. 79, Separate No. 362.

Soliman, M.M., March 1965, "Boundary Flow Considerations in the Design of Wells," *Proceedings of the ASCE, Journal of the Irrigation and Drainage Division*, vol. 91, No. IR1, pp. 159-177.

Theis, C.V., 1935, "Relation Between the Lowering of the Piezometer Surface on the Rate and Duration of Discharge of a Well Using Ground Water Storage," *Transactions of the American Geophysical Union*, vol. 16, pp. 519-524.

Vaadia, Y., and V.H. Scott, January 1958, "Hydraulic Properties of Perforated Well Casing," *Proceedings of the ASCE, Journal of the Irrigation and Drainage Division*, vol. 84, No. IR1, paper 1505.

Zangar, C.N., 1953, "Theory and Problems of Water Percolation," Bureau of Reclamation, Engineering Monograph No. 8, p. 47.

Chapter VII

ARTIFICIAL RECHARGE, ARTIFICIAL STORAGE AND RECOVERY, AND SUBSIDENCE

7-1. Introduction.—Artificial recharge can be defined as any process by which man fosters the transfer of surface water into the ground-water system. Usually, this augmentation of the natural movement of the surface water into the subsurface formations is by some method of construction. Purposes for artificial recharge include:

- Ground-water (well field) management

- Reduction of land subsidence

- Renovation of wastewater

- Improvement of ground-water quality

- Storage during periods of high or excessive surface flow

- Reduction of floodflows

- Increased well yield

- Decrease in size of the areas needed for water-supply systems

- Reduction of saltwater intrusion or leakage of mineralized water

- Increase of streamflow

- Precipitation storage

- Secondary recovery of oil (Pettyjohn, 1981)

Artificial recharge projects also serve as a mechanism for water conservation.

A successful artificial recharge project consists of two critical elements: (1) a dependable supply of acceptable quality water, and (2) a thorough understanding of geologic and subsurface hydrologic conditions. Artificial recharge can greatly enhance water availability when integrated into a basin management program. Artificial recharge of ground water can be accomplished through surface spreading (infiltration) basins, through injection wells, or through shallow shafts which penetrate a cap layer. Infiltration basins can be used where soils are permeable and aquifers are unconfined. Recharge wells can be used in either unconfined or confined aquifers, and shafts supply unconfined aquifers as the shafts terminate above the water table. In all cases, quality of recharge water can greatly affect the success of the project. However, in many cases, reclaimed wastewater has successfully been used for recharge.

Artificial recharge can be used for general aquifer recharge or for temporary storage in specific localities. The latter is generally termed artificial storage and recovery (ASR). Studies indicate that "bulbs" of injected fresh water can be stored in saline aquifers.

In areas of newly constructed reservoirs, recharge of ground water can occur through bank storage as water from the reservoir gradually seeps into formerly unsaturated formations, sometimes causing land damage or structure instability. In the past, this recharge was considered detrimental because it resulted in losses in surface reservoir storage; however, in recent years this recharge has been recognized as a valuable opportunity to integrate water resources management options under certain conditions.

7-2. Surface Spreading.—

(a) General.—Surface spreading generally takes place through temporary (seasonal) or permanent ponds constructed in alluvial or other highly permeable deposits. They may be constructed in-stream or off-stream. In-stream ponds are often designed to store only flood water. Temporary berms may be constructed using earth dams which wash out during floodflows or rubber dams which can be deflated during times of excessively high or low flows. Flashboard dams can also be constructed.

Recharge basins should be located where the water table is of sufficient depth to preclude the intersection of a water-table mound with the basin floor. In addition, the material below the basin should be fairly uniform because perched ground-water mounds can develop above layers of low hydraulic conductivity (Schuh and Shaver, 1988b).

(b) Earth Dikes.—Where earth dikes can be constructed from
river bottom material, they can be constructed relatively
inexpensively. Normal operating flows can be regulated with a
gated bypass pipe installed beneath one end of the dike. Operation
of dikes with bypass valves requires at least daily checking of
water levels and flows with necessary gate adjustments (Lenahan,
1988). Thus, labor costs should be evaluated to determine cost
effectiveness of this construction.

(c) Rubber Dams.—Rubber dams on concrete foundations can be
designed to conform to the flood channel when deflated, so they can
be used for year-round operation. They may be filled with water,
air over water, or air only. The choice of filling medium is related
to dam height and differences in deflation rate and characteristics.
Water-filled dams lower at about the same rate along the entire
length of the dam, resulting in fairly uniform flow across the entire
channel. Air-and-water or air-filled dams usually have a "V" notch
near the center of the dam which causes a concentration of high-
velocity water at this location. This high flow may necessitate the
installation of features to control erosion or hydraulic jump
(Lenahan, 1988).

(d) Flashboard Dams.—A flashboard dam requires a concrete
foundation and vertical guides to hold the flashboards in place.
They have the advantage of being less subject to vandalism than
rubber dams. However, the boards and guides must be removed
prior to the flood season (Lenahan, 1988). Where flash floods are
unpredictable, flashboard dams may not be suitable.

(e) Estimates of Infiltration Rates.—Water infiltrating to the
water table must displace almost an equal volume of air. The rate
of infiltration is influenced by the flow of air ahead of and the
entrapment behind the advancing wetting front, and therefore can
vary with time. The flow of air also influences the shape of the
wetting front. Initial water content of the soil also has a
pronounced effect on the wetting front penetration, even in
relatively deep layered soil profiles.

Most recharge areas concentrate a large volume of infiltrating
water in a small area. As a result, a ground-water mound
develops beneath the area. As the recharge starts, the mound
begins to grow; when the recharge ceases, the mound disintegrates
as the water spreads throughout the aquifer (Fetter, 1988).

This growth and disintegration of the mound can be described using math or a digital computer model; it can also be used to evaluate the impact of the recharge on the water table (Fetter, 1988).

The presence of a ground-water mound below the recharge facility will not affect the infiltration rate provided the water-table mound and associated capillary fringe remain below the bottom of the recharge basin (Schuh and Shaver, 1988b).

Under conditions where the water table is 5 to 10 meters (15 to 30 feet) below ground surface, the wetting front caused by artificial recharge is sharp and migrates downward with time. The water content behind the wetted front may not be at its highest possible value, but it is essentially constant behind the front for most of the profile. As the front migrates downward, it also moves laterally, and as it approaches the water table, its width, $2B'$ is greater than the width, $2B$, of the basin. Under these conditions, the infiltration rate, I, can be calculated by the Green and Ampt formula:

$$I = K\frac{H_c + H + z_f}{z_f} \qquad\qquad 7\text{-}1$$

where:

K = unsaturated hydraulic conductivity at a given water content
H_c = effective capillary drive (a measure of the soil capillary pull expressed as an equivalent depth of water)
H = water depth in the basin
z_f = depth of the sharp wetting front

When H is 0.3 meter (1 foot) or more, H_c is insignificant, and even for shallower basin depth, H_c quickly becomes negligible compared to z_f. While infiltration continues, recharge does not occur until the wetting front reaches the water table. For practical purposes, the soil can be assumed to be saturated and the saturated hydraulic conductivity can be used (Morel-Seytoux, 1985). When the water supply to the basin floor is discontinued, infiltration ceases, but the water in storage in the vadose zone continues downward, although usually at a slower rate. The wetting profile is also redistributed. Changing infiltration rates caused by clogging can be determined using a flagged grid in the recharge basin to visually map water advance with time (Patch and Schuh, 1993).

Research indicates that organic mats can be effective in increasing infiltration rates of turbid water and in actually increasing the unsaturated hydraulic conductivity of the soil. However, fines may penetrate more deeply into the soil when organic mats are used, which could require deeper and more expensive renovation of the basin at a later time (Schuh and Shaver, 1988a, and Schuh, 1991).

(f) Effect of Water Depth.—Infiltration rate may not always be related to water depth. Conditions at the site as well as water quality must be evaluated to determine optimum depth of water. If the ground-water table is above the bottom of the basin, an increase in water depth can increase infiltration rates because of the increased gradient from the basin outward. However, where the ground-water table is located below the base of the basin, applying Darcy's law shows that little change occurs in infiltration rate (Bouwer, 1988).

Equation 7-1, shows that as z_f increases, the effect of water depth decreases and eventually becomes negligible. However, where the soil has materials of low hydraulic conductivity overlying material of higher hydraulic conductivity, z_f is limited by the depth of the low-conductivity material (Schuh and Shaver, 1988b).

Where a well-developed clogging layer overlies the entire wetted perimeter, the entire head caused by increased water is dissipated over the entire clogging layer, and flow through this layer increases in relation to head, other factors being equal. However, problems can develop as a result of increased depth.

Seepage rate across a clogging layer can be decreased by compaction of the clogging layer caused by an increase in head. In addition, increased water depth may reduce the rate of turnover of water in the basin, contributing to algal growth. Increase in algal concentration can increase uptake of carbon dioxide from the water for photosynthesis, increasing the pH of the water. At a pH of 9 or 10, calcium carbonate precipitates out and accumulates on the bottom, further reducing infiltration rates (Bouwer, 1988).

(g) Basin Management Techniques.—Basin management
techniques to optimize recharge and minimize replacement
expenses include: (1) natural methods such as rainfall, shrink-
swell, freeze-thaw; (2) cleaning or removal of soil materials;
(3) tillage of basin floor; (4) filtration; (5) flocculation of sediment;
and (6) management of the ponded depth (Schuh and Shaver,
1988b). Wet-dry cycles are often used as a management tool to
reduce clogging and maintain infiltration rates. After the surface
material has dried out, scraping, disking, or both, may be used to
remove surficial fines, algal or other organic material, or mineral
deposits.

7-3. Injection Wells.—In contrast to surface water, artificial
injection wells are generally more expensive to construct and
maintain but can tap deep aquifers and do not require as much
space.

Injection wells generally require very careful construction
because of the proclivity to clogging. Careful design and
installation of filter pack are especially important because clogging
of the filter pack or borehole wall can greatly reduce injection
potential and cause increases in head at the well. The well screen
open area and screen length should be optimal (about twice as long
as for a withdrawal well pumping the same volume of water)
(Driscoll, 1986). In many ASR projects, a single well is used
alternately for injection and pumping. This procedure aids in
flushing out any fines that might be injected into the filter pack or
aquifer. The reversal in direction in flow in dual-use wells could
lead to a decrease in pore space, which would lower the
permeability of the aquifer in the immediate vicinity of the well.
However, this situation rarely occurs, and its influence on the rise
in the injection head is small (Huisman and Olsthoorn, 1983).

In projects involving only injection, periodic pumping may be
necessary to remove fines. Such wells may be constructed with a
two-way bypass pipe to allow recharge, sampling, and
redevelopment. Huisman and Olsthoorn (1983) report that
backpumping for 5 to 15 minutes at about the same capacity as the
injection rate removes about 80 percent of the increase in the
injection head, and pumping at 3 to 5 times the injection rate
removes about another 10 percent of the head.

Injection wells should be designed to be as efficient as a typical
high-capacity well. Many of the same assumptions used to derive
the equations describing pumping wells are applicable in

calculating the pressure buildup over time in injection wells (Driscoll, 1986). The pressure buildup cone in an injection well is opposite of the drawdown cone in a pumping well.

Clogging can also occur in injection wells because of:

- Entrained air in the recharge water

- Microbial growth in a well

- Chemical reactions between the recharge water and the native ground water

- Chemical reaction between the recharge water and the aquifer material

- Ionic reactions resulting in dispersion of clay particles

- Swelling of colloids in a sand and gravel aquifer

- Iron precipitation

- Biochemical changes in the recharge water and ground water involving iron- or sulfate-reducing bacteria

- Differences in temperature of recharge water compared to aquifer water (O'Hare et al., 1986)

Control of entrained air can generally be done by altering the injection procedure. Removal of chemical deposits and microbial material may require brushing or jetting the screen or chemical cleaning (see chapter XVI).

Where recharge into confined aquifers is proposed, an evaluation of the effect of injecting large quantities of water into the aquifer must be made. Unless transmissivity of the aquifer is very high, a pressure buildup may occur in the vicinity of the well or well field. If sufficiently high, this pressure could result in inadvertent hydrofracturing of overlying or underlying confining beds which could lead to leakage from or to the aquifer.

7-4. Conjunctive Wells.—A conjunctive well is screened in both a shallow unconfined aquifer and a deeper aquifer. When ground water is pumped from the deeper aquifer, the potentiometric surface is lowered and water can drain from the shallow aquifer into the deeper aquifer (O'Hare et al., 1986).

Use of such a method requires careful analysis of potential contamination. Also, State or local laws may preclude mixing of two aquifers.

7-5. Shafts.—A shaft is a well-like opening, usually made by dry excavation, which terminates above the water table. It can be used where a low-permeability, near-surface layer exists which would make a recharge basin infeasible, or where land costs for recharge basins would be prohibitive. Shafts require clean water because the only way to clean the recharge surface is to remove more sand from the cavity. Shafts with cavities mined at the bottom are effective ways to recharge clarified runoff water from playas (Hauser, 1988).

7-6. Horizontal Wells.—Where a thin aquifer is located under a low-permeability near-surface layer, horizontal wells may be preferable to shafts. Construction of horizontal recharge wells is similar to that for horizontal drains (section 13-5). Horizontal wells or drains have also been used to disperse recharged water in a shallow aquifer. Dispensing the recharged water over a larger area effectively increases the storage volume of the aquifer (Bureau of Reclamation, 1992).

7-7. Aquifer Storage and Recovery.—Aquifer storage and recovery involves injecting treated drinking water into a suitable storage zone during times when available water supply and treatment capacity exceed the system demands. When demand exceeds conventional supply or treatment capacity, the stored water is pumped out, chlorinated, and pumped into the distribution system (Pyne, 1988). This method can considerably reduce capital costs for additional surface storage or water treatment facilities which would otherwise be required to meet needs in times of high demand.

Three principal criteria govern the site-specific feasibility of ASR (Pyne, 1988):

- A seasonal variation in water supply, water demand, or both, typically with the ratio of maximum to average day demand exceeding 1.3.

- A reasonable scale of water facilities capacity, generally anaverage demand of 11,000 m^3 (3 mgd) or greater.

- A suitable storage zone, generally one with a transmissivity greater than 185 m^2/day (2,000 ft^2/day), leakance less than 2 x 10^{-3}/day, and total dissolved solids less than 4,000 mg/L.

Although most ASR projects are located in fresh water aquifers, injection into deep saline aquifers is also used. Water pumped into such aquifers does not immediately mix with the native ground water, but forms a bubble or cylindrical-shaped body of injected water around the well. Experiments have shown that recovery of injected water is 35 to 75 percent, with retrieval efficiency increasing with increasing number of injection and retrieval cycles (O'Hare et al., 1986.)

Because the water injected for ASR has generally been chlorinated, concern has existed regarding the fate of chlorinated disinfection byproducts (DBP's), particularly trihalomethanes (THM's). Additional concern has been expressed over the fate of DBP precursors present in the finished water when it is stored in the aquifer and the effect on subsequent DBP formation when the recovered water is rechlorinated. Observations at several ASR sites indicate that THM's and haloacetic acids (HAA's) are removed from chlorinated water during aquifer storage. Elimination of residual chlorine during aquifer storage results in biodegradation of biodegradable organic material (BOM) and lowers the DBP formation potential of the stored water (Singer et al., 1993).

7-8. Use of Reclaimed Wastewater for Recharge.—In using wastewater to recharge aquifers, many engineering issues need to be addressed, such as:

- The quality of the wastewater source
- Storage prior to treatment
- Specifications of treatment processes and design criteria
- Process redundancy requirements
- Parameters affecting plant process control and operation
- Storage of treated (reclaimed) water prior to use
- Operation and maintenance criteria

The principal barrier to ground-water recharge with reclaimed wastewater appears to be neither technical nor economic, but institutional (Asano, 1985). The source of water for artificial recharge includes precipitation, flood or other surplus water, imported water, and reclaimed water. Most of the increasing attention has been focused on the use of reclaimed municipal wastewater. However, because of the increasing concern that low concentration of stable organics and heavy metals may cause long-term health effects, and because of the potential presence of pathogenic organisms in reclaimed wastewater, recharge operations with reclaimed wastewater normally entail further treatment following conventional secondary treatment (Asano, 1985). Health considerations are the governing factor in use of reclaimed wastewater for recharge. Constituents of concern include trace organics, inorganics (particularly heavy metals), microorganisms, and radionuclides. The source of wastewater may affect feasibility of recharge. Areas where industry supplies a significant portion of wastewater may be unsuitable because of the probability of relatively high concentrations of heavy metals and organic chemicals.

Soil-aquifer treatment (SAT) can be an effective means of providing final treatment for recharged wastewater. Studies by Idelovitch and Michail (1985) indicated that if soil conditions are favorable, SAT is usually very effective in removing organic compounds, detergents, phosphorus, and a number of metals; however, nitrogen and boron may not be effectively removed. The sodium absorption ratio (SAR) may increase because of recharge.

Planning for recharge using wastewater is usually considerably more complicated than for using fresh water. State and local agencies may have regulations for recharge of municipal wastewater. Legal questions as to ownership of the wastewater may arise, particularly where a central treatment plant serves several communities or where discharged water has previously been discharged to a stream. In addition, underground flow may differ in direction from surface flow, and recharged water could possibly flow out of the basin of intended use. Extensive investigations may be necessary before approval for recharge projects using reclaimed wastewater can be obtained from the agencies involved. Studies should provide good information on degree of dilution from natural ground water, residence time of reclaimed water underground, unsaturated zone flow, and ground-water quality. Contingency plans should be developed to specify alternative measures to supply water in the event the mixed water is determined to be unsuitable for human consumption. Extensive chemical monitoring of preinjection and in situ ground water will

be required. Considerable public relations work may be necessary
to assure the public that they will not be endangered by the
proposed activities.

Reclaimed water should be thoroughly disinfected prior to
injection or percolation. Ozonation can be very effective without
producing the byproducts produced by chlorination. However,
dechlorination shortly after adding chlorine to the final plant
effluent may minimize the formation of chlorinated hydrocarbons
(California, State of, 1987). Recharge using injection wells may
require considerably more treatment than that required for
recharge in basins because of the detrimental effects of clogging on
wells. Locations for recharge ponds should be selected to be
distant from drinking water-supply wells.

7-9. Effects of Water Chemistry.—Effectiveness of artificial
recharge is highly dependent upon water chemistry, including
turbidity, microbial activity, ion-exchange reactions, and
precipitating constituents such as calcium carbonate, pH, and
temperature.

(a) Turbidity.—Even a small amount of turbidity can result in
clogging of the basin floor and walls or the filter or aquifer
material surrounding an injection well. Although fines settling in
a recharge basin can generally be scraped off during rehabilitation
work, fines clogging a recharge well generally result in permanent
damage and even destruction (Hauser, 1988).

(b) Microbial activity.— Primary causes of biological clogging are:
(1) biomass formation, (2) solid microbial byproducts, and (3) gases
formed during photosynthesis and respiration (Schuh and Shaver,
1988b). Disinfecting the influent water may not be effective in
reducing microbial activity because resistant populations may
develop. Mineral precipitates, particularly iron and manganese,
may develop beneath the surface oxidized zone because of
anaerobic respiration or the formation of H_2S during the reduction
of sulphur. Intermittent rest periods which allow for
decomposition of solid microbial products and for venting of gases
may reduce or eliminate microbial clogging. Oxygen production by
algae may cause diurnal decreases in basin hydraulic conductivity,
although in some cases oxygen may aid in renovating the basin
and increasing recharge. Gases formed by anaerobic respiration
may also impede flow. Denitrification can also cause clogging
(Schuh and Shaver, 1988b).

Proper management and operation of recharge basins can significantly reduce losses in hydraulic conductivity caused by clogging. Bacterial clogging is more apt to be a problem in injection wells where bacteria are more difficult to detect and bacterial clogging of aquifer or filter material as well as screen can occur. Although chlorination may be effective, in some cases it may actually compound the problem, and other methods, such as sulfamic acid or even proprietary treatment, may be necessary. Methods of treatment are more fully described in chapter XVI.

(c) Precipitation.—Generally, mineral precipitation is not a problem in recharge basins (Schuh and Shaver, 1988b). However, in some areas calcium carbonate may cause clogging of injection wells. Acid or mechanical treatment may be necessary to restore the open area of screen (see chapter XVI).

7-10. Recordkeeping.—Accurate records are essential in artificial recharge projects to aid in evaluation of effectiveness of the project, provide information on water in storage, protect against ground-water contamination, and protect the owner in the event of a lawsuit. Monitor wells for water levels and water quality should be used both within and, if possible, adjacent to the site. Monitor wells should extend through the aquifer being recharged unless they are excessively deep. If an underlying or overlying aquifer is present, it may be advantageous to monitor that aquifer also, particularly where treated wastewater is being recharged. The total number of monitor wells will depend on conditions at the site, quality of the recharging water, and proposed use of recharged water.

7-11. Governmental Regulations.—A number of States have developed, or are developing, rules and regulations for injection or extraction of artificially stored water. Requirements may involve water quality, spacing of wells, location of injection wells relative to domestic water-supply wells, and monitoring well installation and analysis. Using reclaimed wastewater has a number of effects on water quality as well as on water right issues.

7-12. Modeling Techniques.—Numerical modeling is not generally used for recharge basins because the equations involved are fairly simple, and the variations of hydraulic conductivity with different percentages of saturation of soil makes numeric models very complicated and results open to question. However, numeric modeling can be a valuable tool in evaluating well-field recharge. A number of computer models have been developed which can be

used to estimate effects of recharge from well fields. MODFLOW
has been widely applied to such studies; however, many
proprietary models are also readily available. A three-dimensional
model (e.g., MODFLOW) has also been developed for recharge
problems.

7-13. Subsidence.—Subsidence is the sinking of a large area of
the earth's crust, usually because of the compaction of fine-grained
sediments in the aquifer system resulting from head decline, and is
not noticed by most residents because it occurs so gradually and
over a broad area. Compaction and subsidence are directly related
to a change in effective stress, which is caused by a decrease in
water levels. A permanent reduction of artesian pressure will
cause some compaction of fine-grained sediments considerably
above and below the aquifer (Davis and DeWiest, 1966).

Because of the slow drainage of fine-grained deposits, subsidence
at a particular time is more closely related to past water-level
change than to current change (e.g., in California, ground-water
withdrawals increased greatly until large imports of surface water
through various canals occurred, but even though water levels in
the area started to rise, the rate of subsidence began to decrease
3 years later).

The depositional environments at various subsidence sites are
varied, but the one common feature is that a thick sequence of
unconsolidated or poorly consolidated sediments form an
interbedded aquifer-aquitard system (Freeze and Cherry, 1979). It
should come as no surprise to find that the process of aquitard
drainage leads to the aquitard compaction, just as the process of
aquifer drainage leads to compaction of the aquifer. Because the
compressibility of the clay is 1 to 2 orders of magnitude greater
than the compressibility of sand, the compaction of the aquitard is
much greater than that of the aquifer, but because the hydraulic
conductivity of the clay may be several orders of magnitude less
than that of the sand, the compaction process is much slower in
the aquitards than in the aquifers (Freeze and Cherry, 1979).
Once the aquitard is dewatered, the effective rebound is only about
one-tenth the amount that it has been compressed (Freeze and
Cherry, 1979).

Extensive problems can be caused by subsidence. Ground-water
storage capacity is permanently destroyed because the voids are
collapsed, which results in reduction in specific yield. Elevations

determined by expensive surveys must be re-established periodically and topographic maps must be redrawn. Flow in canals and rivers may become sluggish because of the already low gradients by subsidence. If a well extends below the major zones of subsidence, the bottom of the casing will remain stationary while the overlying material settles downward, the resulting stress will commonly collapse the casing. Piles for buildings and other structures (canals, bridges, pipelines, highways, etc.) often remain stationary as the ground surface subsides (Davis and DeWiest, 1966).

Several types of subsidence exist in California, where most of the subsidence in the United States has occurred. Subsidence caused by decline of water levels, extraction of oil and gas, oxidation of peaty sediments, and hydrocompaction have been documented (Prokopovich and Marriott, 1983). For Federal Central Valley Project canals in the State, the approximate cost of design modifications and rehabilitation due to subsidence amounts to some $41 million in 1983 (Prokopovich and Marriott, 1983).

Land subsidence caused by ground-water withdrawal began in the San Joaquin Valley in the mid-1920's and locally exceeded 8.4 meters (28 feet) by 1970. By 1977, subsidence reached a maximum of 8.9 meters (29.6 feet) in western Fresno County (Ireland et al., 1984). Subsidence rates increased greatly until surface water was imported through major canals and aqueducts in the 1950's and late 1960's.

Irrigation canals are frequently affected by land subsidence; therefore, monitoring subsidence is essential for operation, maintenance, and possible rehabilitation of canals. An inexpensive technique to measure subsidence along canals was described by Prokopovich and Hall (1983). Canals can be divided into pools by check structures which are designed to control or stop waterflow by the manipulation of check gates. The depth of the water in the canal is measured at the upstream and downstream ends of all pools with check gates completely closed. The difference of depth at the downstream and upstream ends of each pool corresponds to the invert gradient of the pool and is constant in stable areas. Changes reflect the amount of subsidence during the time interval between consecutive sets of measurements (Prokopovich and Hall, 1983).

Texas has also had severe subsidence from ground-water overdrafts in the Houston-Galveston area. As much as 3 meters

(10 feet) of subsidence occurred between 1906 and 1978, but
2.7 meters (9 feet) of this subsidence occurred between 1943 and
1978 (Gabrysch, 1982). Other subsidence problems exist in
Arizona, Wyoming, Montana, Idaho, Nevada, and Washington. In
Arizona, mining ground water has created a fault, and subsidence
faults are also found in Idaho, Nevada, Texas and California
(Coates, 1981). The United States is not the only country afflicted
with subsidence from ground-water pumping; Mexico City, Venice,
and cities in England and Japan have also experienced problems
(Coates, 1981).

Predictive simulation models can be developed to relate possible
pumping patterns in the aquifer-aquitard system to the subsidence
rate that will result (Freeze and Cherry, 1979). Analytical flow
models can be used to analyze time-dependent increases in
effective pressure or ultimate and time-dependent subsidence or
rebound.

7-14. Bibliography.—

Asano, Takashi, 1985, *Artificial Recharge of Groundwater*,
Butterworth Publishers, pp. 3-17.

Bouwer, Herman, 1988, "Systems for Artificial Recharge of Ground
Water," Proceeding of the International Symposium, American
Society of Civil Engineers, Anaheim, California, pp. 2-12.

Bureau of Reclamation, 1992, "Oakes Test Area Study Program,
Draft Interim Status Report."

California, State of, November 1987, "Report of the Scientific
Advisory Panel on Groundwater Recharge with Reclaimed
Wastewater," State Water Resources Control Board, Department
of Water Resources and Department of Health Services,
November 1987.

Coates, D.R., 1981, *Environmental Geology*, John Wiley & Sons.

Davis, S.N. and R.J.M. DeWiest, 1966, *Hydrogeology*, John Wiley &
Sons.

Driscoll, F.G., 1986, *Groundwater and Wells*, Johnson Division,
St. Paul, Minnesota, 1986.

Fetter, C.W., 1988, *Applied Hydrogeology*, 2d edition, Charles E. Merrill Publishing Company.

Freeze, R.A. and J.A. Cherry, 1979, *Groundwater*, Prentice-Hall, Inc., Englewood Cliffs, New Jersey.

Gabrysch, R.K., 1982, "Ground-Water Withdrawals and Land-Surface Subsidence in the Houston-Galveston Region, Texas, 1909-80," U.S. Geological Survey Open-File Report 82-571.

Hauser, Victor L., 1988, "Ground Water Recharge through Wells," Proceeding of the International Symposium, American Society of Civil Engineers, Anaheim, California, pp. 97-106.

Huisman, L. and T.N. Olsthoorn, 1983, *Artificial Groundwater Recharge*, Pitman Advanced Publishing Program.

Idelovitch, Emanuel and Medy Michail, 1985, "Groundwater Recharge for Wastewater Reuse in the Dan Region Project: Summary of Five-Year Experience, 1977-1981," In: *Artifical Recharge of Groundwater*, Butterworth Publishers, pp. 481-506 (Takashi Asano, editor, 1985).

Ireland, R.L., J.F. Poland, and F.S. Riley, 1984, "Land Subsidence in the San Joaquin Valley, California, as of 1980," U.S. Geological Survey Professional Paper 437-I, 1984.

Lenahan, Earl L., 1988, "Conjunctive Use in the Niles Cone, California," *Proceedings of the International Symposium*, American Society of Civil Engineers, Anaheim, California, pp. 495-504.

Morel-Seytoux, Hubert J., "Conjunctive Use of Surface and Ground Waters," In: *Artificial Recharge of Groundwater*, Butterworth Publishers, pp. 35-67 (Takashi Asano, editor, 1985).

O'Hare, M.P., D.M. Fairchild, P.A. Hajali, and L.W. Canter, 1986, *Artificial Recharge of Ground Water: Status and Potential in the Contiguous United States*, Lewis Publishers, Inc., 1986.

Patch, Jon and W.M. Schuh, 1993, "Condition of the Forest River Colony Artificial Recharge Basin Test," Unpublished Report of the North Dakota State Water Commission.

Pettyjohn, Wayne A., 1981, "Introduction to Artificial Ground-Water Recharge," Environmental Protection Agency Office of Research and Development, Robert S. Kerr Environmental Research Laboratory.

Pyne, R. David G., 1988, "Aquifer Storage Recovery: A New Water Supply and Ground Water Recharge Alternative," Proceedings of the International Symposium, American Society of Civil Engineers, Anaheim, California, pp. 107-121.

Prokopovich, N.P. and M.J. Marriott, 1983, "Cost of Subsidence to the Central Valley Project, California," *Bulletin of the Association of Engineering Geologists*, vol. XX, No. 3, pp. 325-332.

Prokopovich, N.P. and H.J. Hall, 1983, "An Inexpensive Technique to Measure Subsidence Along Canals," *Bulletin of the Association of Engineering Geologist*, vol. XX, No. 3, pp. 317-323.

Schuh, W.M. and R.B. Shaver, 1988a, "Hydraulic Effect of Turbid Infiltration through a Shallow Basin," *Proceedings of the International Symposium*, American Society of Civil Engineers, Anaheim, California, pp. 85-96.

_____, 1988b, "Feasibility of Artificial Recharge to the Oakes Aquifer, Southeastern North Dakota: Evaluation of Experimental Recharge Basins," North Dakota Water Commission, Water Resources Investigations Report 7.

_____, 1991, "Effects of an Organic Mat Filter on Artificial Recharge with Turbid Water," *Water Resources Research*, vol. 27, No. 6. Demonstration Program, Interim Report.

Singer, Philip C., R. David G. Pyne, Mallikarjun AVS, Cass T. Miller, and Carolyn Mojonnier, November 1993, "Examining the Impact of Aquifer Storage and Recovery on DBPs," *Research and Technology Jour. AWWA*, pp. 85-94.

Chapter VIII

PUMPING TESTS TO DETERMINE
AQUIFER CHARACTERISTICS

8-1. Methods for Estimating Approximate Values of Aquifer Characteristics.—In practically all ground-water investigations, data are required on the aquifer characteristics, transmissivity, storativity, and boundaries. Several methods of making such tests with various degrees of accuracy are available.

When inventorying existing wells, the data collected often include yield and drawdown, from which specific capacity values may be determined. Section 9-17 discusses methods of estimating the transmissivity of aquifers from such specific capacity data. The procedure is basic but must be used with judgment because well yield depends on several factors, some of which are not readily determinable. When using this method, similarly constructed wells should be grouped together and actually tested for yield and drawdown, if possible.

When wells are equipped with meters or weirs, discharge can be measured easily. Even when not metered, the discharge of wells yielding less than 400 liters (100 gallon) a minute can be readily measured with sufficient accuracy by using a calibrated bucket or drum and a stopwatch. Most wells are not metered, and those having larger discharges are the most significant. Several convenient methods of measuring approximate well discharge with a minimum of equipment are described in the Bureau of Reclamation *Water Measurement Manual* (1981).

Static water levels in wells in the vicinity of the test should be measured during the test and after wells have been shutdown for some time, preferably 12 hours or more. If this condition cannot be realized, the status of such wells should be recorded.

Other methods of estimating approximate values of permeability, transmissivity, and sometimes storativity, include bail tests, slug tests, and analyses of cyclic pumping or natural ground-water fluctuations. They are described by Ferris et al. (1962) and Lohman (1972). However, these methods are either of limited applicability or the results are of questionable accuracy. They are mentioned only as possible alternatives when other methods are not available.

8-2. Controlled Pumping Tests to Determine Aquifer Characteristics.—The most accurate, reliable, and commonly used method of determining aquifer characteristics is by controlled aquifer pumping tests. Before performing such a pumping test, personnel should be acquainted with the contents of chapter IX of this manual. The number of pumping tests required is determined largely by the size of the area, the uniformity and homogeneity of the aquifer or aquifers involved, and known or suspected boundary conditions. One test is usually accurate for a small area, but in an extensive area, several tests may be necessary. A reasonably sophisticated test may cost from $2,000 to $10,000 (in 1995 dollars), not including the cost of the well or treatment of discharge waters, so every effort should be directed toward obtaining a maximum amount of accurate and reliable data.

8-3. Types of Aquifers.—The investigations discussed in chapters II and IX will permit determination of the type of aquifer or aquifers and the interrelationships which may be involved at a particular test site. These factors should be considered in planning pumping tests.

(a) Unconfined Aquifers.—Relatively thin aquifers located at shallow depths are readily tested because test wells and observation wells may be drilled economically to fully penetrate the aquifer. Observation wells may be located short distances from the pumped well; thus, field measurements are easily and quickly made. Long-term pumping is generally not required to obtain usable drawdown measurements. The testing for deeper thin aquifers is similar except for the increased cost of the deeper holes and pump setting.

Problems arise when the aquifer is excessively thick, 30 meters (100 feet) or more, and the water table is deep. Existing wells may not fully penetrate the aquifer, and the cost of drilling fully penetrating observation wells and a test well may not be economically justifiable.

When a pumping well does not fully penetrate an unconfined aquifer, the distorted flow pattern to the well is accentuated and the distance to observation wells should be adjusted accordingly (see section 8-4).

(b) Confined Aquifers.—A confined aquifer will often be overlain by an unconfined aquifer from which it is separated by a confining layer. The unconfined aquifer should always be cased off in the

pumping and observation wells. If the confined aquifer is not excessively thick, the well should be screened for the entire thickness of the aquifer. The nearest observation well should be located at least 7.5 meters (25 feet) from the pumping well and should penetrate and be screened in the upper 10 percent of the aquifer at a minimum. The water level in the pumping well should not be allowed to fall below the bottom of the upper confining bed during an aquifer test.

In a thick, confined aquifer where drilling a fully penetrating well would not be economical, a similar relationship holds regarding the location and depth of observation wells as described above for an unconfined aquifer. However, the area of influence in an artesian aquifer expands more rapidly than in an unconfined aquifer, and the distance to the nearest observation well is not as critical from the standpoint of pumping time to obtain measurable drawdown.

In a confined aquifer, partial penetration by a discharging well may be compensated for by spacing the observation wells an adequate distance from the pumping well, but the relationship of twice the aquifer thickness times the square root of the permeability ratios described in section 8-4 applies. Another method is to use two piezometers at each distance with the closest pair located at least one-half the aquifer thickness from the pumped well. One of the piezometers in each pair is open to the upper 10 percent of the aquifer and the other to the lower 10 percent. The average of the drawdowns in each pair of piezometers is used as the effective drawdown at each distance. If the pumping well is screened through the midsection of the aquifer, the same arrangement may be used as was described previously for an unconfined aquifer under similar conditions, or a piezometer point may be set at the same elevation as the midpoint of the screen.

(c) *Composite and Leaky Aquifers.*—Many areas are underlain by an unconfined aquifer and one or more confined aquifers. The confining layers may vary from practically impermeable to moderately permeable as compared to the aquifers. In the latter case, interchange of water between aquifers may occur depending on the pressure differences which exist among them. Under such conditions, each aquifer should be pump tested separately. The pumping well should be cased through the untested sections and should be screened through the entire thickness of the tested aquifer. Observation wells and piezometers should be set to

conform with the design of the pumping well. In testing such
aquifers, the semilog straight-line plots of time or distance against
drawdown to determine the length of the test may not apply, and if
the test is run sufficiently long, the plot may become a line of zero
drawdown. The field data are analyzed as described in section 9-8.

(d) Delayed Drainage.—Consideration should also be given to the
nature of the aquifer and the probable effect of delayed drainage
on a pumping test (see section 5-10). To estimate the nature of the
aquifer materials, the well logs and sample cuttings should be
carefully examined. From this examination, the minimum planned
time for a test should be estimated on the basis of the following
tabulation:

Minimum pumping time recommended for aquifer test

Predominant aquifer material	Minimum pumping time (hours)
Silt and clay	170
Fine sand	30
Medium sand and coarser materials	4

In many instances, economic and other factors will rule out tests
as long as 170 hours, so less than ideal test results may have to
suffice.

**8-4. Selection and Location of Pumping Wells and
Observation Wells.**—If an existing well is to be used for a test,
the well should ideally closely conform to the requirements for
aquifer testing. Also, the log data on types of construction and
performance characteristics of other wells in the area should be
examined. Other nearby wells may be suitable as observation
wells, but in most cases, additional observation wells will have to
be drilled.

The pumping well and observation wells should be located, if at
all possible, to conform to known or suspected boundaries,
including deep percolation from irrigation. The wells should be
located far enough away from the boundaries to permit recognition
of drawdown trends before the boundary conditions influence the
drawdown readings (see sections 5-5 and 5-11). If more than one
boundary is present, the effects of the first one should be relatively
stable before the influence of the second becomes effective.

Conversely, a study may involve an estimate of induced seepage from a stream or body of surface water as a result of pumping from a nearby well. In such a study, the well may be placed relatively close to the recharge boundary, and one or more observation wells should penetrate into the bed of the surface-water body.

In selecting or locating sites for observation wells, an effort should be made to meet ideal conditions. If a partially penetrating well is used in the ideal aquifer, the observation wells should be located at a minimum distance equal to 1-1/2 to 2 times the aquifer thickness from a partially penetrating pumping well. This configuration will result in a flow pattern equivalent to that of a fully penetrating well. Any well with an 85-percent or more open or screened hole in the saturated thickness may be considered as fully penetrating. If the aquifer is vertically anisotropic, r ideally should be:

$$r > 1.5M\left(\frac{K_r}{K_z}\right)^{1/2} \qquad\qquad 8\text{-}1$$

where:

r = distance from the center of the pumping well to the observation well

M = thickness of the aquifer

K_r = horizontal permeability

K_z = vertical permeability

In reality, this relationship may be of limited value because a reliable and economical method of determining vertical and horizontal permeability is unavailable, aquifer thickness may be unknown, and prolonged pumping would be required. Suggested alternatives for observation well locations are presented in chapter IX. When laying out observation well locations, consideration should be given to the proposed duration of the test and the probable magnitude of the transmissibility and storativity. If estimates can be made of S and T, the drawdowns at various distances and at increasing times, the time when u (defined in chapter IX) will be less than 0.01, and the probable length of time required for the test can be estimated (see sections 9-3 and 9-5). The drawdown at any point in the area of influence will increase with time and the rate of discharge. Conversely, the greater the diffusivity factor, $\alpha = S/T$, the slower the rate of expansion of the area of influence.

Although any number of observation wells may be used, the recommended minimum number is four—three on a line passing through the center of the pumped well and one on a line normal to the previously mentioned line and passing through the pumped well. The distance from the pumped well to the nearest observation well, and the spacing between observation wells, involve consideration of the ideal conditions, how the test conditions conform to ideal conditions, desirable adjustments to compensate for departures from the ideal, and feasible locations for the wells in the field.

If a well must be drilled specifically for testing, the design should reflect whether it is to be purely a test well to be abandoned after the test or whether it would fit into the final plan as a production well. In the former case, the least costly construction commensurate with the purpose should be followed. In the second, a pilot hole may be necessary, and good well design from the standpoint of efficiency, long well life, and desired yield should be followed.

8-5. Disposal of Discharge.—When planning the test, the method and place of disposal of discharge from the well should be determined. The discharge from the pumped well should be transported some distance from the well for convenience and comfort during the test. However, other factors of greater importance must be considered.

If the aquifer is unconfined and the unsaturated materials overlying the aquifer are relatively permeable, the discharge should be transported by pipeline to an existing drain beyond the probable area of influence that will develop during the test. Otherwise, the deep percolation may be recirculated, and the test may be adversely affected.

If the aquifer is unconfined, the water table lies at depths of 30 meters (100 feet) or more, and the overlying materials are of low permeability, an existing ditch or drain that will remove the flow rapidly from the area may be used safely. Regardless of the point of discharge, determining the ultimate fate of the water is important so that no public or private property is subjected to damage. Discharge from a confined aquifer may be treated similarly.

If the water may be contaminated, it must be stored, sampled and treated if necessary.

8-6. Preparations for Pumping Test.—If possible, tests should be run when heavy rains would least likely occur. Infiltration and deep percolation of precipitation may adversely affect a test.

For a few days before starting the test, water levels in the pumping well and observation wells should be measured at about the same time each day to determine whether a measurable trend exists in ground-water levels. If such a trend is apparent, a curve of the change in depth versus time should be prepared and used to correct the water levels read during the test.

In areas of severe winter climate where the frostline may extend to depths of a meter or more, pumping tests should be avoided during the winter where the water table is less than about 3 to 4 meters from the surface. Under some circumstances, the frozen soil acts as a confining bed, and when combined with leaky aquifer and delayed storage characteristics, may make the results of the test unreliable. Also, during periods of increasing or decreasing frost level, the water table may fluctuate significantly in response to the freezing or thawing.

If the aquifer is confined, barometric changes may affect water levels in wells (see section 9-13). An increase in barometric pressure may cause a decrease in the water levels, and a decrease in pressure may cause an increase in water levels. If water levels and barometric pressures are measured several times daily for at least 4 days prior to running a test and both measurements are expressed in meters (feet) of water (25 millimeters mercury = 0.346 meter of water and 1 inch mercury = 1.134 feet of water), a plot may be made correlating the two measurements. The slope of the straight line of closest fit through each set of measurements for each well will give the barometric efficiency of each hole. If a relationship is recognized, barometric readings should be made at the same time as water-level measurements during the test and the required correction should be made to the measured water levels.

A day or two before the test, the well should be tested for several hours for yield and drawdown, operation of the discharge measuring equipment, general operating conditions, and approximate best rates of discharge for the test. At the same time, water levels should be measured in the observation wells to assure response.

The measuring point at the pumped well and all observation holes should be selected, marked clearly with paint, and the elevations determined by leveling, if required.

The distance and bearing from the center of the pumping well to the center of all observation wells should be measured to the nearest meter or foot. The distance and bearing to the closest point of any nearby boundary, such as a lake, stream, or other discharging wells, should also be measured.

8-7. Instrumentation and Equipment Required for a Test.—The following items should be available for use in the test:

- An orifice, weir, flowmeter, or other type of water measuring device that will measure accurately in the range of the discharges expected (Bureau of Reclamation, 1981; Purdue University, 1949) (see section 8-9).

- Depth-to-water measuring devices as described in section 3-4, excluding air lines which are not sufficiently accurate for use in pumping tests. Transducers attached to automation data loggers are increasingly used in pumping tests and are recommended because of the greater accuracy and frequency of measurements and reduction in personnel requirements. If electric sounders are used, extra batteries, waterproof electrical tape, and other supplies should be available for servicing them on the job. The number of measuring devices available for the test should equal the number of observation wells and the pumping well, plus one extra for use as a spare. Continuous water stage recorders are very useful if the recording rate is compatible with the drawdown rate.

If an electric sounder is used for determining water levels, a similar method for recording changes is as follows:

  – Install a notched board about 1 meter (3 feet) long with
    the notched end slightly overlapping the top of the
    casing and the other end supported to maintain a
    horizontal position.

  – Cover the board with paper.

  – Just before the pumping test starts, set the probe so it
    is just at the water level and lay the cable along the
    board. Make a mark on the cable near the far end of
    the board for pumping, at the well end for recovery,
    and make a mark on the paper at this point.

  – Record drawdown or recovery by inserting or
    withdrawing the probe so it is just hitting the water.
    Mark on the paper the location of the cable mark,
    noting on the paper the time elapsed since pumping
    started or stopped.

  – After completion of the test, measure the distances
    from the initial point and make a table of drawdown
    versus time. This method enables one person to record
    data for several wells except in the very initial stages
    of the test when measurements are made very
    frequently.

• A tachometer or revolution counter if an internal-
  combustion engine is used for power.

• Steel tapes, graduated in millimeters (hundredths of a foot).

• A thermometer with a range between 0° and 50 °C (32 and
  120 °F) if water temperature is an important factor.

• Synchronized watches for all observers.

• Log-log 3- by 5-cycle and semilog 3-cycle graph paper,
  rulers, pencils, and forms for recording measurements.

• A gate valve on the pump discharge pipe to control the
  discharge.

• A barometer or recording barograph if the test is in an
  artesian aquifer.

- A stopwatch.

- A carpenter's level if an orifice is used.

- Two or more 1-liter (1 quart) water sample bottles.

If oil is present on the water, a cork should be inserted in the lower end of the pipe. The cork can be dislodged by air pressure or by pouring water in the pipe. Drawdown measurements are made in the pipe, which protects the probe from oil on top of the water and dampens turbulence caused by vibration of the pump and permits more accurate measurements.

8-8. Running a Pumping Test.—Immediately before starting the pump, the water levels should be measured in all observation wells and in the pumped well to determine the static water levels upon which all drawdowns will be based. These data and the time of measurement should be recorded.

The instant of starting the pump should be recorded as the zero time of the test. Wells will ordinarily show a slow decline in discharge with time as the drawdown increases. This decline may be compensated for by limiting the discharge by partially closing the valve at the start of the test and opening it slightly when measurements of discharge show a recognizable decline. The objective is to maintain a constant pump discharge throughout the test. A maximum variation of about 5 percent in the discharge should be the goal.

The well discharge should be controlled to keep it as constant as possible after the initial excess discharge has been stabilized. The discharge can be controlled by either regulating the valve (preferable) or, if an internal-combustion engine is used for power, by changing the speed of the engine. The tone or rhythm of an internal-combustion engine provides an aural check of performance. If the tone changes suddenly, the discharge should be checked immediately and proper adjustments should be made to the gate valve or to the engine speed if necessary.

During a pumping test to determine aquifer characteristics, water levels in the pumping well and observation holes should be measured to give at least 10 observations of drawdown within each log cycle of time. Adherence to the time schedule should not be at the expense of accuracy in the drawdown measurements. A suggested scheduling measurement is as follows:

0 to 10 minutes:	1, 1.5, 2, 2.5, 3.25, 4, 5, 6.5, 8, and 10 minutes
10 to 100 minutes:	10, 15, 20, 25, 30, 40, 50, 65, 80, and 100 minutes
100 minutes to completion:	1- to 2-hour intervals

During the early part of the test, sufficient manpower should be available to have at least one person at each observation well and at the pumping well. After the first 2 hours, two people are usually sufficient to continue the test.

It is important, particularly in the early part of the test, to record with maximum accuracy the time at which readings are taken. The foregoing time schedule should be followed as closely as possible, but if for some reason the schedule is missed, the actual time of taking the reading should be recorded. Estimating drawdown readings to fit the schedule may lead to erroneous results. Readings in the pumping well and observation wells need not be taken simultaneously as long as the schedule is generally followed and readings are recorded at the exact time taken. A 20-millimeter (3/4-inch) inside-diameter or larger pipe should be installed in the pumping well from above the pump base to the top of the pump bowls.

8-9. Measurement of Discharge.—Practically any type of device designed for measurement of flow in pressure conditions can be used to measure pump discharge when the discharge pipe is running full. In some instances, an "L" may need to be inserted on the pipeline with a riser extending above the elevation of the pump discharge, or the discharge end may need to be elevated to keep the pipe full at all times. Commercially available flowmeters with totalizing registers are commonly used, particularly for discharges less than about 760 liters (200 gallon) per minute. Meters should be calibrated and checked for accuracy within the discharge and pressure ranges to which they will be subjected, and they should be installed in conformance with the manufacturer's recommendations. The rate of discharge is usually determined by measuring the time required to discharge a given volume; however, an instantaneous flow indicator is preferable. The *Water Measurement Manual* (Bureau of Reclamation, 1981) gives data on venturi, flow nozzle, and orifice meters. Some meters of the required capacity may be available in the field offices or laboratories of the Bureau of Reclamation.

Where discharge is free flowing from the discharge pipe into a ditch or canal, weirs as described in the *Water Measurement Manual* (Bureau of Reclamation, 1981) may be used.

The most commonly used device for measuring discharge during a pumping test is probably the free discharge pipe orifice. When used in conjunction with a pipeline, the orifice may be placed at the end of the pipeline or it may discharge into a tank or reservoir which feeds the pipeline. The latter arrangement is usually more convenient because measurement and adjustment of discharge can then be made in the immediate vicinity of the well. Figures 8-1 and 8-2 illustrate pipe orifice arrangement and details.

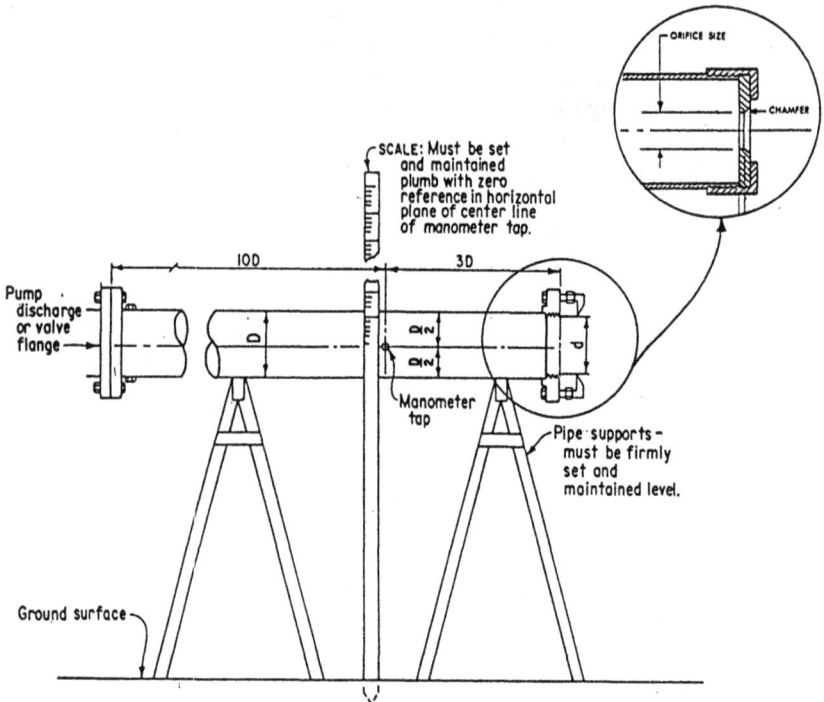

Figure 8-1.—Free discharge pipe orifice.

Numerous combinations of pipe and orifice sizes and applicable tables are available. The free-discharging orifice developed by Layne & Bowler, Inc., and tested at Purdue University (1949), is well suited to field use. Construction details for this device are shown on figure 8-2. Table 8-1 gives the specifications of the various sizes of steel pipe used in the Layne & Bowler orifice

Orifice plate

DETAIL A

Bevel at any angle

Width not greater than 2mm ($\frac{1}{16}$ in.)

p

ORIFICE OPENING

1.3 to 1.6 m (4 to 5 ft.) of 25 to 40mm (1 to 1$\frac{1}{2}$ in.) diameter clear glass or plastic tube

Rubber stopper

Brass tube

125 to 150mm (5 to 6in.) of 6 to 13mm ($\frac{1}{4}$ to $\frac{1}{2}$ in.) i.d. hose to slip over tap.

3 to 6mm ($\frac{1}{8}$ to $\frac{1}{4}$ in.) Manometer Tap. This face of tap must be filed smooth with inside of pipe with no projections or depressions

SECTION B-B
MANOMETER DETAILS

Manometer

B

B

H

> 3D

A

D

TYPICAL PIPING ARRANGEMENT

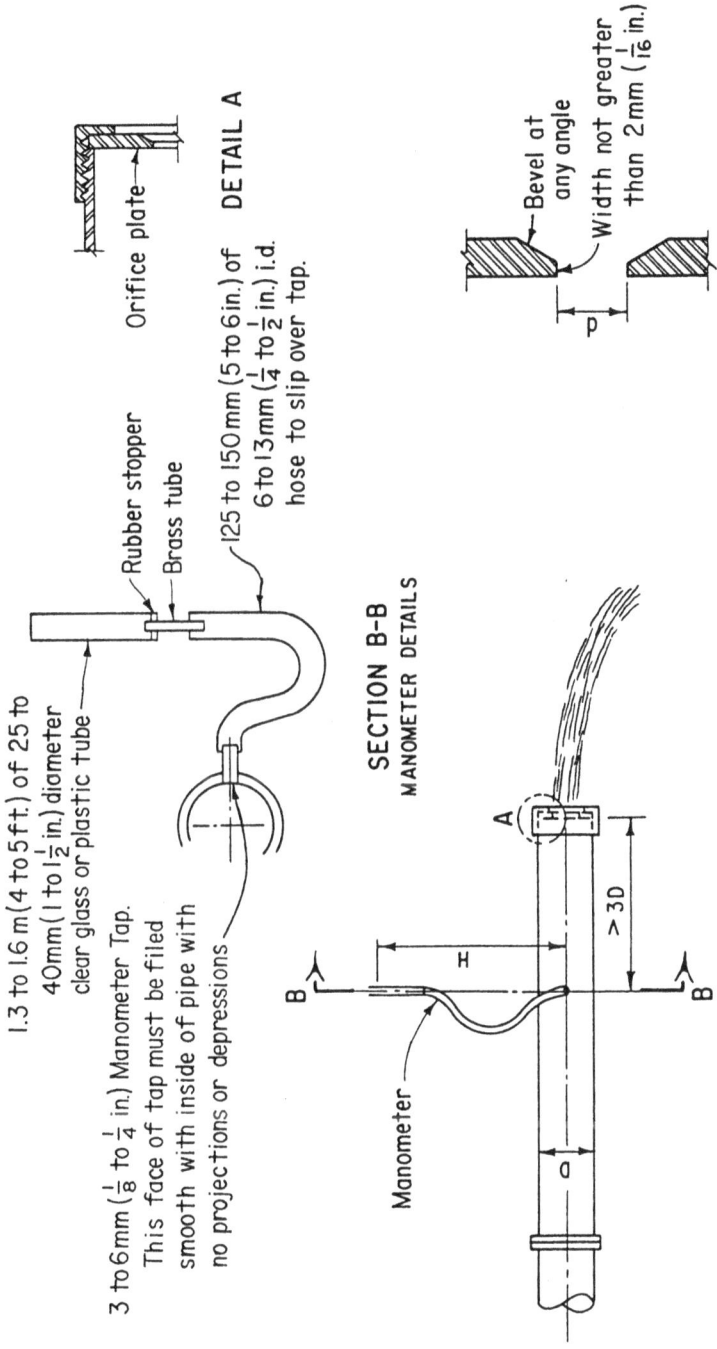

Figure 8-2.—Pipe orifice details.

meters. If constructed and operated as described below, the pipe orifice combinations shown in tables 8-2a and 8-2b and on figure 8-3 permit an accuracy of ±2 percent. Flow rates may be read directly from either the table or figure.

Table 8-1.—Pipe recommended for free discharge orifice use

Nominal pipe size (inches)	Outside diameter (inches)	Inside diameter (inches)	Wall thickness (inches)	Class	Schedule No.	Weight per foot (plain) (ends) (pounds)
4	4.500	4.026	0.237	Std.	40	10.79
6	6.625	6.065	0.280	Std.	40	18.97
8	8.625	8.071	0.277	—	30	24.70
10	10.750	10.192	0.279	—	—	31.20
12	12.750	12.090	0.330	—	30	73.77

It should be emphasized that the tables and figures shown here apply only to a free-discharging orifice, in which the flow exiting the orifice falls freely in the air. Many references provide tables for in-line orifices. These devices have a section of pipe and a second manometer tap downstream from the orifice. Data developed for in-line orifice plates should not be used with the free discharging orifice shown on figure 8-2.

Orifices may be machined from threaded pipe caps or from 5- to 6-millimeter (3/16- to 1/4-inch) steel plate stock and attached to the pipe by thread protectors or similar devices. The plates should be carefully machined as true circles to automatically center in the pipe when attached. The orifice should be accurately machined to the specified diameter and centered in the plate.

The downstream edge of the orifice should be beveled at an angle of about 45° but leaving a root of uniform width of 2 millimeters (1/16 inch) or less on the upstream side. The upstream edge of the orifice should be sharp, clean, and free from rust and any pits or other imperfections. Indices consisting of two lines normal to each other and passing through the center of the orifice are commonly inscribed on the downstream face of the orifice plates to assist in centering them on the end of the pipe. The *Hydraulic Handbook* published by Fairbanks Morse Pump Division (1977) contains tables and descriptions for a similar orifice plate, but with a 1/8-inch root on the orifice opening. If the available orifice plate is of this type, the Fairbanks Morse handbook should be consulted.

Table 8-2a.—Orifice tables (SI Metric)

(For measurement of water in liters per minute through pipe orifices with free discharge.
The following table is based on the original calibration by Purdue University.)

Head in millimeters	4" pipe 3" orifice	6" pipe 3" orifice	6" pipe 4" orifice	6" pipe 5" orifice	8" pipe 4" orifice	8" pipe 5" orifice	8" pipe 6" orifice	10" pipe 6" orifice	10" pipe 7" orifice	12" pipe 8" orifice	12" pipe 9" orifice	12" pipe 10" orifice	Head in millimeters
25										1,467	1,617	2,545	25
50										2,074	2,286	3,599	50
75										2,540	2,800	4,408	75
100										2,933	3,233	5,090	100
125	402									3,280	3,615	5,691	125
150	440	315	664	1,558			1,760			3,593	3,960	6,234	150
175	475	340	717	1,683			1,901			3,881	4,277	6,733	175
200	508	364	767	1,800		1,133	2,032		2,481	4,148	4,573	7,198	200
225	539	386	813	1,909	686	1,202	2,156	1,679	2,632	4,400	4,850	7,635	225
250	568	407	857	2,012	723	1,267	2,272	1,769	2,774	4,638	5,112	8,048	250
275	596	427	899	2,110	759	1,329	2,383	1,856	2,910	4,864	5,362	8,440	275
300	622	446	939	2,204	792	1,388	2,489	1,938	3,039	5,081	5,600	8,816	300
325	648	464	978	2,294	825	1,445	2,591	2,017	3,163	5,288	5,829	9,176	325
350	672	481	1,015	2,381	856	1,499	2,688	2,094	3,283	5,488	6,049	9,522	350
375	696	498	1,050	2,464	886	1,552	2,783	2,167	3,398	5,681	6,261	9,856	375
400	719	515	1,085	2,545	915	1,603	2,874	2,238	3,509	5,867	6,467	10,180	400
425	741	530	1,118	2,623	943	1,652	2,962	2,307	3,617	6,047	6,666	10,493	425
450	762	546	1,150	2,699	970	1,700	3,048	2,374	3,722	6,223	6,859	10,797	450
475	783	561	1,182	2,773	997	1,747	3,132	2,439	3,824	6,393	7,047	11,093	475
500	803	575	1,213	2,845	1,023	1,792	3,213	2,502	3,923	6,559	7,230	11,381	500
525	823	590	1,243	2,916	1,048	1,836	3,293	2,564	4,020	6,721	7,408	11,662	525
550	843	603	1,272	2,984	1,073	1,880	3,370	2,624	4,115	6,879	7,583	11,937	550
600	880	630	1,328	3,117	1,120	1,963	3,520	2,741	4,298	7,185	7,920	12,467	600
650	916	656	1,383	3,244	1,166	2,043	3,664	2,853	4,473	7,479	8,243	12,976	650
700	951	681	1,435	3,367	1,210	2,120	3,802	2,961	4,642	7,761	8,555	13,466	700
750	984	705	1,485	3,485	1,253	2,195	3,935	3,065	4,805	8,033	8,855	13,939	750
800	1,016	728	1,534	3,599	1,294	2,267	4,065	3,165	4,963	8,297	9,145	14,396	800
850	1,047	750	1,581	3,710	1,334	2,337	4,190	3,263	5,116	8,552	9,427	14,839	850
900	1,078	772	1,627	3,817	1,372	2,404	4,311	3,357	5,264	8,800	9,700	15,269	900
950	1,107	793	1,672	3,922	1,410	2,470	4,429	3,449	5,408	9,041	9,966	15,688	950
1,000	1,136	814	1,715	4,024	1,446	2,534	4,544	3,539	5,549	9,276	10,225	16,095	1,000
1,050	1,164	834	1,757	4,123	1,482	2,597	4,656	3,626	5,686	9,505	10,477	16,493	1,050
1,200	1,244	891	1,879	4,408	1,585	2,776	4,978	3,876	6,078	10,162	11,200	17,631	1,200
1,350										10,778	11,880	18,701	1,350
1,500										11,361	12,523	19,713	1,500
1,650										11,916	13,134	20,675	1,650

Note: Only use discharge values within the specified range. Discharge values not shown are outside the range of accuracy for the constant "K."

Table 8-2b.—Orifice tables (U.S. Customary)

(For measurement of water in gallons per minute through pipe orifices with free discharge.
The following table is based on the original calibration by Purdue University.)

Head in inches	4" pipe	6" pipe			8" pipe			10" pipe		12" pipe			Head in inches
	3" orifice	3" orifice	4" orifice	5" orifice	4" orifice	5" orifice	6" orifice	6" orifice	7" orifice	8" orifice	9" orifice	10" orifice	
5	107									873	963	1,516	5
6	117	84	177	415						957	1,055	1,660	6
7	127	91	191	448						1,033	1,139	1,793	7
8	135	97	204	479	183	302	469			1,105	1,218	1,917	8
9	144	103	217	508	193	320	506	447	661	1,172	1,292	2,033	9
10	151	108	228	536	202	337	541	471	701	1,235	1,362	2,143	10
11	159	114	240	562	211	354	574	494	739	1,296	1,428	2,248	11
12	166	119	250	587	220	370	605	516	775	1,353	1,491	2,348	12
13	172	124	260	611	228	385	635	537	809	1,408	1,552	2,444	13
14	179	128	270	634	236	399	663	558	842	1,462	1,611	2,536	14
15	185	133	280	656	244	413	690	577	874	1,513	1,668	2,625	15
16	191	137	289	678	251	427	716	596	905	1,562	1,722	2,711	16
17	197	141	298	699	258	440	741	614	935	1,611	1,775	2,794	17
18	203	145	306	719	265	453	765	632	963	1,657	1,827	2,875	18
19	209	149	315	739	272	465	789	650	991	1,703	1,877	2,954	19
20	214	153	323	758	279	477	812	666	1,018	1,747	1,925	3,031	20
21	219	157	331	776	286	489	834	683	1,045	1,790	1,973	3,106	21
22	224	161	339	795	292	501	856	699	1,071	1,832	2,019	3,179	22
23	229	164	346	813	298	512	877	715	1,096	1,873	2,065	3,250	23
24	234	168	354	830	311	523	898	730	1,121	1,914	2,109	3,320	24
25	244	175	368	864	322	544	918	760	1,145	1,992	2,195	3,456	25
26	253	181	382	897	334	565	937	789	1,191	2,067	2,278	3,586	26
28	262	188	396	928	345	585	976	816	1,236	2,139	2,358	3,712	28
32	271	194	409	958	355	604	1,013	843	1,280	2,210	2,436	3,834	32
34	279	200	421	988	365	622	1,048	869	1,322	2,278	2,510	3,952	34
36	287	206	433	1,017	375	640	1,082	894	1,362	2,344	2,583	4,067	36
38	295	211	445	1,044	385	658	1,116	919	1,402	2,408	2,654	4,178	38
40	303	217	457	1,072	395	675	1,148	942	1,440	2,470	2,723	4,286	40
42	310	222	468	1,098	404	692	1,180	966	1,478	2,531	2,790	4,392	42
44	317	227	479	1,124	413	708	1,210	988	1,514	2,591	2,856	4,496	44
46	324	232	490	1,149	422	724	1,240	1,011	1,550	2,649	2,920	4,597	46
48	331	237	500	1,174		739	1,269	1,032	1,585	2,706	2,983	4,696	48
54							1,298		1,619	2,870	3,164	4,980	54
60							1,326			3,026	3,335	5,250	60
66										3,173	3,653	5,506	66
72										3,314		5,751	72

Note: Only use discharge values within the specified range. Discharge values not shown are outside the range of acceptable accuracy for the constant "K."

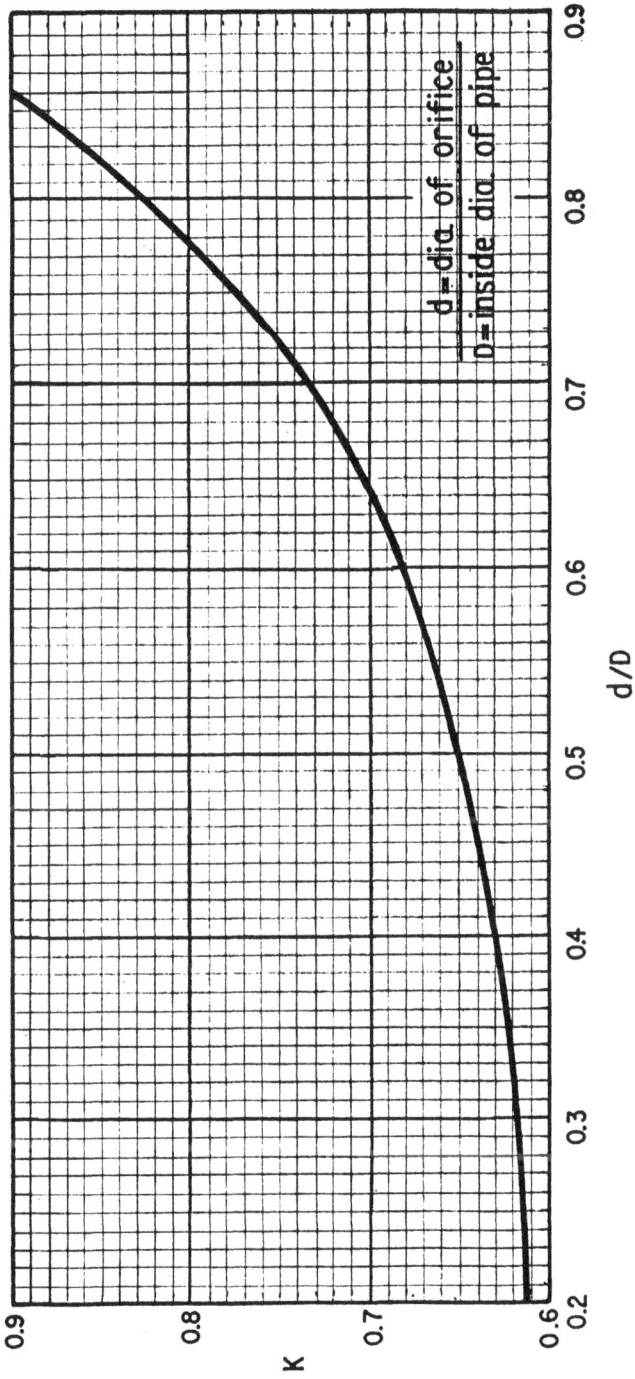

Figure 8-3.—Pipe orifice discharge equation.

To ensure accurate measurements, the pipe orifice assemblies must be installed as follows:

- The manometer tube tap must be located at least 3 pipe diameters from the orifice plate and accurately located on the horizontal diameter of the pipe.

- The manometer tube tap must be located at least 10 pipe diameters upstream from an elbow, valve, reducer, or similar fitting.

- The manometer tap fitting should have an inside diameter of 3 to 6 millimeters (1/8 to 1/4 inch) and must be smooth and flush with the inside surface of the pipe.

- The pipe must be truly horizontal.

- The pipe must be full of water at all times, and the water must fall freely from the orifice into the air without any obstruction.

- Before each measurement, the bottom of the pipe immediately behind the orifice plate should be cleaned of sand or other debris.

- The interior of the pipe should be clean, smooth, and free of grease.

- The manometer hose and gauge should be free of air bubbles whenever a reading is being made.

- Manometer readings should not register less than 25 millimeters (1 inch) greater than the inside radius of the pipe or greater than 1,500 millimeters (60 inch). If readings are more or less than these values, the orifice size should be changed.

- No leaks should exist between the pump head and the orifice plate.

Usual practice is to securely anchor a 2-meter (6-foot) long scale, reading in millimeters (inch), in a true vertical position with the zero point accurately located at the centerline of the manometer tap.

A 1.5- or 2-meter (5- or 6-foot) length of plastic tubing of 6- to 12-millimeters (1/4- to 1/2-inch) inside diameter should be attached to the manometer tap (figure 8-2). Clear tubing should be used and held against the scale when it is to be read. A recommended method is to use a 0.3- to 0.6-meter (1- to 2-foot) length of 25-millimeter (1-inch) diameter clear glass or plastic tube with a hollow rubber plug with a glass or brass tube at the bottom to which the smaller diameter hose is attached. This arrangement dampens the surging commonly associated with many pumps and permits easier and more accurate readings. If regular surging is evident in the tube, the range of such surging should be noted, and the mean of the readings should be taken.

Good practice includes lowering the hose and tube below the manometer tap and allowing water to flow through it for a short time to clear all air bubbles and sand from the system before making a reading. When lighting conditions are poor, reading the gauge may be facilitated by adding a few drops of vegetable coloring or cooking dyes to the clear tube just before making a reading.

When very large discharges are to be measured, or the range of discharges to be measured exceeds that of a single orifice, an arrangement using two pipe orifices as shown on figure 8-4 is suggested. This arrangement sometimes permits ready changing of orifice plates by diverting the flow through one pipe without shutting down the pump. The sum of the readings from both orifices can be used to determine larger discharges.

At times, orifices of the sizes specified in tables 8-2a and 8-2b or figure 8-3 may not be available, and those furnished by the contractor or others must be used. The condition and installation of such meters should conform to the requirements previously stated. In addition, the d/D ratio (diameter ratio) should be between 0.4 and 0.85, where d is the orifice diameter and D is the exact inside diameter of the pipe. For these cases, figure 8-4 gives the general orifice discharge equation and shows the variation of the discharge coefficient, C, as a function of the diameter ratio. The diameter ratio is the only significant influence on C within the range of pipe sizes and discharges covered by tables 8-2a and 8-2b. However, for smaller flow rates, the discharge coefficient also varies as a function of the Reynolds number, a parameter relating viscous and inertial flow forces. To use figure 8-4, one should perform the following check:

Let: Q_{gpm} = discharge, gallons per minute (may have to estimate if not already known)

D_{inches} = pipe diameter, inches

If:

$$10{,}000\left(\frac{Q_{gpm}}{D_{inches}}\right) \geq 1.0$$

then figure 8-4 can be used.

If the situation does not pass this check, one may be able to use a smaller pipe and orifice size to increase the ratio of Q_{gpm}/D_{inches}. For discharges less than 100 gallons per minute, the use of the pipe orifice discharge meter is generally impractical. In these cases, the flow rate can be determined by timing the filling of a 55-gallon oil drum with the top removed. For very small discharges, a 5-gallon bucket may be used.

In addition to the previously discussed relatively accurate methods of measuring discharge, a number of trajectory methods, (such as the California pipe and the coordinate method, which give good to fair approximations of discharge when conditions are favorable), can be found in the *Water Measurement Manual* (Bureau of Reclamation, 1981). These methods are not recommended where high accuracy is required; however, they are useful in reconnaissance and similar surveys where a general idea of the range of capacities of wells is desired. They require little equipment and are relatively easy to perform.

8-10. Determining Duration of a Test.—The duration of a test is determined by the adequacy of the data in the form of curves obtained from plotting time versus drawdown, distance versus drawdown, or both relationships, and possibly economic factors related to costs of pumping and monitoring the test.

The time-drawdown graph (see section 9-5(b)) is a plot of drawdown against the log of time since pumping began. This graph is most simply made on semilog paper with drawdown

1. Nominal 250mm (10in) thread protector.
2. Gasket, 265mm (10.45in) o.d., 250mm (10in) i.d.
3. Orifice plate of 10mm (⅜in) stock. 265mm (10.45in) o.d. by 200mm (8in)i.d. and a 265mm (10.45in) o.d. by 150mm (6in) i.d. 3 to 6mm (⅛ to ¼in) i.d. tap
4. Minimum 2.7m (9ft) length of 270mm (10¾in) o.d. pipe with 7mm (0.279in) wall. Weight per foot (plain ends) of 14kg (31.20lbs) and threaded both ends.
5. Tapped and threaded 250mm (10in) screw flange.
6. 250mm (10in) gate valve.
7. 250mm (10in) 90° flanged elbow.
8. 250mm (10in) flanged tee.
9. 250mm (10in) flanged discharge from pump.
10. Nominal 200mm (8in) thread protector.
11. Gasket, 209mm (8.35in) o.d., 200mm (8in) i.d.
12. Orifice plate of 10mm (⅜in) stock. 209mm (8.35in) o.d. by 100mm (4in) i.d. and a 209mm (8.35in) o.d. by 150mm (6in) i.d.
13. Minimum 2.1m (7ft) length of 215mm (8⅝in) o.d. pipe with 7mm (0.277in) wall. Weight per meter (ft)(plain ends) of 11.12kg (24.70lbs) and threaded both ends.
14. Tapped and threaded 200mm (8in) screw flange.
15. 200mm (8in) gate valve.
16. 250mm (10in) by 8⅝in) flanged reducer taper elbow.
17. For convenience, two 250mm (10in) valves could be used between the tee and the elbows rather than at the discharge end of elbows.

Figure 8-4.—Double pipe orifice.

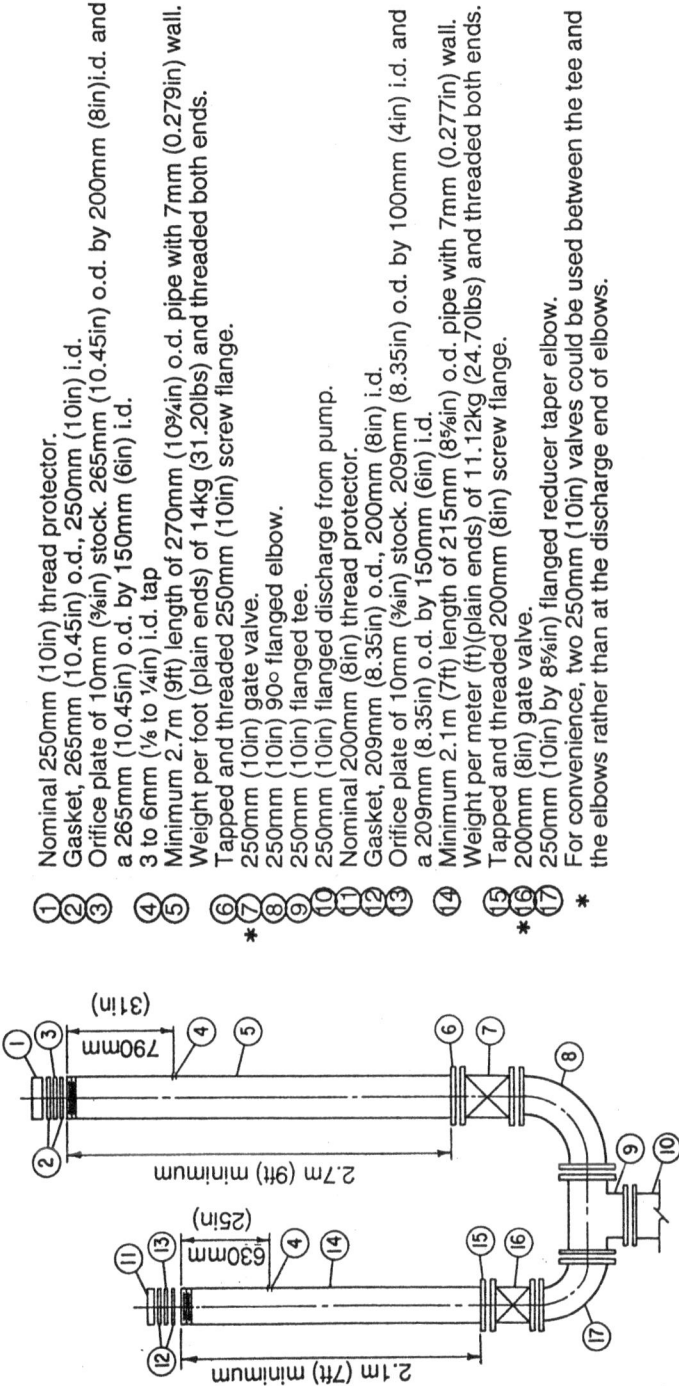

plotted on the arithmetic scale and time plotted on the log scale. A graph should be made for the pumping well and for each observation well included in the test.

The distance-drawdown graph (see section 9-5(a)) is a plot of drawdowns occurring simultaneously in each observation well against the distance on log scale from the observation wells to the pumping well. The time selected would usually be the longest available or the last reading taken unless a boundary has been encountered. This graph can be used for determining duration of test only when three or more observation wells are included in the test because at least three points are needed to establish and verify a straight line.

In a relatively simple test, the time-drawdown plotted points for each observation hole fall initially on a curve which, with time, approximates a straight line within the limits of plotting. When the straight-line condition is attained, continued pumping will result in measured points which will fall on an extension of the straight line. The straight-line plot will be attained earliest for the pumping well, then for nearby observation wells, and at later times for more distant wells. When three or more drawdowns, measured at hourly intervals in the most distant hole, fall on the line, the time-drawdown conditions have been met. In an artesian aquifer, the straight-line condition is reached quite rapidly, but in an unconfined aquifer, the condition develops more slowly. Where three or more observation wells are available, a distance-drawdown graph is made as a check before pumping is stopped because sometimes this graph shows that the test should be continued. If pumping has continued long enough, these plotted points will also fall on a straight line within the limits of plotting. When this condition is reached, the test may be stopped. The time for approximate straight-line plotting conditions to be reached may range from 2 hours to as much as 3 weeks, but usually a satisfactory test can be completed within 48 hours.

When a partially penetrating pumping well and observation wells are used, a preliminary estimate of transmissivity and storativity should be made during the test from the straight-line plots of drawdown against time on the log scale. The test ideally should be continued until the value of:

$$\frac{S}{4Tt} estimated \ for \ each \ hole \ is \ less \ than \frac{0.1}{M}$$

where:

r = distance from pumping well to observation hole, meters (feet)

S = storativity, dimensionless

T = transmissivity, m²/t (ft²/t)

t = time

M = aquifer thickness, meters (feet)

This relationship is discussed in detail in section 9-10 of this manual.

The measurements from some tests may show an irregular curve or dispersion which does not appear to approach the straight-line condition even after several hours of pumping. If this condition persists for more than 24 hours past the minimum estimated pumping time (see section 8-3(d)), pumping may be stopped, and recovery measurements may be made. Study of conditions and results may show that a good test cannot be obtained or that a longer test may be required.

If boundaries are suspected, a test may be run for a longer period to determine whether the boundary effects will become apparent.

8-11. Recovery Test for Transmissivity.—When the pump is stopped after running the pump-out test, the drawdown and time at which it was shutdown are recorded. Measurement of water level is immediately initiated in the pumped well and in all observation wells, if any. The same procedure and time pattern is followed as at the beginning of a pumping test (see section 8-8). As in the pumping test, the time and depth to water are noted for each measurement. The recovery usually will not return to the original static water level within a reasonable length of time; so, when several measurements at 1-hour intervals show less than 30 millimeters (0.1 foot) difference in recovery, measurements may be discontinued. A good check on the transmissivity value can be made of recovery in the pumped well and of transmissivity and storativity from recovery measurements in the observation wells (see section 9-7).

Transmissivity, but not storativity, can be approximated by the following procedure using only a pumping well. The static water level is measured and the time recorded before starting the pump discharging at a uniform rate. Measurements of the drawdown in the well are continued until the plot of drawdown versus log t falls on a straight line over a 3-hour period. The well must fully penetrate the aquifer to obtain a reasonably accurate value for T by this method. The accuracy obtained will be less than the accuracy obtained from the pump-out test using observation wells. Accuracy will be best for homogeneous, isotropic conditions and will lessen as these conditions deteriorate and as boundary conditions increase in effect. The pump is then stopped, and the remainder of the procedure is the same as described for a recovery test after the pump is stopped. Section 9 provides a more indepth discussion of recovery tests.

8-12. Bibliography.—

Bureau of Reclamation, 1981, *Water Measurement Manual*, 2d edition, Denver, Colorado.

Fairbanks Morse Pump Division, 1977, *Hydraulic Handbook*, 10th edition, Colt Industries, Kansas City, Kansas.

Ferris, J.G., D.B. Knowels, R.H. Brown, and R.W. Stallman, 1962, "Theory of Aquifer Tests," U.S. Geological Survey Water-Supply Paper 1536-E.

Lohman, S.W., 1972, "Ground-Water Hydraulics," U.S. Geological Survey Professional Paper 708.

Purdue University, 1949, "Measurement of Water Flow through Pipe Orifice with Free Discharge," Engineering School, Lafayette, Indiana.

Chapter IX

ANALYSIS OF DISCHARGING WELL AND OTHER TEST DATA

9-1. Background Data.—Quantitative data on hydraulic characteristics of aquifers including transmissivity, storativity, and boundary conditions are essential to the understanding and solution of aquifer problems and the proper evaluation and utilization of ground-water resources. Field tests provide the most reliable method of obtaining these data. Such tests involve the removal from or the addition of water to a well and subsequent observation of the reaction of the aquifer to the change. The normal change in the water level is the creation of a cone-like zone of depression or buildup surrounding the well. This cone is unique in shape and lateral extent and is dependent primarily on the time since the start of testing, volume or rate of water withdrawn or added, and the hydraulic characteristics of the aquifer.

Analyses of results of systematic observations of water level and of other test data yield values of aquifer characteristics. The extent and reliability of these analyses are dependent on features of the test including duration of test, number of observation wells, and method of analysis. Two general types of analyses are available for determination of aquifer characteristics: (1) steady state or equilibrium methods which yield values of transmissivity and related permeability, and (2) transient or nonequilibrium methods which also yield storativity and boundary conditions. The principal difference between the two methods is that the transient method permits analysis of ground-water conditions which change with time and involve storage, whereas the steady-state method does not.

Although transmissivity values can be obtained from the pumping well, determination at strativity requires use of observation wells. Observation wells also provide information on variability of transmissivity in heterogeneous aquifers. In addition, in certain aquifer conditions (leaky aquifer), it may be necessary to utilize observation wells to obtain acceptable results from aquifer tests.

Test analyses also require an understanding and appreciation of the hydrologic and geologic setting of the aquifer. Conditions that should be known include: location, character, and distance of nearby bodies of surface water; depth, thickness, and stratigraphic conditions of the aquifer; and construction details of the test well and of observation wells, if used. However, despite knowledge of all apparent conditions that tend to influence a test, deviation of aquifer conditions from the ideal on which analyses are based and imperfections of the testing procedure generally rule out precise results.

9-2. Steady-State Equations.—The Theim-Forchheimer or equilibrium equations (Bentall, 1963) are based on the following assumptions:

- Aquifer is homogeneous, isotropic, and of uniform thickness.

- The discharging well penetrates and receives water through the entire thickness of the aquifer.

- Transmissivity and hydraulic conductivity or permeability are constant at all times and at all locations.

- Discharging has continued for a sufficient duration for the hydraulic system to reach a steady state.

- Flow to the well is horizontal, radial, and laminar, and originates from a circular open water source with a fixed radius and elevation which surrounds the well.

- Rate of discharge from the well is constant.

The equilibrium equations (see figure 9-1), which yield values of hydraulic conductivity and transmissivity, were used for years as the only equations available for the analysis of discharging well tests. The general test procedure is to simultaneously pump from a test well at a constant, known rate and to periodically measure the drawdown in two or more nearby observation wells. The test is normally continued until the plot (for each observation well) of time since start of pumping plotted on log scale, versus drawdown plotted on arithmetic scale, falls on a straight line. Another check on adequacy of duration of the test consists of a similar straight-line plot of distance versus drawdown at the same period of time in three or more observation wells. Often, because of the absence of one or more of the ideal aquifer conditions on which the analyses are based, straight-line plots do not develop within a reasonable time limit of pumping. Nevertheless, the test results may be adequate for the purposes intended.

$$Q = \frac{2\pi KM(h_e - h_w)}{\log_e r_e/r_w}$$

Q = Discharge of the well
K = Coefficient of permeability
M = Thickness of artesian
     aquifer
T = Transmissivity = KM
h_e = Piezometric pressure at
     circumference of area
     of influence
h_w = Piezometric pressure
     at well
r_e = Radius of area of
     influence
r_w = Radius of well
$\log_e$ = Natural log = ln

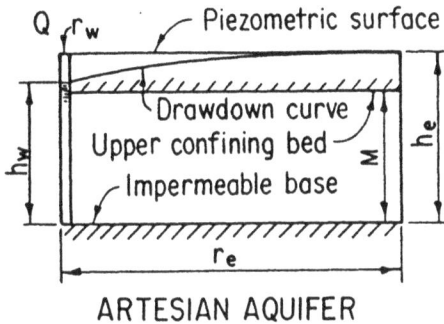

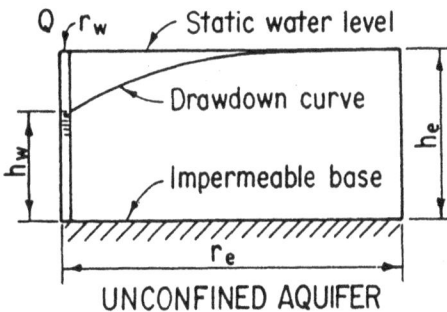

ARTESIAN AQUIFER

$$Q = \frac{\pi K (h_e^2 - h_w^2)}{\log_e r_e/r_w}$$

Q = Discharge of the well
K = Coefficient of permeability
h_e = Saturated thickness of
     aquifer at circumference
     of area of influence
h_w = Saturated thickness
     of aquifer at
     well
r_e = Radius of area of
     influence
r_w = Radius of well
$\log_e$ = Natural log = ln

UNCONFINED AQUIFER

Figure 9-1.—Application of the equilibrium equations.

Rearranged for determination of hydraulic conductivity, K, the equilibrium equations are:

For a confined (artesian) aquifer,

$$K = \frac{Qln\ (r_2/r_1)}{2\pi M\ (s_1 - s_2)} \qquad \text{9-1}$$

and

$$T = \frac{Qln\ (r_2/r_1)}{2\pi(s_1 - s_2)} \qquad \text{9-2}$$

For an unconfined aquifer,

$$K = \frac{Qln\ (r_2/r_1)}{\pi(h_2^2 - h_1^2)} \qquad \text{9-3}$$

and

$$T = \frac{QMln\ (r_2/r_1)}{\pi(h_2^2 - h_1^2)} \qquad \text{9-4}$$

where:

ln	= natural log = (common log x 2.303)
K	= permeability or hydraulic conductivity, L/t
Q	= discharge of the test well, L^3/t
M	= saturated thickness of the aquifer, L
T	= KM = transmissivity of the aquifer, L^2/t
L	= unit length
t	= unit time
$r_1, r_2, . \ r_n$	= horizontal distances from centerline of the test well to centerline of observation wells 1, 2, . .n ., L
$h_1, h_2, . \ h_n$	= saturated thicknesses or piezometric heads of the aquifer at distances $r_1, r_2, . . r_n$ from the test well, L
$s_1, s_2, . \ s_n$	= drawdown in observation wells at distances $r_1, r_2, . . r_n$ from the test well, L

Based on the previous equations, values of K and T can be computed using values of drawdown, s, at the same time in two or more observation wells located at different distances from the test well. For example, pumping test data on three observation wells from table 9-1 are shown plotted on figure 9-2. Table 9-1 presents only a partial record of the test data.

The three companion curves on figure 9-2 represent plots of time versus drawdown for the three observation wells shown in table 9-1 where:

Q = 2.70 to 2.75 ft$_3$/s
M = 50 ft
At t = 960 minutes in observation wells 1, 2, and 3, respectively
s_1 = 1.89 ft at r_1 = 100 ft
s_2 = 1.36 ft at r_2 = 200 ft
s^3 = 0.80 ft at r_3 = 400 ft
h_1 = M-s_1 = 48.11 ft
h_2 = M-s_2 = 48.64 ft
h_3 = M-s_3 = 49.20 ft

By using equation 9-3, and assuming an unconfined aquifer,

$$(K = \frac{161.8\ln\frac{200}{100}}{\pi[(48.64)^2 - (48.11)^2]} = 0.70ft/min), \; and$$

$T = KM$ = (0.215)(15.24) = 3.27 m^2/min
      = (0.70)(50) = 35 ft^2/min

Two precautions are recommended when using the equilibrium equations: (1) in an unconfined aquifer, drawdowns exceeding 10 percent of the aquifer thickness should not be used in the calculations, and (2) measurements should not be used at points where the slope of the cone of depression exceeds 15 degrees.

Basic assumptions of the equilibrium equations provide that the test well is fully penetrating, has 100-percent open hole (or screen), and that flow to the well is horizontal. In many tests, these conditions are not met; however, as discussed in section 6-2, the distribution of flow to a well approaches that of the assumed horizontal condition at a distance from the well equal to approximately 1.5 times the thickness of the aquifer. Accordingly, to minimize effects of convergent flow on test results, the nearest observation well should be located at least 1.5 times the aquifer

Table 9-1.—Tabulated discharging well test data for solution of
equilibrium and nonequilibrium equations

Project: Sioux Flats   Discharge measured by: 1 foot Parshall flur
Feature: Wide Gap damsite   Drawdown measured by: Electrical sounder
Location: NW 1/4 sec. 4, T. 59 N., R. 13 E.   Reference point: North side of casing collar

Pump test No. 1, test well

Date	Time	Depth to water, ft	Drawdown s, ft	Gauge reading, ft	Discharge, ft³/s	Remarks
5-16	0840	60.99				
5-17	0830	61.01				
5-18	0845	61.00				
5-19	0820	60.98				
	0840	[1]60.99	0.0			Pump started
	0900	72.30	11.3	0.79	2.70	
	1000	72.60	11.6	.79	2.70	
	1100	72.80	11.8	.79	2.70	Pump off 5-21 0730
	1155	72.80	11.8	.80	2.75	
	1255	72.80	11.8	.80	2.75	Avg. Q = 2.7 f
	1355	73.00	12.0	.80	2.75	
	1455	73.20	12.2	.80	2.75	
	1555	73.20	12.2	.80	2.75	M = 50 ft
	1655	73.20	12.2	.80	2.75	
	1800	73.20	12.2	.80	2.75	
	1856	73.30	12.3	.80	2.75	
	1948	73.40	12.4	.80	2.75	
	2057	73.40	12.4	.80	2.75	
	2203	73.40	12.4	.80	2.75	
	2300	73.60	12.6	.80	2.75	
	2358	73.50	12.5	.80	2.75	
5-20	0104	73.60	12.6	.80	2.75	
	0204	73.60	12.6	.80	2.75	
	0259	73.60	12.6	.80	2.75	
	0400	73.80	12.8	.80	2.75	
	0501	74.00	13.0	.80	2.75	
	0602	73.90	12.9	.80	2.75	
	0702	73.90	12.9	.80	2.75	
	0759	73.80	12.8	.80	2.75	
	0855	73.80	12.8	.80	2.75	
	0955	73.80	12.8	.80	2.75	
	1055	73.80	12.8	.80	2.75	

[1] Static water level.

Table 9-1.—Tabulated discharging well test data for solution of
equilibrium and nonequilibrium equations - continued

Pump test No. 1, observation well No. 1, r = 100 ft

Date	Time	Depth to water, ft	Drawdown s, ft	t, min	r^2/t, ft²/min	Remarks
5-16	0845	60.43				
5-17	0825	60.45				
5-18	0840	60.43				
5-19	0815	60.42				
	0841	60.42	0.00			Pump started at 0840
	0845	60.50	.08	5	2,000	
	0850	60.64	.22	10	1,000	Pump off 5-21 at
	0855	60.74	.32	15	670	0730
	0900	60.83	.41	20	500	
	0905	60.90	.48	25	400	
	0910	60.96	.54	30	333	M = 50 ft
	0920	61.06	.64	40	250	
	0930	61.14	.72	50	200	
	0940	61.20	.78	60	170	
	0950	61.27	.85	70	140	
	1000	61.32	.90	80	125	
	1010	61.36	.94	90	110	
	1020	61.40	.98	100	100	
	1030	61.44	1.02	110	91	
	1040	61.47	1.05	120	83	
	1140	61.62	1.20	180	56	
	1240	61.73	1.31	240	42	
	1340	61.83	1.41	300	33	
	1440	61.90	1.48	360	28	
	1540	61.96	1.54	420	24	
	1640	62.01	1.59	480	21	
	1740	62.05	1.63	540	19	
	1840	62.09	1.67	600	17	
	1940	62.14	1.72	660	15	
	2040	62.17	1.75	720	14	
	2240	62.26	1.84	840	12	
5-20	0040	62.31	1.89	960	10	
	1845	62.59	2.17	2,045	4.9	

Table 9-1.—Tabulated discharging well test data for solution of
equilibrium and nonequilibrium equations - continued

Project: Sioux Flats   Drawdown measured by: Electric
Feature: Wide Gap damsite   sounder
Location: NW 1/4 sec. 4, T. 29 N., R. 13 E   Reference point: East side of casing
                                                  collar

Pump test No. 1, observation well No. 2, r = 200 ft

Date	Time	Depth to water, ft	Drawdown s, ft	t, min	r^2/t, ft^2/min	Remarks
5-16	0835	58.41				
5-17	0820	58.39				
5-18	0820	58.40				
5-19	0810	58.41				
	0838	[1]58.41				Pump started at 0840
	0847	58.41	0.00	7	5,720	
	0852	58.44	.03	12	3,332	Pump off 5-21 at
	0857	58.48	.07	17	2,352	0730
	0902	58.52	.11	22	1,820	
	0907	58.56	.15	27	1,480	$M = 50$ ft
	0912	58.59	.18	32	1,252	
	0922	58.66	.25	42	952	
	0932	58.72	.31	52	768	
	0942	58.77	.36	62	644	
	0952	58.81	.40	72	556	
	1002	58.85	.44	82	488	
	1012	58.89	.48	92	436	
	1022	58.92	.51	102	392	
	1032	58.95	.54	112	357	
	1042	58.98	.57	122	328	
	1142	59.12	.71	182	220	
	1242	59.22	.81	242	165	
	1342	59.30	.89	302	132	
	1442	59.38	.97	362	110	
	1542	59.44	1.03	422	94	
	1642	59.49	1.08	482	83	
	1742	59.53	1.12	542	72	
	1842	59.57	1.16	602	66	
	1942	59.61	1.20	662	60	
	2042	59.64	1.23	722	55	
	2242	59.73	1.32	842	44	
5-20	0042	59.77	1.36	962	40	
	1845	60.01	1.65	2,045	20	

[1] Static water level.

Table 9-1.—Tabulated discharging well test data for solution of
equilibrium and nonequilibrium equations - continued

Project: Sioux Flats

Feature: Wide Gap damsite

Location: NW 1/4 sec. 4, T. 59 N., R. 13 E

Drawdown measured by: "Popper"

Reference point: East side of casing collar

Pump test No. 1, observation well No. 3, r = 400 ft

Date	Time	Depth to water, ft	Drawdown s, ft	t, min	r^2/t, ft²/min	Remarks
5-16	0850	58.47				
5-17	0830	58.48				
5-18	0835	58.48				
5-19	0820	58.47				
	0838	[1]58.47				Pump on at 0840
	0855	58.47	0.00	15	10,720	Pump off 5-21 at
	0900	58.47	.00	20	8,000	0730
	0905	58.47	.00	25	6,400	
	0910	58.47	.00	30	5,280	
	0915	58.48	.01	35	4,640	M = 50 ft
	0920	58.49	.02	40	4,000	
	0930	58.50	.03	50	3,200	
	0940	58.52	.05	60	2,720	
	0950	58.54	.07	70	2,240	
	1000	58.55	.08	80	2,080	
	1010	58.57	.10	90	1,760	
	1020	58.59	.12	100	1,600	
	1030	58.61	.14	110	1,456	
	1040	58.62	.15	120	1,328	
	1050	58.64	.17	130	1,232	
	1150	58.73	.26	190	848	
	1250	58.81	.34	250	640	
	1350	58.87	.40	310	512	
	1450	58.93	.46	370	432	
	1550	58.98	.51	430	368	
	1650	59.02	.55	490	320	
	1750	59.06	.59	550	288	
	1850	59.10	.63	610	256	
	1950	59.12	.65	670	240	
	2050	59.15	.68	730	224	
	2250	59.22	.75	850	192	
5-20	0050	59.27	.80	970	160	
	1845	59.54	1.07	2,045	78	

[1] Static water level.

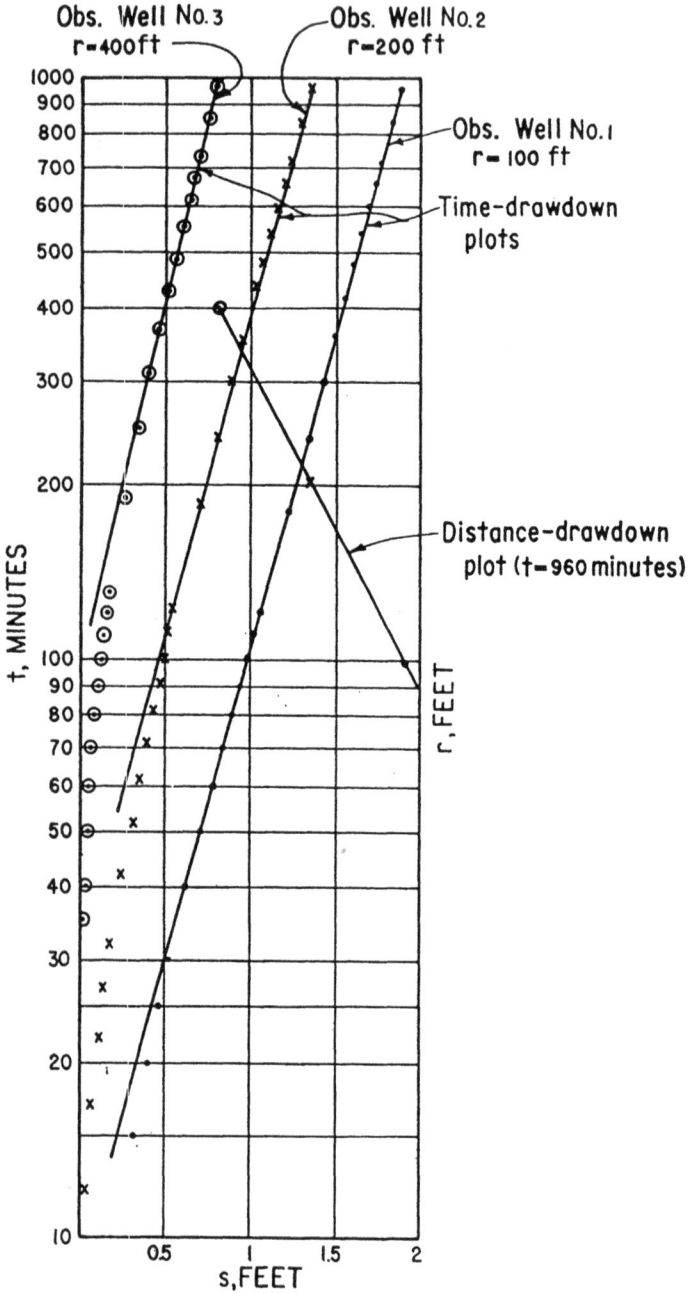

Figure 9-2.—Plots of water levels in three
observation wells during a pumping test.

thickness from the test well unless a large aquifer thickness provides for distances that are unreasonably large. In this event, pairs of piezometers located at reasonable distances from the well may be substituted. Piezometers should be set in both the lower and upper 15 percent of the aquifer and the drawdowns in these piezometers should be averaged for computational purposes. Where this is not feasible, observation wells with screened or open hole zones duplicating those of the test well may be substituted. These procedures are empirical, but they serve to minimize errors caused by vertical flow convergence resulting from partially penetrating wells. Mathematical methods of correcting for flow convergence have been developed, but their usefulness may be questionable when applied to field conditions (DeWiest, 1965; Ferris, Knowles, Brown, and Stallman, 1962; Hantush, July 1961, September 1961; Jacob, 1945; Kruseman and DeReder, 1970; Muskat, 1946; and Rouse, 1949).

9-3. Transient Equations.—Transient equations permit analysis of aquifer conditions that vary with time and involve storage. The assumptions on which the equations are based include:

- Aquifer is confined, horizontal, homogeneous, isotropic, of uniform thickness, and of infinite areal extent

- Pumping well is of infinitesimal diameter and fully pene- trates the aquifer. Flow to the well is radial, horizontal, and laminar

- All water comes from storage in the aquifer within the area of influence and is released from storage instantaneously with decline in pressure

- Transmissivity and storativity of the aquifer are constant in time and space.

The transient equations are directly applicable to confined condi- tions and are suitable for use, with limitations, in unconfined aquifers. These limitations are related to the percentage of drawdown in observation wells as related to the total aquifer thickness. If the drawdown exceeds 25 percent of the aquifer thickness, the transient equations should not be used. However, if the percentage is less than 10, little error is introduced. For values between 10 and 25 percent, the following correction factor derived by C.E. Jacob (Bentall, 1963) should be applied:

$$s' = s - \frac{s^2}{2M}$$ 9-5

where:

  s = measured drawdown in an observation well
  M = saturated thickness of the aquifer prior to pumping
  s' = corrected drawdown

Figure 9-3 uses data from table 9-2 to illustrate this method of correction and its application to semilog plots of drawdown versus time and drawdown versus distance. Table 9-2 presents only a partial record of the test data.

Deviations of many aquifers from assumptions of uniform thickness, horizontally, and homogeneity, are usually minor or are averaged; therefore, such deviations do not seriously affect the accuracy of aquifer test results. Also, most aquifers are anisotropic; but if the test well and observation wells are fully penetrating, the effect on drawdowns is minor. However, if the aquifer is strongly anisotropic and the test well is not fully penetrating, drawdowns in observation wells may be misleading because only that portion of the aquifer actually penetrated by the well contributes flow to the well. In such cases, use of total screen or open hole length as a substitute for total aquifer thickness will give reasonably reliable results, but the storativity computed from such data is usually too low.

The assumption that aquifers are infinitely extensive is never realized because all have boundaries. Nevertheless, most regional aquifers may be considered extensive when related to the timeperiod during which pumping tests are normally run. However, if boundaries are close to the test well, straightforward application of transient equations may yield unreliable results. This is often the situation in pumping tests conducted for Bureau of Reclamation purposes. These tests are often located in river valleys where bedrock boundaries, layered deposits, and stream recharge can greatly affect drawdown. In such cases, methods as described in section 9-11 should be used.

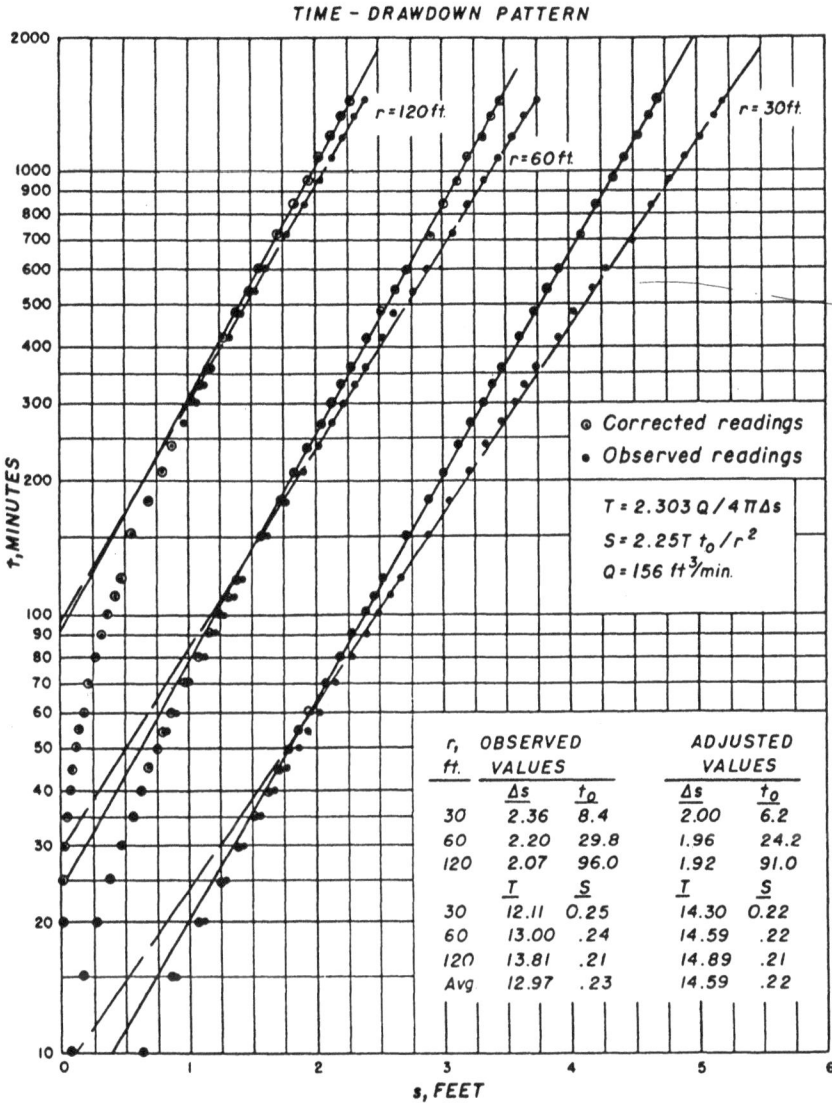

TIME - DRAWDOWN PATTERN

$T = 2.303 Q / 4\pi\Delta s$

$S = 2.25 T t_0 / r^2$

$Q = 156$ ft^3/min.

r, ft.	OBSERVED VALUES Δs	t_0	ADJUSTED VALUES Δs	t_0
30	2.36	8.4	2.00	6.2
60	2.20	29.8	1.96	24.2
120	2.07	96.0	1.92	91.0
	T	S	T	S
30	12.11	0.25	14.30	0.22
60	13.00	.24	14.59	.22
120	13.81	.21	14.89	.21
Avg	12.97	.23	14.59	.22

Figure 9-3.—Effects of correcting drawdown readings in a free aquifer (sheet 1 of 2).

The fact that a well is not fully penetrating may be partially compensated for by design and location of observation wells as described in section 6-2. The diameter of the well, unless it is large and the discharge small, is seldom an adverse factor in interpretation of aquifer tests.

DISTANCE — DRAWDOWN PATTERN
(t= 1440 MIN)

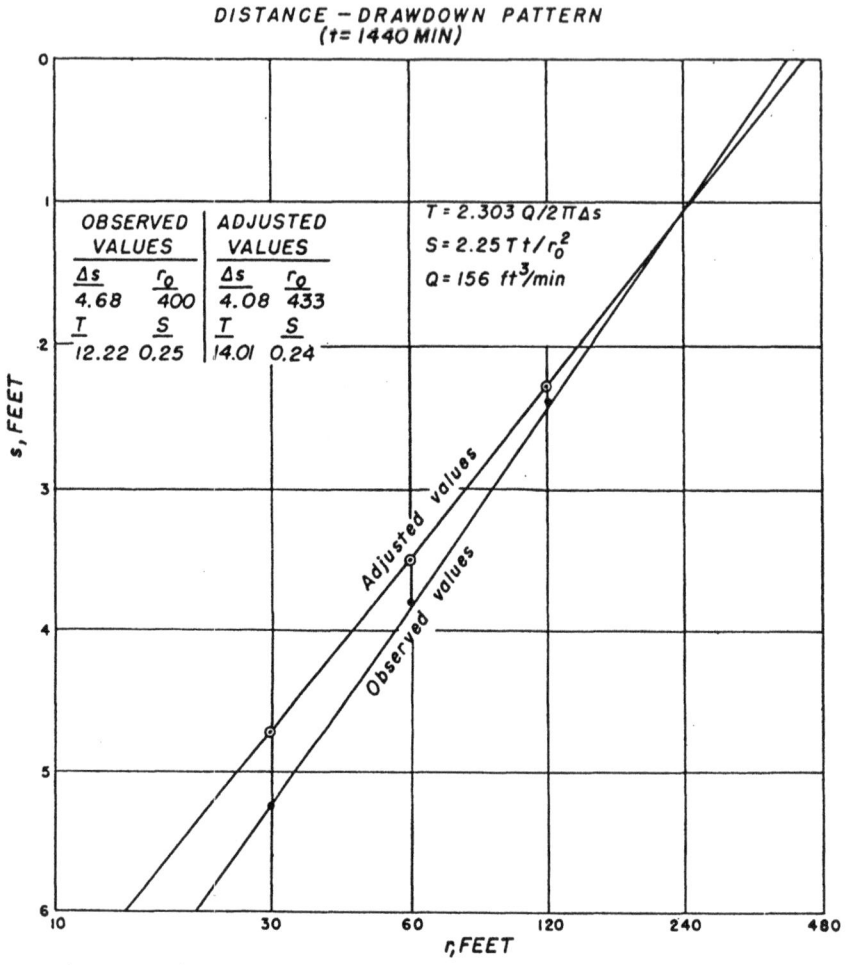

(ΔS values between one intercept on
log scale of r=30' to r=300')

Figure 9-3.—Effects of correcting drawdown
readings in a free aquifer (sheet 2 of 2).

The assumption that all water pumped from a well comes from storage within the aquifer is seldom realized because few, if any, aquifers are totally isolated. In addition to storage within the aquifer, water may originate as leakage from overlying or underlying aquifers or from recharge, precipitation, irrigation, or hydraulically connected bodies of surface water. Methods of treating leaky aquifers and recharge as boundary conditions are given in sections 9-8 and 9-11, respectively.

Table 9-2.—Tabulated discharging well test data for obtaining drawdown corrections

Project: Mesa
Feature: Drainage investigation
Location: 30 feet north of pumped well

Drawdown measured by: "Popper"
Reference point: North side of casing
   collar

Pump test No. 1, observation well No. 1, r = 30 ft

Date	Time	Elapsed time, min	Depth to water, ft	Drawdown s, ft	s^2, ft^2	$s^2/2M$, ft	s', ft	Remarks
12-6	0800	0	15.45	0	0	0	0	Pump on
	0802	2	15.49	0.04	0.0016	0	0.04	
	0804	4	15.64	.19	.0361	0	.19	
	0806	6	15.80	.35	.1225	0	.35	
	0808	8	15.96	.51	.2601	0.01	.50	
	0810	10	16.08	.63	.3969	0.01	.62	Pump off 12-7
	0815	15	16.35	.90	.8100	.02	.88	at 0800
	0820	20	16.56	1.11	1.232	.02	1.09	
	0825	25	16.73	1.28	1.638	.03	1.25	
	0830	30	16.88	1.43	2.045	.04	1.39	
	0835	35	17.01	1.56	2.434	.05	1.51	Q = 156 ft^3/min
	0840	40	17.12	1.67	2.789	.05	1.62	
	0845	45	17.22	1.77	2.133	.06	1.71	Avg. M = 26 ft
	0850	50	17.31	1.86	3.460	.07	1.79	
	0855	55	17.39	1.94	3.764	.07	1.87	$2M$ = 52 ft
	0900	60	17.47	2.02	4.080	.08	1.94	$s' = s\text{-}s^2/2M$
	0910	70	17.61	2.16	4.666	.09	2.07	
	0920	80	17.73	2.28	5.198	.10	2.18	
	0930	90	17.85	2.40	5.760	.11	2.29	
	0940	100	17.95	2.50	6.250	.12	2.38	
	0950	110	18.04	2.59	6.708	.13	2.46	
	1000	120	18.12	2.67	7.129	.14	2.53	
	1030	150	18.33	2.88	8.294	.16	2.72	
	1100	180	18.51	3.06	9.364	.18	2.88	
	1130	210	18.66	3.21	10.304	.20	3.01	
	1200	240	18.80	3.35	11.223	.22	3.13	

Table 9-2.—Tabulated discharging well test data for obtaining
drawdown corrections - continued

Project: Mesa
Feature: Drainage investigation
Location: 60 feet north of pumped well

Drawdown measured by: "Popper"
Reference point: North side of
　casing collar

Pump test No. 1, observation well, No.2, r = 60 ft

Date	Time	Elapsed time, min	Depth to water, ft	Drawdown s, ft	s^2, ft^2	$s^2/2M$, ft	s', ft	Remarks
12-6	0800	0	18.10	0	0	0	0	Pump on
	0802	2	0	0	0	0	0	
	0804	4	0	0	0	0	0	
	0806	6	18.12	0.02	0	0	0.02	
	0808	8	18.14	.04	0	0	.04	
	0810	10	18.18	.08	0.01	0	.08	Pump off 12-7 at
	0815	15	18.27	.17	.03	0	:17	0800
	0820	20	18.27	.27	.07	0	.27	
	0825	25	18.47	.38	.14	0	.38	
	0830	30	18.57	.47	.22	0	.47	
	0835	35	18.65	.55	.30	0.01	.54	Q = 156 ft^3/min
	0840	40	18.73	.63	.40	.01	.62	
	0845	45	18.78	.68	.46	.01	.67	Avg. M = 26 ft
	0850	50	18.86	.76	.58	.01	.75	
	0855	55	18.94	.84	.71	.01	.83	$2M$ = 52 ft
	0900	60	19.00	.90	.81	.02	.88	$s' = s \text{-} s^2/2M$
	0910	70	19.10	1.00	1.99	.02	.98	
	0920	80	19.22	1.11	1.23	.02	1.09	
	0930	90	19.30	1.20	1.44	.03	1.17	
	0940	100	19.38	1.28	1.64	.03	1.25	
	0950	110	19.46	1.36	1.85	.04	1.32	
	1000	120	19.52	1.42	2.02	.04	1.38	
	1030	150	19.81	1.61	2.59	.05	1.56	
	1100	180	19.87	1.77	3.13	.06	1.71	
	1130	210	20.01	1.91	3.65	.07	1.84	
	1200	240	20.12	2.02	4.08	.08	1.94	

Table 9-2.—Tabulated discharging well test data for obtaining
drawdown corrections - continued

Project: Mesa
Feature: Drainage investigation
Location: 120 feet north of pumped well

Drawdown measured by: "Popper"
Reference point: North side of casing
collar

Pump test No. 1, observation well, No. 3, r = 120 ft

Date	Time	Elapsed time, min	Depth to water, ft	Drawdown s, ft	s^2, ft^2	$s^2/2M$, ft	s', ft	Remarks
12-6	0800	0	17.95	0	0	0	0	Pump on
	0802	2	17.95	0	0	0	0	
	0804	4	17.95	0	0	0	0	
	0806	6	17.95	0	0	0	0	
	0808	8	17.95	0	0	0	0	
	0810	10	17.95	0	0	0	0	Pump of 12-7 at 0800
	0815	15	17.95	0	0	0	0	
	0820	20	17.96	0.01	0	0	0.01	
	0825	25	17.97	.02	0	0	.02	
	0830	30	17.99	.04	0	0	.04	
	0835	35	18.00	.05	0	0	.05	Q = 156 ft³/min
	0840	40	18.02	.07	0	0	.07	Avg. M = 26 ft
	0845	45	18.05	.10	0.01	0	.10	
	0850	50	18.07	.12	0.01	0	.12	
	0855	55	18.10	.15	.02	0	.15	$2M$ = 52 ft
	0900	60	18.13	.18	.03	0	.18	$s' = s - s^2/2M$
	0910	70	18.16	.21	.04	0	.21	
	0920	80	18.22	.27	.07	0	.27	
	0930	90	18.27	.32	.10	0	.32	
	0940	100	18.32	.37	.14	0	.37	
	0950	110	18.38	.43	.18	0	.43	
	1000	120	18.42	.47	.22	0	.47	
	1030	150	18.55	.60	.36	0.01	.59	
	1100	180	18.65	.70	.49	.01	.69	
	1130	210	18.76	.81	.66	.01	.80	
	1200	240	18.85	.90	.81	.02	.88	

Also, the assumption that water is discharged from storage instantaneously with a decline in head is seldom realized, especially in a free aquifer. Normally there is a lag caused by slow drainage and, as a consequence, the apparent storativity increases with time and approaches a constant value.

For most tests that are run long enough for drawdowns to reach apparent stability, the storage value determined at that time is sufficiently accurate and reliable for most applications. If drainage is unusually slow, Boulton's analysis (see section 9-10) (Boulton, 1955, 1964) should be applied. Although transmissivity and storativity are not constant everywhere within most aquifers, pumping tests tend to average out these values.

The most widely used transient equation (Jacob, 1940; Theis, 1935) is that by Theis:

$$s = \frac{Q}{4\pi T} \int_{\frac{r^2 s}{4Tt}}^{\infty} \frac{e^{-u} du}{u} \qquad\qquad 9\text{-}8$$

Where:

s = drawdown in an observation well located at a given radius from the test well at a specific time, L

Q = uniform discharge from the well, L^3/t

T = transmissivity of the aquifer, L^2/t

r = radius of the observation well, L

S = storativity of the aquifer (nondimensional)

t = time since start of pumping, t

$$u = \frac{r^2 s}{4Tt} \text{ (nondimensional)} \qquad\qquad 9\text{-}7$$

9-4. Type Curve Solutions of the Transient Equation.—
Since T appears twice in equation 9-6, a mathematical solution for each individual problem becomes tedious. Theis provided a graphical method of solution that gives satisfactory results when applied to tests in aquifers which conform approximately to ideal conditions.

The integral expression in the Theis equation is given by the series:

$$\int_u^\infty \frac{e^{-u}}{u} du = W(u) = -0.5772 - \ln u + u - \frac{u^2}{2 \cdot 2!} + \frac{u^3}{3 \cdot 3!} - \frac{u^4}{4 \cdot 4!} \ldots \qquad 9\text{-}8$$

where $W(u)$ is the well function or exponential integral of u

Equation 9-6 may be rewritten:

$$s = \frac{QW(u)}{4\pi T} \qquad 9\text{-}9$$

Values of $W(u)$ for u values of 10^{-15} to 9.9 are given in table 9-3 where the value of u is expressed as some number (N) between 1 and 9.9, multiplied by 10 with a range of appropriate exponents. For example, if u has a value of 0.0027, this is 2.7 x 10^{-3}. The value of $W(u)$ is found by reading across the table opposite $N = 2.7$ and under the column headed N x 10^{-3}. The value of $W(u)$ is 5.3400. More complete tabulations of the $W(u)$ function are presented by the U.S. Geological Survey (Kazmann, 1941) and the National Bureau of Standards (1940).

Equations 9-9 and 9-7, respectively:

$$s = \frac{Q}{4\pi T} W(u)$$

and

$$u = \frac{r^2 s}{4Tt}$$

may be expressed in common logarithmic form:

$$\log s = \left[\log \frac{Q}{4\pi T}\right] + \log W(u) \qquad 9\text{-}10$$

$$\log \frac{r^2}{t} = \left[\log \frac{4T}{S}\right] + \log u \qquad 9\text{-}11$$

Table 9-3.—Value of W(u) for values of u between 10^{-15} and 9.9

N/u	$N\times10^{-15}$	$N\times10^{-14}$	$N\times10^{-13}$	$N\times10^{-12}$	$N\times10^{-11}$	$N\times10^{-10}$	$N\times10^{-9}$	$N\times10^{-8}$	$N\times10^{-7}$	$N\times10^{-6}$	$N\times10^{-5}$	$N\times10^{-4}$	$N\times10^{-3}$	$N\times10^{-2}$	$N\times10^{-1}$	N
1.0	33.9616	31.6590	29.3564	27.0538	24.7512	22.4486	20.1460	17.8435	15.5409	13.2383	10.9357	8.6332	6.3315	4.0379	1.8229	0.2194
1.1	33.8662	31.5637	29.2611	26.9585	24.6559	22.3533	20.0507	17.7482	15.4456	13.1430	10.8404	8.5379	6.2363	3.9436	1.7371	.1860
1.2	33.7792	31.4767	29.1741	26.8715	24.5689	22.2663	19.9637	17.6611	15.3586	13.0560	10.7534	8.4509	6.1494	3.8576	1.6595	.1584
1.3	33.6992	31.3966	29.0940	26.7914	24.4889	22.1863	19.8837	17.5811	15.2785	12.9759	10.6734	8.3709	6.0695	3.7785	1.5889	.1355
1.4	33.6251	31.3225	29.0199	26.7173	24.4147	22.1122	19.8096	17.5070	15.2044	12.9018	10.5993	8.2968	5.9955	3.7054	1.5241	.1162
1.5	33.5561	31.2535	28.9509	26.6483	24.3458	22.0432	19.7406	17.4380	15.1354	12.8328	10.5303	8.2278	5.9266	3.6374	1.4645	.1000
1.6	33.4916	31.1890	28.8864	26.5838	24.2812	21.9786	19.6760	17.3735	15.0709	12.7683	10.4657	8.1634	5.8621	3.5739	1.4092	.08631
1.7	33.4309	31.1283	28.8258	26.5232	24.2206	21.9180	19.6154	17.3128	15.0103	12.7077	10.4051	8.1027	5.8016	3.5143	1.3578	.07465
1.8	33.3738	31.0712	28.7686	26.4660	24.1634	21.8606	19.5583	17.2557	14.9531	12.6505	10.3479	8.0455	5.7446	3.4581	1.3089	.06471
1.9	33.3197	31.0171	28.7145	26.4119	24.1094	21.8068	19.5042	17.2016	14.8990	12.5964	10.2939	7.9915	5.6906	3.4050	1.2649	.05620
2.0	33.2684	30.9658	28.6632	26.3607	24.0581	21.7555	19.4529	17.1503	14.8477	12.5451	10.2426	7.9402	5.6394	3.3547	1.2227	.04890
2.1	33.2196	30.9170	28.6145	26.3119	24.0093	21.7067	19.4041	17.1015	14.7989	12.4964	10.1938	7.8914	5.5907	3.3069	1.1829	.04261
2.2	33.1731	30.8705	28.5679	26.2653	23.9628	21.6602	19.3576	17.0550	14.7524	12.4498	10.1473	7.8449	5.5443	3.2614	1.1454	.03719
2.3	33.1286	30.8261	28.5235	26.2209	23.9183	21.6157	19.3131	17.0106	14.7080	12.4054	10.1028	7.8004	5.4999	3.2179	1.1099	.03250
2.4	33.0861	30.7835	28.4809	26.1783	23.8758	21.5732	19.2706	16.9680	14.6654	12.3628	10.0603	7.7579	5.4575	3.1763	1.0762	.02844
2.5	33.0453	30.7427	28.4401	26.1375	23.8349	21.5323	19.2298	16.9272	14.6246	12.3220	10.0194	7.7172	5.4167	3.1365	1.0443	.02491
2.6	33.0060	30.7035	28.4009	26.0983	23.7957	21.4931	19.1905	16.8880	14.5854	12.2828	9.9802	7.6779	5.3776	3.0983	1.0139	.02185
2.7	32.9683	30.6657	28.3631	26.0606	23.7580	21.4554	19.1528	16.8502	14.5476	12.2450	9.9425	7.6401	5.3400	3.0615	.9849	.01918
2.8	32.9319	30.6294	28.3268	26.0242	23.7216	21.4190	19.1164	16.8138	14.5113	12.2087	9.9061	7.6038	5.3037	3.0261	.9573	.01686
2.9	32.8968	30.5943	28.2917	25.9891	23.6865	21.3839	19.0813	16.7788	14.4762	12.1736	9.8710	7.5687	5.2687	2.9920	.9309	.01482
3.0	32.8629	30.5604	28.2578	25.9552	23.6526	21.3500	19.0474	16.7449	14.4423	12.1397	9.8371	7.5348	5.2349	2.9591	.9057	.01305
3.1	32.8302	30.5276	28.2250	25.9224	23.6198	21.3172	19.0146	16.7121	14.4095	12.1069	9.8043	7.5020	5.2022	2.9273	.8815	.01149
3.2	32.7984	30.4958	28.1932	25.8907	23.5880	21.2855	18.9829	16.6803	14.3777	12.0751	9.7726	7.4703	5.1706	2.8965	.8583	.01013
3.3	32.7676	30.4651	28.1625	25.8599	23.5573	21.2547	18.9521	16.6495	14.3470	12.0444	9.7418	7.4395	5.1399	2.8668	.8361	.008939
3.4	32.7378	30.4352	28.1326	25.8300	23.5274	21.2249	18.9223	16.6197	14.3171	12.0145	9.7120	7.4097	5.1102	2.8379	.8147	.007891
3.5	32.7088	30.4062	28.1036	25.8010	23.4985	21.1959	18.8933	16.5907	14.2881	11.9855	9.6830	7.3807	5.0813	2.8099	.7942	.006970
3.6	32.6806	30.3780	28.0755	25.7729	23.4703	21.1677	18.8651	16.5625	14.2599	11.9574	9.6548	7.3526	5.0532	2.7827	.7745	.006160
3.7	32.6532	30.3506	28.0481	25.7455	23.4429	21.1403	18.8377	16.5351	14.2325	11.9300	9.6274	7.3252	5.0259	2.7563	.7554	.005448
3.8	32.6266	30.3240	28.0214	25.7188	23.4162	21.1136	18.8110	16.5085	14.2059	11.9033	9.6007	7.2985	4.9993	2.7306	.7371	.004820
3.9	32.6006	30.2980	27.9954	25.6928	23.3902	21.0877	18.7851	16.4825	14.1799	11.8773	9.5748	7.2725	4.9735	2.7056	.7194	.004267
4.0	32.5753	30.2727	27.9701	25.6675	23.3649	21.0623	18.7598	16.4572	14.1546	11.8520	9.5495	7.2472	4.9482	2.6813	.7024	.003779
4.1	32.5506	30.2480	27.9454	25.6428	23.3402	21.0376	18.7351	16.4325	14.1299	11.8273	9.5248	7.2225	4.9236	2.6576	.6859	.003349
4.2	32.5265	30.2239	27.9213	25.6187	23.3161	21.0136	18.7110	16.4084	14.1058	11.8032	9.5007	7.1985	4.8997	2.6344	.6700	.002969
4.3	32.5029	30.2004	27.8978	25.5952	23.2926	20.9900	18.6874	16.3848	14.0823	11.7797	9.4771	7.1749	4.8762	2.6119	.6546	.002633
4.4	32.4800	30.1774	27.8748	25.5722	23.2696	20.9670	18.6644	16.3619	14.0593	11.7567	9.4541	7.1520	4.8533	2.5899	.6397	.002336
4.5	32.4575	30.1549	27.8523	25.5497	23.2471	20.9446	18.6420	16.3394	14.0368	11.7342	9.4317	7.1295	4.8310	2.5684	.6253	.002073
4.6	32.4355	30.1329	27.8303	25.5277	23.2252	20.9226	18.6200	16.3174	14.0148	11.7122	9.4097	7.1075	4.8091	2.5474	.6114	.001841
4.7	32.4140	30.1114	27.8088	25.5062	23.2037	20.9011	18.5985	16.2959	13.9933	11.6907	9.3882	7.0860	4.7877	2.5268	.5979	.001635
4.8	32.3929	30.0904	27.7878	25.4852	23.1826	20.8800	18.5774	16.2748	13.9723	11.6697	9.3671	7.0650	4.7667	2.5065	.5848	.001453
4.9	32.3723	30.0697	27.7672	25.4646	23.1620	20.8594	18.5568	16.2542	13.9516	11.6491	9.3465	7.0444	4.7462	2.4871	.5721	.001291

Table 9.3 — Value of W(u) for values of u between 10^{-15} and 9.9 - continued

N\u	$N \times 10^{-15}$	$N \times 10^{-14}$	$N \times 10^{-13}$	$N \times 10^{-12}$	$N \times 10^{-11}$	$N \times 10^{-10}$	$N \times 10^{-9}$	$N \times 10^{-8}$	$N \times 10^{-7}$	$N \times 10^{-6}$	$N \times 10^{-5}$	$N \times 10^{-4}$	$N \times 10^{-3}$	$N \times 10^{-2}$	$N \times 10^{-1}$	N
5.0	32.3521	30.0495	27.7470	25.4444	23.1418	20.8392	18.5366	16.2340	13.9314	11.6289	9.3263	7.0242	4.7261	2.4679	.5598	.001148
5.1	32.3323	30.0297	27.7271	25.4246	23.1220	20.8194	18.5168	16.2142	13.9116	11.6091	9.3065	7.0044	4.7064	2.4491	.5478	.001021
5.2	32.3129	30.0103	27.7077	25.4051	23.1026	20.8000	18.4974	16.1948	13.8922	11.5896	9.2871	6.9850	4.6871	2.4306	.5362	.0009086
5.3	32.2939	29.9913	27.6887	25.3861	23.0835	20.7809	18.4783	16.1758	13.8732	11.5706	9.2681	6.9659	4.6681	2.4126	.5250	.0008086
5.4	32.2752	29.9726	27.6700	25.3674	23.0648	20.7622	18.4596	16.1571	13.8545	11.5519	9.2494	6.9473	4.6495	2.3948	.5140	.0007198
5.5	32.2568	29.9542	27.6516	25.3491	23.0465	20.7439	18.4413	16.1387	13.8361	11.5336	9.2310	6.9289	4.6313	2.3775	.5034	.0006409
5.6	32.2388	29.9362	27.6336	25.3310	23.0285	20.7259	18.4233	16.1207	13.8181	11.5155	9.2130	6.9109	4.6134	2.3604	.4930	.0005708
5.7	32.2211	29.9185	27.6159	25.3133	23.0108	20.7082	18.4056	16.1030	13.8004	11.4978	9.1953	6.8932	4.5958	2.3437	.4830	.0005085
5.8	32.2037	29.9011	27.5985	25.2959	22.9934	20.6908	18.3882	16.0856	13.7830	11.4804	9.1779	6.8758	4.5785	2.3273	.4732	.0004532
5.9	32.1866	29.8840	27.5814	25.2789	22.9763	20.6737	18.3711	16.0685	13.7659	11.4633	9.1608	6.8588	4.5615	2.3111	.4637	.0004039
6.0	32.1698	29.8672	27.5646	25.2620	22.9595	20.6569	18.3543	16.0517	13.7491	11.4465	9.1440	6.8420	4.5448	2.2953	.4544	.0003601
6.1	32.1533	29.8507	27.5481	25.2455	22.9429	20.6403	18.3378	16.0352	13.7326	11.4300	9.1275	6.8254	4.5282	2.2797	.4454	.0003211
6.2	32.1370	29.8344	27.5318	25.2293	22.9267	20.6241	18.3215	16.0189	13.7163	11.4138	9.1112	6.8092	4.5122	2.2645	.4366	.0002864
6.3	32.1210	29.8184	27.5158	25.2133	22.9107	20.6081	18.3055	16.0029	13.7003	11.3978	9.0952	6.7932	4.4963	2.2494	.4280	.0002555
6.4	32.1053	29.8027	27.5001	25.1975	22.8949	20.5923	18.2898	15.9872	13.6846	11.3820	9.0795	6.7775	4.4806	2.2346	.4197	.0002279
6.5	32.0898	29.7872	27.4846	25.1820	22.8794	20.5768	18.2742	15.9717	13.6691	11.3665	9.0640	6.7620	4.4652	2.2201	.4115	.0002034
6.6	32.0745	29.7719	27.4693	25.1667	22.8641	20.5616	18.2590	15.9564	13.6538	11.3512	9.0487	6.7467	4.4501	2.2058	.4036	.0001816
6.7	32.0595	29.7569	27.4543	25.1517	22.8491	20.5465	18.2439	15.9414	13.6388	11.3362	9.0337	6.7317	4.4351	2.1917	.3959	.0001621
6.8	32.0446	29.7421	27.4395	25.1369	22.8343	20.5317	18.2291	15.9265	13.6240	11.3214	9.0189	6.7169	4.4204	2.1779	.3883	.0001448
6.9	32.0300	29.7275	27.4249	25.1223	22.8197	20.5171	18.2145	15.9119	13.6094	11.3068	9.0043	6.7023	4.4059	2.1643	.3810	.0001293
7.0	32.0156	29.7131	27.4105	25.1079	22.8053	20.5027	18.2001	15.8976	13.5950	11.2924	8.9899	6.6879	4.3916	2.1508	.3738	.0001155
7.1	32.0015	29.6989	27.3963	25.0937	22.7911	20.4885	18.1860	15.8834	13.5808	11.2782	8.9757	6.6737	4.3775	2.1376	.3668	.0001032
7.2	31.9875	29.6849	27.3823	25.0797	22.7771	20.4746	18.1720	15.8694	13.5668	11.2642	8.9617	6.6598	4.3636	2.1246	.3599	.00009219
7.3	31.9737	29.6711	27.3685	25.0659	22.7633	20.4608	18.1582	15.8556	13.5530	11.2504	8.9479	6.6460	4.3500	2.1118	.3532	.00008239
7.4	31.9601	29.6575	27.3549	25.0523	22.7497	20.4472	18.1446	15.8420	13.5394	11.2368	8.9343	6.6324	4.3364	2.0991	.3467	.00007364
7.5	31.9467	29.6441	27.3415	25.0389	22.7363	20.4337	18.1311	15.8286	13.5260	11.2234	8.9209	6.6190	4.3231	2.0867	.3403	.00006583
7.6	31.9334	29.6308	27.3282	25.0257	22.7231	20.4205	18.1179	15.8153	13.5127	11.2102	8.9076	6.6057	4.3100	2.0744	.3341	.00005886
7.7	31.9203	29.6178	27.3152	25.0126	22.7100	20.4074	18.1048	15.8022	13.4997	11.1971	8.8946	6.5927	4.2970	2.0623	.3280	.00005263
7.8	31.9074	29.6049	27.3023	24.9997	22.6971	20.3945	18.0919	15.7893	13.4868	11.1842	8.8817	6.5798	4.2842	2.0503	.3221	.00004707
7.9	31.8947	29.5921	27.2895	24.9869	22.6844	20.3818	18.0792	15.7766	13.4740	11.1714	8.8689	6.5671	4.2716	2.0386	.3163	.00004210
8.0	31.8821	29.5795	27.2769	24.9744	22.6718	20.3692	18.0666	15.7640	13.4614	11.1589	8.8563	6.5545	4.2591	2.0269	.3106	.00003767
8.1	31.8697	29.5671	27.2645	24.9619	22.6594	20.3568	18.0542	15.7516	13.4490	11.1464	8.8439	6.5421	4.2468	2.0155	.3050	.00003370
8.2	31.8574	29.5548	27.2523	24.9497	22.6471	20.3445	18.0419	15.7393	13.4367	11.1342	8.8317	6.5298	4.2346	2.0042	.2996	.00003015
8.3	31.8453	29.5427	27.2401	24.9376	22.6350	20.3324	18.0298	15.7272	13.4246	11.1220	8.8195	6.5177	4.2226	1.9930	.2943	.00002699
8.4	31.8333	29.5307	27.2282	24.9256	22.6230	20.3204	18.0178	15.7152	13.4126	11.1101	8.8076	6.5057	4.2107	1.9820	.2891	.00002415
8.5	31.8215	29.5189	27.2163	24.9137	22.6112	20.3086	18.0060	15.7034	13.4008	11.0982	8.7957	6.4939	4.1990	1.9711	.2840	.00002162
8.6	31.8098	29.5072	27.2046	24.9020	22.5995	20.2969	17.9943	15.6917	13.3891	11.0865	8.7840	6.4822	4.1874	1.9604	.2790	.00001936
8.7	31.7982	29.4957	27.1931	24.8905	22.5879	20.2853	17.9827	15.6801	13.3776	11.0750	8.7725	6.4707	4.1759	1.9498	.2742	.00001733
8.8	31.7868	29.4842	27.1816	24.8791	22.5765	20.2739	17.9713	15.6687	13.3661	11.0636	8.7610	6.4593	4.1646	1.9393	.2694	.00001552
8.9	31.7755	29.4729	27.1703	24.8678	22.5652	20.2626	17.9600	15.6574	13.3548	11.0523	8.7497	6.4480	4.1534	1.9289	.2648	.00001390
9.0	31.7643	29.4618	27.1592	24.8566	22.5540	20.2514	17.9488	15.6462	13.3437	11.0411	8.7386	6.4368	4.1423	1.9187	.2602	.00001245
9.1	31.7533	29.4507	27.1481	24.8455	22.5429	20.2404	17.9378	15.6352	13.3326	11.0300	8.7275	6.4258	4.1313	1.9087	.2557	.00001115
9.2	31.7424	29.4398	27.1372	24.8346	22.5320	20.2294	17.9268	15.6243	13.3217	11.0191	8.7166	6.4148	4.1205	1.9007	.2513	.000009988
9.3	31.7315	29.4290	27.1264	24.8238	22.5212	20.2186	17.9160	15.6135	13.3109	11.0083	8.7058	6.4040	4.1098	1.8888	.2470	.000008948
9.4	31.7209	29.4183	27.1157	24.8131	22.5105	20.2079	17.9053	15.6028	13.3002	10.9976	8.6951	6.3934	4.0992	1.8791	.2429	.000008018
9.5	31.7103	29.4077	27.1051	24.8025	22.4999	20.1973	17.8948	15.5922	13.2896	10.9870	8.6845	6.3828	4.0887	1.8695	.2387	.000007185
9.6	31.6998	29.3972	27.0946	24.7920	22.4895	20.1869	17.8843	15.5817	13.2791	10.9765	8.6740	6.3723	4.0784	1.8599	.2347	.000006439
9.7	31.6894	29.3869	27.0843	24.7817	22.4791	20.1765	17.8739	15.5713	13.2688	10.9662	8.6637	6.3620	4.0681	1.8505	.2308	.000005771
9.8	31.6792	29.3766	27.0740	24.7714	22.4688	20.1663	17.8637	15.5611	13.2585	10.9559	8.6534	6.3517	4.0579	1.8412	.2269	.000005173
9.9	31.6690	29.3664	27.0639	24.7613	22.4587	20.1561	17.8535	15.5509	13.2483	10.9458	8.6433	6.3416	4.0479	1.8320	.2231	.000004637

The bracketed values in equations 9-10 and 9-11 are constants for any given test. Logarithmic plots of test data for s versus r^2/t will be similar to a logarithmic plot of $W(u)$ versus u, which is referred to as a type curve. If a test data curve is superimposed on the type curve while keeping the coordinate axes of the sheets parallel and the test data curve shifted to the point of best fit on the type curve, the displacements of common points of the test data curve and the type curve will be equal to the constants in brackets as shown in figure 9-4. The displacement values are used in equations 9-7 and 9-9, respectively, to solve for S and T:

$$S = \frac{4Tut}{r_2}$$

$$T = \frac{Q}{4\pi s} W(u)$$

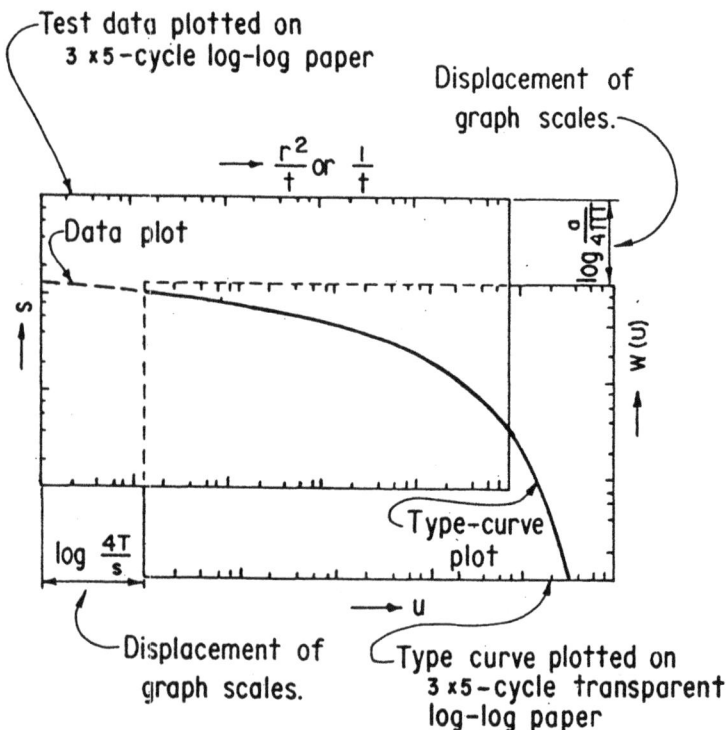

Figure 9-4.—Superimposition of the type curve on test data for graphic solution of the nonequilibrium equation.

A type curve of u versus $W(u)$ has been prepared on 3- by 5-cycle logarithmic paper (figure 9-5). **A full-scale drawing of figure 9-5 has been placed in a pocket at the back of this manual**. Field data plotted on 3- by 5-cycle logarithmic paper can be used as an overlay on this drawing to determine the best fit on the type curve. Table 9-1 (section 9-2) shows a portion of the recorded data from a pumping test, and figures 9-6, 9-7, and 9-8 show plots of the data.

There are two procedures for solving the nonequilibrium equation using the type curve of figure 9-5 (Lohman, 1972; Prickett, 1965; Schict, 1972):

(a) Time-Drawdown Solution.—The time-drawdown data curve can be prepared by plotting s in an observation well against the reciprocal of t or against r^2/t on log-log paper and fitting this curve to the type curve as shown on figure 9-6. An easier procedure which avoids the need for calculating reciprocals is to plot s against t (figure 9-7) in a similar manner and fit this data plot against a type curve which has been reversed end-for-end. This end-for-end switch of the type curve is the equivalent of plotting the reciprocal of $1/t$ or t on the data curve. The data curve plot of each observation well is matched to the type curve while keeping the axes of the two sheets parallel. The curves may match only in a given segment because results from the test may depart from the ideal conditions on which the type curve is based. Such factors as unconfined aquifer conditions, boundaries, and leakage result in departure from the type curve. Any common index point, usually where u and $W(u)$ are equal to 1 to simplify computations, on the overlapping sheets is marked and the values of u, $W(u)$, t, and s are recorded. Transmissivity and storativity are then computed using these values in the following equations:

$$T = \frac{Q}{4\pi s} W(u) \qquad\qquad 9\text{-}12$$

$$S = \frac{4Tu}{\dfrac{r^2}{t}} \text{ or } \frac{4Ttu}{r^2} \qquad\qquad 9\text{-}13$$

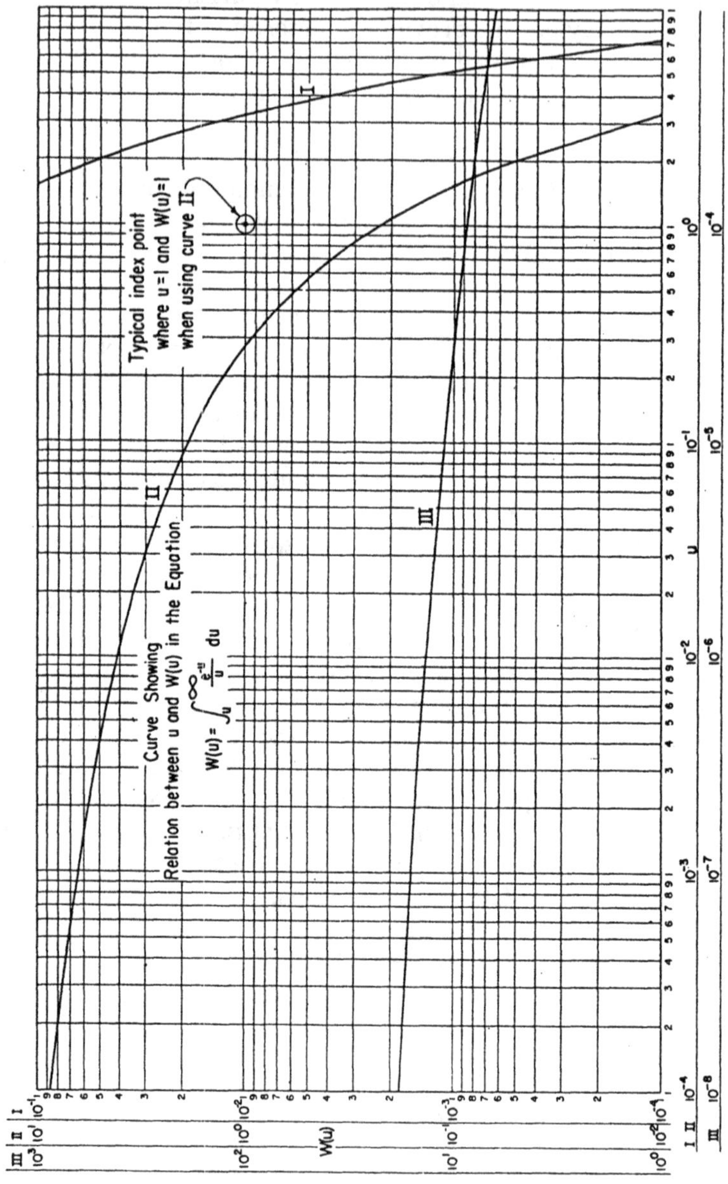

Figure 9-5.—Type curve resulting from the plotting of u versus $W(u)$.

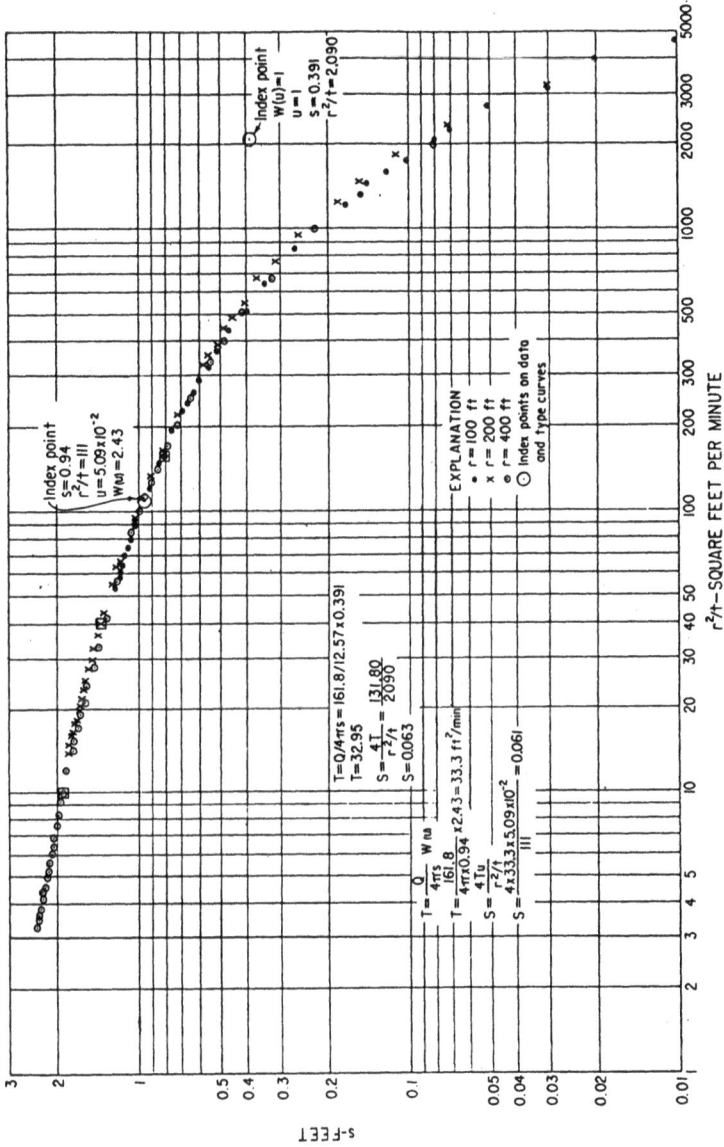

Figure 9-6.—Type curve solution of the nonequilibrium equation using s versus r^2/t.

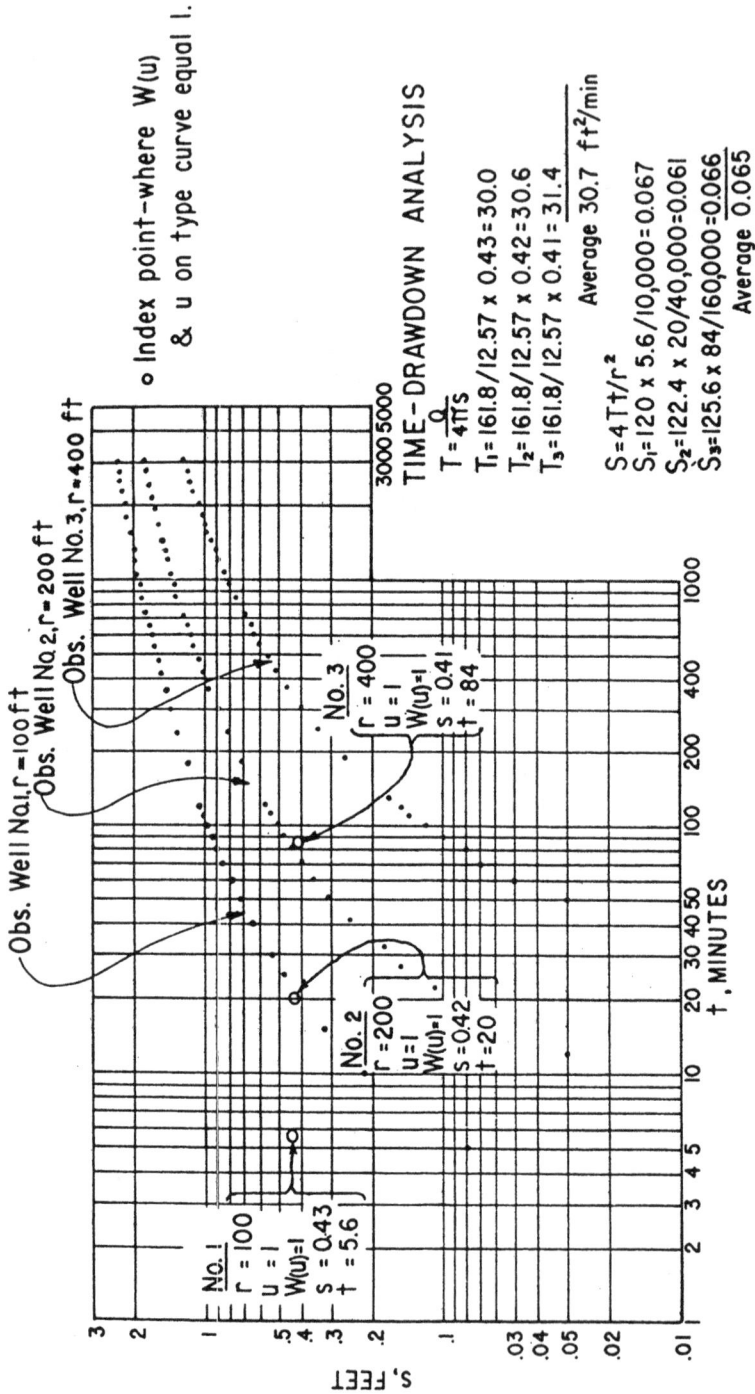

Figure 9-7.—Type curve solution of the nonequilibrium equation using s versus t.

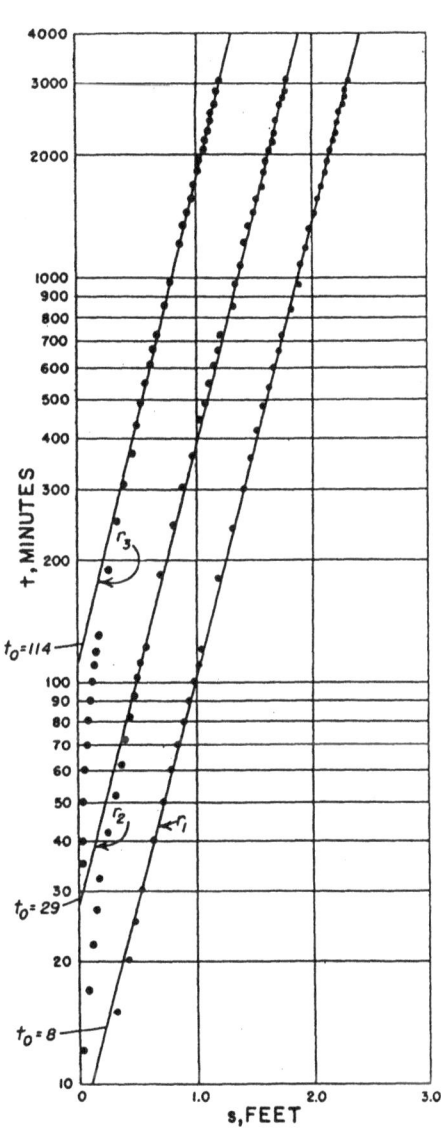

TIME - DRAWDOWN ANALYSIS

JACOB'S APPROXIMATION

$Q = 1210 \ gal/min = 161.8 \ ft^3/min$

	Δs	t_o
$r_1 = 100'$	$1.90 - 1.0 = 0.90$	8
$r_2 = 200'$	$1.38 - 0.49 = 0.89$	29
$r_3 = 400'$	$1.68 - 0.80 = 0.88$	114

$T = 2.303 Q / 4\pi \Delta s$

$S = 2.25 T t_o / r^2$

$\dfrac{2.303 Q}{4\pi} = \dfrac{372.6}{12.5664} = 29.65$

$T_1 = \dfrac{29.65}{0.90} = 33.0 \ ft^2/min$

$T_2 = \dfrac{29.65}{0.89} = 33.3 \ ft^2/min$

$T_3 = \dfrac{29.65}{0.88} = 33.7 \ ft^2/min$

$AVG. \ T = 33.3 \ ft^2/min$

$S_1 = 2.25 \times 33.0 \times 8/10,000 = 0.060$

$S_2 = 2.25 \times 33.3 \times 29/40,000 = 0.055$

$S_3 = 2.25 \times 33.7 \times 114/160,000 = 0.054$

Average $= 0.056$

Figure 9-8.—Straight-line solution of the nonequilibrium equation (sheet 1 of 2).

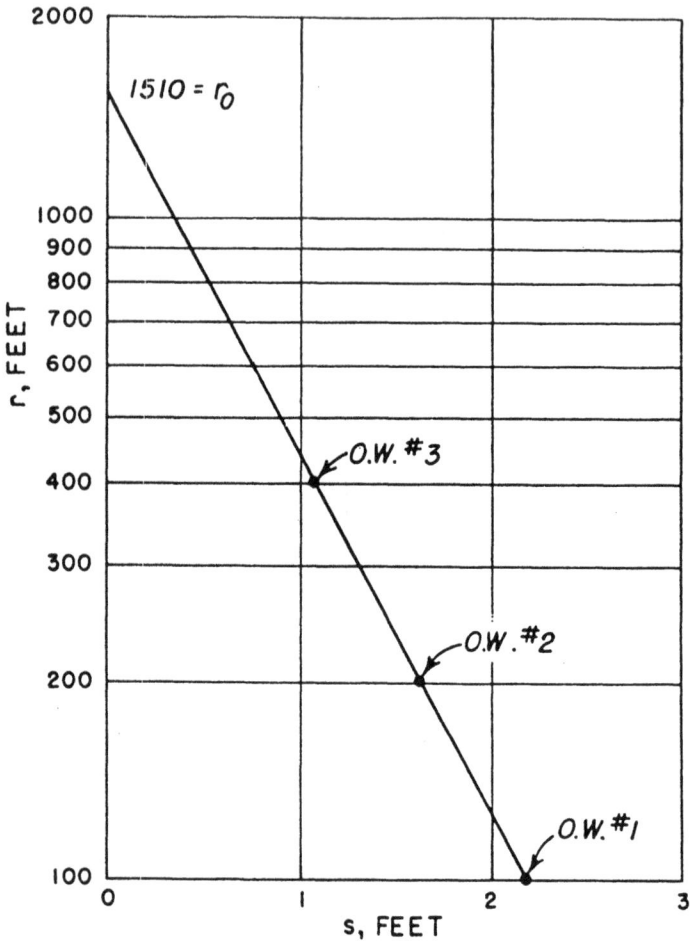

DISTANCE-DRAWDOWN ANALYSIS

JACOB'S APPROXIMATION

$Q = 1210 \, gal/min = 161.8 \, ft^3/min$

$t = 2045 \, minutes$

$\Delta s = 2.17 - 0.33 = 1.84 \, ft$

$r_0 = 1510 \, ft$

$T = 2.303Q / 2\pi \Delta s$

$T = 372.6 / 2\pi \times 1.84 = 32.2 \, ft^2/min$

$S = 2.25Tt / r_0^2 = 0.065$

Figure 9-8.—Straight-line solution of the nonequilibrium equation (sheet 2 of 2).

(b) *Distance-Drawdown Solution.*—The data curve is prepared by plotting the drawdowns measured at the same time in three or more observation wells against the distances squared divided by time, s versus r^2/t. The three points enclosed in squares on figure 9-6 illustrate this procedure. The three points should be fitted to the type curve, the values of s, r^2/t, $W(u)$, and u obtained, and the values of T and S computed using equations 9-12 and 9-13.

A more tedious but reliable analysis involves plotting s versus r^2/t for each drawdown measurement in each observation well as shown in figure 9-6. The use of computers greatly facilitates these computations. This method enables better determination of departures from the type curve which may indicate aquifer conditions that are less than the ideal (Bentall, 1963; Brown, 1953; Bruin and Hudson, 1955; Davis and DeWiest, 1966; Heath and Trainer, 1968; Rouse, 1949; UNESCO, 1967; UPO Johnson Division, 1966; Walton, 1962, 1970).

9-5. Jacob's Approximation Solutions for the Nonequilibrium Equation.—In addition to solution by the type curve methods previously described, a method developed by Cooper and Jacob (1946) permits an approximate solution to the Theis nonequilibrium equation using a straight-line graphical approach, which is simple and may offer advantages over the type curve methods.

When the value of u in equation 9-8 is less than 0.01 (i.e., when r^2/t becomes very small), the terms following the first two terms of the series equation for $W(u)$ may be neglected and equation 9-9 may then be approximated by:

$$s = \frac{Q}{4\pi T} \left[\ln \left(\frac{1}{u} \right) - 0.5772 \right] \qquad \text{9-14}$$

which can be reduced to:

$$s = \frac{Q}{4\pi T} \left(\ln \frac{2.25Tt}{r^2 S} \right) \qquad \text{9-15}$$

by solving for the value of u necessary to make s equal to zero.

Converting to common logarithms, equation 9-15 may be rewritten:

$$s = \frac{2.303Q}{4\pi T}\left(\log \frac{2.25Tt}{r^2S}\right) \qquad\qquad 9\text{-}16$$

Inasmuch as a plot of the drawdown s versus the logarithm of distance r or time t is a straight line, two simple semilog graphical methods can be used to solve for transmissivity and storativity.

(a) *Distance-Drawdown Solution.*—In this method, drawdowns taken at the same time in each of three or more observation wells are plotted against the distance on a log scale of each observation well from the test well as shown in figure 9-8. The straight-line portion of this curve is projected to cover at least one log cycle and the zero drawdown axis. Transmissivity is calculated by:

$$T = \frac{2.303Q}{2\pi \Delta s} \qquad\qquad 9\text{-}17$$

where:

Δs is the difference in drawdown over one log cycle.

Storativity is determined by projecting the straight line to the zero drawdown interception which defines the distance r_o, and substituting the value of r_o into:

$$S = \frac{2.25Tt}{r_o^2} \qquad\qquad 9\text{-}18$$

Figure 9-8 shows an example of a distance versus drawdown solution.

(b) *Time-Drawdown Solution.*—In this method the drawdown in each observation well since pumping began is plotted against time on the log scale. The straight-line portion of the plot is projected to intercept one or more log cycles and the zero drawdown axis.

where:

$$u = \frac{r^2S}{4Tt} \;\; and \;\; u' = \frac{r^2S}{4Tt'}$$

t is the time since pumping started, and t' is the time since pumping stopped. Q, T, S, and r have been previously defined in equation 9-6.

The value:

$$\frac{r^2S}{4Tt'}$$

decreases as t' increases. Therefore, the recovery equation can be written in the form:

$$T = \frac{2.303Q}{4\pi s'} \log\left(\frac{t}{t'}\right) \qquad\qquad 9\text{-}22$$

To solve equation 9-21 graphically using semilog paper, data from table 9-4 are plotted on figure 9-9 where t/t' is plotted on the log scale against residual drawdown s' on the arithmetic scale. When t' becomes large, the plot of the observed data should fall on a straight line. The slope of the line gives the value of the quantity,

$$\frac{\log\dfrac{t}{t'}}{s'}$$

in equation 9-22. The value of t/t' is usually chosen over one log cycle so that $\log t/t'$ is unity and then equation 9-22 becomes:

$$T = \frac{2.303Q}{4\pi \Delta s'} \qquad\qquad 9\text{-}23$$

where $\Delta s'$ is the change in residual drawdown over one log cycle t/t'. The storativity cannot be determined because the effective radius of the test well cannot be determined. The recovery method may give a slightly high value of transmissivity in unconfined aquifers, but it is reasonably accurate when used with data from an artesian (confined) aquifer. In areas where boundary conditions

Table 9-4.—Tabulated data from a discharging and recovering well test

Project: Las Vegas　　　　　Discharge measured by: Parshall flume
Feature: Pichaco Dam　　　　Drawdown measured by: M scope
　　　　　　　　　　　　　　Reference point: North side of casing collar

		Pump test No. 1			Test well		
Date	Time	time, min	Depth to water, ft	Drawdown, ft	Gauge reading, ft	Discharge, ft³/s	Remarks
5-16	0840		60.99				
5-17	0820		61.01				
5-18	0845		61.00				
5-19	0820		60.98				
	0840	0	60.99				Pump started
	0843	3	71.2	10.2			Q = 162.9 ft³/min
	0848	8	71.2	10.6			
	0853	13	71.3	10.8			
	0900	20	72.3	11.3	0.79	2.70	
	1000	80	72.6	11.6	.79	2.70	
	1100	140	72.8	11.8	.79	2.70	
	1155	195	72.8	11.8	.80	2.75	
	1255	255	72.8	11.8	.80	2.75	
	1355	315	73.0	12.0	.80	2.75	
	1455	375	73.2	12.2	.80	2.75	
	1555	435	73.2	12.2	.80	2.75	
	1655	495	73.2	12.2	.80	2.75	
	1800	560	73.2	12.2	.80	2.75	
	1856	616	73.3	12.3	.80	2.75	
	1958	668	73.4	12.4	.80	2.75	
	2057	737	73.4	12.5	.80	2.75	
	2200	800	73.5	12.5	.80	2.75	Pump off

Table 9-4.—Tabulated data from a discharging and recovering well test - continued

Recovery of test well

Date	Time	t, min	t', min	t/t'	Depth to water, ft	Residual drawdown, ft	Remarks
5-19	2200	800	0	0.0	73.5	12.5	Pump off
	2203	803	3	268.0	41.0±	+20.0	
	2208	808	8	101.0	56.0±	+5.0	
	2213	813	13	62.5	60.5	+0.5	
	2220	820	20	41.0	62.49	1.5	
	2320	880	80	11.0	61.99	1.0	
5-20	0020	940	140	6.7	61.79	0.80	
	0115	995	195	5.1	61.68	.69	
	0215	1,055	255	4.1	61.58	.59	
	0315	1,115	315	3.5	61.50	.51	
	0415	1,175	375	3.1	61.48	.49	
	0515	1,235	435	2.8	61.45	.46	
	0615	1,295	495	2.6	61.37	.38	
	0720	1,360	560	2.4	61.33	.34	
	0816	1,416	616	2.3	61.32	.33	
	0908	1,418	668	2.2	61.32	.33	
	1017	1,527	727	2.1	61.21	.22	
	1120	1,600	800	2.0	61.21	.22	

Table 9-4.—Tabulated data from a discharging and
recovering well test - continued

Pump test No. 1, observation well No. 1, r = 100 ft

Date	Time	Time, min	Depth to water, ft	Drawdown, ft	Remarks
5-16	0845		61.20		
5-17	0825		61.21		
5-18	0840		61.21		
5-19	0815		61.20		
	0835		61.20		
	0840		61.20		Pump started
	0845	5	61.28	0.08	
	0850	10	61.42	.22	
	0855	15	61.55	.33	
	0900	20	61.61	.41	
	0905	25	61.70	.50	
	0910	30	61.75	.55	
	0920	40	61.86	.66	
	0930	50	61.93	.73	
	0940	60	62.00	.80	
	0950	70	62.06	.86	
	1000	80	62.12	.92	
	1010	90	62.16	.96	
	1020	100	62.20	1.00	
	1030	110	62.24	1.04	
	1040	120	62.27	1.07	
	1140	180	62.44	1.24	
	1240	240	62.55	1.35	
	1340	300	62.65	1.45	
	1440	360	62.72	1.52	
	1540	420	62.79	1.59	
	1640	480	62.85	1.65	
	1740	540	62.91	1.71	
	1840	600	62.93	1.73	
	1940	660	62.97	1.77	
	2046	720	63.01	1.81	
	2200	800	63.06	1.86	Pump off

Table 9-4.—Tabulated data from a discharging and
recovering well test - continued

Recovery of observation well

Date	Time	t, min	t', min	t/t'	Depth to water, ft	Residual drawdown, ft
5-19	2200	800	0	0.0	63.06	1.86
	2205	805	5	161.0	62.98	1.78
	2210	810	10	81.0	62.84	1.64
	2215	815	15	54.3	62.73	1.53
	2220	820	20	41.0	62.65	1.45
	2225	825	25	33.3	62.57	1.37
	2230	830	30	27.7	62.52	1.32
	2240	840	40	21.0	62.52	1.22
	2250	850	50	17.0	62.35	1.15
	2300	860	60	14.3	62.29	1.09
	2310	870	70	12.4	62.23	1.03
	2320	880	80	11.0	62.17	0.97
	2330	890	90	9.88	62.14	.94
	2340	900	100	9.00	62.10	.90
	2350	910	110	8.27	62.07	.87
5-20	2400	920	120	7.67	62.05	.85
	0100	980	180	5.44	61.90	.70
	0200	1,040	240	4.33	61.81	.61
	0300	1,100	300	3.67	61.74	.54
	0400	1,160	360	3.22	61.69	.49
	0500	1,220	420	2.90	61.66	.46
	0600	1,280	480	2.67	61.60	.40
	0700	1,340	540	2.48	61.56	.36
	0800	1,400	600	2.33	61.56	.36
	0900	1,460	660	2.21	61.54	.34
	1000	1,520	720	2.11	61.51	.31
	1120	1,600	800	2.00	61.49	.29

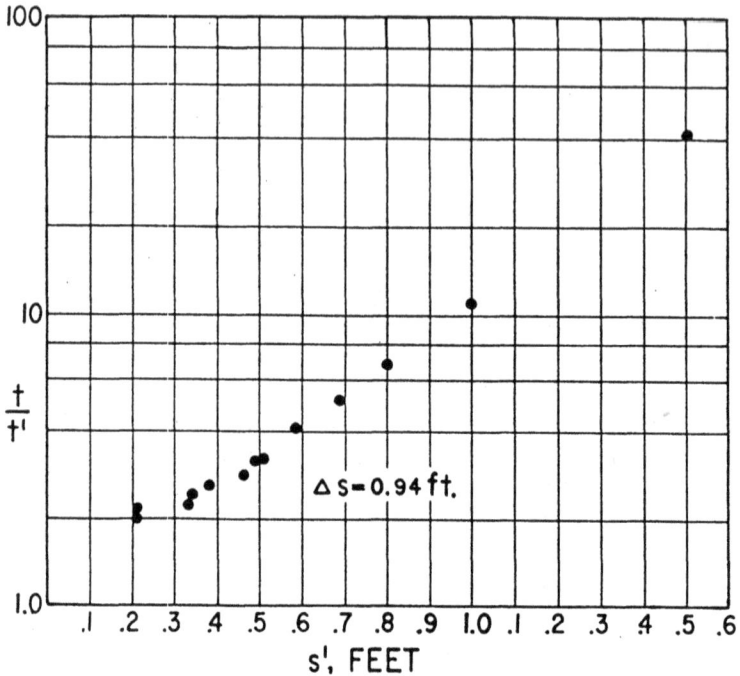

$$T = \frac{2.303Q}{4\pi\Delta s'} = \frac{(2.303)(162.9)}{(4\pi)(0.94)} = 31.7 \text{ ft}^2/\text{min}$$

Figure 9-9.—Recovery solution for
transmissivity in a discharging well.

are known or suspected, the recovery method should be used with
caution because of the difficulty in separating the influence of
boundaries.

The recovery method may be used to analyze the recovery of
observation wells to determine T and S. Values of t/t' are plotted
on semilog paper against residual drawdowns (see figure 9-10 and
table 9-4). The value of T is determined as described for recovery
of a test well. Storativity can be estimated from recovery data in
an observation well by using the equation:

$$S = \frac{2.25Tt'/r^2}{\log^{-1}[S_p-s')/\Delta(s_p-s')]} \qquad \text{9-24}$$

where:

s_p = pumping period drawdown projected to time t'
s' = residual drawdown at time, t'
$(s_p - s')$ = recovery at time, t'

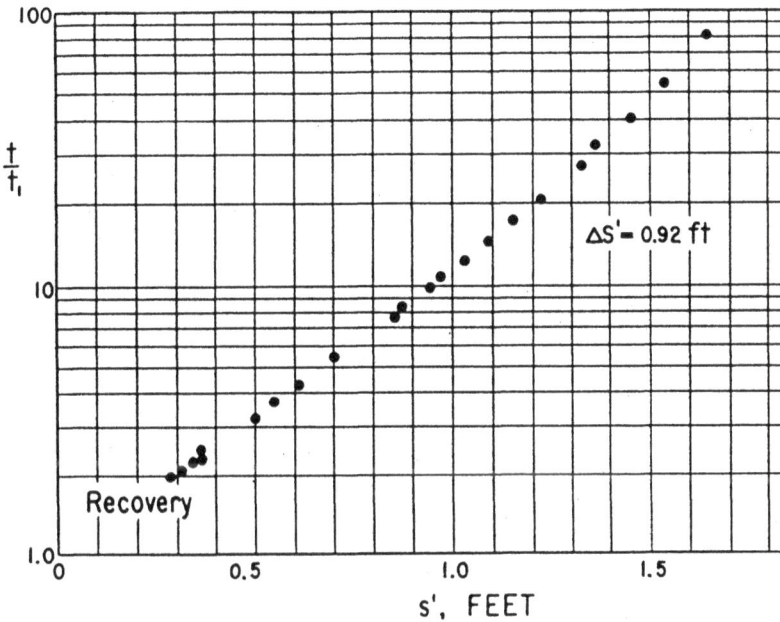

Figure 9-10.—Recovery solution for transmissivity and storativity in an observation well.

$$T = \frac{2.303\ Q}{4\pi\Delta S'} = \frac{(2.303)\ (162.9)}{4\pi\ (0.92)} = 32.1\ \text{ft}^2/\text{min}$$

$$S = \frac{2.25\ (Tt'/r^2)}{\log^{-1}[(S_p-S')/\Delta(S_p-S')]} = \frac{(2.25)(32.1)\left(\frac{600}{10,000}\right)}{\log^{-1}(1.75/0.99)} = 0.07$$

Note: Semilog plots necessary to determine $(S_p-S') = 1.75$ and $\Delta(S_p-S') = 0.99$ not shown.

9-8. Leaky Aquifer Solutions.—Under sufficient head, even apparently impermeable geologic materials will transmit water, and confining layers enclosing artesian aquifers are no exception. Where two or more aquifers are separated by a confining layer, pumping from one aquifer may disturb the mutual hydraulic balance and result in an increase or decrease in leakage between the aquifers. Such leakage is a boundary condition. Theoretically, the area of influence of a discharging well expands until leakage into the aquifer induced by the well equals the well discharge. At this point, the area of influence stabilizes and the drawdown becomes constant with time. Conversely, if the discharge from a well in an aquifer that is losing water by leakage balances the amount of leakage, the area of influence will stabilize.

In 1946, Jacob (1946) published a mathematical solution to the problem involving a single confined aquifer overlain by a leaky confining bed above which was an unconfined aquifer. In later work by Glover, Moody, and Tapp (1954) and Glover (1960) of the Bureau of Reclamation, simplified methods of analysis using a family of type curves were developed. These curves, shown on figure 9-11, should be superimposed on a log-log plot of drawdown versus time from three observation wells (figure 9-12) taken from a tabulation of field measurements and data (table 9-5). Table 9-5 presents only a partial record of the test data. This fitting of curves is similar to the procedure described earlier for using the Theis-type curve. The values of s and t are applied to the equations:

$$u = \frac{\dfrac{s}{Q}}{2\pi KM} \qquad\qquad 9\text{-}25$$

When $u = 1$, equation 9-25 can be written:

$$s = \frac{Q}{2\pi KM}$$

Also, since $T = KM$, equation 9-25 can be written:

$$T = \frac{Qu}{2\pi s}$$

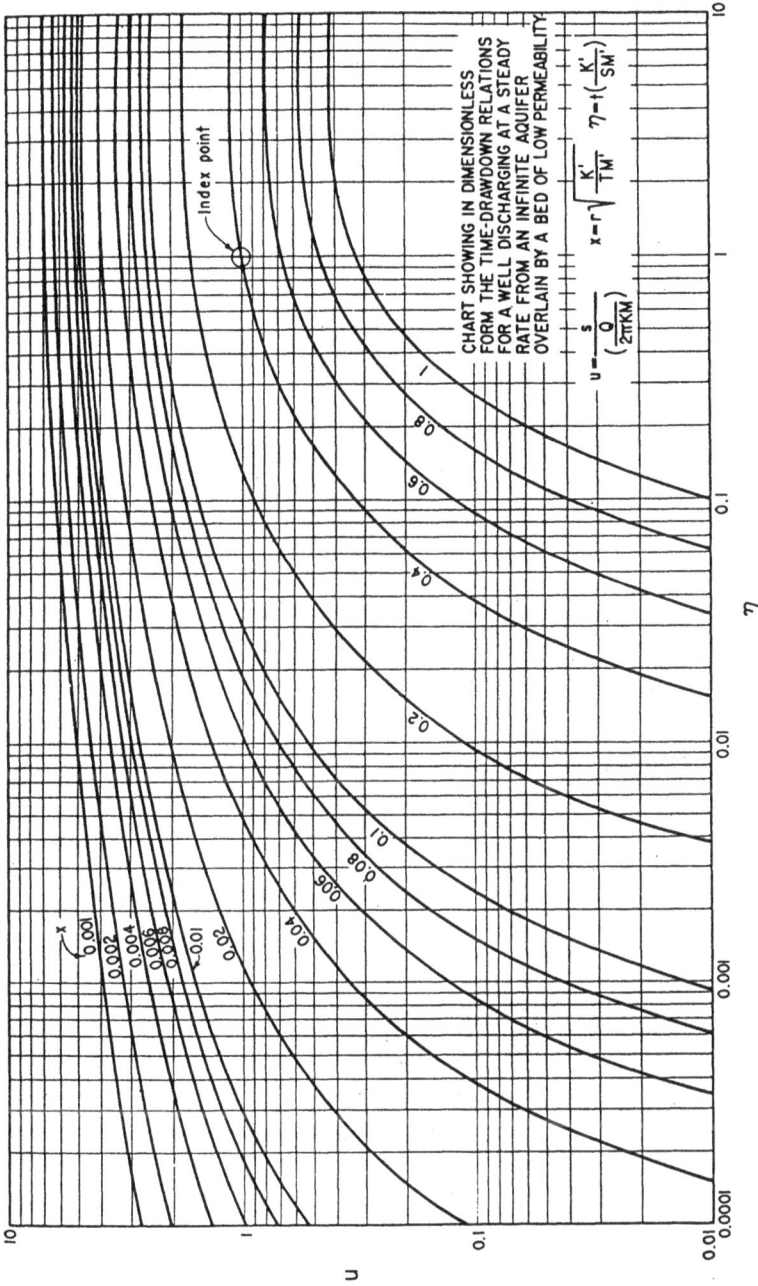

Figure 9-11.—Leaky aquifer type curves.

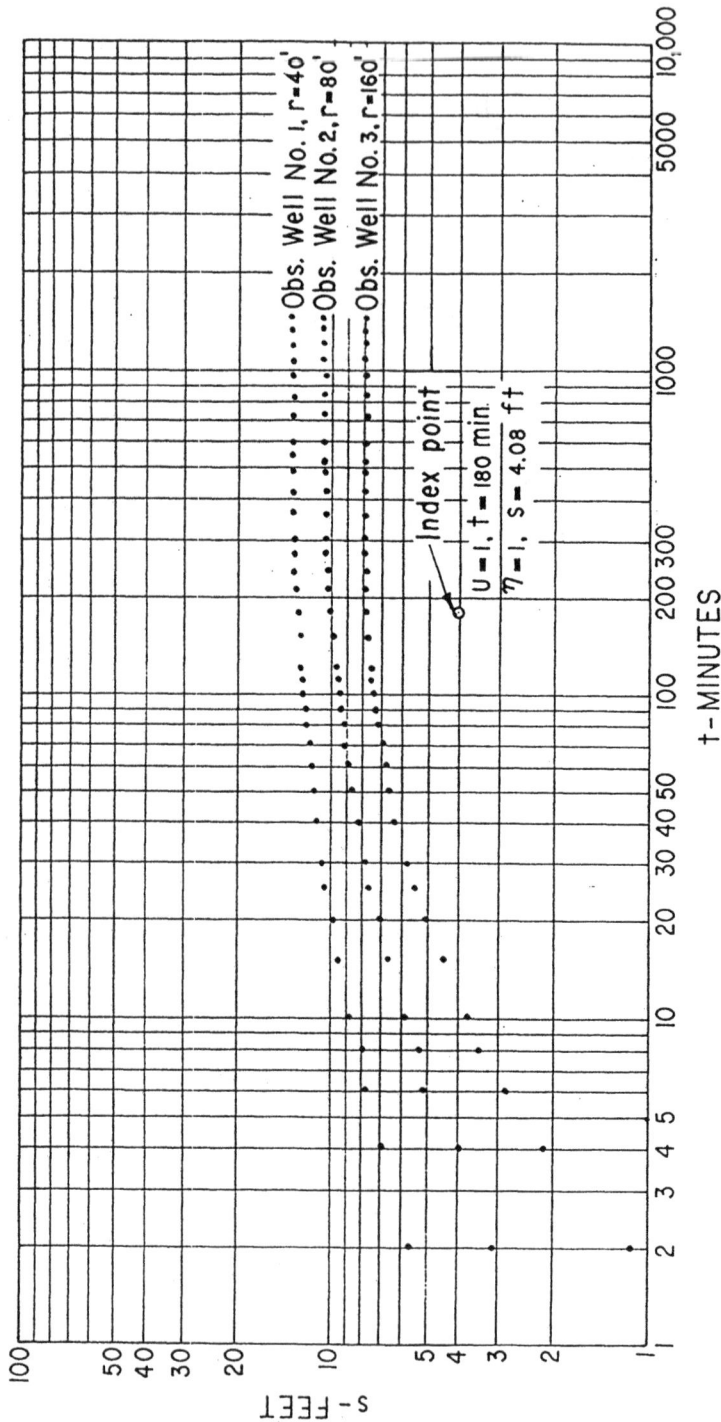

Figure 9-12.—Leaky aquifer type curve solution for transmissivity, storativity, and permeability of the leaky bed.

Table 9-5.—Tabulated discharging well test data for determining characteristics of a leaky aquifer

Project: Texas Hill

Feature: Salt Flat drainage program

Location: 200 feet north SW corner, sec. 10, T. 20 N., R. 5 W.

Drawdown measured by: "Popper"

Distance from pumped well: 40 feet north

Pump test No. 1, observation well No. 1

Date	Time	Elapsed time, min	Depth to water, ft	Drawdown s, ft	Remarks
8-24	1050		26.59	0	
8-25	1055		26.58	0	
8-26	1045		26.59	0	Static water level
	1100	0	26.59	0	Pump on
	1102	2	32.24	5.65	
	1104	4	33.55	6.96	
	1106	6	34.31	7.72	
	1108	8	34.59	8.00	
	1110	10	35.30	8.71	Pump off 8-28 at 1600
	1115	15	36.06	9.47	
	1120	20	36.58	9.99	
	1125	25	36.94	10.35	
	1130	30	37.29	10.70	
	1140	40	37.73	11.14	Average discharge:
	1150	50	38.05	11.46	= 4,488 gal/min
	1200	60	38.21	11.62	= 600 ft³/min
	1210	70	38.45	11.86	
	1220	80	38.61	12.02	
	1230	90	38.85	12.26	
	1240	100	38.92	12.33	
	1250	110	38.96	12.37	
	1300	120	39.00	12.41	
	1330	150	39.28	12.69	
	1400	180	39.44	12.85	
	1430	210	39.68	13.09	
	1500	240	39.72	13.13	
	1530	270	39.84	13.25	
	1600	300	39.92	13.33	
	1700	360	39.96	13.37	
	1800	420	40.00	13.41	

Table 9-5.—Tabulated discharging well test data for determining
characteristics of a leaky aquifer - continued

Project: Texas Hill   Drawdown measured by: "Popper"
Feature: Salt Flat drainage program   Distance from pumped well: 80 feet north
Location: 240 feet north SW corner, sec. 10,
T. 20 N., R. 5 W.

Pump test No. 1, observation well No. 2

Date	Time	Elapsed time, min	Depth to water, ft	Drawdown s, ft	Remarks
8-24	1045		26.54	0	
8-25	1050		26.54	0	
8-26	1040		26.54	0	Static water level
	1100	0	26.54	0	Pump on
	1102	2	29.64	3.10	
	1104	4	30.56	4.02	
	1106	6	31.59	5.05	
	1108	8	31.83	5.29	
	1110	10	32.51	5.97	Pump off 8-28 at 1600
	1115	15	33.26	6.72	
	1120	20	33.70	7.16	
	1125	25	34.14	7.60	
	1130	30	34.50	7.96	
	1140	40	34.90	8.36	Average discharge:
	1150	50	35.17	8.63	= 4,488 gal/min
	1200	60	35.45	8.91	= 600 ft³/min
	1210	70	35.73	9.19	
	1220	80	35.85	9.31	
	1230	90	36.01	9.47	
	1240	100	36.09	9.55	
	1250	110	36.17	9.63	
	1300	120	36.29	9.75	
	1330	150	36.49	9.95	
	1400	180	36.61	10.07	
	1430	210	36.73	10.19	
	1500	240	36.81	10.27	
	1530	270	36.89	10.35	
	1600	300	36.93	10.39	
	1700	360	36.93	10.34	
	1800	420	36.96	10.42	

Table 9-5.—Tabulated discharging well test data for determining
characteristics of a leaky aquifer - continued

Project: Texas Hill
Feature: Salt Flat drainage program
Location: 320 feet north SW corner, sec. 10,
  T. 20 N., R. 5 W.

Drawdown measured by: "Popper"
Distance from pumped well: 160 feet
  north

Pump test No. 1, observation well No. 3

Date	Time	Elapsed time, min	Depth to water, ft	Drawdown s, ft	Remarks
8-24	1040		26.60	0	
8-25	1045		26.61	0	
8-26	1035		26.60	0	
	1100	0	26.60	0	Pump on
	1102	2	27.71	1.11	
	1104	4	28.75	2.15	
	1106	6	29.46	2.86	
	1108	8	30.06	3.46	
	1110	10	30.38	3.78	Pump off 8-28 at 1600
	1115	15	31.18	4.58	
	1120	20	31.69	5.09	
	1125	25	32.09	5.49	
	1130	30	32.45	5.85	
	1140	40	32.97	6.37	Average discharge:
	1150	50	32.24	6.64	= 4,488 gal/min
	1200	60	33.40	6.80	= 600 ft³/min
	1210	70	33.56	6.96	
	1220	80	33.76	7.16	
	1230	90	33.96	7.36	
	1240	100	34.04	7.44	
	1250	110	34.12	7.52	
	1300	120	34.16	7.56	
	1330	150	34.24	7.64	
	1400	180	34.48	7.88	
	1430	210	34.52	7.92	
	1500	240	34.56	7.96	
	1530	270	34.56	7.96	
	1600	300	34.56	7.96	
	1700	360	34.55	7.95	
	1800	420	34.56	7.96	

Similarly,

$$\eta = t \left(\frac{K'}{SM'} \right)$$

9-26

When $\eta = 1$, equation 9-26 can be written:

$$t = \frac{SM'}{K'}$$

The terms of the above equations are as defined on figure 9-13.

Each of the family of curves has an x value as noted on figure 9-11. When the closest fit is found between the data curve and the type curve, and the values of s and t have been obtained, the x value of the type curve to which the fit is made is noted. If the fit falls between curves, the x value is interpolated. The value x is related to r, K″, T, and M′ by:

$$x = r \sqrt{\frac{K'}{TM'}}$$

9-27

or

$$\frac{K'}{M'} = T \left(\frac{x}{r} \right)^2$$

Since T, x, and r are known, the value of the leakage factor K'/M' can be calculated from equation 9-27, and then S is calculated by solving equation 9-26 when $\eta=1$. If M' is approximately known, K' can be calculated.

Using data from table 9-5 and figures 9-11, 9-12, and 9-13, some sample calculations for material in this section could be:

Q = pump discharge = 600 ft³/min or 10 ft³/s
x_1 = x value for type curve which fits the plotted s-t relation for observation well No. 1 = 0.04
x_2 = x value for well No. 2 = 0.08
x_3 = x value for well No. 3 = 0.16
M' = 6.1 meters (20 feet)

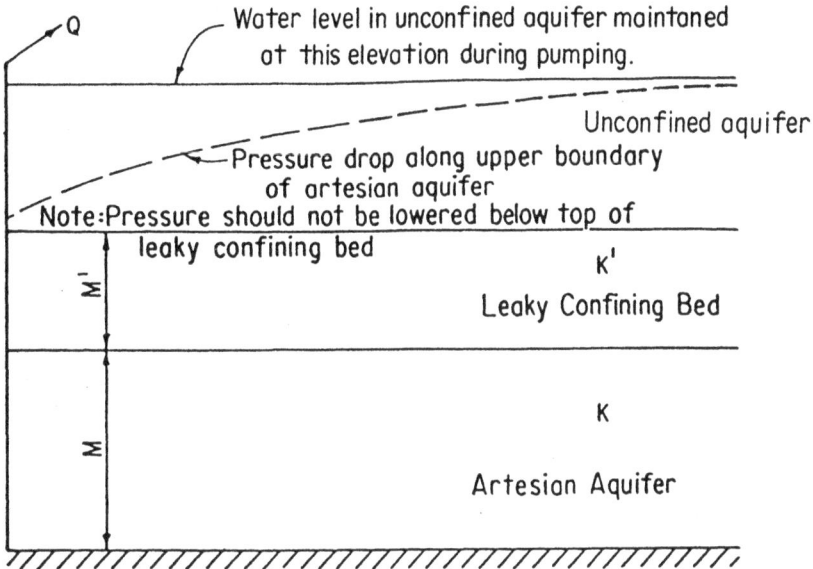

Q

Water level in unconfined aquifer maintoned at this elevation during pumping.

Unconfined aquifer

Pressure drop along upper boundary of artesian aquifer

Note: Pressure should not be lowered below top of leaky confining bed

K'

Leaky Confining Bed

M'

K

M

Artesian Aquifer

K — Hydraulic conductivity of ortesion oquifer.----------L/t

M — Thickness of artesian oquifer---------------------L

T — KM Transmissivity of artesian aquifer------------L^2/t

K' — Vertical permeability of leaky confining bed -------L/t

M' — Thickness of leaky confining bed------------------L

$\dfrac{K'}{M'}$ — Leakance---1/t

Q — Discharge from well-----------------------------L^3/t

S — Storativity of artesian aquifer, dimensionless

Figure 9-13.—Leaky aquifer definitions.

At the index point shown on figure 9-12:

$u = 1$, $\eta = 1$, $t = 180$ min, and $s = 1.24$ meters (4.08 feet)

$$T = \frac{Qu}{2\pi s} = \frac{600(1)}{2\pi(4.08)} = 23.4 \; ft^2/min$$

$$x_1 = r_1\sqrt{\frac{K'}{TM'}} \quad or \quad \frac{K'}{M'} = \frac{Tx_1^2}{r_1^2}$$

$$\frac{K'}{M'} = \frac{(23.4)(0.04)^2}{(40)^2} = 2.34 \times 10^{-5} \; min^{-1}$$

$$S = \frac{tK'}{M'} = 180 \; (2.34 \times 10{-5}) = 0.004$$

If the aquifer is overlain and underlain by other aquifers from which it is separated by confining layers, the assumed M' value may be erroneous, and it may be impossible to separate the contribution from each aquifer. Then, the leakage factor computed is a combined value of both layers.

In unconfined aquifers, delayed drainage may cause the plot of s versus t on log-log paper to appear similar to data plotted from a leaky aquifer. In extreme cases of delayed drainage, several weeks of pumping may be required to differentiate between the two conditions. Such a lengthy testing period generally cannot be justified from an economic standpoint, so analysis must be based on judgment of short-term test results, known and inferred aquifer conditions, and knowledge of well construction details (Glover, 1964; Glover, Moody, and Tapp, 1954; Hantush, 1956, 1960, 1961, 1961, and 1959; Hantush and Jacob, 1955; Walton, 1970).

Additionally, the presence of the lateral boundaries may further influence the drawdown curves so that a reliable interpretation of aquifer conditions and calculation of aquifer characteristics are virtually impossible.

9-9. Constant Drawdown Solutions.—Aquifer test procedures discussed thus far have relied on the use of constant discharge and variable drawdown of a well. Under some conditions, such as with a flowing artesian well, it is simpler to test the well by permitting it to discharge at a variable rate but with constant drawdown. This can be done by shutting off discharge until the pressure inside the well stabilizes and then permitting resumption of discharge. During this discharge period, the rate of discharge is recorded periodically.

Jacob and Lohman (1952) derived equations for tests of this type, including a straight-line approximation solution to determine T and S:

$$T = \frac{2.303}{\cfrac{4\pi\Delta\left(\dfrac{S_w}{Q}\right)}{\Delta\,\log\left(\dfrac{t}{r_w^2}\right)}} \qquad \text{9-28}$$

where:

T	= transmissivity, L^2/t
s_w	= constant drawdown, L, (the difference in meters (feet) of water between the static head and the top of the casing or center of the discharge valve)
Q	= weighted average discharge during a timed interval, L^3/t
t	= elapsed time since start of test, t
r_w	= radius of the well, L
$\Delta\,(S_w/Q)$	= change in the ratio S_w/Q over a time period, $1/L^2$
$\Delta\,\log\,(t/r_w^2)$	= change in base 10 logarithm over a time period

Table 9-6 shows the tabulated data from a constant drawdown test and the straight-line method of solution. In this table, column (1) shows the times at which measurements were made, column (2) shows the average rate of discharge during the time interval shown in column (3), column (4) shows the total discharge during the time interval, column (5) shows the elapsed time since start of discharge, column (6) is the constant drawdown divided by rate of discharge in column, and column (7) is time shown in column (5) divided by the square of the well radius.

In the straight-line method of solution, the values for S_w/Q are plotted on an arithmetic scale against values of t/r_w^2 for the corresponding time on a log scale. A straight line is then fitted through the points as shown on figure 9-14.

Table 9-6.—Constant drawdown test data
from Lohman (1965)

Field data for flow test on Artesia Heights well near
Grand Junction, Colorado, September 22, 1948
(Valve opened at 10:29 a.m. S_w = 92.33 ft; r_w = 0.276 ft)
Data from Lohman (1965, tables 6 and 7, well 28)

Time of observation	Rate of flow, gal/min	Flow interval, min	Total flow during interval, gal	Time since flow started, min	$\dfrac{S_w}{Q}$ ft min/gal	$\dfrac{t}{r_w^2}$ min/ft²
10:30 a.m.	7.28	1	7.28	1	12.7	13.1
10:31 a.m.	6.94	1	6.94	2	13.3	26.3
10:32 a.m.	6.88	1	6.88	3	13.4	39.4
10:33 a.m.	6.28	1	6.28	4	14.7	52.6
10:34 a.m.	6.22	1	6.22	5	14.8	65.7
10:35 a.m.	6.22	1	6.22	6	15.1	78.8
10:37 a.m.	5.95	2	11.90	8	15.5	105
10:40 a.m.	5.85	3	17.55	11	15.8	145
10:45 a.m.	5.66	5	28.30	16	16.3	210
10:50 a.m.	5.50	5	27.50	21	16.8	276
10:55 a.m.	5.34	5	26.70	26	17.3	342
11:00 a.m.	5.34	5	26.70	31	17.3	407
11:10 a.m.	5.22	10.5	54.81	41.5	17.7	345
11:20 a.m.	5.14	9.5	48.83	51	18.0	670
11:30 a.m.	5.11	10	51.10	61	18.1	802
11:45 a.m.	5.05	15	75.75	76	18.3	999
12:00 p.m.	5.00	15	75.00	91	18.5	1,190
12:12 p.m.	4.92	12	59.04	103	18.8	1,354
12:22 p.m.	4.88	10	53.68	113	18.9	1,485
Total[1]		114	596.98			

[1] 596.98 gal/114 min = 5.23 gal/min, weighted average discharge.

The change in value of S_w/Q is determined over one log cycle of t/r_w^2 and is set equal to $\Delta\,(S_w/Q)$. In the example on figure 9-14, $\Delta\,S_w/Q$ = 18.40 - 15.38 = 3.02 feet per minute per gallon. Measurements must be converted to consistent units and inserted into equation 9-28 to give T in square feet per minute as follows:

$$T = \frac{2.303}{4\pi\Delta\left(\dfrac{S_w}{Q}\right) / \Delta\log\left(\dfrac{t}{r_w^2}\right)} = \frac{2.303}{\dfrac{4\pi(22.59)}{1}} = 0.008 ft^2/min$$

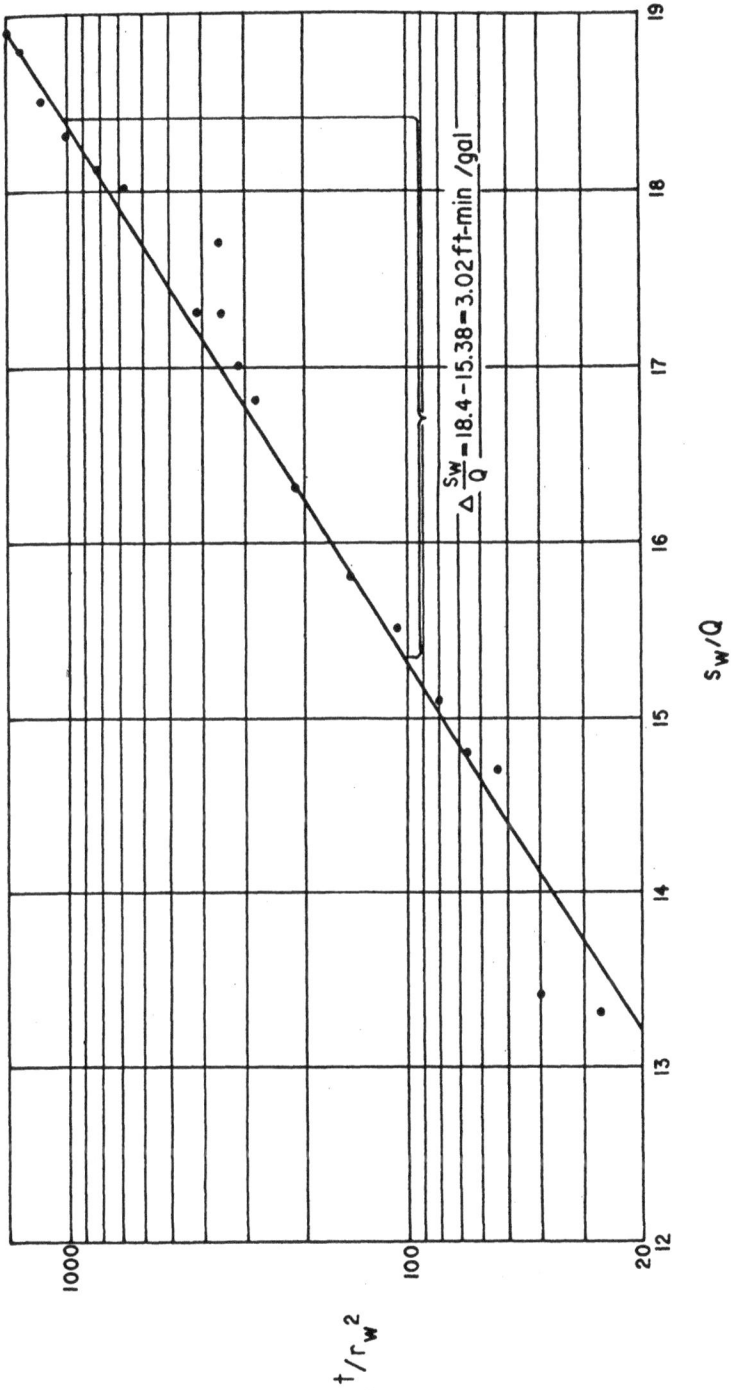

$$\Delta \frac{s_w}{Q} = 18.4 - 15.38 = 3.02 \, \text{ft-min /gal}$$

Figure 9-14.—Constant drawdown solution for transmissivity and storativity.

If r_w is known, the straight line on the plot can be extended to intercept S_w/Q, and S can be computed from the equation:

$$S = 2.25T\left(\frac{t}{r_w^2}\right)$$

Transmissivity, T, can be roughly checked by recovery measurements made after the discharge has been stopped. Equations 9-22 and 9-23 and the weighted average discharge per period of discharge are used with the recovery data.

9-10. Delayed Drainage Solutions.—The early response of an unconfined aquifer to a discharging well is especially dependent on the degree of isotropy present. Tests of short duration in such aquifers may be unreliable because of the delayed drainage effect produced by anisotropy. The usual plots of log t versus log s may show a steep initial slope, then a flat segment followed by another slope. The general shape of the curve is that of an elongated letter S (see table 9-7 and figure 9-15). The early part of the curve may be influenced by several or all of the following factors:

- Changing storativity caused by delayed drainage

- Expansion of water below the water table resulting from reduction in pressure

- Vertical flow components

- Thinning of the saturated zone as drawdown increases

- Observation well lag

- Aquifer heterogeneity

Where these factors are present, a log-log plot of s versus t from an observation well appears similar to plots of test data from a leaky aquifer except that the drawdown seldom stabilizes with time, as may be the case with the leaky aquifer.

The early part of a data curve, influenced by delayed drainage, may permit a good fit to the Theis type curve and give a reasonably reliable value for T, but S may be so small as to be in the artesian range. As pumping continues, the area of influence and the drawdown decrease in growth at a logarithmic rate, and the delayed drainage tends to catch up with these portions of the

Table 9-7.—Adjusted field data for delayed yield analysis on Fairborn, Ohio, well from Lohman (1972)

Time since pumping began t, min	Corrected drawdowns s, ft	Time since pumping t, min	Corrected drawdowns s, ft	Time since pumping t, min	Corrected drawdowns s, ft
0.165	0.12	2.65	0.92	80	1.28
.25	.195	2.80	.93	90	1.29
.34	.255	3.00	.94	100	1.31
.42	.33	3.50	.95	120	1.36
.50	.39	4.00	.97	150	1.45
.58	.43	4.50	.975	200	1.52
.66	.49	5.00	.98	250	1.59
.75	.53	6.00	.99	300	1.65
.83	.57	7.00	1.00	350	1.70
.92	.61	8.00	1.01	400	1.75
1.00	.64	9.00	1.015	500	1.85
1.08	.67	10.00	1.02	600	1.95
1.16	.70	12.00	1.03	700	2.01
1.24	.72	15.00	1.04	800	2.09
1.33	.74	18.00	1.05	900	2.15
1.42	.76	20.00	1.06	1,000	2.20
1.50	.78	25.00	1.08	1,200	2.27
1.68	.82	30.00	1.13	1,500	2.35
1.85	.84	35.00	1.15	2,000	2.49
2.00	.86	40.00	1.17	2,500	2.59
2.15	.87	50.00	1.19	3,000	2.66
2.35	.90	60.00	1.22		
2.50	.91	70.00	1.25		

response, resulting in the relatively flat portion of the data curve. Eventually, the various factors reach a balance and the final slope develops from which reliable values of T and S may be computed using the Theis nonequilibrium solution. This solution, however, may require pumping for an excessively long period and involves a tedious trial-and-error process.

Boulton (1955, 1964) developed equations for treating delayed drainage influence; Prickett (1965) and Stallman (1963) improved on Boulton's work to make it more amenable to practical application. Neuman (1975) developed type curves for delayed drainage. The abbreviated form of Boulton's equations are:

$$s = \frac{Q}{4\pi T} W\left(u_{\alpha y}\frac{r}{M}\right) \qquad \text{9-29}$$

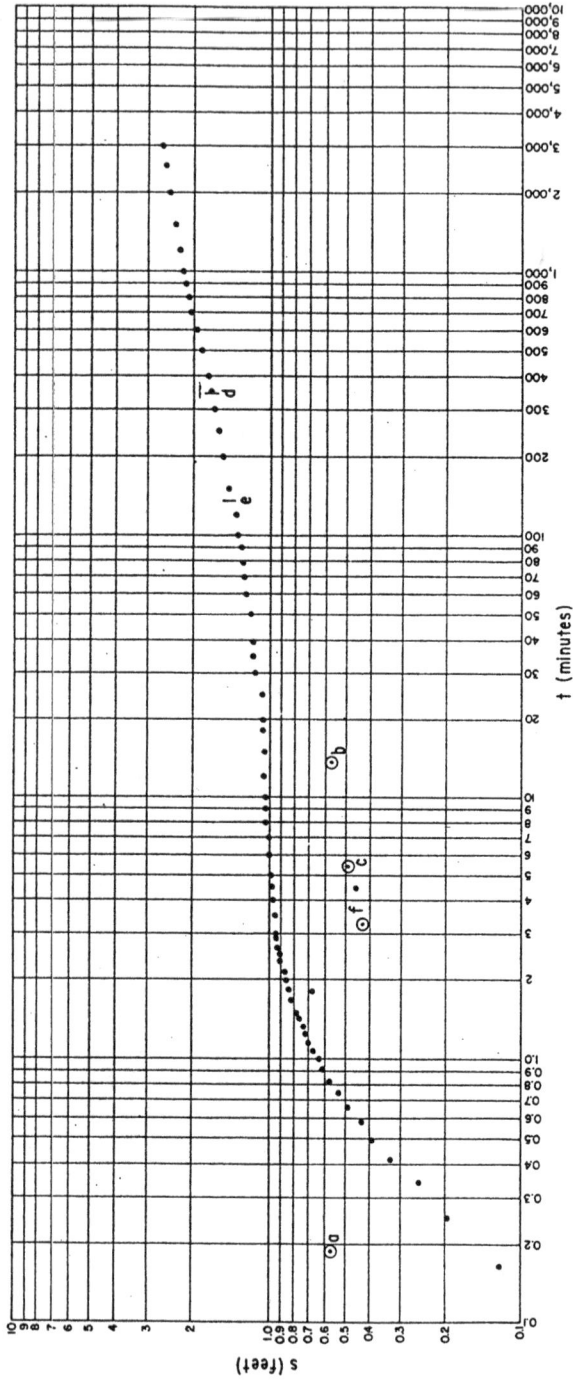

Figure 9-15.—Delayed drainage type curve solution for transmissivity and storativity (from table 9-7).

$$u_\alpha = \frac{r^2 S}{4Tt}$$

9-30

$$uy = \frac{r^2 S_y}{4Tt}$$

9-31

$$d = \frac{\left(\dfrac{r}{M}\right)^2 \dfrac{1}{u\alpha}}{4t} \quad type\ A\ curve$$

9-32

$$d = \frac{\left(\dfrac{r}{M}\right)^2 \dfrac{1}{uy}}{4t} \quad tupe\ B\ curve$$

9-33

where in consistent units:

s	= drawdown at time t and at a distance r, L
Q	= rate of discharge, L^3/t
T	= transmissivity, L^2/t
$W\ (u_{\alpha y}\ r/M)$	= well function of u when η tends to infinity
t	= time since start of pumping, t
r	= distance of observation well from test well, L
S	= early time coefficient of storage (nondimensional)
S_y	= true specific yield or coefficient of storage (nondimensional)
d	= reciprocal of the delay index, t
$\eta = (S+S_y/S)$	= (nondimensional)
M	= aquifer thickness, L

On figure 9-16, the type curve consists of two families of curves. Type A curves are shown to the left of the r/M values, and type Y curves are shown to the right of the r/M values.

The type A curves are applicable to early time pumping and type Y curves to later time pumping when response of the aquifer is in accord with the Theis nonequilibrium assumptions. Values for $1/u_\alpha$ are shown at the top of the type curves, and $1/u_y$ values are at the bottom.

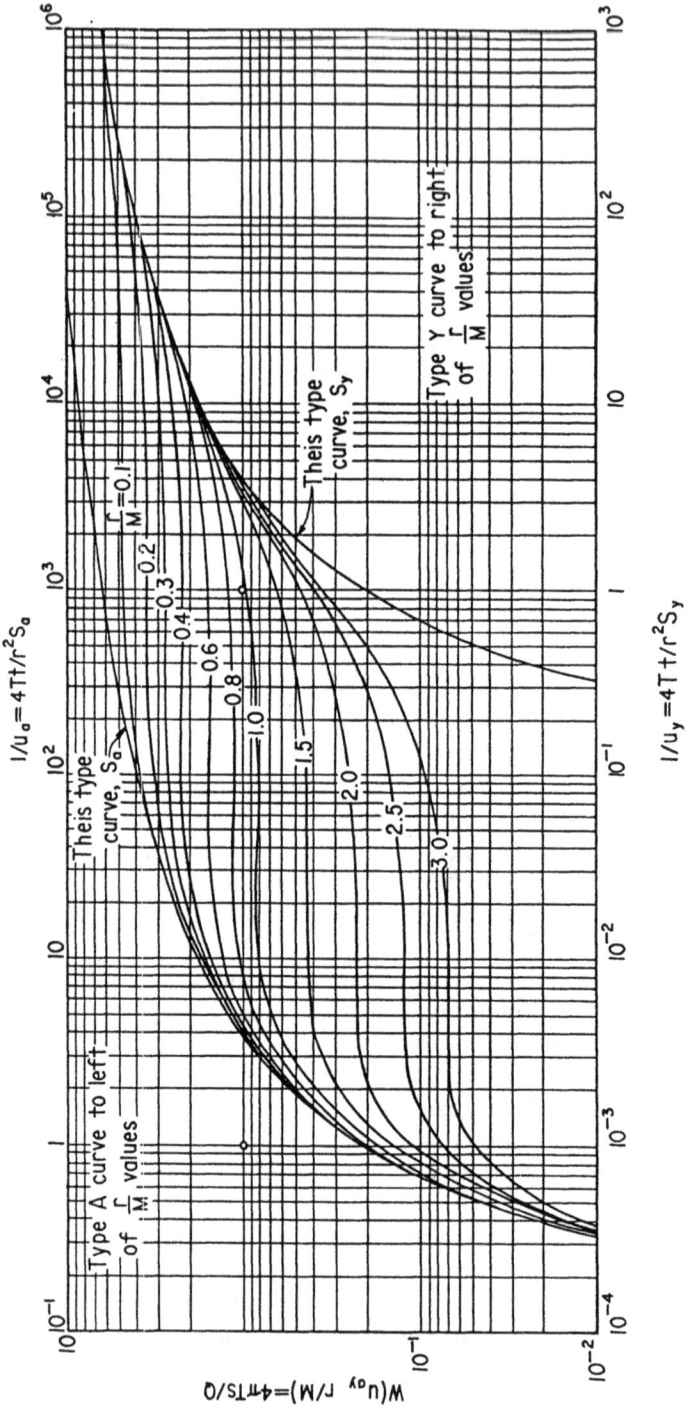

Figure 9-16.—Delayed drainage type curves.

The method of using the type curves for finding values of T, S, S_y, η, and d, and the delay index (i.e., the reciprocal of d, will now be discussed).

The field data s and t (table 9-7) are plotted on log-log paper (figure 9-15) to the same scale as the type curves. This time drawdown curve is analogous to two of the families of unconfined aquifer-type curves shown on figure 9-16. The field data curve is superimposed on the type curves, while keeping the axes of both sheets parallel, and moved horizontally and vertically until the best possible fit is found for type A curves. If necessary, interpolate a position and r/M value between the type curves. The r/M value of the match-type curve is noted. In this match position, a point on the intersection of the major axes of the type curve is selected and the corresponding point on the data curve is marked. The point selected may be anywhere on the type curve provided it overlies part of the data curve. The coordinates of the common match points are s, $1/u_\alpha$, $W\ (u_{\alpha y}\,r/M)$, and t. These values are substituted in equation 9-29 to determine T. The value for S is then calculated using the determined value of T and the values of $1/u_\alpha$ and t from the match points using equation 9-30.

The curves are then moved horizontally with respect to each other, and as much as possible of the late field data curve is matched to the type Y curve. The type Y curve should have the same r/M value as was used when matching to the type A curve. A similar match of the field data curve to the type curve is made. The coordinates of the match point s, $1/u_{\alpha y}$, $W\ (u_{\alpha y}\,r/M)$, and t are substituted in equations 9-29 and 9-31 to determine T and S_y. The value of T is calculated using equation 9-29 and coordinates $W\ (u_{\alpha y}\ r/M)$, and s. This T value should be similar to that obtained in matching the early time data. The value of S_y is determined using equation 9-31, the calculated value of T, and coordinates $1/u_v$ and t from the later time match.

The reciprocal of the delay index, d, is calculated by substituting the r/M value with the later time, match point coordinates $1/u\alpha$ and t in equation 9-32.

$$\text{If } \eta \ = \ \frac{S+S_v}{S})6.5,$$

Values of the aquifer characteristics determined are probably of acceptable reliability whether Jacob's corrections for drawdown and partial penetration have or have not been made. If $\eta < 6.5$ and corrections have not been made, they should be computed, a new data curve prepared, and the aquifer characteristics recalculated. If the recalculation results in η>6.5, the new values are probably sufficiently reliable for most applications. However, if η still remains less than 6.5, Boulton's equation (equation 6) (in Prickett, 1965) should be used. As an example of a delayed drainage solution, figure 9-15 was matched to the A-type curve on figure 9-16. Where:

$$\frac{1}{u_\alpha} = 1 \quad and \quad W\left(u_{\alpha y}\frac{r}{M}\right) = 1$$

on the type curve, $s = 0.56$ foot and $t = 0.18$ minute (point a on figure 9-15). In addition, the test data showed $Q = 144.4$ ft^3/min and $r = 73$ feet. Inserting the appropriate values in equation 9-29:

$$0.56 = \frac{144.4X1}{12.57XT} \quad and \quad T = 20.5 \ ft^2/min$$

Referring to equation 9-30

$$u_\alpha = \frac{r^2S}{4Tt}$$

when:

$u_\alpha = 1$, $r = 73ft$, $t = 0.18$ min, and $T = 20.5$ ft^2/min, $then$:

$$S = \frac{4X20.5X0.18}{73^2X1} = 0.003$$

Moving the data curve to the right on the type curve to the best latetime match where $s = 0.56$ foot (point b on figure 9-15) gives a value of 13.8 min for:

$$W\left(u_{\alpha y}\frac{r}{M}\right) = 1 \quad and \frac{1}{u_y} = 1$$

on the type curve. Inserting the appropriate values in equation (29) does not change the value of T, but using equation 9-31:

$$S_y = \frac{4X20.5X13.8}{73^2X1} = 0.21$$

9-11. Determination of Aquifer Boundaries.—The equilibrium equation is based on the concept of an aquifer of infinite areal extent, which obviously does not exist. Finite boundaries (section 2-9) in one or more directions complicate application of the equation. Suitably located image wells serve to simulate hydraulically the flow regime caused by such boundaries and may permit the hydraulic system to be analyzed as being in an aquifer of infinite areal extent.

Although most boundaries are not abrupt nor do they follow a straight line, it is usually possible to treat them as abrupt changes along a straight line. The more distant the boundary from the discharging well, the less the magnitude of influence will be on the drawdown and the longer the time will be before drawdown is influenced. Figures 9-17 and 9-18 show boundary relationships to image wells.

Figure 9-18 illustrates a recharge boundary where an aquifer is bounded on one side by a fully penetrating stream. As the cone of depression expands and eventually encounters the boundary, both the rate of expansion of the cone and the rate of drawdown in the discharging well are slowed. With continuation of pumping, the cone of depression expands along the stream until the well receives a major contribution from the stream.

To analyze the influence of a recharge boundary, a recharge image well is hypothetically located on the opposite side of the boundary from the real discharging well and at an equal distance from the boundary. The image well recharges water to the aquifer at the same rate as the real well discharges. Consequently, the water-level buildup by the image well cancels the drawdown at the boundary resulting from the real well. This satisfies the limits of the problem.

Under actual field conditions, the stream channel is seldom fully penetrating and hydraulic continuity between the stream and aquifer may be restricted by partial plugging of the streambed by fine materials. However, such deviations from ideal conditions merely tend to cause a shifting of the boundary farther from the discharging well.

Similarly, in the case of an impermeable boundary, figure 9-17 shows a fault which has brought impermeable material in contact with an aquifer. In such a situation, there is no flow across the boundary. A discharging image well is hypothetically located an equal distance on the opposite side of the boundary from the real discharging well. The image well is assumed to discharge at the

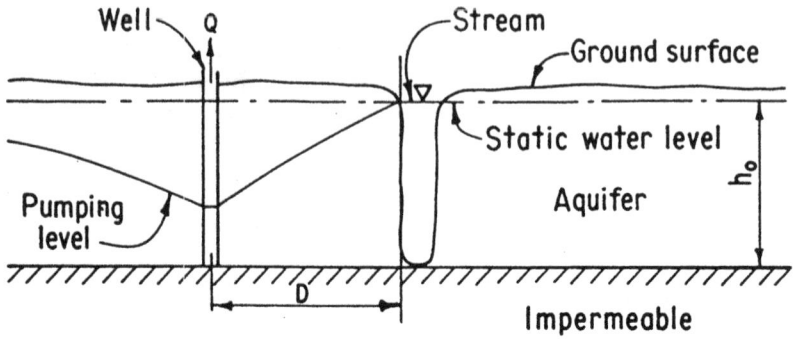

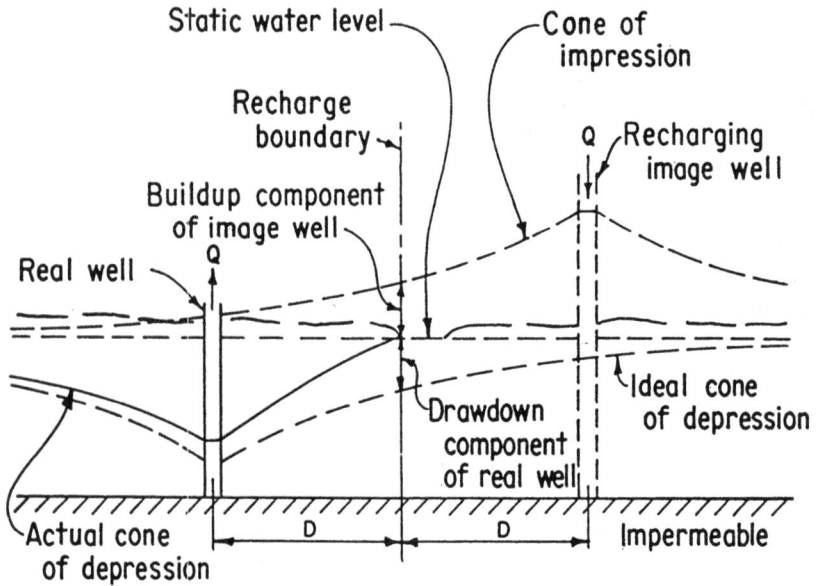

(after U.S. Geological Survey)

Figure 9-17.—Relationship of an impermeable
boundary and an image well.

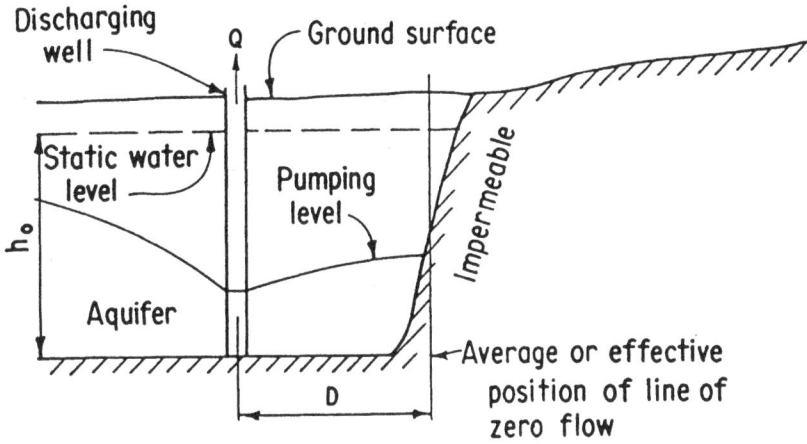

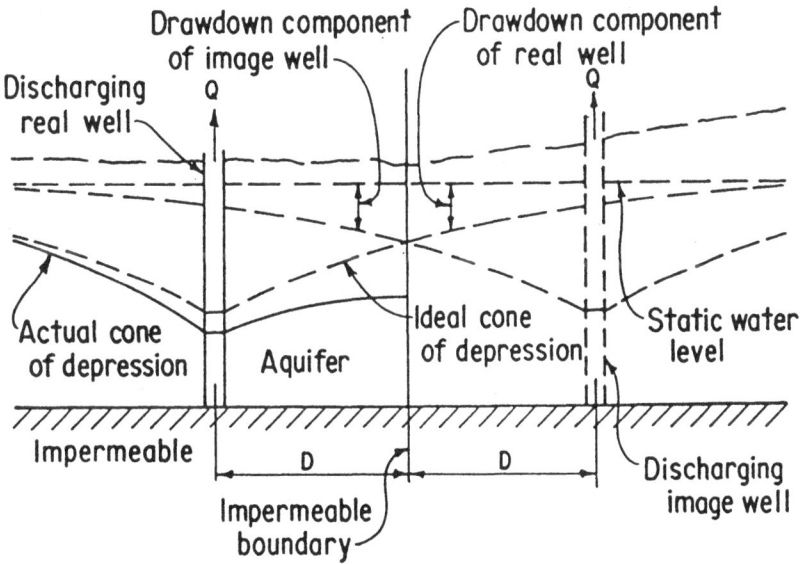

(After U.S. Geological Survey)

Figure 9-18.—Relationship of a recharge
boundary and an image well.

same rate as the real well. Theoretically, a ground-water divide forms along the line of the boundary as the real and image well cones of depression intercept. This satisfies the limits of the problem.

To solve problems involving image wells, the principle of superposition as described in section 5-20 is used. This method provides that when two or more cones of depression interfere or overlap, the resultant drawdown can be determined by adding algebraically the individual drawdowns involved at the point of intersect. Figures 9-17 and 9-18 show the theoretical, individual drawdowns of the real and image wells and the resultant or actual drawdowns.

In many instances, sufficient geologic and hydrologic information is available to permit anticipation of the direction and distance to boundaries from a well. Where such information is not available, the influence of the well must be analyzed. The resultant cones of depression have typical shapes, depending on the type of boundary, which may aid in analyzing the boundaries. Analysis by either the Theis type curve solution (section 9-4) or Jacob's approximation (section 9-5) will provide recognition criteria and estimated distance and direction of boundaries.

The approximation method is advantageous for most analyses because changes in slope of a straight line are usually easier to recognize than in a log-log curve. Some data curves may present problems in differentiating between positive boundary, leaky aquifer, or delayed drainage conditions (Ferris, 1948; Ferris, Knowles, Brown, and Stallman, 1962; and Walton, 1970). Lohman (1972) presents a suite of curves for recharging and discharging wells which permit calculation of transmissivity or storativity using log-log plots (see figure 9-19).

(a) Boundary Location by Type Curve Analyses.—Figure 9-19 shows the relationship between a type curve and actual data from observation wells for a recharge boundary; figure 9-20 shows similar relationships for an impermeable boundary. These figures were plotted from tabulated data that are partially reproduced in tables 9-8 and 9-9. Figure 9-19 shows s versus t plotted on log-log paper, whereas figure 9-20 shows s versus 1/t plotted on log-log paper to show the two different methods plotted. Either method can be used.

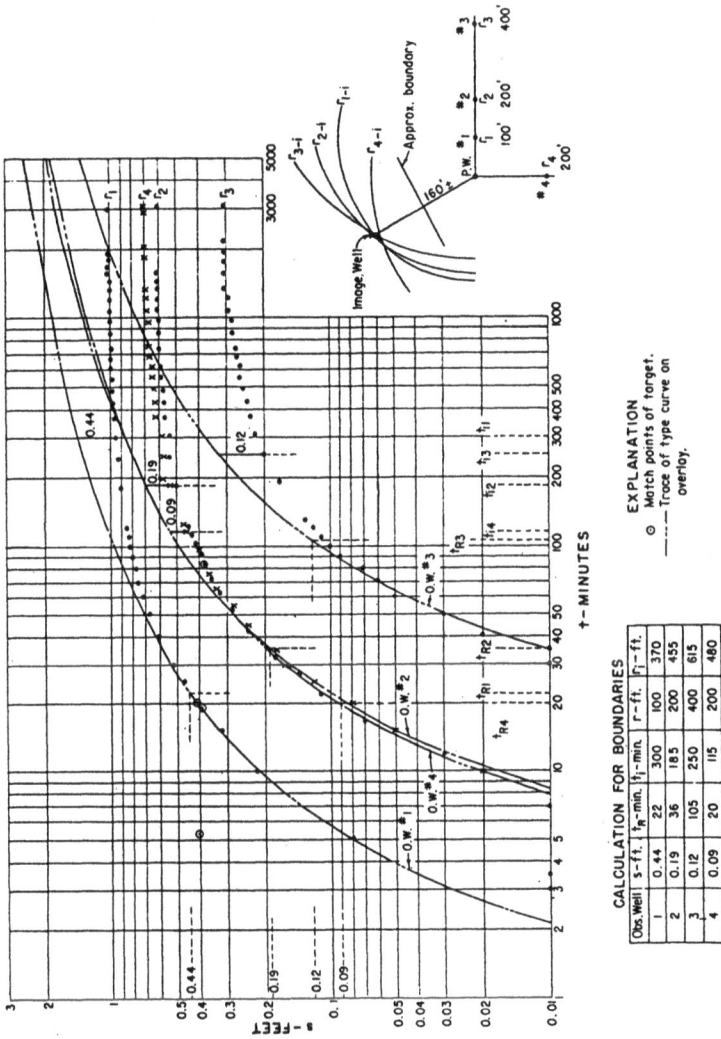

Figure 9-19.—Recharge boundary location by type curve analysis.

CALCULATION FOR BOUNDARIES

Obs.Well	r-ft.	s-ft.	W_ti-min	t_i-min	V_Rx-min	t_x-min	r_i-ft
1	100	0.41	.00385	260	0.05	20	360
2	200	0.40	.0028	357	0.014	72	445
3	400	0.10	.0042	238	0.0120	84	673
4	200	0.25	.0037	270	0.022	45	490

EXPLANATION
⊙ Match points of target.
---- Trace of type curve on overlay.

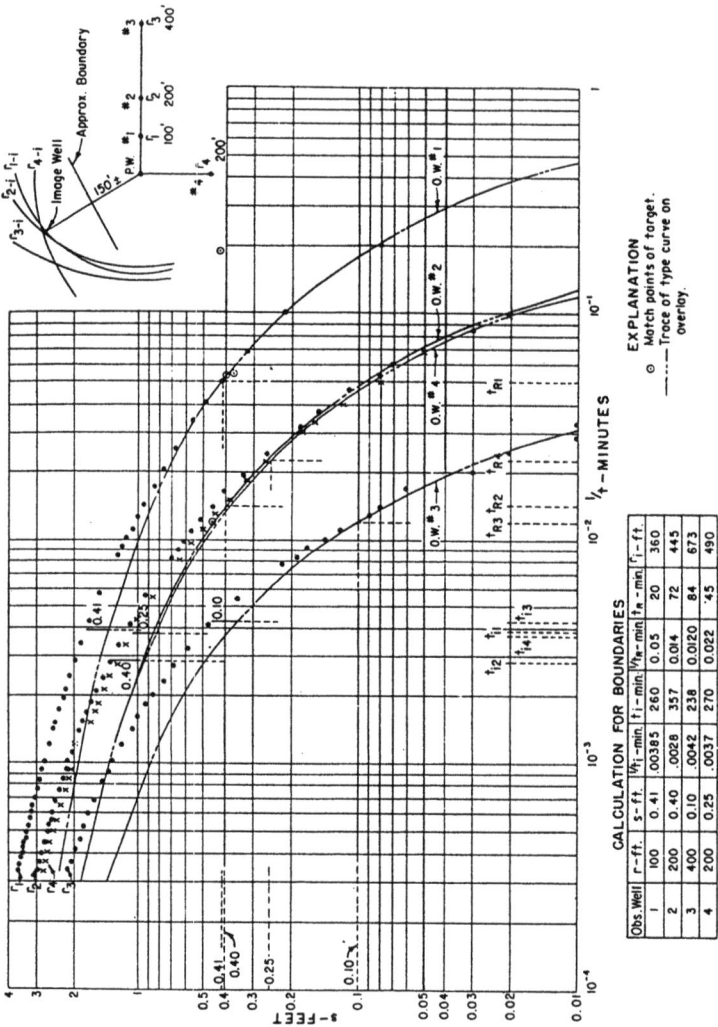

Figure 9-20.—Impermeable boundary location by type curve analysis.

Table 9-8.—Tabulated discharging well test data for determining a recharge boundary

Project: Tongue Valley
Feature: Drainage tests, Area 1
Location: 1,280 ft east of SW corner, sec. 10,
T. 25 N., R. 16 W.

Discharge measured by: 7- by 10-inch orifice
Drawdown measured by: Electrical sounder
Reference point: North side of casing collar

Pump test No. 1, pumping well

Date	Time	Depth to water, ft	Drawdown s, ft	Elapsed time, min	Manometer reading, in	Discharge, gal/min	Remarks
2-14	1130	38.42					
2-15	1120	38.47					
2-16	1135	38.53					
2-17	1125	38.59					
2-18	0710	[1] 38.64					
	0805	38.64	0.00	0			Pump started
	0807	42.36	3.72	2	34.5	1,230	
	0809	42.67	4.03	4	34.0	1,220	
	0811	42.85	4.21	6	34.0	1,220	
	0813	42.97	4.33	8	34.0	1,220	
	0815	43.09	4.45	10	33.7	1,214	Pump off 2-20 at
	0820	43.27	4.63	15	33.7	1,214	1115
	0825	43.39	4.75	20	33.4	1,208	
	0830	43.49	4.85	25	33.5	1,210	
	0835	43.55	4.91	30	33.5	1,210	Average discharge 1,210 gal/min
	0845	43.67	5.03	40	33.4	1,208	
	0855	43.75	5.11	50	33.4	1,208	
	0905	43.80	5.16	60	33.6	1,212	
	0915	43.85	5.21	70	33.5	1,210	
	0925	43.89	5.25	80	33.4	1,208	
	0935	43.90	5.26	90	33.5	1,210	
	0945	43.94	5.30	100	33.5	1,210	
	1005	43.97	5.33	120	33.5	1,210	
	1105	44.03	5.39	180	33.5	1,210	
	1205	44.06	5.42	240	33.6	1,212	
	1305	44.09	5.45	300	33.7	1,214	
	1405	44.12	5.48	360	33.8	1,216	Adjusted discharge value
	1505	44.15	5.51	420	33.5	1,210	
	1605	44.14	5.50	480	33.5	1,210	
	1705	44.14	5.50	540	33.5	1,210	

[1] Static water level.

Table 9-8.—Tabulated discharging well test data for
determining a recharge boundary - continued

Project: Tongue Valley   Drawdown measured by:
Feature: Drainage tests, Area 1   "Popper"
Location: 100 ft north of pumping well, sec. 10, T. 25   Reference point: West side
  N., R. 16 W.   of casing collar

Pump test No. 1, observation hole No. 1, $r = 100$ ft

Date	Time	Depth to water, ft	Drawdown s, ft	Correction[1]	Corrected drawdown, ft	Elapsed time, min	Remarks
2-14	1125	38.39					
2-15	1115	38.44					
2-16	1130	38.50					
2-17	1120	38.56					
2-18	0705	[2] 38.61					
	0805	38.61	0.00			0	Pump started
	0810	38.69	.08			5	
	0820	38.83	.22			10	
	0825	38.93	.32			15	
	0830	39.02	.41			20	
	0835	39.08	.47			25	Pump off 2-20 at 1115
	0845	39.14	.53			30	
	0855	39.22	.61			40	
	0905	39.28	.67			50	
	0915	39.32	.71			60	
	0925	39.36	.75			70	
	0935	39.39	.78			80	
	0945	39.41	.80			90	
	0955	39.43	.82			100	
	1005	39.44	.83			110	
	1105	39.46	.85			120	
	1205	39.50	.89	0.01	0.90	180	
	1305	39.54	.93	.01	.93	240	
	1405	39.55	.94	.01	.95	300	
	1505	39.55	.94	.02	.96	360	
	1605	39.56	.95	.02	.97	420	
	1705	39.59	.96	.02	.98	480	
	1805	39.59	.96	.02	.98	540	
	1905	39.57	.96	.03	.99	600	
	2005	39.57	.96	.03	.99	660	

[1] Static water-level rising.
[2] Static water level.

Table 9-9.—Tabulated discharging well test data for
determining an impermeable boundary

Project: Dry Lake
Feature: Playa Reservoir
Location: SW 1/4, sec. 10, T. 4 N.,
  R. 21 E.

Discharge measured by: 7- by
  10-inch orifice
Drawdown measured by: Electrical
  sounder
Reference point: North side of casing
  collar

Pump test No. 1, pumping well

Date	Time	Depth to water, ft	Drawdown s, ft	Manometer reading, in	Discharge, gal/min	Elapsed time, min	Remarks
7-15	0810	30.90					
7-16	0815	30.91					
7-17	0810	30.90					
7-18	0840	[1] 30.90					
	0900					0	Pump started
	0902	34.20	3.30	34.0	1,220	2	Pump off 7-21
	0904	34.90	4.00	34.5	1,230	4	at 0900
	0906	35.00	4.10	34.0	1,220	6	
	0908	35.20	4.30	34.0	1,220	8	
	0910	35.30	4.40	33.7	1,214	10	Average Q = 1,220 gal/min 161.8 ft³/min
	0915	35.40	4.50	33.7	1,214	15	
	0920	35.60	4.70	33.4	1,208	20	
	0925	35.70	4.80	33.5	1,210	25	
	0930	35.80	4.90	33.5	1,210	30	
	0940	35.90	5.00	33.5	1,210	40	
	0950	36.00	5.10	33.5	1,210	50	
	1000	36.10	5.20	33.5	1,210	60	
	1010	36.20	5.30	33.5	1,210	70	
	1020	36.30	5.40	33.5	1,210	80	
	1030	36.40	5.50	33.5	1,210	90	
	1040	36.50	5.60	33.5	1,210	100	
	1100	36.60	5.70	33.4	1,208	120	
	1200	36.90	6.00	33.4	1,208	180	
	1300	37.10	6.20	33.5	1,210	240	
	1400	37.30	6.40	33.6	1,212	300	
	1500	37.40	6.50	33.6	1,212	360	
	1600	37.50	6.60	33.5	1,210	420	
	1700	37.60	6.70	33.5	1,210	480	
	1800	37.70	6.80	33.5	1,210	540	
	1900	37.80	6.90	33.5	1,210	600	
	2000	37.90	7.00	33.5	1,210	660	
	2100	38.00	7.10	33.4	1,208	720	
	2200	38.01	7.20	33.5	1,210	780	

[1] Static water level.

Table 9-9.—Tabulated discharging well test data for
determining an impermeable boundary - continued

Project: Dry Lake   Drawdown measured by: "Popper"
Feature: Playa Reservoir   Reference point: North side of casing
Location: SW 1/4, sec. 10, T. 4 N., R. 21 E.   collar

Pump test No. 1, observation well No. 1, r = 100 ft north of pumped well

Date	Time	Depth to water, ft	Drawdown s, ft	Elapsed time, min	Remarks
7-15	0815	30.87			
7-16	0820	30.87			
7-17	0815	30.88	0.00		
7-18	0830	[1] 30.88			
	0900			0	Pump started
	0905	30.96	.08	5	
	0910	21.10	.22	10	
	0915	31.20	.32	15	
	0920	31.29	.41	20	Pump off 7-21 at 0900
	0925	31.37	.49	25	
	0930	31.44	.56	30	
	0940	31.55	.67	40	
	0950	31.65	.77	50	
	1000	31.73	.85	60	
	1010	31.83	.95	70	
	1020	31.89	1.01	80	
	1030	31.96	1.08	90	
	1040	32.02	1.14	100	
	1050	32.08	1.20	110	
	1100	32.13	1.25	120	
	1200	32.39	1.51	180	
	1300	32.58	1.70	240	
	1400	32.75	1.87	300	
	1500	32.87	1.99	360	
	1600	32.98	2.10	420	
	1700	33.08	2.20	480	
	1800	33.16	2.28	540	
	1900	33.24	2.36	600	
	2000	33.34	2.46	660	
	2100	33.38	2.50	720	
	2300	33.51	2.63	840	
7-19	0100	33.65	2.77	960	

[1] Static water level.

To determine the distance to a boundary from a discharging well, the data, curve is fitted to the type curve as shown on figures 9-19 and 9-20. Then, successive steps to read the data are: (1) difference in drawdown between the type curve and the data curve, s_i, is noted for any time, t_i; (2) this difference is spotted on the s axis and the time, t_R, noted where the s_i value intersects the trace of the type curve. The values of t_i and t_R are the times of equal drawdown resulting from the image well and real well, respectively. On figure 9-19, for observation well No. 1 at a radius of 30 meters (100 feet) from the test well, s_i = 0.134 meter (0.44 foot) when t_i = 300 minutes. Transferring s_i = 0.134 meters (0.44 foot) to the s axis of the graph and reading across to the type curve, t_R = 22 minutes. The time values are related by:

$$\frac{r_i^2}{t_i} = \frac{r_R^2}{t_R} \qquad\qquad 9\text{-}34$$

The distance r_R from the real well to the closest observation well being known, then:

$$r_i = \sqrt{\frac{(100)^2 X 300}{22}} = 370 \; feet$$

This process is followed for each observation well, and a circle is drawn on a scaled plan of the test site layout using a radius equal to r_i distance computed for each observation well. The intersection of the circles marks the approximate location of the image well. If the observation wells lie in a straight line, especially if normal to the boundary, the circles may be nearly tangent to each other and the location of the image well may be weakly determined. This problem can be minimized by offsetting at least one observation well, as well No. 4 was offset in the plan views of figures 9-19 and 9-20.

To locate the boundary, a line is drawn between the real well and the image well. The boundary is located at the midpoint and is normal to the line. Values of T and S may be calculated using the same procedures explained previously. However, the values of t and s must be taken from the early portion of the data curve that coincides with the type curve before the deviation caused by boundary conditions is evident.

Either a leaky roof aquifer or a number of boundaries near the test area will occasionally cause the data curve to depart from the type curve in a manner that makes analysis difficult. In such instances, analysis by the leaky roof method should be tried if the curves appear similar. If this is not satisfactory, analysis by either the straight-line approximation or delayed storage method may permit a reliable interpretation.

(b) Boundary Location by Straight-Line Approximation.—Equation 9-16 in section 9-5 on the straight-line approximation method of solution shows that the slope of the line of the semilog graph is dependent only upon the rate of pumping and the transmissivity of the aquifer. In a pumping test, the discharge is held constant and the transmissivity is assumed to be constant. The plot of drawdown in an observation well versus time on a log scale initially follows a curve which changes gradually to a straight line as pumping continues.

When the area of influence of a discharging well reaches an impermeable boundary, the rate of drawdown will be doubled. This results from the addition of the equal influence caused by the hypothetical image well to the influence of the real well. Influence from a second impermeable boundary will triple the drawdown. A plot of drawdown versus time on a log scale shows straight sections of lines (or legs) each having a slope that is an approximate multiple of the initial slope. The transition from each leg to another is not sharp, but follows a curve.

If the influence from two or more boundaries located at approximately equal distances reaches an observation well almost simultaneously, the drawdown data will plot on a path of increasing curvature, and the straight-line portion of the plot may be three or more times steeper than the initial slope. Approximate locations of image wells may be obtained by drawing tangents with double or triple (or one-half or one-third) slope values to the initial slope of the data.

To determine the location of an image well by using the straight-line approximation method, the various legs of the curve are extended through the plotted points that fall on a straight line. Figure 9-21 shows an impermeable boundary determination. The time, t_i, is determined for a certain drawdown difference between the second leg and the extension of the first leg. This point should not be chosen where the plotted points are on a curve. In the

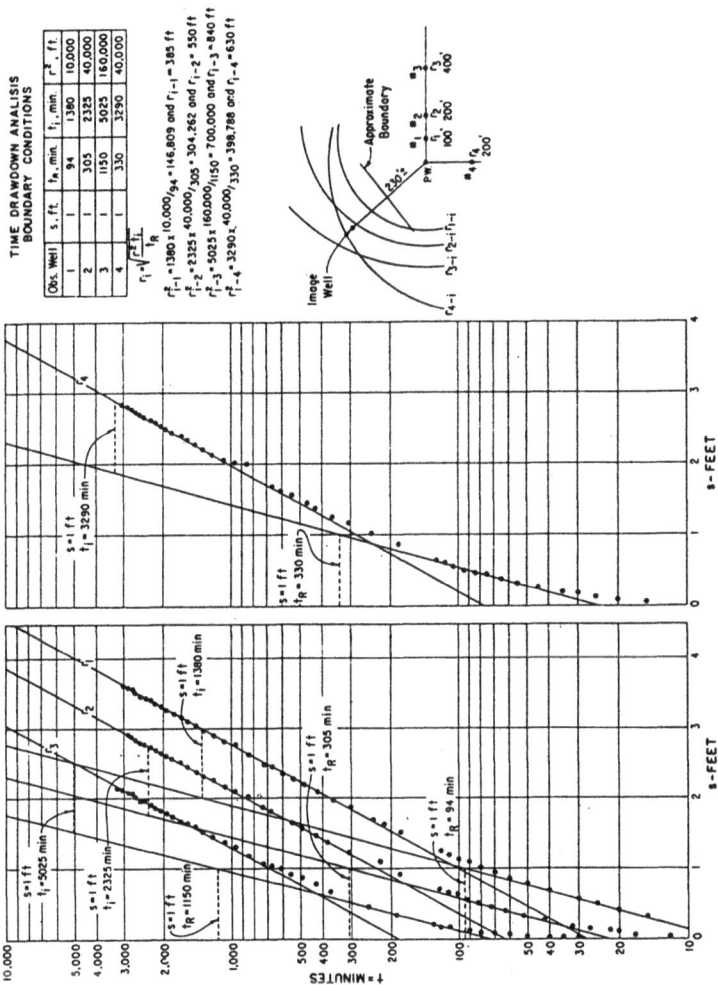

Figure 9-21.—Impermeable boundary determination by straight-line analysis.

example on figure 9-21, the location is chosen where $s = 0.3$ meter (1 foot). This drawdown of 0.3 meter (1 foot) is then located on the straight line of the first leg and the time, t_R, noted. If a third leg appears, the same procedure is followed in determining the amount of drawdown between the second and third legs, but referring the time and amount of drawdown to the fifth leg. Since the distance from the discharging well to each observation well is known, the distance to the image well causing the same amount of drawdown (or recharge for a recharge boundary) can be determined by the relationship shown in the previous equation 9-34:

$$\frac{r_i^2}{t_1} = \frac{r^2_R}{t_R}$$

In figure 9-19 for the observation well at a radius r_i of 30 meters (100 feet) and a drawdown $s = 0.3$ meter (1 foot), the values of t are: $t_i = 1{,}380$ minutes and $t_R = 94$ minutes. From this, the radius of the image well from the observation well was calculated as 117.3 meters (385 feet), and the boundaries can now be located.

(c) *Multiple Boundaries.*—Where two or more boundaries are present, image wells should be added as shown on figure 9-22 to maintain the condition of no flow across an impermeable boundary and no drawdown along a recharge boundary. Figure 9-22 also illustrates a discharging well between impermeable and recharge boundaries. This figure shows that when the first two image wells have been located, a repetitious pattern of image well spacings and discharge-recharge characteristics are present which permit the location and type of additional image wells. Theoretically, image wells extend to infinity in either direction, but in practice, the number is limited to that which results in an acceptable effect at the boundaries. The analysis then consists of analyzing the system of real and image wells as though they were in an ideal infinite aquifer. Bentall, 1963b; Ferris, 1948; UNESCO, 1967; and Walton, 1962 and 1970, discuss various complex boundary conditions and their treatment.

9-12. Interference and Well Spacing.—When a well discharges, a cone of depression is formed with its axis and lowest point at the well. As discharging continues, the circumference of the cone extends farther and farther from the well and the cone continues to deepen. Theoretically, the cone continues to expand to infinity; however, under actual conditions this is not the case. Recharge may balance discharge, thereby stabilizing the cone both

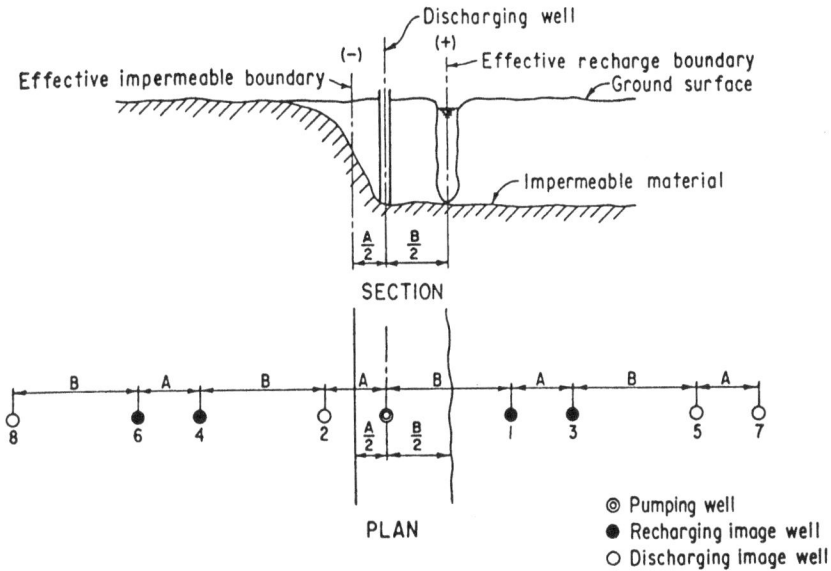

Figure 9-22.—Use of multiple image wells for
boundary determination.

horizontally and vertically, and the rate and amount of drawdown
at some distance becomes too small to measure. This distance may
be anywhere from several hundred feet to several miles, depending
on time and rate of discharge and aquifer characteristics. The
distance to the point of negligible drawdown (radius of influence)
can be calculated by assuming a very small value for s and using
the Theis nonequilibrium equation 9-6 in section 9-3. The values
of Q, t, S, and T must be known to use the equation.

Wells in a well field designed for water supply should be spaced
as far apart as possible so their areas of influence will have a
minimum of interference with each other. Because drawdown
interference is additive, the capacity of wells with intersecting
areas of influence will be reduced or drawdowns increased.

The solid lines on figure 9-23 show drawdown in wells A and B,
assuming that only one well is discharging at a time and that the
aquifer is infinite. When both wells are discharging at the same
time, drawdown curves between the wells would be additive and
their combined drawdown is shown by the dotted lines. The
drawdown curve with both wells discharging can be estimated for
any point (see line segment 0-0 on figure 9-23). Drawdown caused
by well A is shown by line segment 1-2, well B drawdown is

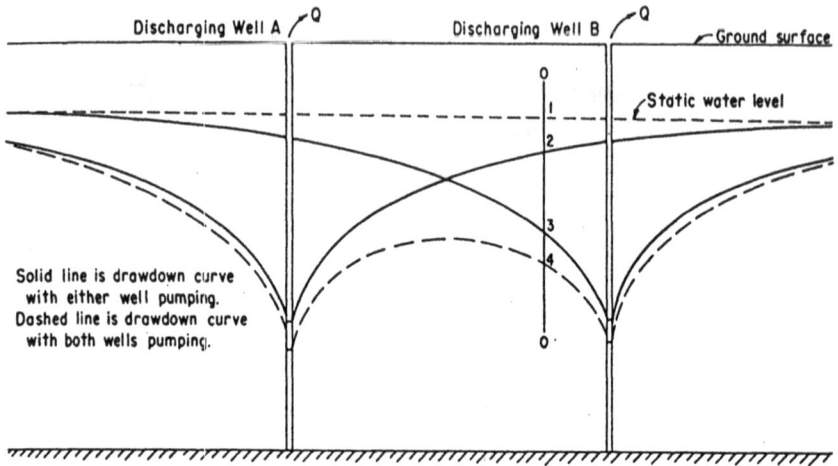

Figure 9-23.—Interference between discharging wells.

segment 1-3, and the drawdown caused by both wells is shown by line segment 1-4, or segment 1-3 plus segment 1-2. This is the principle of superposition or mutual interference and applies in well hydraulics whether the wells are real or images. For drainage wells, it may be desirable to lay out the well spacing to intentionally cause interference, thereby increasing the drainage effect at the midpoints between wells. In any case, the curve of the cone of depression should be known either by calculation or field measurement so that proper spacing of wells can be established (Bentall, 1963a, 1963b; Ferris, Knowles, Brown, and Stallman, 1962; Hantush, 1964; Lang, 1961; Moody, 1955a, 1955b; Remson, McNeary, and Randolph, 1961; Walton, 1970).

9-13. Barometric Pressure and Other Influences on Water Levels.—Water levels in wells in many artesian aquifers respond to changes in atmospheric pressure. An increase in atmospheric pressure causes the water level to decline, and a decrease causes the water level to rise. The barometric efficiency of an aquifer may be expressed:

$$BE = \frac{s_\omega}{s_b} X100$$

where:

BE = the barometric efficiency (nondimensional)
s_w = the water-level change, L
s_b = the barometric pressure change, L

Barometric efficiency can be estimated by plotting the water-level changes as ordinates and barometric pressure changes as abscissas on rectangular coordinate paper as shown on figure 9-24. The slope of a straight line fitted through the plotted points is the barometric efficiency which may be as high as 80 percent. In conducting discharging well tests in artesian aquifers when drawdowns are expected to be small, records of water levels and barometric pressures should be kept for several days prior to the start of the test to determine the influence of pressure changes on water levels. Barometric readings are continued during the test and measured drawdowns corrected accordingly by applying the barometric efficiency (table 9-10). For example, a well which showed a 50-percent barometric efficiency would have a water rise of 0.015 meter (0.05 foot) for each decrease of 0.030 meter (0.10 foot) in barometric pressure measured in head of water and conversely. Such values should be added to or subtracted from the measured drawdowns to eliminate the influence of atmospheric pressure changes.

Similar fluctuations in water levels may result from ocean and earth tidal fluctuations, earthquakes, and passing trains (Ferris, Knowles, Brown, and Stallman, 1962; Todd, 1959, 1980).

9-14. Well Performance Tests.—The tests described thus far are intended primarily to determine aquifer characteristics, including transmissivity, storativity, and boundary conditions. Similar tests are also conducted to determine well characteristics. The two principal well characteristics of well performance, yield, and drawdown are measures of the capacity of the well to produce water; both should be taken into account when capacity is considered. More specifically, the tests are conducted for the following reasons: (1) to determine general adequacy of development prior to completion; (2) upon completion to determine general capacity, establish a baseline for later tests, and to determine correct pump capacity and setting; and (3) to determine deterioration of the well following a period of use.

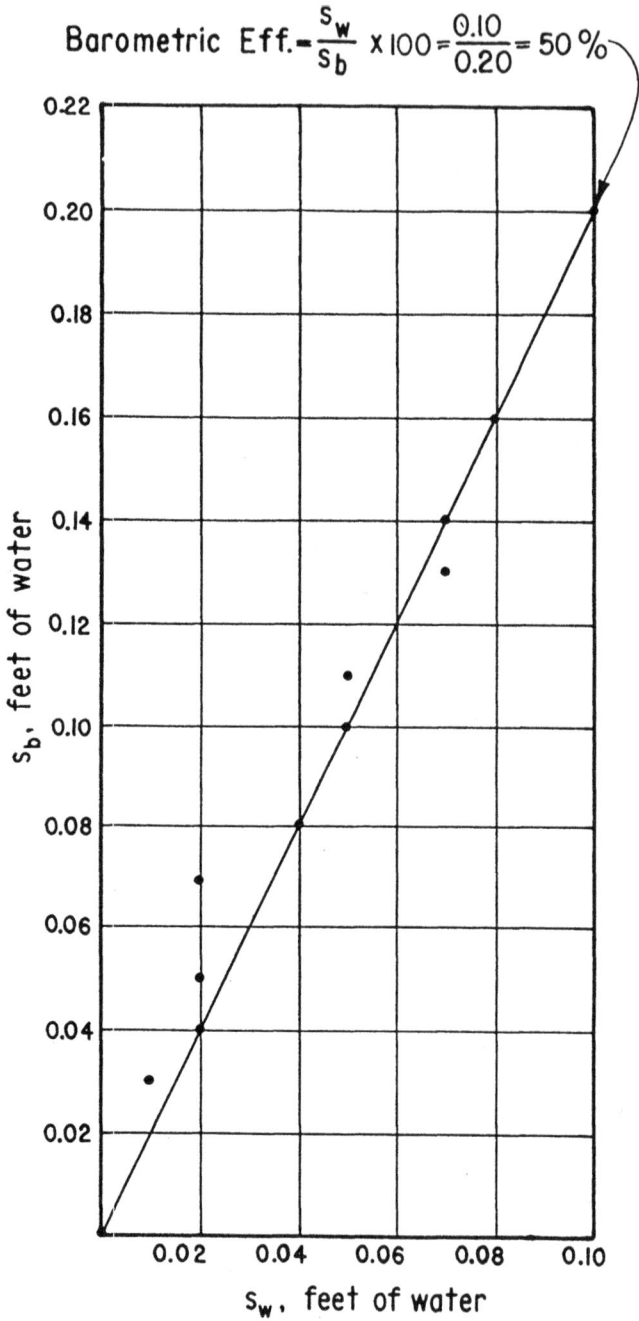

Figure 9-24.—Relationship of water levels and
barometric pressure.

Table 9-10(a).—Water level and barometric pressure data (SI metric)

Time	Millimeters of mercury	Meters of water	Change in barometric pressure, m	Summation of change in barometric pressure, m	Depth to water level, m	Water level elevation, m	Change in water level, m	Summation of change in water level, m
0800	753.62	10.245	0.000	0.000	11.414	1604.254	0.000	0.000
0900	753.11	10.242	-.007	-.007	11.418	+.257	+.004	+.004
1000	752.86	10.238	-.004	-.011	11.421	+.259	+.003	+.007
1100	752.60	10.235	-.003	-.014	11.421	+.259	+.000	+.007
1200	752.04	10.228	-.007	-.021	11.427	+.266	+.006	+.013
1300	751.33	10.218	-.010	-.031	11.430	+.269	+.003	+.016
1400	750.82	10.212	-.005	-.036	11.436	+.275	+.006	+.022
1500	749.30	10.190	-.022	-.044	11.445	+.284	+.009	+.031
1600	750.06	10.201	+.011	-.033	11.439	+.278	+.006	+.037
1700	750.52	10.208	+.007	-.026	11.436	+.275	-.003	+.034
1800	750.82	10.212	+.004	-.022	11.436	+.275	-.003	+.031
1900	751.33	10.218	+.007	-.016	11.430	+.269	-.006	+.025
2000	751.84	10.225	+.007	-.009	11.421	+.259	-.009	+.016

Note: 1 mm of mercury = 0.0136 m of water.

Table 9-10(b).—Water level and barometric pressure data (U.S. customary)

Time	Inches of mercury	Feet of water	Change in barometric pressure, ft	Summation of change in barometric pressure, ft	Depth to water level, ft	Water level elevation, ft	Change in water level, ft	Summation of change in water level, ft
0800	29.67	33.65	0.00	0.00	37.45	5263.30	0.00	0.00
0900	29.65	33.62	-0.03	-0.03	37.46	+0.31	+0.01	+0.01
1000	29.64	33.61	-0.01	-0.04	37.47	+0.32	+0.01	+0.02
1100	29.62	33.60	-0.01	-0.05	37.47	+0.32	+0.00	+0.04
1200	29.61	33.57	-0.03	-0.08	37.49	+0.34	+0.02	+0.04
1300	29.58	33.54	-0.03	-0.11	37.50	+0.35	+0.01	+0.05
1400	29.56	33.52	-0.02	-0.13	37.52	+0.37	+0.02	+0.07
1500	29.50	33.45	-0.07	-0.20	37.55	+0.40	+0.03	+0.10
1600	29.53	33.49	+0.04	-0.16	37.53	+0.38	+0.02	+0.08
1700	29.55	33.51	+0.02	-0.14	37.52	+0.37	-0.01	+0.07
1800	29.56	33.52	+0.01	-0.13	37.52	+0.37	-0.00	+0.07
1900	29.58	33.54	+0.03	-0.10	37.50	+0.35	-0.02	+0.05
2000	29.60	33.57	+0.03	-0.07	37.47	+0.32	-0.03	+0.02

Note: 1 inch of mercury = 1.134 feet of water.

(a) Step Tests.—Step tests are used principally to determine the comparative specific capacity (i.e., yield versus drawdown, of a well at different rates of discharge). Usually the tests are started at a low step, such as 25 percent of design capacity, and increased in three or four incremental steps, such as 50, 75, 100, and 125 percent of capacity. Ideally, a recovery period should follow each step, but the costs of such a procedure often cannot be justified. More commonly, the entire test is run without pause. The first step is continued until the water level reaches approximate stabilization. This usually is reached within a period ranging from 1 to 4 hours. Subsequent steps are run for the same length of time as the first.

The specific capacity of wells declines with increasing discharge and length of pumping time. The decline caused by increasing discharge is usually small in wells in artesian aquifers, but may be large in wells in free aquifers. If a step test of a new well shows increasing specific capacity with increasing discharge, it probably indicates that the well is continuing to develop and that the original development was inadequate.

In the determination of most favorable capacity, if an exact design capacity is not necessary, a plot of drawdown versus yield for each step may show a point at which yield is optimum. Also, such plots, when combined with constant yield test data, will yield data on correct pump settings.

The yield of most wells declines with use because of general deterioration of the well through buildup of encrustation on screens, plugging of aquifers and gravel packs, and other similar factors. If a step test was conducted upon completion of the well, the running of a duplicate test may yield data on the extent of the deterioration and clues as to the nature.

Analyses of data from step tests of discharging wells have been used by Jacob (1947) and Rorabaugh (1953) to determine the efficiency of wells. Mogg (1968) contends, however, that determination of well efficiency should properly be based on the theoretical specific capacity of a well which is a function of transmissivity of the aquifer. This requires test data from a fully penetrating test well and an observation well.

(b) Constant Yield Tests.—For wells of low capacity and
intermittent operation, such as domestic or stock wells, a bailing or
pumping test of several hours' duration may be adequate to
determine the yield and correct pump capacity and setting.

Wells of larger capacity which must operate for prolonged periods
should be tested at a rate approximating the intended capacity for
a long enough period to simulate actual production pumping conditions. Normally, unless aquifer conditions are simple and uniform
or are well known, such a test should be continued at a uniform
rate for a minimum of 72 hours.

9-15. Streamflow Depletion by a Discharging Well.—Where
an aquifer is hydraulically connected to a stream or lake, discharge
from a nearby well will result in depletion of the surface water
body. It is often important to know, because of possible water
rights conflicts, the timing and extent of such depletion. A method
of estimating surface-water depletion resulting from operation of a
nearby well has been developed by Glover and Balmer (1954) in an
analysis made by the images method. Figure 9-25 is a graph from
which the portion of the well discharge contributed by a stream
can be estimated. In this graph, q is the discharge of the well
drawn from the stream; Q is the total well discharge; X is the
distance from the well to the stream, or more accurately, the
computed distance to the recharge boundary; T is the
transmissivity; t is the time since discharge began; and S is the
specific yield. If values of X, Q, T, and S are known, the
percentage of water pumped from the well that originates from the
stream can be estimated for any time. Similarly, by assuming a
small value such as 0.01 (it cannot be zero) on the abscissa, an
estimate may be made of the time at which practically all water
pumped would come from the stream. In an unconfined aquifer,
the drawdown in the well should not be less than 50 percent of the
original saturated thickness of the aquifer (Glover, 1960, 1964,
1968; Glover and Balmer, 1954; Hantush, 1964, 1965; Moulder and
Jenkins, 1969; Rorabaugh, 1951; Walton, 1963, 1966, 1967, 1980).

**9-16. Estimates of Future Pumping Levels and Well
Performance.**—In well design, an estimate of minimum pumping
levels during the prospective well life may be necessary to provide
for an adequately deep pump setting and head and power
requirements. Initially, hydrographs of water levels in nearby
wells are examined to determine long-term ground-water trends
and seasonal fluctuations. Potential additional ground
development is evaluated and, if possible, the effects of such

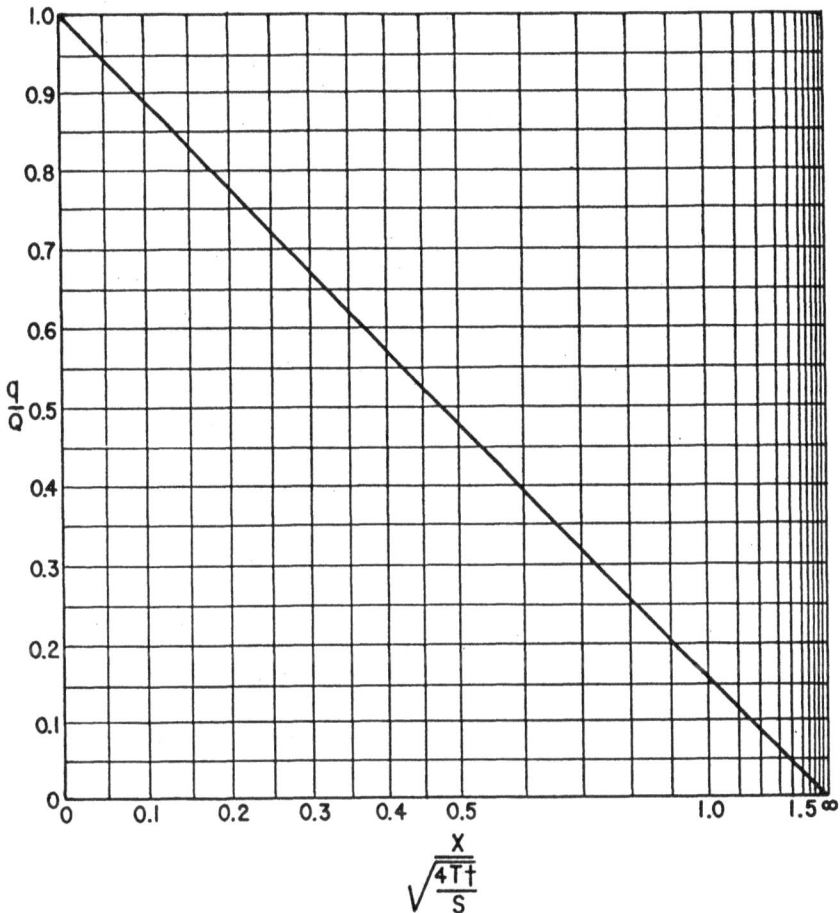

q — Flow that comes from the stream
Q — Discharge of the well
x — Distance from the well to the stream or the
      computed distance to a recharge boundary
T — Transmissivity
t — Time from beginning of pumping
s — Specific yield

Figure 9-25.—Analysis of streamflow depletion by a well.

development on future ground levels estimated. Also, in the case of free aquifers, the effect of decrease in aquifer thickness on transmissivity is estimated from the relationship $T = KM$. Using the estimated or determined value of S, the value of T (corrected as necessary), and estimated pumping schedules, the maximum levels are estimated which might be encountered in a well in which the drawdown in the well exceeds the theoretical drawdown by 20 to 30 percent. This 20- to 30-percent factor compensates for losses incurred in the well. After completion and testing of a well, step test analyses are used to refine the original estimate of minimum water levels. The same procedure can be used on existing or rehabilitated wells.

Common problems for analysis are to predict long-term performance of a well or group of wells, or the reaction of an aquifer to various distribution of wells and rates of pumping. The computed values of T and S may be substituted in the various equations based on the nonequilibrium equation to arrive at solutions to such problems. For example, it may be necessary to determine the drawdown in a well pumping at a given rate for a given period of time. S, T, and r (radius of the well) would be known. This permits solving for u. Using this value of u, we can find $W(u)$ from figure 9-5 or table 9-3 in section 9-4. Substituting the value for $W(u)$ in equation 9-9, we can then solve for s. The value obtained for s will be the theoretical drawdown just outside the well casing. The drawdown inside the well will be greater, depending on the efficiency of the well. The drawdown at any distance from the well can be found by setting a value for r. When more than one well is involved, the drawdown at any point is equal to the sum of the drawdowns for each well at that point.

Other uses of equations 9-7 and 9-9 would be to determine:

- length of time a well can be pumped before drawdown reaches a given value

- rate at which a well can be pumped for a given period before the top of the screen is exposed or the pump breaks suction

- distance from a well at which drawdown reaches a selected value in a given time. This is especially useful in locating drainage wells or determining interference between wells.

Determining the feasible yield from a well or well field requires an economic analysis involving the amount of water needed, specific capacity of the wells, the cost of power, and the cost of wells and pumps.

9-17. Estimating Transmissivity from Specific Capacity.— For some investigations, especially those of a reconnaissance nature, those covering large areas, or investigations that are limited by funds, estimates of approximate transmissivity values may be acceptable. The drilling of test and observation wells and the running of tests may be unnecessary or unjustified if there are existing wells in the area. Limited data on the yields and drawdown of such wells may be used to calculate approximate transmissivity values from specific capacity data.

In the previously discussed methods of determining transmissivity by discharging well tests, two essential terms in the analyses are the well discharge, Q, and drawdown, s, or more specifically Q/s. In a closely controlled test, Q and s are related to values of time, distance, and storativity in an observation well to obtain a transmissivity value that is as accurate and reliable as possible. However, within limits, these three factors may be ignored and the simple Q/s relationship used to determine approximate transmissivity values. Figure 5-4 in section 5-3 shows the general relationship between specific capacity, in gallons per minute of discharge versus feet of drawdown, and transmissivity. If the specific capacity values are based on pumping periods of several hours or more, the transmissivity values tend to be conservatively low because the inefficiency of the well probably overshadows the effects of drawdown increasing with time.

Figure 5-4 may also be used to determine the value of specific capacity when the transmissivity is known. In all instances, however, it should be understood that the values from figure 5-4 are approximations only. Walton (1970) gives a more thorough analysis of the specific capacity method of determining transmissivity.

9-18. Flow Nets.—Forchheimer (1930) evolved a graphical solution for complex ground flow problems. His methods are used primarily for analysis of flow under and around foundations, through dams and similar structures at which relatively uniform conditions prevail. He considered that such problems could be analyzed by flow nets in a two-dimensional flow system through a unit cross section. In 1937, Arthur Casagrande (1937) published

an English discussion of the principle and application of flow nets
(Ferris, Knowles, Brown, and Stallman, 1962; Peattie, 1956; and
Walton, 1970).

Solutions are based on the law of continuity, which states that
the volume of water flowing into a saturated element of soil is
equal to the volumes flowing out of it, and on Darcy's law, where
boundaries are fixed and flow is steady. The two-dimensional flow
can be represented by orthogonal families of curves. One set of
curves represents flow lines and the other equipotential lines. The
two sets of curves intersect each other at right angles. Since the
number of lines is infinite, there is no unique way of constricting a
flow net.

The equipotential lines, which are drawn one unit apart,
represent lines of an equal water or piezometric level. The flow
lines are drawn to represent equal flow rates through the zone
bounded by the lines.

Boundaries on all flow nets are either equipotential or flow lines.
Flow lines start at right angles to recharge boundaries and
impermeable boundaries are represented by flow lines.
Theoretically, there can be no flow across a flow line.

The two families of lines are drawn to form a net of orthogonal
squares which are not usually squares, but possess the property
that the ratio of the sum of the opposite sides of each square
bounded by two flow lines and two equipotential lines is unity and
the lines intersect at right angles. If η_q equals the number of flow
tubes, η_e, the number of equipotential drops, and h, the head
difference between the inlet and outlet of the net, the flow Q is
represented by:

$$Q = Kh\frac{\eta_q}{\eta_e} \qquad\qquad 9\text{-}36$$

where:

η_q/η_e is known as the shape factor and K is the permeability.

Forchheimer's solution can be applied to qualitative analysis and
sometimes to quantitative determination of ground flow
distribution where conditions are uniform, flow is steady, and

sufficient data are available to permit drawing a reasonably accurate flow net. However, the basic assumptions are seldom met, so the Forchheimer method must be used with care.

The ground gradient, i, can be estimated by measuring the distance, L, between equipotential lines h_1, and h_2. By using the gradient equation,

$$i = \frac{h_1 - h_2}{L}$$

and the area A of a plane normal to the direction of flow which can be estimated from the width and thickness of the flow zone, the flow rate can be calculated using Darcy's law: $Q = KiA$.

Figure 9-26 is a water-table elevation contour map on which flow lines have been sketched. A ground mound caused by deep percolation is shown near the upper, right corner. Water flows out radially from the mound. The distance between adjacent flow lines, W, increases with distance from the mound and with flattening gradient. Flow lines and equipotential lines form an orthogonal network of cells, the size of which increase as the gradient decreases. Theoretically, the flow through any one cell equals the flow through any other cell.

Since there can be no flow across a flow line and if the aquifer is uniformly thick,

$$Q_1 = Q_2 = K_1 W_1 i_1 = K_2 W_2 i_2 \text{ or}$$

$$\frac{K_1}{K_2} = \frac{W_2 i_2}{W_1 i_1} \qquad\qquad 9\text{-}37$$

9-19. Drainage Wells.—Drainage wells differ little from conventional water-supply wells except that the main objective is lowering of the water table rather than water supply.

Where conditions are favorable, a method developed by Hantush (1964) may be useful in determining the feasibility of using drainage wells, general features, layout of facilities, and operating

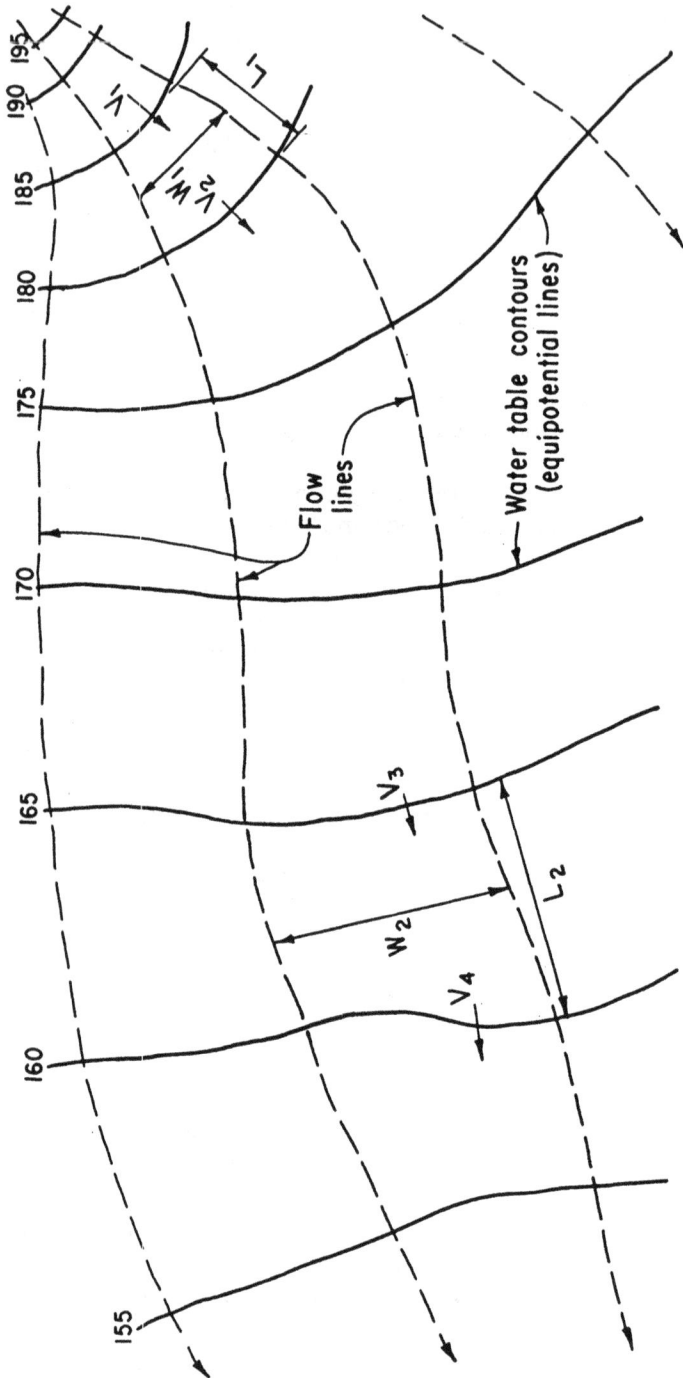

Figure 9-26.—Flow net analysis.

criteria. Although the equation is steady state, averaging transient conditions during long-term pumping permits its use for initial estimates of required well discharge and spacing.[1]

Hantush's equations are (see figure 9-27):

$$f_o = \frac{K}{W}\left(\frac{h_e}{r_w}\right)^2\left[1-\left(\frac{h_w}{h_e}\right)^2\right]$$ 9-38

$$\log\left(\frac{r_e}{r_w}\right) = 0.464 \; \log \, f_o - 0.157$$ 9-39

$$r_e = \frac{M}{d}$$ 9-40

$$Q = \pi W r_e^2$$ 9-41

where, using consistent units:

f_o = a pure number whose value depends on the aquifer characteristics and geometry of the well

W = rate of uniform deep percolation, L/t

K = hydraulic conductivity or permeability of the aquifer, L/t

h_e = maximum permissible thickness at the center of each cell with an area drained by a network of wells, L (the maximum thickness of the saturated aquifer between the dewatered root zone and the base of the aquifer)

r_w = effective radius of a well, L

h_w = saturated thickness at a pumping well which completely penetrates the aquifer, L

r_e = radius of a circle circumscribing the diversion area of each of a large number of equally spaced wells, L

M = spacing of the wells, L

d = dimensionless constant which depends on the grid on which wells are located (for a rectangular grid, it is $\sqrt{2} = 1.414$ and for an equilateral grid it is $\sqrt{3} = 1.732$)

Q = steady discharge of an individual well in the grid, L^3/t

---

[1] These equations were developed for a single well with an impervious boundary at radius r_e. They do not allow for well interference and are therefore conservative.

RECTANGULAR GRID   EQUILATERAL GRID

SECTION A-A

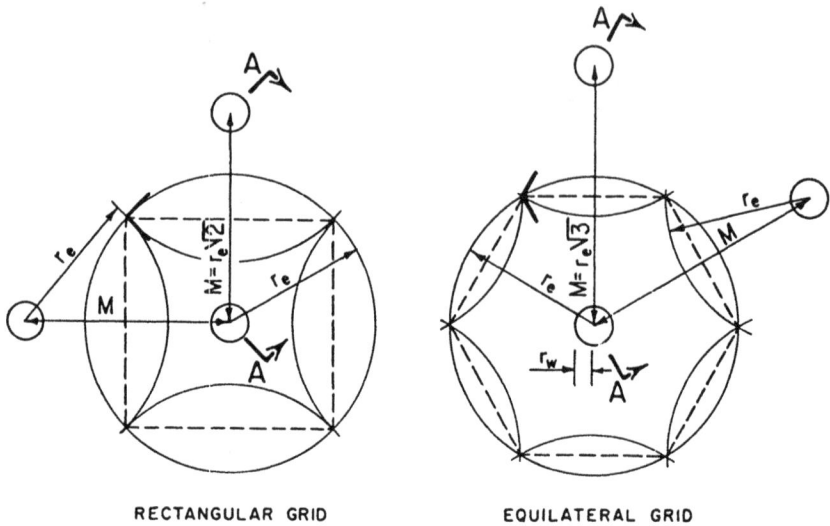

Figure 9-27.—Analysis of drainage well requirements.

The Hantush analysis gives a preliminary estimate of the drainage requirements. Further refinement may be necessary, depending on the reliability of the data and complexity of the hydrologic system.

9-20. Bibliography.—

Bentall, R. (compiler), 1963a, "Methods of Determining Permeability, Transmissibility, and Drawdown," U.S. Geological Survey Water-Supply Paper 1536-I.

_____, 1963b, "Short Cuts and Special Problems in Aquifer Test," U.S. Geological Survey Water-Supply Paper 1545-C.

Boulton, N.S., December 1951, "The Flow Pattern Near a Gravity Well in a Uniform Water-Bearing Medium," Journal of the Institution of Civil Engineers (London), vol. 36, No. 10, pp. 534-550.

_____, August 1954, "The Drawdown of the Water Table Under Non-Steady Conditions Near a Pumped Well in an Unconfined Formation," Institution of Civil Engineers Proceedings (London), vol. 3, part III, No. 2, Paper No. 5979, pp. 564-579.

_____, 1955, "Unsteady Radial Flow to a Pumped Well Allowing for Delayed Yield for Storage," International Association of Scientific Hydrology, Assemblee Generale de Roma, Tome II, Publication No. 37, Rome, pp. 473-477.

_____, November 1963, "Analysis of Data from Nonequilibrium Pumping Tests Allowing for Delayed Yield from Storage," Institution of Civil Engineers Proceedings (London), vol. 26, Paper No. 6693, pp. 469-482, also, "Discussion of . . ." in vol. 28, pp. 603-610, August 1964.

Brown, R.H., 1953, "Selected Procedures for Analyzing Aquifer Test Data," *Journal of the American Water Works Association*, vol. 45, No. 8, pp. 844-866.

Bruin, J., and H.E. Hudson, Jr., 1955, "Selected Methods for Pumping Test Analysis," Illinois State Water Survey Report of Investigation No. 25, Urbana.

Casagrande, A., 1937, "Seepage Through Dams," *Journal of the New England Water Works Association*, vol. 51, pp. 131-172.

Chow, V.T., June 1952, "On the Determination of Transmissibility and Storage Coefficient from Pumping Test Data," Transactions of the American Geophysical Union, vol. 33, No. 3, pp. 397-404.

Cooper, H.H., Jr., and C.E. Jacob, August 1946, "A Generalized Graphical Method for Evaluating Formation Constants and Summarizing Well-Field History," Transactions of the American Geophysical Union, vol. 27, No. IV, pp. 526-534.

Davis, S.N., and R.J.M. DeWiest, 1966, "Hydrogeology," John Wiley & Sons, New York.

DeWiest, R.J.M., 1965, "Geohydrology," John Wiley & Sons, New York.

Ferris, J.G., 1948, "Ground Water Hydraulics as a Geophysical Aid," Michigan Department of Construction, Technical Report No. 1, Lansing.

Ferris, J.G., D.B. Knowles, R.H. Brown, and R.W. Stallman, 1962, "Theory of Aquifer Tests," U.S. Geological Survey Water-Supply Paper 1536-E.

Forchheimer, Philipp, 1930, "Hydraulik," 3d edition, B.G. Teubner Verlagsgesellschaft, Berlin.

Glover, R.E., March 1960, "Studies of Ground Water Movement," Bureau of Reclamation Technical Memorandum 657.

_____, 1960, "Ground Surface Water Relationship," Water Research Conference, Colorado State University, Fort Collins.

_____, R.E., 1964, "Ground-Water Movement," Bureau of Reclamation, Engineering Monograph No. 31.

_____, 1968, "The Pumping Well," Colorado State University Agricultural Experiment Station, Technical Bulletin No. 100, Fort Collins.

_____, and G.G. Balmer, June 1954, "River Depletion Resulting from Pumping a Well near a River," Transactions of the American Geophysical Union, vol. 35, No. 3, pp. 468-470.

_____, Moody, W.T., and W.N. Tapp, June 11, 1954, "Till Permeabilities as Estimated from the Pump-Test Data Obtained During the Irrigation Well Investigations," memorandum to Drainage and Ground Engineer, Bureau of Reclamation.

"Ground Water and Wells," 1966, UOP Johnson Division, Driscoll, Fletcher G., (principal author and editor), 1986, "Ground Water and Wells," 2d edition, UOP Johnson Division, St. Paul, Minn.

Hantush, M.S., April 1955, "Steady Three-Dimensional Flow to a Well in a Two-Layered Aquifer," Transactions of the American Geophysical Union, vol. 36, No. 2, pp. 286-292.

_____, December 1956, "Analysis of Data from Pumping Tests in Leaky Aquifers," Transactions of the American Geophysical Union, vol. 37, No. 6, pp. 702-714.

_____, August 1959, "Nonsteady Flow to Flowing Wells in Leaky Aquifers," *Journal of Geophysical Research*, vol. 64, No. 8, pp. 1043-1052.

_____, November 1960, "Modification of the Theory of Leaky Aquifers," *Journal of Geophysical Research*, vol. 65, No. 11, pp. 3713-3725.

_____, 1961, "Tables of the Functions W(UB)," New Mexico Institution of Mining and Technology, Professional Paper No. 103, Socorro.

_____, 1961, "Tables of the Functions W(UB)," New Mexico Institution of Mining and Technology, Professional Paper No. 104, Socorro.

_____, July 1961, "Drawdown Around a Partially Penetrating Well, Proceedings of the American Society of Civil Engineers, *Journal of the Hydraulics Division*, vol. 87, No. HY4, pp. 83-98.

_____, September 1961, "Aquifer Tests on Partially Penetrating Wells," Proceedings of the ASCE, *Journal of the Hydraulics Division*, vol. 87, No. HY5, pp. 171-195.

_____, March 1962, "Drainage Wells in Leaky Water-Table Aquifers," Proceedings of the ASCE, *Journal of the Hydraulics Division*, vol. 88, No. HY2, pp. 123-137.

_____, March 1964, "Supplement to Peterson's Design of Replenishment Wells," Proceedings of the ASCE, *Journal of Irrigation and Drainage Division*, vol. 90, No. IR1, part 1, pp. 67-76.

_____, June 15, 1964, "Depletion of Storage, Leakage, and River Flow by Gravity Wells in Sloping Sand," *Journal of Geophysical Research*, vol. 69, No. 12, pp. 2551-2560.

_____, June 15, 1965, "Wells near Streams with Semipervious Beds," *Journal of Geophysical Research*, vol. 70, No. 12, pp. 2829-2838.

_____, 1966, "Wells in Homogeneous Anisotropic Aquifers," Water Resources Research, vol. 2, No. 2, pp. 273-279.

_____, January 15, 1966, "Analysis of Data from Pumping Tests in Anisotropic Aquifers," *Journal of Geophysical Research*, vol. 71, No. 2, pp. 421-426.

_____, and C.E. Jacob, February 1955, "Non Steady Radial Flow in an Infinite Leaky Aquifer," Transactions of the American Geophysical Union, vol. 36, No. 1, pp. 93-100.

Heath, R.C., and F.W. Trainer, "Introduction to Ground Hydrology," John Wiley & Sons, New York, 1968.

Skeat, W.0. (editor), 1969, "Manual of British Water Engineering Practice," W. Heiner & Sons, Cambridge.

Jacob, C.E., 1940, "On the Flow of Water in an Elastic Artesian Aquifer," Transactions of the American Geophysical Union, Reports and Papers, vol. 21, pp. 574-586.

_____, 1945, "Adjustment for Partial Penetration of a Pumping Well" U.S. Geological Survey Open File Report.

_____, April 1946, "Radial Flow in a Leaky Artesian Aquifer," Transactions of the American Geophysical Union, vol. 27, No. II, pp. 198-208.

_____, 1947, "Drawdown Test to Determine the Effective Radius of an Artesian Well," Transactions of the American Society of Civil Engineers, vol. 112, pp. 1047-1070.

_____, and S.W. Lohman, August 1952, "Nonsteady Flow to a Well of Constant Drawdown in an Extensive Aquifer," Transactions of the American Geophysical Union, vol. 33, No. 4, pp. 559-569.

Kazmann, R.G., December 1941, "Inverse Tables of the Exponential Integral," U.S. Geological Survey, Water Resources Branch, Division of Ground Water.

_____, 1965, "Modern Hydrology," Harper & Row, New York. Kruseman, G.P., and N.A. de Ridder, 1970, "Analysis and Evaluation of Pumping Test Data," International Institute for Land Reclamation and Improvement, Book II, Wageningen, Netherlands.

Lang, S.M., 1961, "Methods for Determining the Proper Spacing of Wells in Artesian Aquifers," U.S. Geological Survey Water-Supply Paper 1545-B.

Lennox, D.H., January/February 1969, "Reader's Comment on Step Down Tests," Drillers Journal, UOP Johnson Division, St. Paul, Minnesota.

_____, November 1966, "Analysis and Application of the Step-Drawdown Test," Proceedings of the American Society of Civil Engineers, *Journal of the Hydraulics Division*, vol.92, No. HY6.

Lohman, S.W., 1972, "Ground Water Hydraulics," U.S. Geological Survey Professional Paper 708.

Lohman, S.W., 1965, "Geology and Artesian Water Supply of the Grand Junction Area Colorado," U.S. Geological Survey Professional Paper 451.

Meinzer, 0.E., 1932, "Outline of Methods for Estimating Ground Supplies," U.S. Geological Survey Water-Supply Paper 638-C.

Mogg, J.L., July/August 1968, "Step Drawdown Test Needs Critical Review," Drillers Journal, UOP Johnson Division, pp. 3-8, 11.

Moody, W.T., March 11, 1955a, "Determination of Minimum Drawdown Within an Array of Wells," Memorandum to T.P. Ahrens, Bureau of Reclamation.

_____, February 16, 1955b, "Determination of the Drawdown at the Center of an Array of Wells," memorandum to T.P. Ahrens, Bureau of Reclamation.

Moulder, E.A., and C.T. Jenkins, March 1969, "Analog Digital Models of Stream Aquifer Systems," U.S. Geological Survey, Water Resources Division, Open File Report.

Muskat, Morris, 1946, "The Flow of Homogeneous Fluids through Porous Media," J.W. Edwards, Ann Arbor, Michigan.

National Bureau of Standards, 1940, "Tables of Sine, Cosine, and Exponential Integrals," vols. 1 & 2.

Norris, S.E., and R.E. Fidler, July 1966, "Use of Type Curves Developed from Electric Analog Studies of Unconfined Flow to Determine the Vertical Permeability of an Aquifer at Piketown, Ohio," Ground Water, vol. 4, No. 3, pp. 43-48.

Papadopulos, I.S., October 15, 1966, "Nonsteady Flow to Multiaquifer Wells," Journal of Geophysical Research, vol. 71, No. 20, pp. 4791-4797.

Peattie, K.R., January 1956, "A Conducting Paper Technique for the Construction of Flow Nets," Civil Engineering and Public Works Review (GB), vol. 51, No. 595, pp. 62-64.

Prickett, T.A., July 1965, "Type-Curve Solution to Aquifer Tests Under Water Table Conditions," Ground Water, vol. 3, No. 3, pp. 5-14.

Prickett, T.A., and C.G. Longquist, December 1968, "Comparison Between Analog and Digital Simulation Techniques for Aquifer Evaluation," Illinois State Water Survey Reprint Series No. 114, Urbana.

Remson, I., S.S. McNeary, and J.R. Randolph, 1961, "Water Levels Near a Well Discharging from an Unconfined Aquifer," U.S. Geological Survey Water-Supply Paper 1536-B.

Rorabaugh, M.I., 1951, "Stream Bed Percolation in Development of Water Supplies," International Association of Scientific Hydrology Publication No. 33, Brussels.

_____, December 1953, "Graphical and Theoretical Analysis of Step Drawdown Tests of Artesian Wells," Proceedings of the American Society of Civil Engineers, vol. 79, Separate No. 362.

Rouse, H. (editor), 1950, "Engineering Hydraulics," Proceedings of the Fourth Hydraulics Conference, Iowa Institute of Hydraulic Research, State University of Iowa, Iowa City, June 12-15, 1949.

Schicht, R.J., February 1972, "Selected Methods of Aquifer Test Analysis," Water Resources Bulletin, vol. 8, No. 1 pp. 175-187.

Sheahan, N.T., January/February 1971, "Type-Curve Solution of Step-Drawdown Test," Ground Water, vol. 9, No. 1, pp. 25-29.

Stallman, R.W., 1961, "The Significance of Vertical Flow Components in the Vicinity of Pumping Wells in Unconfined Aquifers," U.S. Geological Survey Professional Paper 424-B.

_____, 1963, "Electric Analog of Three-Dimensional Flow to Wells and Its Application to Unconfined Aquifers," U.S. Geological Survey Water-Supply.

_____, 1965, "Effects of Water Table Conditions on Water Level Changes Near Pumping Wells," Water Resources Research, vol. 1, No. 2, pp. 295-312, Paper 1536-H.

Stone, R.F., January/February 1969, "Reader's Comment on Step Down Tests," *Drillers Journal*, UOP Johnson Division, pp. 10-12.

Theis, C.V., August 1935, "The Relation Between the Lowering of the Piezometric Surface and the Rate and Duration of Discharge of a Well Using Ground-Water Storage," Transactions of the American Geophysical Union, vol. 16, part II, pp. 519-524.

Todd, D.K., 1959, "Ground Water Hydrology," John Wiley & Sons, New York.

Todd, D.K., 1980, "Ground Water Hydrology, 2d edition, John Wiley & Sons, New York.

UNESCO, 1967, "Methods and Techniques of Ground Water Investigations and Developments," Water Resources Series, No. 33, New York.

Walton, W.C., 1962, "Selected Analytical Methods for Well and Aquifer Evaluation," Illinois State Water Survey, Department of Regulation and Education Bulletin No. 49, Urbana.

_____, 1963, "Estimating the Infiltration Rate of a Stream Bed by Aquifer Test," International Association of Scientific Hydrology Publication No. 63, Berkeley, Calif.

_____, 1966, "Effect of Induced Stream Bed Infiltration of Water Levels on Wells During Aquifer Tests," Water Resources Research Center, University of Minnesota, Bulletin No. 2.

_____, 1967, "Recharge from Induced Stream Bed Infiltration Under Varying Water Level and Stream Stage Conditions," Water Resources Research Center, University of Minnesota, Bulletin No. 6.

_____, 1970, "Ground Water Resource Evaluation," McGraw-Hill, New York.

Weeks, E.P., 1964, "Field Methods of Determining Vertical Permeability and Aquifer Anisotropy," U.S. Geological Survey Professional Paper 501-D.

Wenzel, L.K., 1942, "Methods for Determining Permeability of Water-Bearing Materials," U.S. Geological Survey Water-Supply Paper 887.

_____, and A.L. Greenlee, January 1944, "A Method of Determining Transmissibility- and Storage-Coefficients by Tests of Multiple-Well Systems," Transactions of the American Geophysical Union, vol. 24, part II, pp. 547-560.

Chapter X

PERMEABILITY TESTS IN
INDIVIDUAL DRILL HOLES AND WELLS

10-1. General Considerations.—Chapters II, VIII, and IX considered various aspects of pumping tests to determine values of transmissivity, storativity, and boundary conditions of aquifers for use in ground-water inventories, well field design, drainage feasibility, and related activities. Such tests are usually large-scale activities that are time consuming, costly, and require significant manpower. Moreover, they are applicable only to saturated materials.

In many investigations, particularly those involving construction of facilities or analysis of hazardous waste sites, information on low-permeability or unsaturated materials is required. Often, a number of locations need to be tested to provide data on spatial variations in characteristics of subsurface materials.

Laboratory permeability tests of subsurface materials usually are not satisfactory. Test specimens from such materials can seldom, if ever, be obtained in an entirely undisturbed state, and a specimen may represent only a limited portion of the material being investigated.

Tests have been devised that are relatively simple and less costly than aquifer pumping tests. These tests are usually conducted in conjunction with exploratory drilling or existing monitoring wells. They are designed to obtain data relating to possible or existing seepage, uplift pressures, contaminant transport, and similar problems that may occur in low permeability materials.

Exploratory drilling to determine foundation and other conditions is costly and time consuming, but its value is generally recognized. Permeability testing is often an integral part of the operation and the cost of such exploratory drilling. Permeability testing on existing monitor wells may be advantageous to determine characteristics of materials where evaluation of existing data indicates gaps or necessity to confirm previous assumptions. Properly conducted and controlled permeability tests will yield reasonably accurate and reliable data.

The tests described herein show semiquantitative values of permeability. If they are performed properly, however, the values

obtained are sufficiently accurate for most engineering purposes.
Linear, volumetric, pressure, and time measurements should be
made as accurately as available equipment will permit, and gauges
should be checked periodically for accuracy.

The quality of water used in permeability tests is of primary importance. The presence of only a few parts per million of turbidity
or air dissolved in water can plug soil and rock voids and cause
serious errors in test results. Water should be clear and silt free.
To avoid plugging of the soil pores by air bubbles, the use of water
that is a few degrees warmer than the ground temperature of the
test section is a desirable practice.

For some packer tests, pumps of up to 950-liters-per-minute
(250-gallons-per-minute) capacity against a total dynamic head of
50 meters (160 feet) may be required.

If meters and gauges are located in relation to each other as
recommended, the arrangement of pipe, hose, etc., will not
seriously influence the tests. However, in the interest of pumping
efficiency, sharp bends in hose, 90-degree fittings on pipes, and unnecessary changes in pipe and hose diameters should be avoided.

The equations given for computing permeability are applicable
for laminar flow. The velocity at which turbulent flow occurs
depends on the grain size of the materials tested and other factors,
but a safe average figure below which flow would be considered
laminar is about 25 millimeters per second (0.1 foot per second).
Therefore, in tests, if the quotient of the water intake in cubic
units per second divided by the open area of the test section in
square units times the estimated porosity of the tested material is
greater than 0.10, the various given equations may not be
applicable.

In an open hole test, the total open area of the test section is
computed as follows:

$$a = \pi dA + \pi r^2 \qquad\qquad 10\text{-}1$$

where:

a = total open area of the hole face plus the hole bottom
r = radius of the hole
d = diameter of the hole
A = length of the test section of the hole

In a test using perforated casing, the open area of the perforations is computed as follows:

$$a_p = na_s \qquad\qquad 10\text{-}2$$

where:

n = number of perforations
a_p = total open area of perforations
a_s = area of each perforation

If the bottom of the perforated casing is open, this area must be added to the area of the perforations to obtain the total open area.

Where fabricated well screens are used, estimates of open area may require precise measurement of screen component dimensions and the computations based on these measurements. However, information on open area can generally be obtained from the screen manufacturer. Table 11-9 contains information on many commonly used screens.

For purposes of discussion, permeability tests are divided into four types: pressure tests, constant head gravity tests, falling head gravity tests, and slug or bail tests. In pressure tests and falling head gravity tests, one or two packers are used to segregate the test section in the hole. In pressure tests, water is forced into the test section through combined applied pressure and gravity head, although the tests can be performed using gravity head only. In falling head tests, gravity head only is used. In constant head gravity tests, no packers are used, and a constant water level is maintained (Ahrens and Barlow, 1951). Slug or bail tests use only small changes in water level, generally over a short-time period.

10-2. Pressure Permeability Tests in Stable Rock.—
Pressure permeability tests are run using one or two packers to isolate various zones or lengths of drill hole in stable rock. Hole diameters usually do not exceed 87 millimeters (3.5 inches), but larger holes can also be tested if suitable equipment is available. The tests may be run in vertical, angled, or horizontal holes and analyzed if the head and zone relationships can be determined. Pressure tests are often the only practical tests to use when permeability of streambeds or lakebeds must be determined below water.

Compression packers, inflatable packers, leather cups, and similar types of packers have been used for pressure testing. Inflatable packers are usually more economical because they reduce testing time and ensure a tighter seal, particularly in rough-walled or out-of-round holes. The packers are inflated through tubes extending to a cylinder of air or nitrogen at the surface. If a pressure sensing instrument is included, pressure in the test section is sensed by the instrument and is transmitted to the surface by an electrical circuit where it is either read from a register at the surface or is recorded on a chart. Although this arrangement permits an accurate determination of test pressures, other observations outlined in this section should still be made to permit an estimate of permeability should the pressure sensor fail. This double packer arrangement permits successive tests at different depths in a completed hole without having to remove the packer between each test. The pressure sensor can also be adapted where a single packer is used.

(a) *Methods of Testing.*—The most common testing practice by the Bureau of Reclamation is the drilling of about 3 meters (10 feet) of hole and pressure testing the newly drilled section. In rock that tends to ravel or bridge the hole and which must be cemented to permit continuation of drilling, this is the only practical method of testing. In such rock, long lengths of open hole are impractical, and the test sections must be kept short because the tests must be made before the hole is cemented if good test data are to be obtained.

Where the rock is stable and does not require cementing, the following method of testing may offer distinct advantages. The hole is drilled to the total depth without testing. Two inflatable packers 1.5 to 3 meters (5 to 10 feet) apart are mounted near the bottom of the rod or pipe used for making the test. The bottom of the rod or pipe is sealed, and the section between the packers is perforated. The perforations should be at least 6 millimeters (1/4 inch) in diameter, and the total area of all perforations should be greater than two times the inside cross sectional area of the pipe or rod. Tests are made beginning at the bottom of the hole. After each test, the packers are raised the length of the test section and another test made. This procedure is followed until the entire length of the hole has been tested.

(b) *Cleaning Test Sections Before Testing.*—Before each test, the test section should be surged with clear water and bailed out to clean cuttings and drilling fluid from the face of the hole. If the test section is above the water table and will not hold water, water should be poured into the hole during the surging, then bailed out as rapidly as possible. When a completed hole is tested using two packers, the entire hole can be cleaned in one operation. Cleaning the hole is frequently omitted from testing procedures; however, this omission may result in a permeable rock appearing to be impermeable because the hole face is sealed by cuttings or drilling fluid. In such cases, the computed permeability will be lower than the true permeability.

Alternative methods to surging and bailing a drill hole in consolidated formations before pressure testing include the use of a rotating, stiffly bristled brush while washing and jetting with water. An average jet velocity of 45 meters per second (150 feet per second) is desirable. This velocity is approximated by a rate of pumping equal to 5.3 liters per minute per 2-millimeter (1.4 gallons per minute per 1/16-inch) diameter hole in the rod. On completion of jetting, the hole should be blown or bailed out to the bottom, if possible.

(c) *Length of Test Section.*—The length of the test section is governed by the character of the rock, but generally a length of 3 meters (10 feet) is desirable. At times, a good seal cannot be obtained for the packer at the planned elevation because of bridging, raveling, or the presence of fractures. Under these circumstances, the test section length should be increased or decreased or test sections overlapped to ensure that the test is made with well-seated packers. On some tests, a 3-meter section will take more water than the pump can deliver; hence, no back pressure can be developed. When this difficulty occurs, the length of the test section should be shortened until back pressure can be developed, or the falling head test (section 10-4) might be tried.

The test sections should never be shortened to where the ratio D/A is less than 5, where D is the diameter of the hole and A is the length of the test section. Under no circumstances should a packer be set inside the casing when making a test unless the casing has been grouted in the hole. Except under the most adverse condition, the use of test sections greater than 6 meters (20 feet) in length is inadvisable. Longer test sections may not permit sufficient localization of permeable zones and may complicate computations.

(d) *Size of Rod or Pipe to Use in Tests.*—Drill rods are commonly used as intake pipes to make pressure and permeability tests. NX and NW rods can be used for this purpose without seriously affecting the reliability of the test data if the intake of the test section does not exceed 45 to 60 liters per minute (12 to 15 gallons per minute) and the depth to the top of the test section does not exceed 15 meters (50 feet). For general use, 32-millimeter (1-1/4-inch) or larger pipe is more satisfactory. Figures 10-1 through 10-4 show head losses per 3-meter (10-foot) section at various deliveries of water for different sizes of drill rod and 32-millimeter pipe. These figures were compiled from experimental tests. The desirability of using the 32-millimeter pipe, particularly where holes 15 meters (50 feet) or more in depth are to be tested, is obvious from study of the graphs. The couplings on the 32-millimeter (1-1/4-inch) pipe must be turned down to an outside diameter of 45 millimeters (1.8 inch) for use in AX holes.

   (e) *Pumping Equipment.*—Tests are commonly run using a mud pump for pumping the water. Such pumps are generally of the multiple cylinder type with a uniform fluctuation in pressure. Many of these pumps have a maximum capacity of about 95 liters per minute (25 gallons per minute) and, if not in good condition, the capacities may be as small as 64 to 68 liters per minute (17 to 18 gallons per minute). Tests are often difficult, if not impossible, to analyze because such pumps do not have sufficient capacity to develop back pressure in the length of hole being tested. When this happens, the tests are generally reported: "took capacity of pump, no pressure developed." This result does not permit determination of permeability of the material tested other than it is probably high. The fluctuating pressures of multiple cylinder pumps, even when an air chamber is used, are often difficult to read accurately because the high and low readings must be averaged to determine the approximate true effective pressure, a difficulty which may be a source of error. In addition, such pumps occasionally develop instantaneous excessively high pressures which may fracture the rock or blow out a packer.

   Permeability tests made in drill holes ideally should be performed using centrifugal pumps having sufficient capacity to develop back pressure. A pump with a capacity of up to 950 liters per minute (250 gallons per minute) against a total head of 48 meters (160 feet) would be adequate for most testing. Head and discharge of such pumps are easily controlled by changing engine speed or with a control valve on the discharge.

   (f) *Swivels for Use in Tests.*—Swivels used for testing should be selected for minimum head losses.

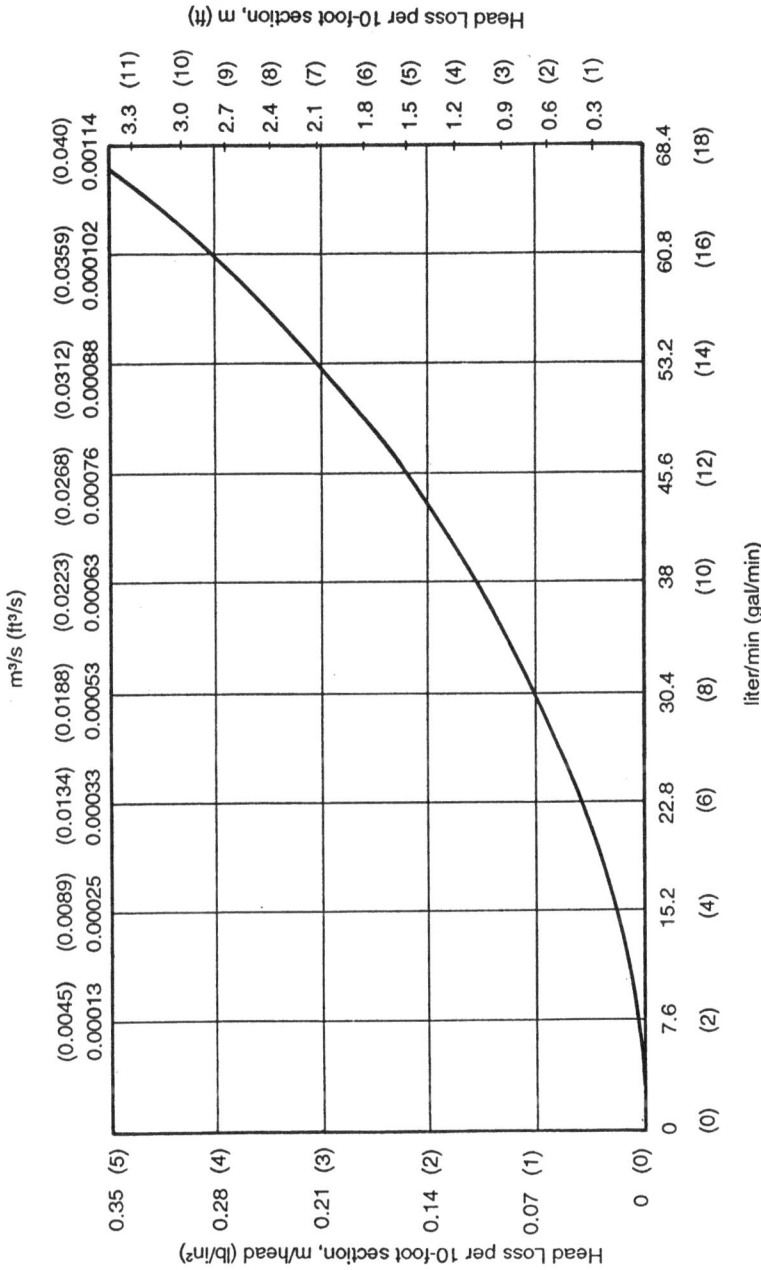

Figure 10-1.—Head loss in a 3-meter (10-foot) section of AX drill rod.

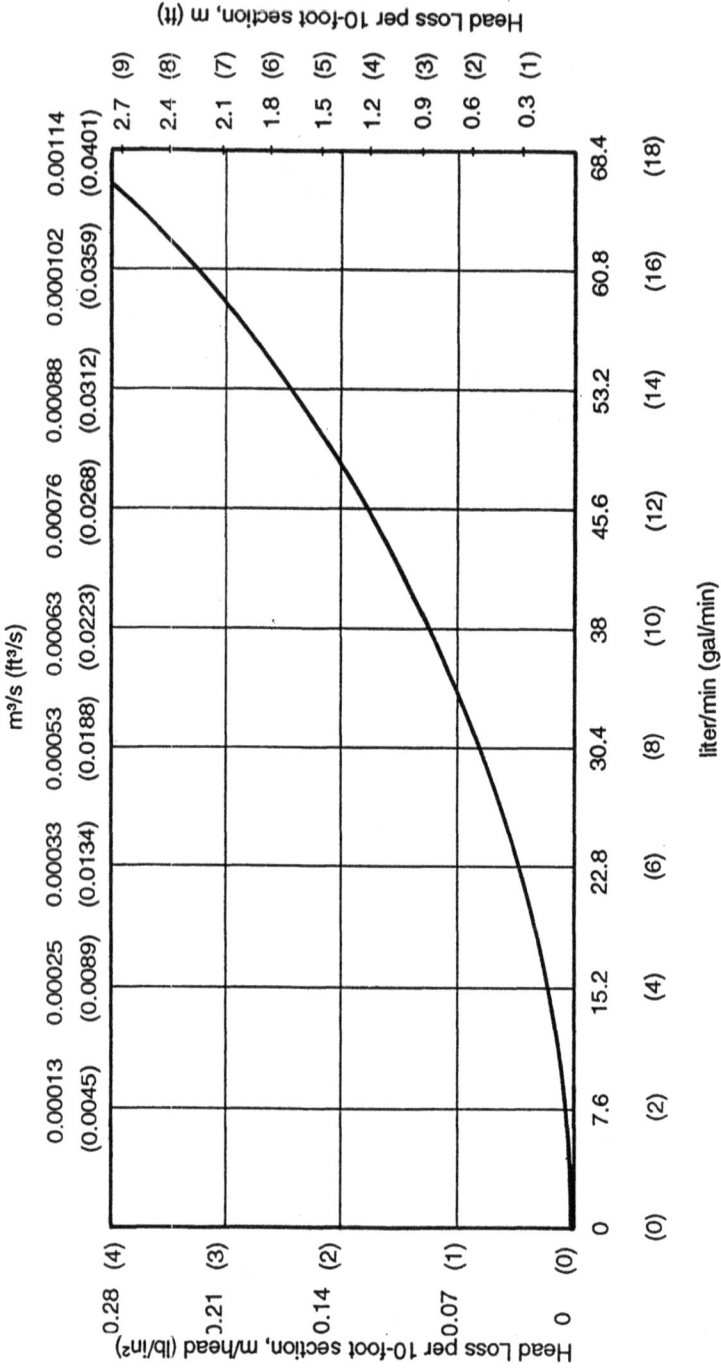

Figure 10-2.—Head loss in a 3-meter (10-foot) section of BX drill rod.

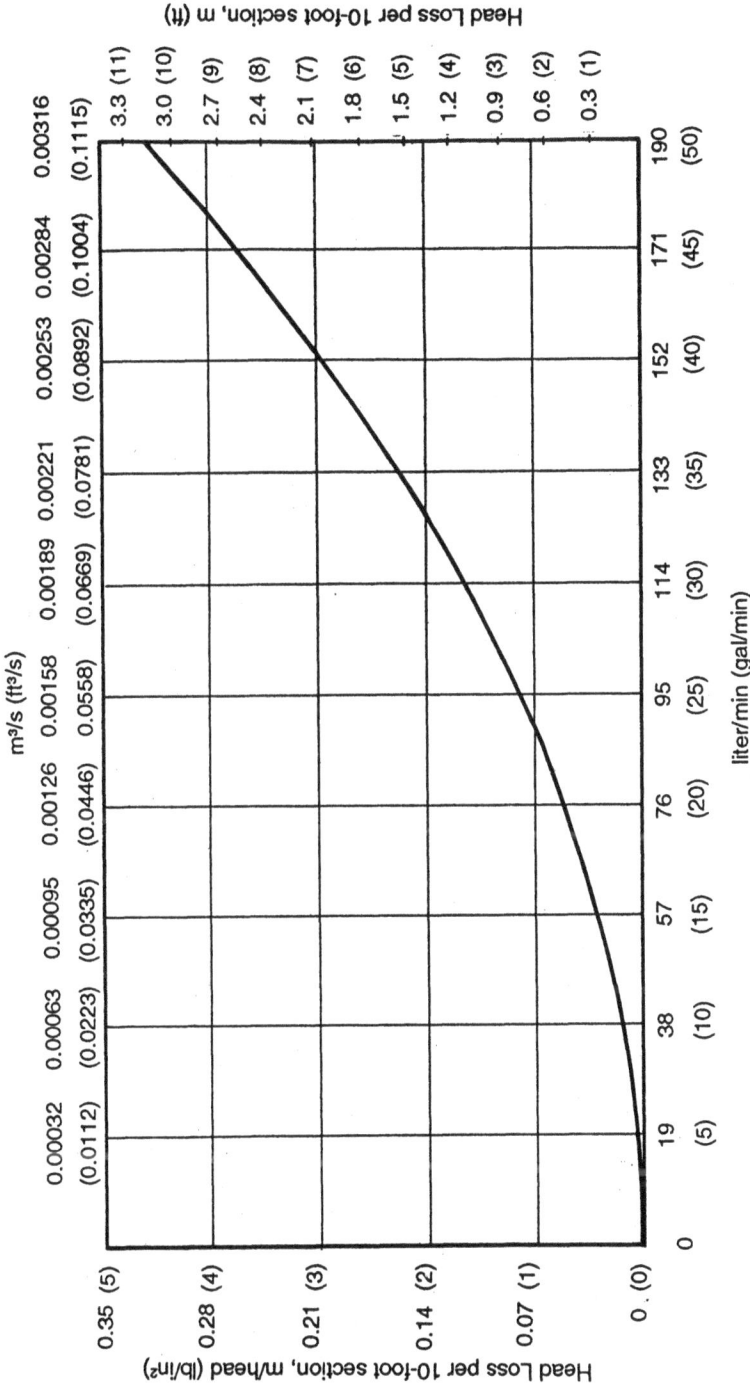

Figure 10-3.—Head loss in a 3-meter (10-foot) section of NX drill rod.

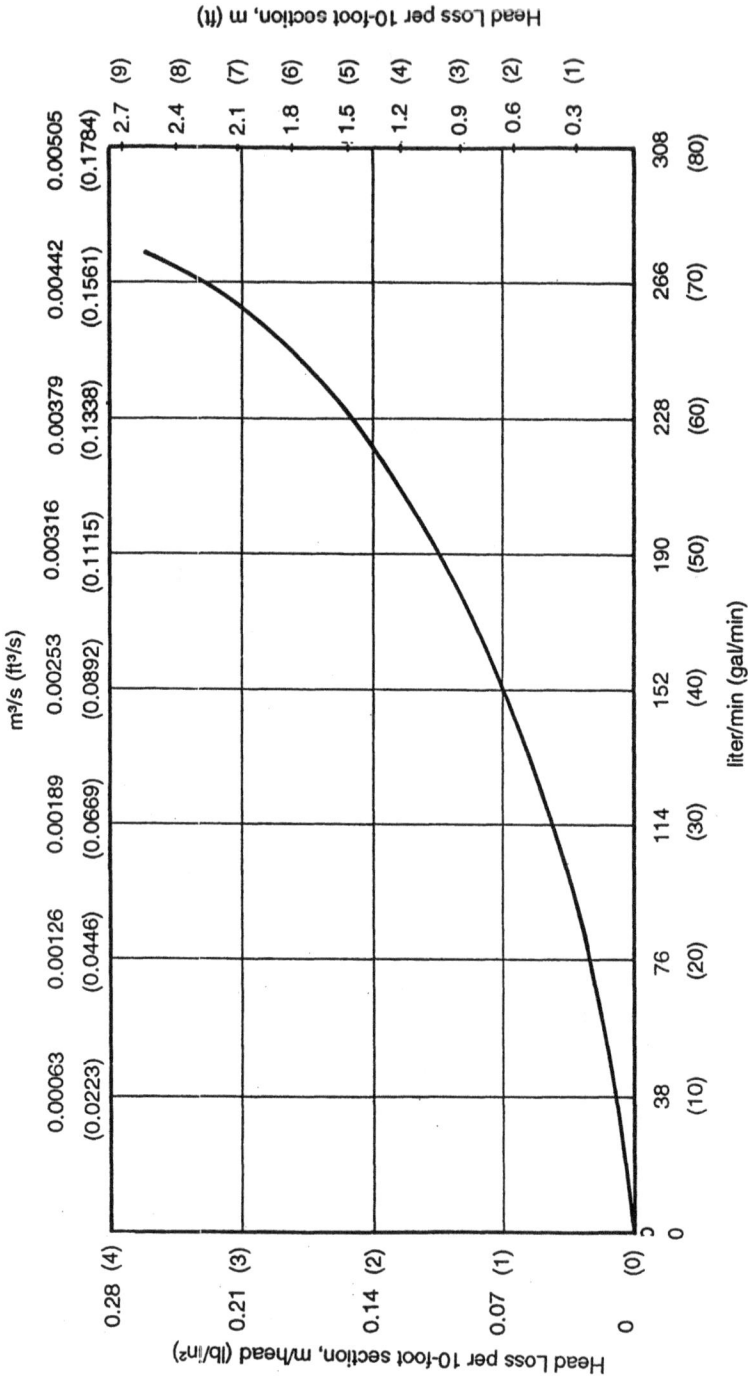

Figure 10-4.—Head loss in a 3-meter (10-foot) section of 32-millimeter (1-1/4-inch) iron pipe.

(g) Location of Pressure Gauge in Tests.—The ideal location for a pressure gauge is near the well head, preferably between the packer and the swivel.

(h) Recommended Watermeters.—Required water deliveries in pressure tests may range from less than 3.8 liters per minute (1 gallon per minute) to as much as 1,500 liters per minute (400 gallons per minute). No one meter is sufficiently accurate at all ranges to be reliably used. Therefore, 2 meters are recommended: (1) a 100-millimeter (4-inch) propeller or impeller-type meter to measure flows between 200 and 1,300 liters per minute (50 and 350 gallons per minute); and (2) a 25-millimeter (1-inch) disk-type meter for flows between 4 and 200 liters per minute (1 and 50 gallons per minute). Ideally, each meter should be equipped with an instantaneous flow indicator as well as a totalizer. Watermeters should be tested frequently to assure reliability.

Inlet pipe adapters should be available for each meter to minimize turbulent inflow. The adapters should be at least 10 times as long as the diameter of the rated size of the meter.

(i) Length of Time for Tests.—The minimum length of time to run a test depends upon the nature of the material tested. Tests should be run until stabilization occurs (i.e., until three or more readings of water intake and pressure taken at 5-minute intervals are essentially equal). In tests above the water table, water should be pumped into the test section at the desired pressure for about 10 minutes in coarse materials or 20 minutes in fine-grained materials before making measurements.

Stability is obtained more rapidly in tests below the water table than in unsaturated material. When multiple pressure tests are made, each pressure theoretically should be maintained until stabilization occurs. This procedure is not practicable in some cases, but good practice requires that each pressure stop be maintained for at least 20 minutes with intake and pressure readings made at 5-minute intervals as the pressure is increased and for 5 minutes as pressure is decreased.

(j) Pressures to be Used in Testing.—Where subsurface conditions for proposed reservoirs or other water-impounding or storage facilities are being investigated, the theoretical minimum pressure used in the test section should equal the head imposed by the maximum reservoir level. However, when tests are made in

locations where the ground surface is below the proposed maximum pool level, the use of such test pressures may be impractical because of the danger of blowouts or fracturing the hole face. Under these conditions, a safe pressure in consolidated rock is 3.4 kilopascals (0.5 lb/in²), or 1.35 meters of water per meter (1.5 feet of water per foot) of depth from the ground surface to the top of the test section. In all other locations, the same criterion is a good rule-of-thumb guide. Tables 10-1(a) and 10-1(b) and 10-2 are provided for converting kilograms per square meter to meters of water (lb/in² to feet of water) and vice versa, respectively.

(k) Arrangement of Equipment.—Recommended arrangement of test equipment starting at the source of water is as follows:

Source of water, suction line, pump, waterline to settling and storage tank or basin, suction line, centrifugal test pump, line to watermeter adapter (if required) or to watermeter, short length of pipe, plug valve, waterline to swivel, sub for gauge, and pipe or rod to packer. All connections should be kept as short and straight as possible with a minimum number of changes in diameter of hose, pipe, etc.

All joints, connections, and hose between the watermeter and the packer or casing should be tight so no water loss occurs between the meter and the test section.

(l) Pressure Permeability Test Methods.—A schematic drawing of the following two methods is shown on figure 10-5.

Method 1: This method is primarily applicable to testing in unconsolidated rock which requires casing as the hole is drilled, although it may be used in stable material if desired. The hole is drilled, the tools are removed, a packer is seated a given distance above the bottom of the hole, water under pressure is pumped into the test section, and the readings are recorded. The packer is then removed, the hole is drilled deeper, the packer is inserted so as to leave the full length of the newly drilled hole within the test section, and the test is repeated.

Method 2: This method is applicable in consolidated rock which is stable and does not require cementing. The hole is drilled to the final depth, cleaned, and blown out or bailed. Two packers, spaced on pipe or drill stem to isolate the desired test section, are used. Tests should

Table 10-1(a).—Conversion of heads of water in meters to hydrostatic pressures in kilopascals

Head (meters)	Hydrostatic pressure in kilopascals (kPa)[1]									
	0	0.1	0.2	0.3	0.4	0.5	0.6	0.7	0.8	0.9
0	—	0.98	1.96	2.94	3.92	4.90	5.88	6.86	7.85	8.83
1	9.81	10.79	11.77	12.75	13.73	14.71	15.69	16.67	17.65	18.63
2	19.61	20.59	21.57	22.85	23.53	24.51	25.49	26.47	27.45	28.44
3	29.42	30.40	31.38	32.36	33.34	34.32	35.30	36.28	37.26	38.24
4	39.22	40.20	41.19	42.17	43.15	44.13	45.11	46.09	47.07	48.05
5	49.03	50.01	50.99	51.97	52.95	53.93	54.91	55.89	56.87	57.86
6	58.84	59.82	60.80	61.78	62.76	63.74	64.72	65.20	66.68	67.66
7	68.64	69.62	70.60	71.58	72.56	73.54	74.53	75.51	76.49	77.47
8	78.45	79.43	80.41	81.39	82.37	83.35	84.33	85.31	86.29	87.27
9	88.25	89.23	90.22	91.20	92.18	93.16	94.14	95.12	96.10	97.08
10	98.06	99.04	100.02	101.00	101.98	102.96	103.94	104.92	105.90	106.89
11	107.87	108.85	109.83	110.81	111.78	112.77	113.75	114.73	115.72	116.70
12	117.68	118.66	119.64	120.62	121.60	122.58	123.56	124.54	125.52	126.50
13	127.48	128.46	129.44	130.42	131.40	132.39	133.37	134.35	135.33	136.31
14	137.29	138.27	139.25	140.23	141.21	142.19	143.17	144.15	145.13	146.11
15	147.09	148.08	149.06	150.04	151.02	152.00	152.98	153.96	154.94	155.92
16	156.90	157.88	158.86	159.84	160.82	161.80	162.78	163.77	164.75	165.73
17	166.71	167.79	168.67	169.65	170.63	171.61	172.59	173.57	174.55	175.53
18	176.51	177.49	178.48	179.46	180.49	181.42	182.40	183.38	184.36	185.34
19	186.32	187.30	188.28	189.26	190.24	191.22	192.20	193.18	194.17	195.15
20	196.13	197.11	198.09	199.07	200.05	201.03	202.01	202.99	203.97	204.95
21	205.93	206.91	207.89	208.87	209.86	210.84	211.82	212.80	213.78	214.76
22	215.74	216.72	217.70	218.68	219.66	220.64	221.62	222.60	223.58	224.57
23	225.55	226.53	227.51	228.49	229.47	230.45	231.43	232.41	233.39	234.37
24	235.35	236.33	237.31	238.29	239.27	240.26	241.24	242.22	243.20	244.18
25	245.16	246.14	147.12	248.10	249.08	250.06	251.04	252.02	253.00	253.78
26	254.96	255.95	256.93	257.91	258.89	259.87	260.85	261.83	262.81	263.79
27	264.77	265.75	266.73	267.71	268.69	269.67	270.65	271.64	272.62	273.60
28	274.58	275.56	276.54	277.52	278.50	279.48	280.46	281.44	282.42	283.40
29	284.38	285.36	286.35	287.33	288.09	289.29	290.27	291.25	292.23	293.21
30	294.19	295.17	296.15	297.13	298.11	299.09	300.07	301.05	302.04	303.02

[1] Based on relationship of 1 meter head of water = 9.80636 kPa.

Table 10-1(b).—Conversion of heads of water in feet to hydrostatic pressures in pounds per inch (weight of water—62.4 pounds per cubic foot)

Head (feet)	Hydrostatic pressure in pounds per cubic foot									
	0	1	2	3	4	5	6	7	8	9
0	—	0.43	0.87	1.30	1.73	2.17	2.60	3.03	3.47	3.90
10	4.33	4.77	5.20	5.64	6.07	6.50	6.94	7.37	7.80	8.24
20	8.67	9.10	9.54	9.97	10.40	10.84	11.27	11.70	12.14	12.57
30	13.00	13.44	13.87	14.31	14.74	15.17	15.61	16.04	16.47	16.91
40	17.34	17.77	18.21	18.64	19.07	19.51	19.94	20.37	20.81	21.24
50	21.67	22.11	22.54	22.98	23.41	23.84	24.28	24.71	25.14	25.58
60	26.01	26.44	26.88	27.31	27.74	28.18	28.61	29.04	29.48	29.91
70	30.34	30.78	31.21	31.65	32.08	32.51	32.95	33.38	33.81	34.25
80	34.68	35.11	35.55	35.98	36.41	36.85	37.28	37.71	38.15	38.58
90	39.01	39.45	39.88	40.32	40.75	41.18	41.62	42.05	42.48	42.92
100	43.35	43.78	44.22	44.65	45.08	45.52	45.95	46.38	46.82	47.25
110	47.68	48.12	48.55	48.99	49.42	49.85	50.29	50.72	51.15	51.59
120	52.02	52.45	52.89	53.32	53.75	54.19	54.62	55.05	55.49	55.92
130	56.36	56.79	57.22	57.66	58.09	58.52	58.96	59.39	59.82	60.26
140	60.69	61.12	61.56	61.99	62.42	62.86	63.29	63.72	64.16	64.59
150	65.02	65.46	65.89	66.33	66.76	67.19	67.63	68.06	68.49	68.93
160	69.36	69.79	70.23	70.66	71.09	71.53	71.96	72.39	72.83	73.26
170	73.69	74.13	74.56	75.00	75.43	75.86	76.30	76.73	77.16	77.60
180	78.03	78.46	78.90	79.33	79.76	80.20	80.63	81.06	81.50	81.93
190	82.36	82.80	83.23	83.67	84.10	84.53	84.97	85.40	85.83	86.27
200	86.70	87.13	87.57	88.00	88.43	88.87	89.30	89.73	90.17	90.60

Table 10-2(a).—Conversion of hydrostatic pressures in kPa to head of water in meters[1]

Pressure (kPa)	Hydrostatic head in meters of water									
	0	1	2	3	4	5	6	7	8	9
0	—	0.10	0.20	0.31	0.41	0.51	0.61	0.71	0.82	0.92
10	1.02	1.12	1.22	1.33	1.43	1.53	1.63	1.73	1.84	1.94
20	2.04	2.14	2.24	2.35	2.45	2.55	2.65	2.75	2.86	2.96
30	3.06	3.16	3.26	3.37	3.47	3.57	3.67	3.77	3.87	3.98
40	4.08	4.18	4.28	4.38	4.49	4.59	4.69	4.79	4.89	5.00
50	5.01	5.20	5.30	5.40	5.51	5.61	5.71	5.81	5.91	6.02
60	6.12	6.22	6.32	6.42	6.53	6.63	6.73	6.83	6.93	7.04
70	7.14	7.24	7.34	7.44	7.55	7.65	7.75	7.85	7.95	8.06
80	8.16	8.26	8.36	8.46	8.57	8.67	8.77	8.87	8.97	9.08
90	9.18	9.28	9.38	9.48	9.59	9.69	9.79	9.89	10.00	10.10
100	10.20	10.30	10.40	10.50	10.60	10.71	10.81	10.91	11.01	11.11
110	11.23	11.32	11.42	11.52	11.62	11.73	11.83	11.93	12.03	12.13
120	12.24	12.34	12.44	12.54	12.64	12.75	12.85	12.95	13.05	13.15
130	13.26	13.36	13.46	13.56	13.66	13.77	13.87	13.97	14.07	14.17
140	14.28	14.38	14.48	14.58	14.68	14.79	14.89	14.99	15.09	15.19
150	15.30	15.40	15.50	15.60	15.70	15.81	15.91	16.01	16.11	16.21
160	16.32	16.42	16.52	16.62	16.72	16.83	16.93	17.03	17.13	17.23
170	17.33	17.44	17.54	17.64	17.74	17.84	17.95	18.05	18.15	18.25
180	18.35	18.46	18.56	18.66	18.76	18.86	18.97	19.07	19.17	19.27
190	19.37	19.48	19.58	19.68	19.78	19.88	19.99	20.09	20.19	20.29
200	20.39	20.50	20.60	20.70	20.80	20.90	21.01	21.11	21.21	21.31

[1] Based on 1 kPa = 0.10197 meter of water.

Table 10-2(b).—Conversion of hydrostatic pressures in pounds per square inch to head of water in feet

Pounds (square inch)	Hydrostatic head in feet of water									
	0	1	2	3	4	5	6	7	8	9
0	—	2.31	4.62	6.93	9.24	11.55	13.86	16.17	18.48	20.79
10	23.10	25.41	27.72	30.03	32.34	34.65	36.96	39.27	41.58	43.89
20	46.20	48.51	50.82	53.13	55.44	57.75	60.06	62.37	64.68	66.99
30	69.30	71.61	73.92	76.23	78.54	80.85	83.16	85.47	87.78	90.09
40	92.40	94.71	97.02	99.33	101.64	103.95	106.26	108.57	110.88	113.19
50	115.50	117.81	120.12	122.43	124.74	127.05	129.36	131.67	133.98	136.29
60	138.60	140.91	143.22	145.43	147.84	150.15	152.46	154.77	157.08	159.39
70	161.70	164.01	166.32	168.63	170.94	173.25	175.56	177.87	180.18	182.49
80	184.80	187.11	189.42	191.73	194.04	196.35	198.66	200.97	203.28	205.59
90	207.90	210.21	212.52	214.83	217.14	219.45	221.76	224.07	226.38	228.69
100	231.00	233.31	235.62	237.93	240.24	242.55	244.86	247.17	249.48	251.79
110	254.10	256.41	258.72	261.03	263.34	265.65	267.96	270.27	272.58	274.89
120	277.20	279.51	281.82	284.13	286.44	288.75	291.06	293.37	295.68	297.99
130	300.00	302.61	304.92	307.23	309.54	311.85	314.16	316.47	318.78	321.09
140	323.40	325.71	328.02	330.33	332.64	334.95	337.26	339.57	341.88	344.19
150	346.50	348.81	351.12	353.43	355.74	358.05	360.36	362.67	364.98	367.29
160	369.60	371.91	374.22	376.53	378.84	381.15	383.46	385.77	388.08	390.39
170	392.70	395.01	397.32	399.63	401.94	404.25	406.56	408.87	411.18	413.49
180	415.80	418.11	420.42	422.73	425.04	427.35	429.66	431.97	434.28	436.59
190	438.90	441.21	443.52	445.83	448.14	450.45	452.76	455.07	457.38	459.69
200	462.00	464.31	466.62	468.93	471.24	473.55	475.86	478.17	480.48	482.79

K=coefficient of permeability, meters (ft) per second under a unit gradient

Q=steady flow into well, m³/$_s$ (ft³/$_s$)

H=h$_1$+h$_2$-L=effective head, m (ft)

h$_1$(above Water table)=distance between Bourdon gage and bottom of hole for method 1 or distance between gage and upper surface of lower packer for method 1, m (ft)

h$_1$(below water table)=distance between gage and water table, m (ft)

h$_2$=applied pressure at gage, 6.89 kPa =0.703 m of water (1 lb/in²=2.307 ft)

L= head loss in pipe due to friction, m (ft);ignore head loss of Q<15 L/min (4 gal/min)in 40mm (1½in) pipe; use length of pipe between gage and top of test section for computations

X=H/T$_u$(100)=percent of unsaturated stratum

A=length of test section, m (ft)

r=radius of test hole, m (ft)

C$_u$= conductivity coefficient for unsaturated materials with partially penetrating cylindrical test wells

C$_s$= conductivity coefficient for semi-spherical flow in saturated materials with partially penetrating cylindrical test wells

U=thickness of unsaturated material, m (ft)

S=thickness of saturated material, m (ft)

T$_u$=U-D+H=distance from water surface in well to water table, m (ft)

D=distance from ground surface to bottom of test section, m (ft)

a= surface area of test section, m² (ft²);area of wall plus area of bottom for method 1; area of wall for method 2

Limitations:
Q/a$_s$, S≥5A, a≥10r, thickness of each packer must be ≥10r in method 2

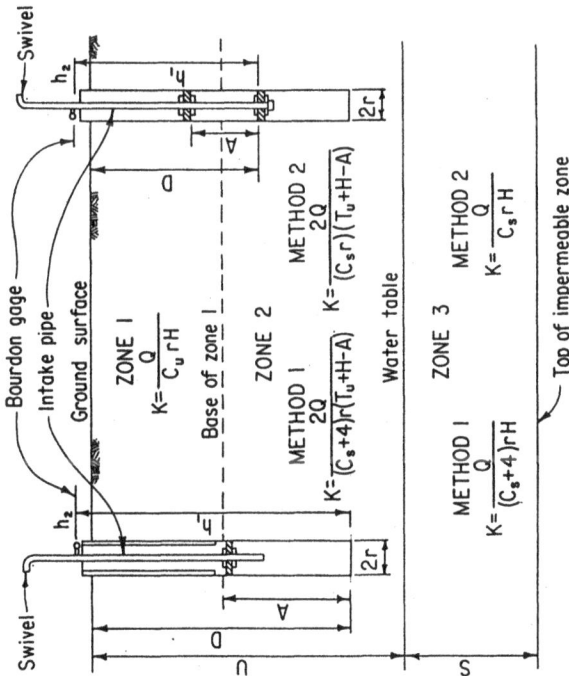

Figure 10-5.—Permeability test for use in saturated or unsaturated consolidated rock (Zangar, 1953).

be started at the bottom of the hole. After each test, the
pipe is lifted a distance equal to the A dimension shown
on figure 10-5, and the test is repeated until the entire
hole is tested.

Data required for computing the permeability may not be available until the hole has encountered the water table or a relatively
impermeable bed. The required data for each test include:

- Radius, r, of the hole, in meters (feet).

- Length of test section A, the distance between the packer
  and the bottom of the hole, method 1, or between the
  packers, method 2, in meters (feet).

- Depth, h, from pressure gauge to bottom of the hole,
  method 1, or from gauge to upper surface of lower packer in
  method 2, in meters (feet). If a pressure sensor is used,
  substitute the pressure recorded in the test section prior to
  pumping for the h value.

- Applied pressure, h_2, at the gauge, in meters (feet), or the
  pressure recorded during pumping in the test section if a
  sensor is used.

- Steady flow, Q, into well at 5-minute intervals, in m³/s
  (ft³/s).

- Nominal diameter in millimeters (inches) and length of
  intake pipe in meters (feet) between the gauge and upper
  packer.

- Thickness, U, of unsaturated material above water table, in
  meters (feet).

- Thickness, S, of saturated material above a relatively
  impermeable bed, in meters (feet).

- Distance, D, from the ground surface to the bottom of the
  test section, in meters (feet).

- The time that the test is started and time measurements
  are made.

- If tests are made in streambeds or lakebeds below water, the effective head is the difference in meters (feet) between the elevation of the free water surface in the pipe and the elevation of the gauge plus the applied pressure.

- If a pressure sensor is used, the effective head in the test section is the difference in pressure before water is pumped into the test section and the pressure readings made during the test.

The following examples show some typical calculations using methods 1 and 2 in the different zones shown on figure 10-5. Figure 10-6 shows the location of the zone 1 lower boundary for use in unsaturated materials. Because the process and not the units is important, only customary units are shown.

Pressure permeability tests examples of methods 1 and 2

Example 1:
  Zone 1 (Method 1)
  Given: $U = 75$ feet, $D = 25$ feet, $A = 10$ feet, $r = 0.5$ foot,
      $h_1 = 32$ feet, $h_2 = 25$ lb/in^2 = 57.8 feet,
      and $Q = 20$ gallons per minute = 0.045 ft^3/s

From figure 10-4: Head loss, L, for a 1-1/4-inch pipe at 20 gallons per minute is 0.76 foot per 10-foot section. If the distance from the Bourdon gauge to the bottom of the pipe is 22 feet, the total head loss, L, is $(2.2)(0.76) = 1.7$ feet.

$H = h_1 + h_2 - L = 32 + 57.8 - 1.7 = 88.1$ feet of effective head,
$T_u = U - D + H = 75 - 25 + 88.1 = 138.1$ feet

$$X = \frac{H}{T_u}\ (100) = \frac{88.1}{138.1}\ (100) = 63.8\%$$

$$\frac{T_u}{A} = \frac{138.1}{10} = 13.8$$

The values for X and T_u/A lie in zone 1 (figure 10-6). To determine the conductivity coefficient, C_u, from figure 10-7:

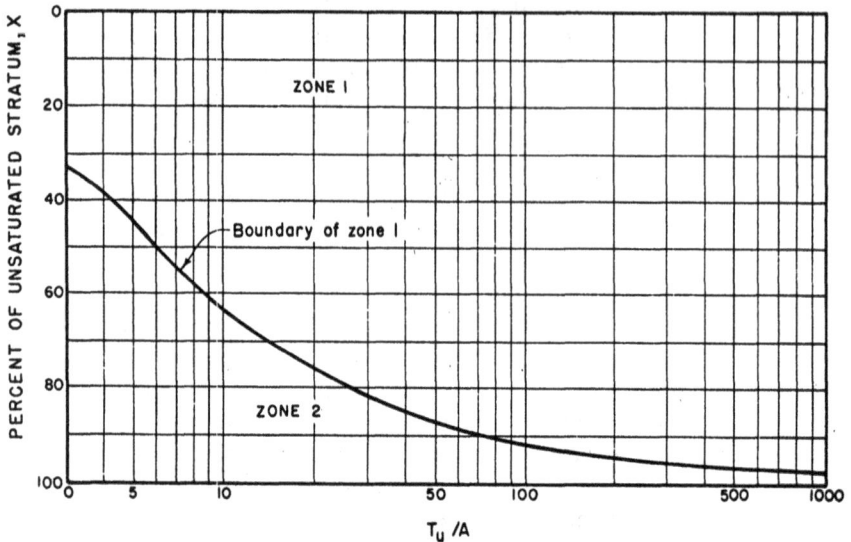

Figure 10-6.—Location of zone 1 lower boundary, for use in
unsaturated materials.

$$\frac{H}{r} = \frac{88.1}{0.5} = 176.2$$

$$\frac{A}{H} = \frac{10}{88.1} = 0.11 \quad also \quad C_u = 62$$

then:

$$K = \frac{Q}{C_u rH} = \frac{0.045}{(62)\,(0.5)\,(88.1)} = 0.000016 \; ft/s$$

Example 2:
  Zone 2
  Given: U, A, r, h_2, Q, and L are as given in example 1
         $D = 65$ feet, and $h_1 = 72$ feet

If the distance from the Bourdon gauge to the bottom of the
intake pipe is 62 feet, the total L is $(6.2)\,(0.76) = 4.7$ feet.

$H = 72 + 57.8 - 4.7 = 125.1$ feet
$T_u = 75 - 65 + 125.1 = 135.1$ feet

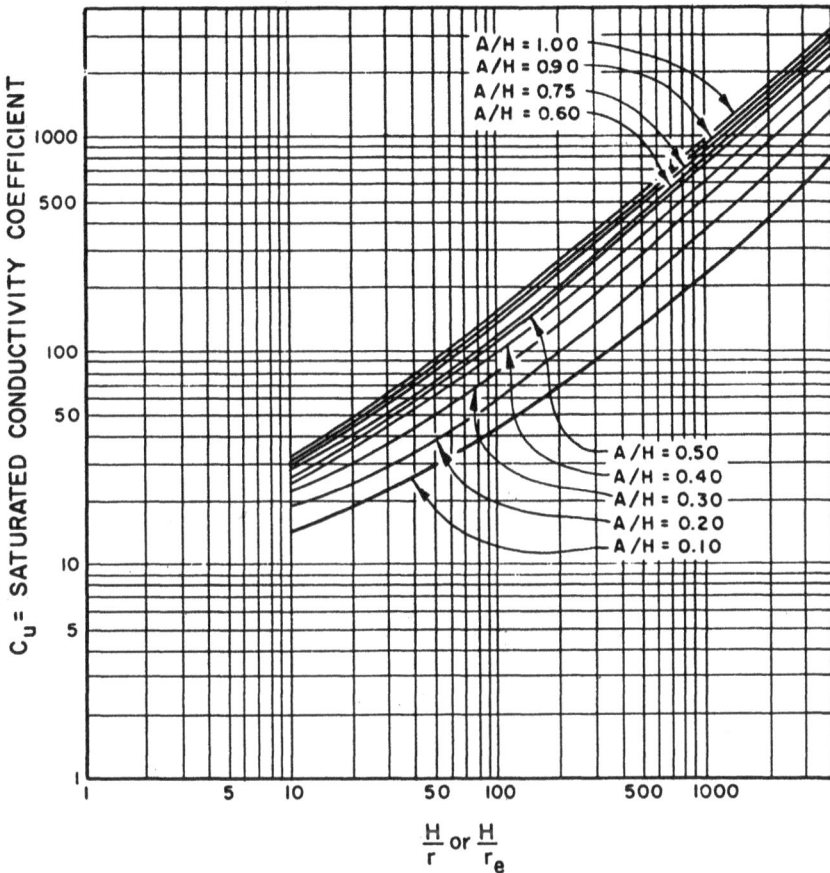

Figure 10-7.—Conductivity coefficients for permeability
determination in unsaturated materials with partially
penetrating cylindrical test wells.

$$X = \frac{125.1}{135.1} (100) = 92.6\% \quad also \frac{T_u}{A} = \frac{135.1}{10} = 13.5$$

The test section is located in zone 2 (figure 10-6). To determine
the conductivity coefficient, C_s, from figure 10-8:

$$\frac{A}{r} = \frac{10}{0.5} = 20 \quad also \quad C_s = 39.5$$

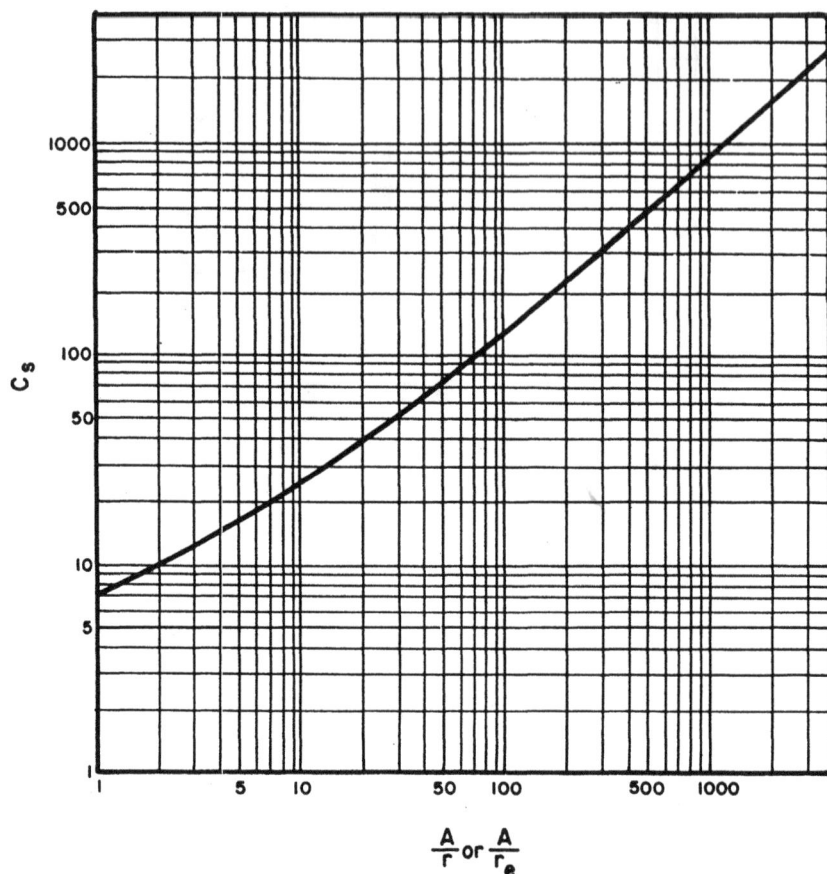

Figure 10-8.—Conductivity coefficients for semispherical flow in saturated materials through partially penetrating cylindrical test wells.

Method 1:

$$K = \frac{2Q}{(C_s + 4)r(T_u + H - A)}$$

$$K = \frac{(2)(0.045)}{(39.5 + 4)(0.5)(135.1 + 125.1 - 10)}$$

10-3

$$K = 0.000016 \ ft/s$$

Method 2:

$$K = \frac{2Q}{(C_s \, r)(T_u + H - A)}$$

10-4

$$K = \frac{(2)(0.045)}{(39.5)(0.5)(135.1 + 125.1 - 10)} = 0.000018 \; ft/s$$

Example 3:
   Zone 3
   Given: $U, A, r, h_2, Q,$ and L are as given in example 1
       $D = 100$ feet, $h_1 = 82$ feet, and $S = 60$ feet

If the distance from the Bourdon gauge to the bottom of the intake pipe is 97 feet, the total L is $(9.7)(0.76) = 7.4$ feet.

$H = 82 + 57.8 - 7.4 = 132.4$ feet

$$\frac{A}{r} = \frac{10}{0.5} = 20 \;\; also \;\; C_s = 39.5 \; from \; figure \; 10\text{-}8$$

Method 1:

$$K = \frac{Q}{(C_s + 4)rH} = \frac{0.045}{(39.5 + 4)(0.5)(132.4)} = 0.000016 \; ft/s$$

Method 2:

$$K = \frac{Q}{C_s rH} = \frac{0.045}{(39.5)(0.5)(132.4)} = 0.000017 \; ft/s \qquad 10\text{-}6$$

Computer programs are available to solve these equations. Computer programs developed by the Bureau of Reclamation are available to the public by contacting the Engineering Geology Group at the Reclamation Service Center in Denver.

(m) Multiple Pressure Tests.—These tests are run in the same manner as other pressure permeability tests except that the pressure is applied in three or more approximately equal steps. For example, if the allowable maximum differential pressure is 621 kPa (90 lb/in^2), the test would be run at pressures of about 207 kPa, 414 kPa, and 621 kPa (30, 60, and 90 lb/in^2).

Each pressure step should be maintained for 20 minutes and intake readings should be made at 5 minute intervals. The pressure is then raised to the next step. On completion of the highest step, the process should then be reversed and the pressure should be maintained for 5 minutes at about the same middle and the lowest pressure steps. A plot of intake against pressure for the five steps in a multiple pressure test may be useful in assessing hydraulic conditions.

Graphs of synthetic test results of multiple pressure tests are plotted on figure 10-9. Synthetic plots have been used in the interest of uniform spacing for illustrative purposes, but the curves are typical of those most often encountered. The results should be analyzed on the basis of confined flow hydraulic principles combined with data obtained from the core or hole logs.

Probable conditions represented by the circled numbers on figure 10-9 are:

(1) Probably very narrow, clean fractures. Flow is laminar, permeability is low, and discharge is directly proportional to head.

(2) Firm, practically impermeable material; fractures are tight. Little or no intake regardless of pressure.

(3) Highly permeable, relatively large open fractures indicated by high rates of water intake and no back pressure. Pressure shown on gauge caused entirely by pipe resistance.

(4) Permeability high with fractures that are relatively open and permeable, but contain filling material which tends to expand on wetting or dislodges and tends to collect in traps that retard flow. Flow is turbulent.

(5) Permeability high, with fracture filling material which washes out, increasing permeability with time. Fractures probably are relatively large. Flow is turbulent.

(6) Similar to (4), but fractures are tighter and flow is laminar.

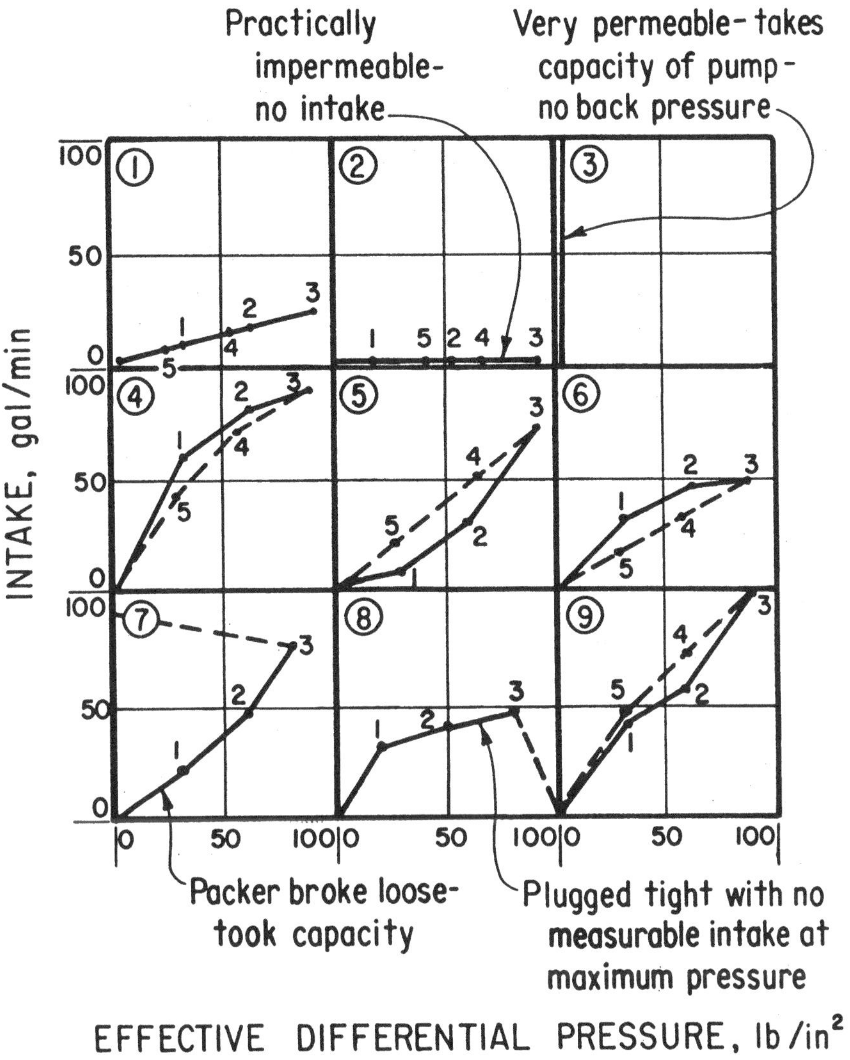

EFFECTIVE DIFFERENTIAL PRESSURE, lb/in²

Figure 10-9.—Plots of simulated, multiple pressure permeability tests.

⑦ Packer failed or fractures are large, flow is turbulent. Fractures have been washed clean; highly permeable. Test takes capacity of pump with little or no back pressure.

(8) Fractures are fairly wide and open but filled with clay gouge material which tends to pack and seal when subject to water under pressure. Takes full pressure with no water intake near end of test.

(9) Open fractures with filling which tend to first block and then break under increased pressure. Probably permeable. Flow is turbulent.

10-3. Gravity Permeability Tests.—Gravity permeability tests are intended primarily for use in unconsolidated or unstable materials and are usually made in larger diameter holes than those used for pressure tests. Gravity tests can be run only in vertical or near-vertical holes. A normal test section length is 1.5 meters (5 feet); however, if the material is stable, will stand without caving or sloughing, is relatively uniform, and sections up to 3 meters (10 feet) long may be tested. Shorter test sections may be used if the length of the water column in the test section is at least five times the diameter of the hole. After each test, the casing for open hole tests is driven to the bottom of the hole and a new test section is opened below it. If perforated casing is used, the pipe may be driven to the required depth and cleaned out, or the hole may be drilled to the required depth, then casing may be driven to the bottom of the hole and the hole may be cleaned out.

(a) Cleaning Test Sections.—Each newly opened test section should be developed by surging and bailing. This procedure should be done slowly and gently so that a large volume of loosely packed material will not be drawn into the hole, but so that compaction caused by drilling will be broken down and some fines will be removed from the formation.

(b) Measurement of Water Levels Through Protective Pipe.—In making gravity tests, insertion of a small-diameter perforated pipe (20 to 40 millimeters [3/4- to 1-1/2-inches]) in the hole is helpful in dampening wave or ripple action on the water surface caused by the inflow of water and also permits more accurate water-level measurements.

In an uncased test section in friable materials liable to wash, the end of the pipe should rest on a 100- to 150-millimeter (4- to 6-inch) cushion of coarse gravel at the bottom of the hole. In more stable material, the pipe may be suspended above the bottom of the hole, but the bottom of the pipe should be located at least 0.6 meter (2 feet) below the top of the water surface maintained in the hole.

Water may also be introduced through the pipe and measurements of water level may be made in the annular space between the pipe and the casing.

(c) *Pumping Equipment and Controls.*—Pressure is not required in the test section which normally has a free water surface. However, pump capacity should be adequate to maintain a constant head during the test.

A problem on many gravity tests is accurate control of the flow of water into the casing. The intake of the test section necessary to maintain a constant head is sometimes so small that inflow cannot be sufficiently limited using a conventional arrangement. A source of error is that many meters register inaccurately at very low flows. To overcome this difficulty, a constant head tank has been developed as shown on figure 10-10.

This tank will deliver 0.2 to 95 liters per minute (0.05 to 25 gallons per minute) of controlled flow. The materials used in its construction are available on most projects or are readily procured from plumbing supply houses, and the tank can be easily assembled by a welder. When the tank is completed, a rating curve must be prepared for it.

A hose delivering 76 to 95 liters per minute (20 to 25 gallons per minute) of water is placed in the tank until overflow begins. The valve is opened to each gradation in turn and the gallon per minute discharge at each opening is measured. Three measurements should be made at each gradation. The chart is prepared plotting the liters per minute (gallons per minute) discharge against the gauge opening. In the field, a close approximation of the discharge can be obtained by reference to the chart when the gauge is opened between gradations. A sample chart, figure 10-11, was prepared for a tank assembled in the research laboratory in Denver.

Some precautions must be observed in using the tank. The hose transmitting water from the tank to the casing should never be attached directly to the valve outlet except as noted in the following paragraph. Water discharging from the valve should fall freely into a hose two or more times the diameter of the valve

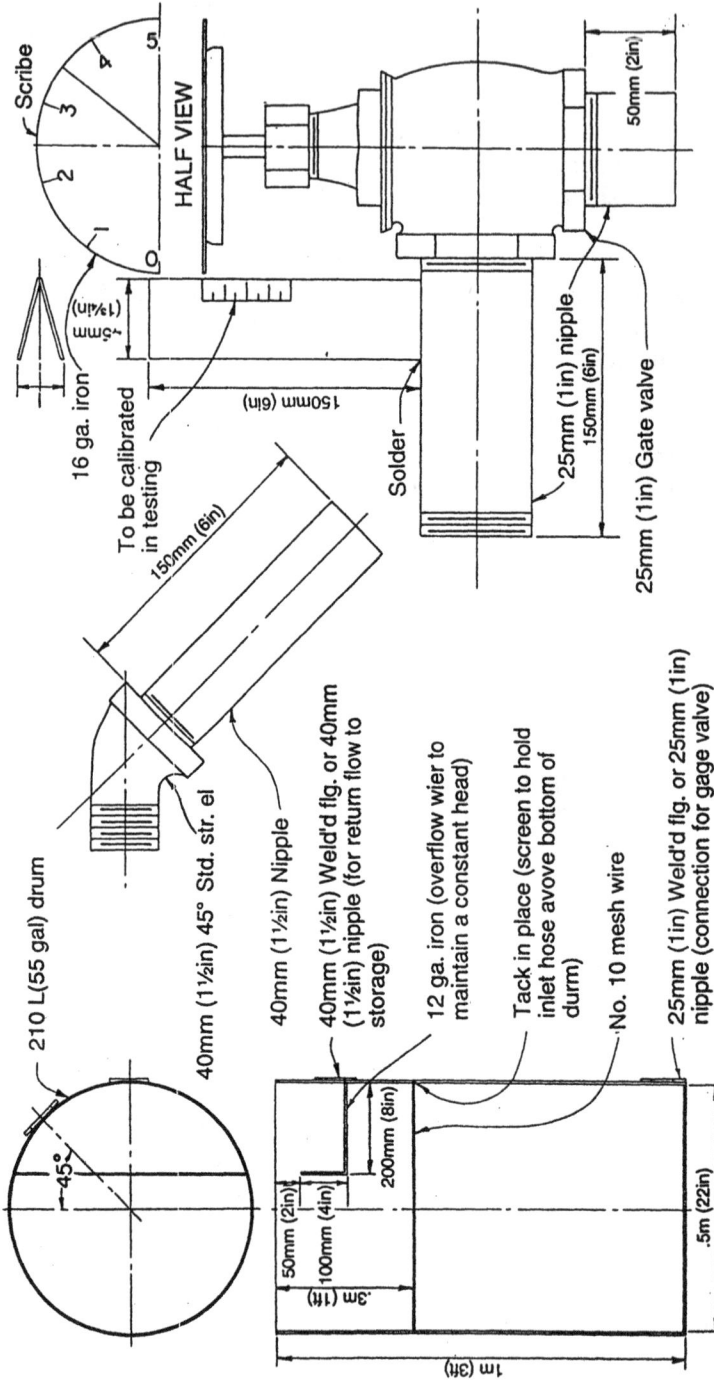

Figure 10-10.—Supply tank for constant head permeability tests.

Gallons per Minute

For 25mm (1in) gate valve with 750 mm (30in) head.

Gage Opening No.

Liters per Minute

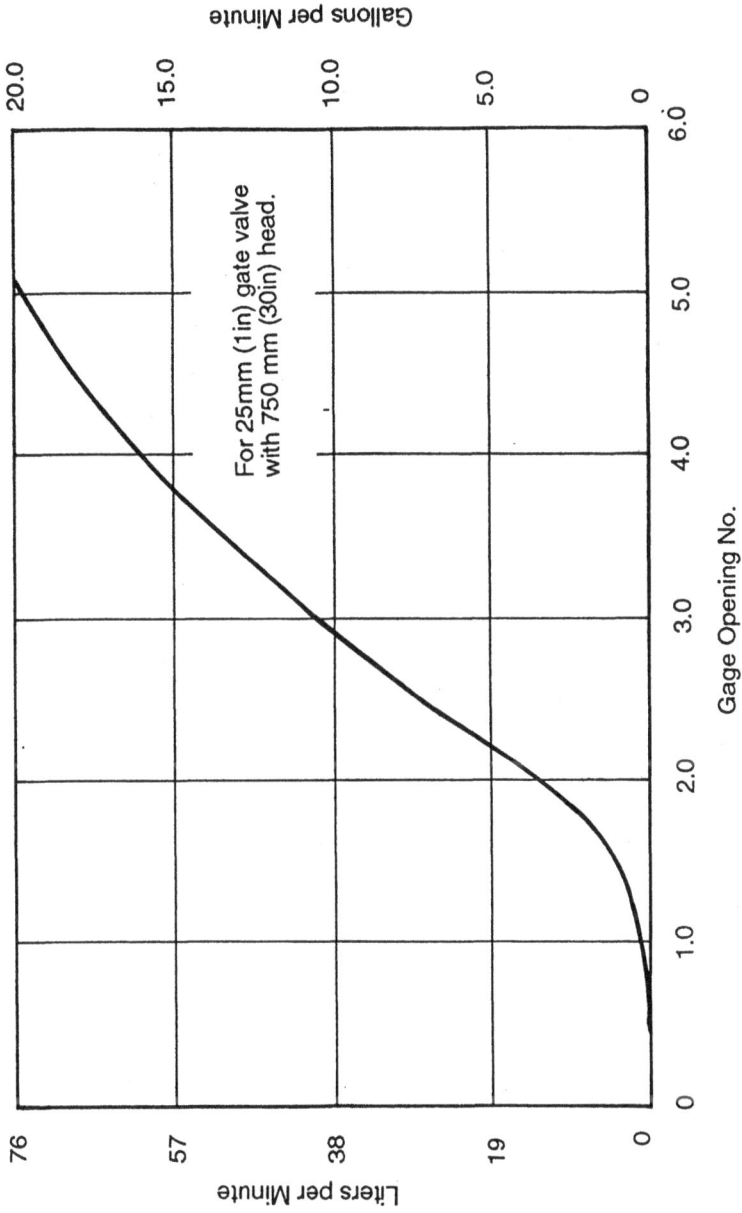

Figure 10-11.—Discharge curve for constant head supply tank.

outlet or into an open tank from which a hose leads to the test hole. Such arrangements give an erratic flow in the hole for a few minutes, but it quickly stabilizes. The crest of the overflow trough must be level to ensure accurate discharge.

For some gravity tests in which the test sections take a very small amount of water, a plug valve is used in the water line from the meter to the casing. A bypass line around the plug valve contains a 12-millimeter (1/2-inch) needle valve. The entire setup is connected directly to the outlet valve of the constant head tank. A 25-millimeter (1-inch) watermeter will not measure accurately the low flows used under such conditions, and after stabilization is obtained, actual flow is determined by the time required to fill a container of known volume.

(d) *Watermeters*.—The watermeter recommendations given in section 10-2(h) are equally satisfactory for use in gravity tests.

(e) *Length of Time for Tests*.—As in pressure tests, establishment of stabilized conditions is of primary importance if good results are to be obtained from gravity tests. Depending on the type of test performed, one of two methods is used. In one method, the inflow of water is controlled until a uniform inflow results in stabilization of water level at a predetermined level. In the other method, a uniform flow of water is introduced into the hole until the water level stabilizes.

(f) *Arrangement of Equipment*.—The recommendations given for equipment arrangement in section 10-2(k) are suitable for use in gravity tests. If used, a constant head tank should be placed so that water flows directly into the casing.

(g) *Gravity Permeability Test (Method 1)*.—For tests in unsaturated and unstable material using only one drill hole, method 1 (figure 10-12) is the most accurate available. Because of mechanical difficulties, this test cannot be economically carried out to a depth greater than about 12 meters (40 feet) when gravel fill must be used in the hole. In performing the test, care should be exercised (after the observation and intake pipes are set) to add gravel in small increments as the casing is pulled back; otherwise, the pipes may become sandlocked in the casing. For tests in unsaturated and unstable material at depths greater than about 12 meters (40 feet), method 2 should be used.

Procedures for various soil conditions are as follows:

(1) Unconsolidated Material.—A 150-millimeter (6-inch) or larger hole is drilled or augered to the depth at which the test is to be made and is then developed gently. A cushion of coarse gravel is placed at the bottom of the hole and the feed pipe (I) and the observation pipe (O) are set in position (figure 10-12). After the pipes are in position, the hole is filled with medium gravel to a depth at least five times the diameter of the hole. If the material will not stand without support, the hole must be cased to the bottom. After casing, the gravel cushion and pipes are put in and the casing is pulled back slowly as medium gravel is fed into the hole. The casing should be pulled back only enough to ensure that the water surface to be maintained in the hole will be below the bottom of the casing. It is advisable to have about 100 millimeters (4 inches) of the gravel fill protrude into the casing.

A metered supply of water is poured into the feed pipe until three or more successive measurements, taken at 5-minute intervals, of the water level through the observation pipe are within ±5 millimeters (0.2 foot). The water supply should be controlled so that the stabilized water level will not be within the casing, but is located more than five times the hole diameter above the bottom of the hole. Adjustment of the flow of water is generally necessary to obtain the desired conditions.

(2) Consolidated Materials.—In consolidated material, or unconsolidated material which will stand without support even when saturated, the gravel fill and casing may be omitted. The use of the coarse gravel cushion is advisable. In all other respects, the test is carried out as in unconsolidated, unstable materials.

(3) General.—Tests should be made at the successive depths selected so that the water level in each test is located at or above the bottom of the hole in the preceding test.

The conductivity coefficients within the limits ordinarily employed in the field can be obtained from figures 10-7 and 10-8. The zone in which the test is made and applicable equations can be found on figures 10-6 and 10-12, respectively.

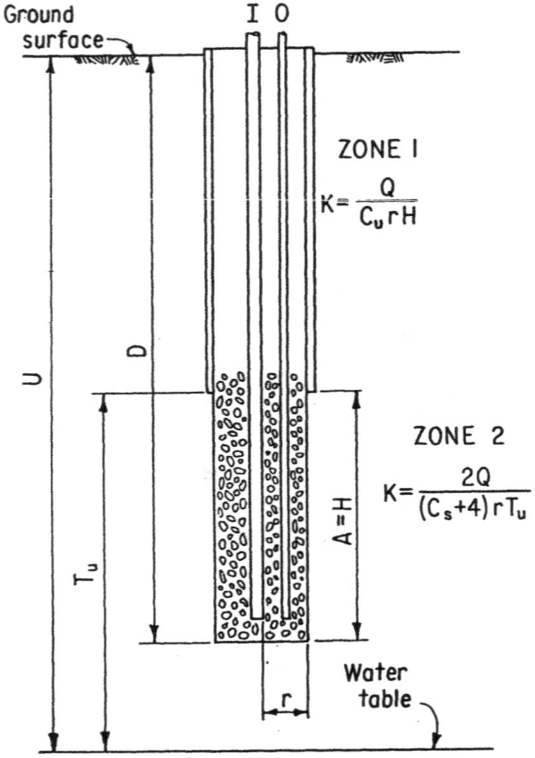
Ground surface
I O
ZONE I
K= Q / C_u r H
ZONE 2
K= 2Q / (C_s+4) r T_u
U
D
T_u
A=H
r
Water table

K = coefficient of permeability, meter per second (ft./sec.) under a unit gradient
Q = uniform flow into well, m^3/s ($ft^3.s$)
r = radius of test section, m (ft.)
H = height of column of water in well, m (ft.)
A = length of test section m (ft.)(for this method, A = H)
C_u and C_s = conductivity coefficients
X = H/T_u (100) = percent of unsaturated stratum
T_u = U - D + H = distance from water surface in well to water table, m (ft.)
U = thickness of unsaturated permeable bed, m (ft.)
D = distance from ground surface to bottom of test section, m (ft.)
I = feed pipe for pouring water into well (a 50mm (2 in.) standard pipe is usually satisfactory)
O = observation pipe (30mm (1¼ in.) o.d. pipe is satisfactory)
a = surface area of test section (area of wall plus area of bottom), m (ft.)
Limitations:
$A \geq 10r$ and $Q/a \geq 0.10$

Figure 10-12.—Gravity permeability test (method 1) (Zangar, 1953).

Data required for computing the permeability may not be available until the hole has penetrated the water table. The required data include:

- Radius of hole, r, meters (feet)

- Depth of hole, D, meters (feet)

- Depth to bottom of casing, meters (feet)

- Depth of water in hole, H, meters (feet)

- Depth to top of gravel in hole, meters (feet)

- Length of test section, A, meters (feet)

- Depth-to-water table, T_u, meters (feet)

- Steady flow, Q, introduced into the hole to maintain a uniform water level, m^3/s (ft^3/s)

- Time test is started and time each measurement is made

Some examples using method 1 are as follows:

Example 4:
Zone 1 (Method 1)
Given: $H = A = 5$ feet, $r = 0.5$ foot, $D = 15$ feet, $U = 50$ feet, and
$Q = 0.10$ ft^3/s
$Tu = U\text{-}D + H = 50 - 15 + 5 = 40$ feet, also $Tu/A = 40/5 = 8$

$$X = \frac{H}{T_u} (100) = \frac{5}{40} (100) = 12.5\%$$

The values for X and T_u/A lie in zone 1 (figure 10-6). To determine the conductivity coefficient C_u, from figure 10-7:

$$\frac{H}{r} = \frac{5}{0.5} = 10, \frac{A}{H} = \frac{5}{5} = 1, \text{ also } C_u = 32$$

From figure 10-12:

$$K = \frac{Q}{C_u rH} = \frac{0.10}{(32)(0.5)(5)} = 0.00125 \ ft/s \qquad 10\text{-}7$$

Example 5:

  Zone 2 (Method 1)

  Given: $H, A, r, U,$ and Q are as given in example 1

      $D = 45$ feet

$$T_u = 50 - 45 + 5 = 10 \ ft \quad also \quad \frac{T_u}{A} = \frac{10}{5} = 2$$

$$X = \frac{5}{10} \ (100) = 50\%$$

Points T_u/A and X lie in zone 2 on figure 10-6.

To determine the conductivity coefficient, C_s, from figure 10-8:

$$\frac{A}{r} = \frac{5}{0.5} = 10 \quad also \quad C_s = 25.5$$

From figure 10-12:

$$K = \frac{2Q}{(C_{s} + 4)rT_u} = \frac{(2)(0.10)}{(25.5 + 4)(0.5)(10)} = 0.00136 \ ft/s \qquad 10\text{-}8$$

(h) Gravity Permeability Test (Method 2).—This method may give erroneous results when used in unconsolidated material because of several uncontrollable factors. However, it is the best of the available pump-in tests for the conditions involved. In most instances, if performed with care, the results obtained by its use are adequate. When permeabilities in streambeds or lakebeds must be determined below water, method 2 is the only practical gravity test available.

A 1.5-meter (5-foot) length of 75- to 150-millimeter (3- to 6-inch) casing is perforated in a uniform pattern. The maximum number of perforations possible, without seriously affecting the strength of the casing, is desirable. The bottom of the perforated section of casing should be beveled on the inside and case hardened so that a cutting edge can be made.

The casing is sunk by drilling or jetting and driving, whichever method will give the tightest fit of the casing in the hole. In poorly compacted material and soils with a nonuniform grain size, development by filling the casing with water to about 1 meter (a few feet) above the perforations and gently surging and bailing is advisable before making the test. A 150-millimeter (6-inch) coarse gravel cushion is poured into the casing, and the observation pipe is set on the cushion.

A uniform flow of water sufficient to maintain the water level in the casing above the top of the perforations is poured into the well. Depth of water measurements are made at 5-minute intervals until three or more measurements are within ±60 millimeters (0.2 foot). The water should be poured through a pipe and measurements made between the pipe and casing. This procedure may be reversed if necessary.

When a test is completed, the casing is sunk an additional 1.5 meters (5 feet) and the test is repeated.

The test may be run in consolidated material using an open hole for the test section. This practice is not recommended because the bottom of the casing under such conditions is seldom tightly fitted in the hole and considerable error may result from seepage upward through the annular space between the casing and the wall of the hole.

Measurements should be made to the nearest 3 millimeters (0.01 foot). The values of C_u and C_s within the limits ordinarily employed in the field can be obtained from figures 10-7 and 10-8. The zone in which the test is made and applicable equations can be found on figures 10-6 and 10-13, respectively.

In making tests in a hole, some data required for computing permeability are not available until the hole has encountered the water table. The recorded data are supplemented by this information as it is determined. The data recorded in each test are as follows:

- Outside radius of casing, meters (feet)

- Length of perforated section of casing A, meters (feet)

- Number and diameter of perforations in length A

K = coefficient of permeability, m (ft.) per second under a unit gradient

Q = steady flow into well, m^3/s ($ft.^3/s$)

H = height of water in well, m (ft.)

A = length of perforated section, m (ft.)

r_i = outside radius of casing (radius of hole in consolidated material), m (ft.)

r_e = effective radius of well = r_i (area of perforations)/(outside area of perforated section of casing); r_1 = r_e in consolidated material that will stand open and is not cased

C_u and C_e = conductivity coefficients

T_u = distance from water level in casing to water table, m (ft.)

a = surface area of test section (area of perforations plus area of bottom), m^2 ($ft.^2$) where clay seal is used at bottom, a = area of perforations

S = thickness of saturated permeable material above an underlying relatively impermeable stratum, m (ft.)

X = H/T_u(100) + percent of unsaturated stratum

U = thickness of unsaturated material above water table, m (ft.)

D = distance from ground surface to bottom of test section, m (ft.)

O = observation pipe, 25 to 32mm (1 to 1¼ in.) pipe

Limitations:

S≥5A, A≥10r, and Q/a≥0.10

Notes:

In zone 3, H is the difference in elevation between the normal water table and the water level in the well. In zones 2 and 3, if a clay seal is placed at the bottom of the casing, the factor 4(r_1/r_e) is omitted from the equations. Where the test is run with "A" as an open hole, r_1/r_e = 1 and [C_s+4(r_1/r_e)]=[C_s+r].

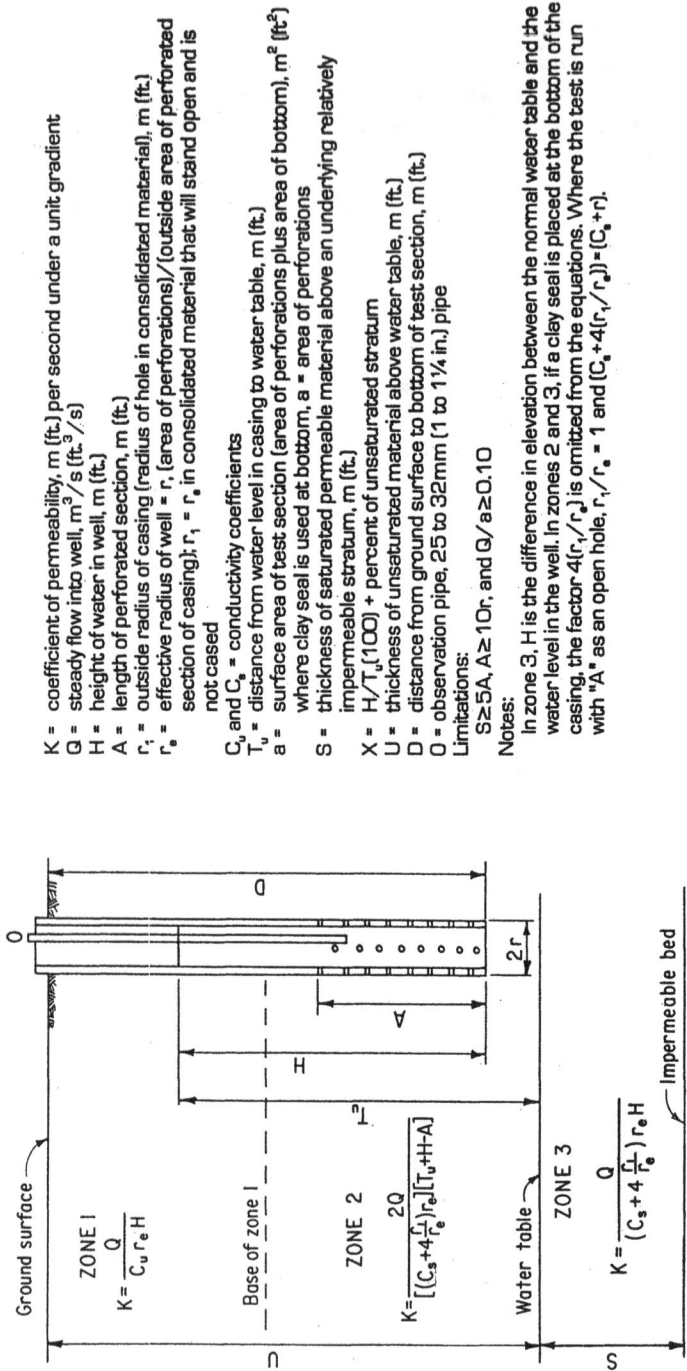

Figure 10-13.—Gravity permeability test (method 2).

- Depth to bottom of hole, D, meters (feet)

- Depth-to-water surface in hole, meters (feet)

- Depth of water in hole, H, meters (feet)

- Depth-to-water table, U, meters (feet)

- Thickness of saturated permeable material above underlying relatively impermeable bed, S, meters (feet)

- Steady flow into well to maintain a constant water level in hole, Q, m³/s (ft³/s)

- Time test is started and time each measurement is made

Some examples using method 2 are as follows:

Example 6:
  Zone 1 (Method 2)
  Given: $H = 10$ feet, $A = 5$ feet, $r_1 = 0.25$ foot, $D = 20$ feet, $U = 50$ feet, $Q = 0.10$ ft³/s and 128 0.5-inch-diameter perforations, bottom of the hole is sealed

Area of perforations $= 128\pi r^2 = 128\pi\ (0.25)^2 = 25.13$ in² $= 0.174$ ft²

Area of perforated section $= 2\pi r_1 A = 2\pi\ (0.25)\ (5) = 7.854$ ft²

$$r_e = \frac{0.174}{7.854}\ (0.25) = 0.00554\ ft$$

$$T_u = U - D + H = 50 - 20 + 10 = 40\ ft \quad also \quad \frac{T_u}{A} = \frac{40}{5} = 8$$

$$X = \frac{H}{T_u}\ (100) = \frac{10}{40}\ (100) = 25\%$$

Points T_u/A and X lie in zone 1 on figure 10-6.

Find C_u from figure 10-7:

$$\frac{H}{r_e} = \frac{10}{0.00554} = 1,805 \ also \ \frac{A}{H} = \frac{5}{10} = 0.5$$

then, $C_u = 1,200$

From figure 10-13:

$$K = \frac{Q}{C_u r_e H} = \frac{0.10}{(1,200)(0.00554)(10)} = 0.0015 \ ft/s \qquad 10\text{-}9$$

Example 7:
Zone 2 (Method 2)
Given: $Q, H, A, r_1, r_e, U, A/H,$ and H/r_e same as example 6
       $D = 40$ feet

$$T_u = 50 - 40 + 10 = 20 \ ft \ also \ \frac{T_u}{A} = \frac{20}{5} = 4$$

$$X = \frac{10}{20} (100) = 50\%$$

Points T_u/A and X lie in zone 2 on figure 10-6.

Find C_s from figure 10-8:

$$\frac{A}{r_e} = \frac{5}{0.00554} = 902 \ also \ C_s = 800$$

From figure 10-13:

$$K = \frac{2Q}{\left[\left(C_s + 4\frac{r_1}{r_e}\right)r_e\right](T_u + H - A)} = \qquad 10\text{-}10$$

$$\frac{0.20}{(5.43)(20 + 10 - 5)} = 0.0015 \ ft/s$$

Example 8:
  Zone 3 (Method 2)
  Given: Q, H, A, r_1, r_e, A/H, H/r_e, U, C_s, and A/r_e are as given in
    example 7
    $S = 60$ feet

From figure 10-13:

$$K = \frac{Q}{\left(C_s + 4\frac{r_1}{r_e}\right)r_eH} = \frac{0.10}{(980.5)(0.00554)(10)} = 0.0018 \; ft/s \qquad 10\text{-}11$$

(i) *Gravity Permeability Test (Method 3)*.—This method is a
combination of the gravity permeability test-methods 1 and 2
which was developed to permit testing under difficult conditions.
It is the least accurate method of testing, but is the only one
available for use under circumstances where method 2 cannot be
used because of the nature of the material (figure 10-14).

In some areas, the material to be tested will be of such character
that a casing that is beveled and case hardened at the bottom will
not stand up under the driving necessary to sink it. This lack of
durability is particularly true in gravelly materials where the
particle size is greater than about 25 millimeters (1 inch). Under
such conditions, method 3 would probably not be satisfactory
because a drive shoe must be used with this method. Use of a
drive shoe causes excessive compaction of the materials and forms
an annular space about the casing, introducing an opportunity for
error in the results. The size of this error is unknown, but if
reasonable care is taken in performing the test, the results will
probably approximate the correct magnitude for the material
tested.

On completion of each test, a 90- to 150-millimeter (3- to 6-inch)
perforated casing is advanced a distance of 1.5 meters (5 feet) or
more by drilling and driving. After each new test section is
developed by surging and bailing, a 150-millimeter (6-inch) gravel
cushion should be placed on the bottom to support the observation
pipe. A uniform flow of water sufficient to maintain the water
level in the casing just at the top of the perforations is then poured
into the well. The water is poured directly into the casing and
measurements are made through the 32-millimeter (1-1/4-inch)
observation pipe. The test should be run until three or more

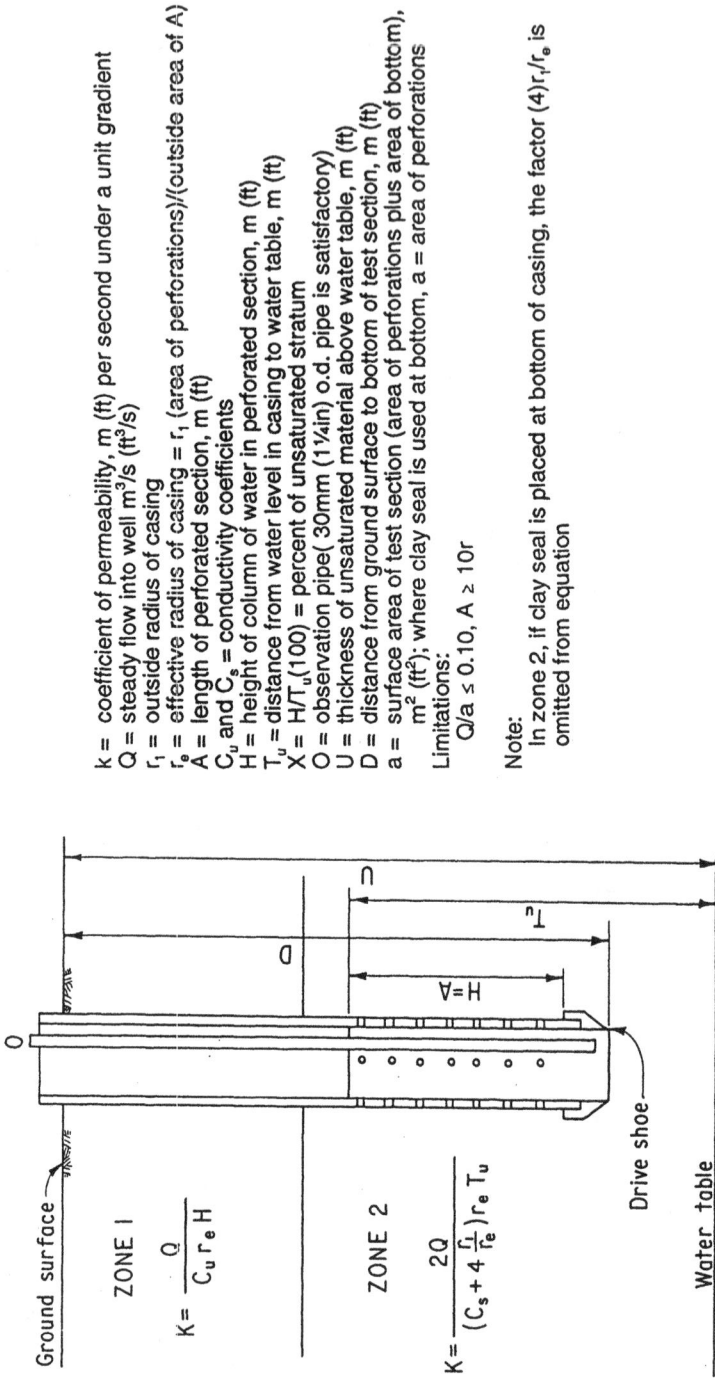

Ground surface

ZONE 1

$$K = \frac{Q}{C_u\, r_e\, H}$$

ZONE 2

$$K = \frac{2Q}{\left(C_s + 4\, \frac{r_i}{r_e}\right) r_e\, T_u}$$

Drive shoe

Water table

k = coefficient of permeability, m (ft) per second under a unit gradient
Q = steady flow into well m³/s (ft³/s)
r_i = outside radius of casing
r_e = effective radius of casing = r_1 (area of perforations)/(outside area of A)
A = length of perforated section, m (ft)
C_u and C_s = conductivity coefficients
H = height of column of water in perforated section, m (ft)
T_u = distance from water level in casing to water table, m (ft)
X = H/T_u(100) = percent of unsaturated stratum
O = observation pipe(30mm (1¼in) o.d. pipe is satisfactory)
U = thickness of unsaturated material above water table, m (ft)
D = distance from ground surface to bottom of test section, m (ft)
a = surface area of test section (area of perforations plus area of bottom), m² (ft²); where clay seal is used at bottom, a = area of perforations

Limitations:
  $Q/a \leq 0.10$, $A \geq 10r$

Note:
  In zone 2, if clay seal is placed at bottom of casing, the factor (4)r_i/r_e is omitted from equation

Figure 10-14.—Gravity permeability test (method 3) (Zangar, 1953).

measurements taken at 5-minute intervals show the water level in the casing to be within ±60 millimeters (0.2 foot) of the top of the perforations.

The values of C_u and C_s, within the limits ordinarily employed in the field, can be obtained from figures 10-7 and 10-8. The zone in which the test is made and applicable equations can be found on figures 10-6 and 10-14, respectively.

The data recorded in each test are as follows:

- Outside radius of casing, r_1, meters (feet)

- Length of perforated section of casing, A, meters (feet)

- Number and diameter of perforations in length A

- Depth to bottom of hole, D, meters (feet)

- Depth-to-water surface in hole, meters (feet)

- Depth of water in hole, meters (feet)

- Depth-to-water table, meters (feet)

- Steady flow into well to maintain a constant water level in hole, Q, m^3/s (ft^3/s)

- Time test is started and time each measurement is made

Example 9:
  Zone 1 (Method 3)
  Given: Q = 10.1 gallon/minute = 0.023 ft^3/s, $H = A$ = 5 feet,
  D = 22 feet, U = 71 feet, T_u = 54.5 feet, r_e = 0.008 foot,
  and r_1 = 1.75 inches = 0.146 foot (nominal 3-inch casing)

$$\frac{T_u}{A} = \frac{54.4}{5} = 10.9$$

$$Also \ X = \frac{H}{T_u} (100) = \frac{5}{54.5} (100) = 9.2\%$$

These points lie in zone 1 (figure 10-6).

Find C_u from figure 10-7:

$$\frac{H}{r_e} = \frac{5}{0.008} = 625 \ also \ \frac{A}{H} = 1$$

then, $C_u = 640$

From figure 10-14:

$$K = \frac{Q}{C_u r_e H} = \frac{0.023}{(640)(0.008)(5)} = 0.0009 \ ft/s$$

Example 10:

Zone 2 (Method 3)

Given: Q, H, A, U, r_e, and r_1 are as given in example 9

$D = 66$ feet and $T_u = 10$ feet

$$\frac{T_u}{A} = \frac{10}{5} = 2 \ also \ X = \frac{5}{10}(100) = 50\%$$

These points lie in zone 2 (figure 10-6).

Find C_s from figure 10-8:

$$\frac{A}{r_e} = \frac{5}{0.008} = 625 \ also \ C_s = 595$$

From figure 10-14:

$$K = \frac{2Q}{\left(C_s + 4\dfrac{r_1}{r_e}\right)r_e T_u} = \frac{(2)(0.023)}{(668)(0.008)(10)} = 0.00086 \ ft/s \qquad \text{10-12}$$

(j) Gravity Permeability Test (Method 4).—Method 4 can be used to advantage in determining the overall average permeability of unsaturated materials above a widespread impermeable layer. However, it does not permit determination of relative permeability variations with depth. The method is actually an application of the steady-state pumping test theory discussed in section 9-2.

An intake well (preferably 150 millimeters [6 inches] or larger) is drilled to a relatively impermeable layer of wide areal extent or to the water table. The well is uncased in consolidated material, but in unconsolidated material, a perforated casing or screen should be set from the bottom to about 1.5 meters (5 feet) below the ground surface. The well should be developed by pouring water into it while surging and bailing prior to the testing for intake capacity.

Before the observation wells are drilled, a test run of the intake well should be made to determine the maximum height, H, of the column of water in the intake pipe above the top of the impermeable stratum that is possible to maintain with available pumping equipment (figure 10-15). The spacing of the observation wells can be determined from this test run. A 25- to 32-millimeter (1- to 1-1/4-inch) observation pipe should be inserted near the bottom of the intake well to facilitate water-level measurements.

A minimum of three observation wells should be installed, by jetting or some other method, to the top of the impermeable stratum. Suitable pipe, perforated for the bottom 3 to 4.5 meters (10 to 15 feet), should be set to the bottom of these wells. The observation wells should be set from the intake well at distances equal to multiples of one-half the height, H, of the water column which it is possible to maintain in the intake well.

The elevations of the top of the impermeable strata or water table in each well are determined, and the test is started. If the saturated thickness is small compared to the height of the water column that can be maintained in the hole, a water table will act as an impermeable layer for purposes of this test. After water has been poured into the intake at a constant rate for an hour, measurements are made of water levels in the observation wells. Measurements are made at 15-minute intervals thereafter, and each set of measurements is plotted on semi-log paper with the square of the height of the water level above the top of the impermeable bed, h^2, against the distance from the intake well to the observation holes, r, for each hole (see figure 10-16). When the plot of a set of measurements permits a straight line to be drawn through the points within the limits of plotting, stable conditions prevail and the permeability may be computed.

The data recorded in each test are as follows:

- Ground elevations at sites of intake well and observation wells

- Elevations of reference points at intake well and observation wells, meters (feet)

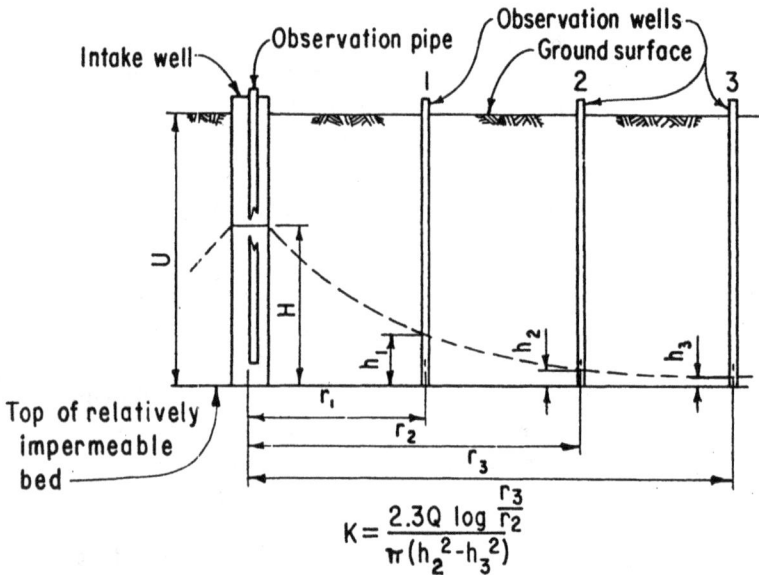

$$K = \frac{2.3Q \log \frac{r_3}{r_2}}{\pi (h_2^2 - h_3^2)}$$

K = coefficient of permeability, m(ft) per second under a unit gradient

Q = uniform flow into intake well, m³/s (ft³/s)

r_1, r_2, and r_3 = distance from intake well to observation holes, m (ft)

h_1, h_2, and h_3 = height of water in observation holes r_1, r_2, and r_3 respectively, above elevation of top of impermeable layer, m(ft)

H = height of column of water in intake pipe above top of impermeable stratum, m(ft)

U = distance from ground surface to impermeable bed, m(ft)

Figure 10-15.—Gravity permeability test (method 4) (Zangar, 1953).

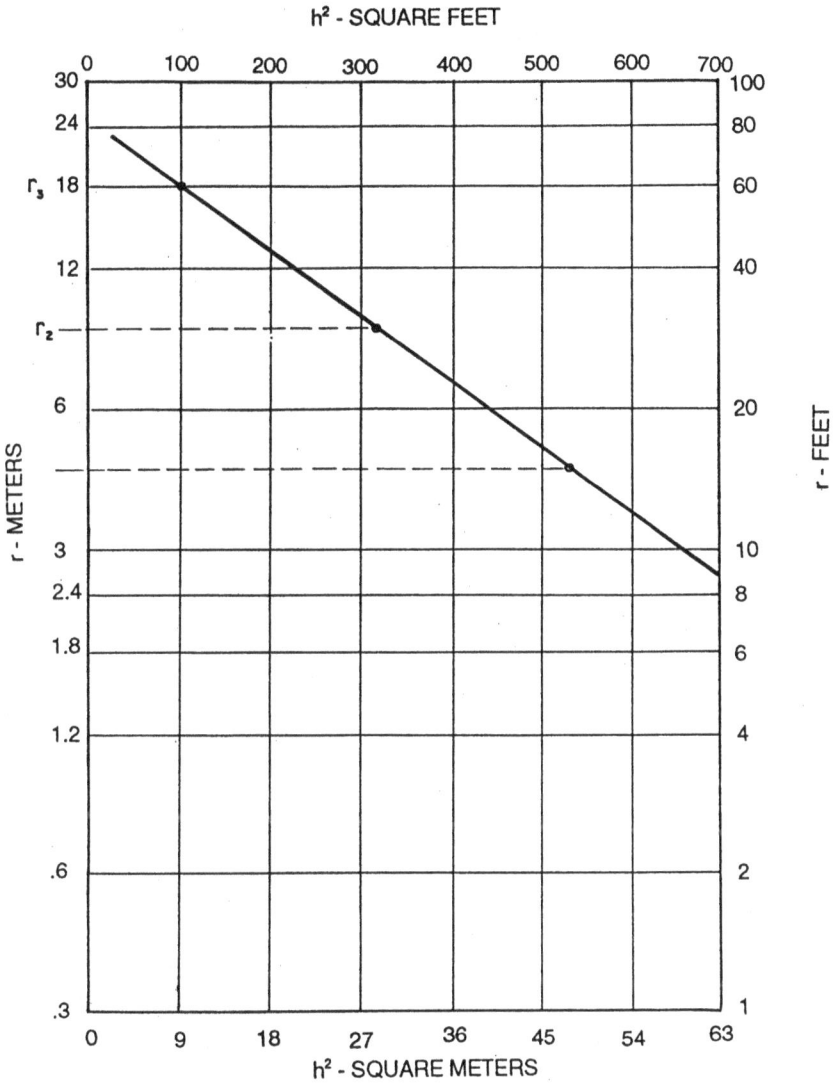

Figure 10-16.—Plot of h^2 versus r for
gravity permeability test (method 4).

- Distances from center of observation wells to center of
  intake well, r_1, r_2, and r_3, meters (feet)

- Elevation of top of impermeable bed at intake well and
  observation wells, meters (feet)

- Depths of water below reference point in intake well and observation wells at 15-minute intervals, meters (feet)

- Uniform flow of water, Q, introduced into well, m^3/s (ft^3/s)

- Time pumping is started and time each measurement is made

Example 11:
  Method 4
  Given: U = 50 feet, Q = 1 ft^3/s, H = 30 feet, r_1 = 15 feet,
      r_2 = 30 feet, r_3 = 60 feet, h_1 = 23.24 feet, h_2 = 17.89 feet,
      h_3 = 10.0 feet, h_1^2 = 540 ft^2, h_2^2 = 320 ft^2, and
      h_3^2 = 100 ft^2

A plot of r against h^2, as shown on figure 10-16, shows that a straight line can be drawn through the plotted points which means that stable conditions exist and the permeability may be computed.

From figure 10-15:

$$K = \frac{2.3\ Q \log \frac{r_3}{r_2}}{\pi\ (h_2^2 - h_3^2)} \qquad 10\text{-}13$$

$\log \dfrac{r_3}{r_2} = 0.3010$, $\log \dfrac{r_3}{r_1} = 0.6021$, *also* $\log \dfrac{r_2}{r_1} = 0.3010$

$$K = \frac{2.3\ Q \log \frac{r_3}{r_2}}{\pi(h_2^2 - h_3^2)} = \frac{2.3\ Q \log \frac{r_3}{r_1}}{\pi(h_1^2 - h_3^2)} = \frac{2.3\ Q \log \frac{r_2}{r_1}}{\pi(h_1^2 - h_2^2)} \qquad 10\text{-}14$$

$$K = \frac{(2.3)(1)(0.3010)}{\pi(220)} =$$

$$\frac{(2.3)(1)(0.6020)}{\pi(440)} = \frac{(2.3)(1)(0.3010)}{\pi(220)} = 0.001\ ft/s$$

10-4. Falling Head Tests.—Falling head tests are used primarily in open holes in consolidated rock. They are performed using inflatable packers identical to those described under pressure testing (section 10-2) and can be used as an alternate method under some circumstances when the pressure transducer or other instrumentation fails. The method of cleaning the hole is the same as that described under pressure testing.

(a) Tests Below the Static Water Level in the Hole.—

- Use inflatable straddle packers, with 3-meter (10-foot) isolated intervals on a 32-millimeter (1-1/4-inch) standard size drop pipe (inside diameter = 35 millimeters [1.38 inches]). Set initially at the bottom of the hole and inflate packers to 2,068 KpA (300 lb/in^2) of differential pressure.

- After packers are inflated, measure water level in drop pipe three or more times at 5-minute intervals until water level stabilizes. Stabilized level will be at the static level of water in the test section.

- After stabilization of level, pour in 8 liters (2 gallons) or more of water as rapidly as possible into the drop pipe. Four liters (1 gallon) of water will raise the water level in a 32-millimeter (1-1/4-inch) pipe 4.1 meters (12.9 feet) if the section is tight.

- Measure water level as soon as possible after water is poured in. Measure initial depth to water and time measurement as soon as possible and at two 5-minute intervals thereafter. If rate of decline exceeds 4.5 meters (15 feet) in 13 minutes, the transmissivity of a 3-meter (10-foot) test section is greater than 18 m^2 per year (200 ft^2 per year), or an average permeability greater than 6 meters per year (20 feet per year).

- The value of transmissivity so determined is only an approximation but is sufficiently accurate for many engineering purposes.

The test is based on an adaptation of the slug method by Ferris and Knowles (1962). The equation for analysis is:

$$T = \frac{V}{2\pi s \Delta t}$$
10-15

where:

T = transmissivity of test section, m²/s (ft²/s)
V = volume of water entering test section in period Δt, m³ (ft³) (300-millimeter [1-foot] decline in 32-millimeter [1-1/4-inch] pipe = 0.000283 m³ [0.01 ft³])
s = decline in water level in period Δt, meters (feet)
Δt = period of time, seconds, between successive water-level measurements (i.e., $t_1 - t_o$, $t_2 - t_1$, etc.)

- If the log indicates the test section is uniform without obvious points of probable concentrated leakage, the average permeability of the test section in meters per second (feet per second) can be estimated from $K = T/A$, where A is the length of the test section in meters (feet). If the log indicates a predominantly impervious test section but with a zone or zones of probable concentrated flow, the average K of the zones can be estimated from $K = T/A'$ where A' is the thickness of the permeable zone or zones in meters (feet).

- After each test, deflate packers, raise test string 3 meters (10 feet), and repeat until entire hole below the static water level has been tested.

(b) *Tests in Unsaturated Materials Above the Water Table.*— Tests above the water table require different procedures and analyses than tests in the saturated zone. Tests made in sections straddling the water table or slightly above it will give high computed values if the equations in section 10-4(a) are used and low computed values if the following equations 10-16 and 10-17 are used. For tests above the water table, the following procedure is used:

- Install a 3-meter (10-foot) straddle packer at the bottom of the hole if the hole is dry or with the top of the bottom packer at the water table if it contains water. Inflate the packer.

- Fill the drop pipe with water to the surface if possible, otherwise to the level permitted by pump capacity.

- Measure water level in the drop pipe and record with time of measurement. Make two or more similar measurements while water-table declines.

- Upon completion of a test, raise the packer 3 meters (10 feet) and repeat this procedure until all of the encased or uncemented hole is tested.

- The equation for analysis of each test section is an adaptation of one derived by Jarvis (1953):

$$K = \frac{r_1^2}{2A\Delta t}\left[\frac{\sinh^{-1}\frac{A}{r_e}}{2}\ln\left(\frac{2H_1 - A}{2H_2 - A}\right) - \ln\left(\frac{2H_1H_2 - AH_2}{2H_1H_2 - AH_1}\right)\right] \qquad 10\text{-}16$$

where:

K = average permeability of the test section, meter per second (ft/s)

A = length of test section, meters (feet)

r_1 = inside radius of drop pipe, millimeters (feet) (17.25 millimeters [0.0575 foot] for 32-millimeter [1-1/4-inch] pipe)

r_e = effective radius of test section, millimeters (feet) (37.5 millimeters [0.125 foot] for a 75-millimeter [3-inch] hole)

Δt = time intervals (t_1-t_0, t_2-t_1), seconds

$\sinh^{-1}$ = inverse hyperbolic sine

$\ln$ = natural logarithm

H = length of water column from bottom of test interval to water surface in standpipe, meters (feet) (H_0, H_1, H_2 lengths at time of measurements t_0, t_1, t_2, etc.)

- For the particular equipment specified, and a 3-meter test section, equation 10-16 may be simplified as follows:

$$K = \frac{1.653\ x\ 10^{-4}}{\Delta t}\left[2.5\ \ln\left(\frac{H_1 - 5}{H_2 - 5}\right) - \ln\left(\frac{H_1H_2 - 5H_2}{H_1H_2 - 5H_1}\right)\right] \qquad 10\text{-}17$$

10-5. Slug Tests.—

(a) Introduction.—Slug testing involves the rapid introduction or removal of small quantities of water, air injection, packer deflation, or other method of causing a rapid rise or lowering of the water level in a well. The time interval for injection or removal must be sufficiently short to be considered instantaneous. Methods which cause an initial rise in water level are generally termed slug tests; methods that cause an initial lowering of the water level are termed bail tests. Both tests create the same effect. During a slug or bail test in a steady-state aquifer, a brief pressure pulse is created at a point in the aquifer and the transient response at that same point is observed and measured. The test measures chiefly the hydraulic conductivity of the aquifer and, with a lesser degree of precision, the storage coefficient (Marsily, 1986).

Slug or bail tests are used to obtain estimates of the aquifer hydraulic conductivity in situations where a pumping test cannot be completed. This test could be appropriate in areas where the aquifer will not yield enough water to conduct a pumping test, where budget constraints preclude full-scale pumping tests on the number of wells for which permeability data are desired, or in areas where disposal of large quantities of water would be a problem, such as hazardous waste sites. In contrast to long-term pumping tests, which provide information on the aquifer over a fairly extensive area, slug tests give information on hydraulic conductivity only for the area in the immediate vicinity of the well. This limit may be an advantage if many tests are run because a better picture of the range in hydraulic conductivities can be obtained. The values for hydraulic conductivities, however, are only estimates and may not accurately characterize the aquifer.

A major concern in conducting slug tests is the assurance that the changes in water level during the test accurately reflect the aquifer characteristics and are not unduly affected by well construction. Thus, unless details of well construction are known, analyzing the test results to give reliable values of hydraulic conductivity will be impossible.

A number of methods exist to analyze slug or bail tests. The choice depends upon the hydrologic conditions as well as other factors. Choice of the slug test procedure and method of analysis thus requires an evaluation of the geologic and hydrogeologic conditions at the test site in addition to consideration of well size and construction.

(b) *Conducting the Slug Test.*—The choice of causing a rise or fall in the water table depends upon the purpose of the test and the conditions at the site. Where the water level is shallow and clean water is readily available, water injection or bailing is often the easiest method. One limitation of the accuracy is that the initial direct pulse of injection water is followed by a lesser amount of water flowing down the inside walls of the well (Black, 1978). Causing a rapid rise by displacing water in the well by dropping a pipe or weight or injecting air may be preferable at remote sites where clean water may not be readily available. The displacement or air injection method may also be desirable at locations where the water level is deep and rapid injection or removal of water may be difficult, at hazardous waste sites where disposal of any water removed is difficult, or at well sites that are being used for chemical sampling.

Prior to introducing the slug, the well bore or test cavity needs to be as clean as possible to remove anything that will impede movement of water from the cavity to the surrounding aquifer material. The cavity length and diameter need to be recorded as accurately as possible. The static water level should then be carefully measured and recorded. Water-level measurements must begin immediately following introduction of the slug. Where the water level changes occur slowly, measurements may be made by hand using a water-level indicator. However, the most reliable method is to use a pressure transducer connected to an automatic data logger. This method permits much more frequent and precise measurements and leads to greater reliability of estimates of hydraulic conductivity.

(c) *Analysis.*—Several slug test analysis methods are commonly used. These methods include Hvorslev (1951), Bouwer and Rice (1989), and Cooper, Bredehoeft, and Papadopulos (1967). Wang et al. (1977) and Barker and Black (1983) have developed methods for analyzing slug tests in fissured aquifers. A number of other investigators have modified these methods or developed other methods. Computer programs are available from various sources to analyze data (e.g. Dawson and Istok [1991]).

(d) *Hvorslev.*—The simplest method of analysis is that of Hvorslev (1951) and Chirlin (1989). This analysis assumes a homogeneous, isotropic, infinite medium in which both soil and water are incompressible. It neglects wellbore storage and thus may not be as accurate where a gravel pack is present. A sketch of the geometry of this test and the method of analysis are shown on figure 10-17. This method may be used where the slug test is conducted below an existing water table.

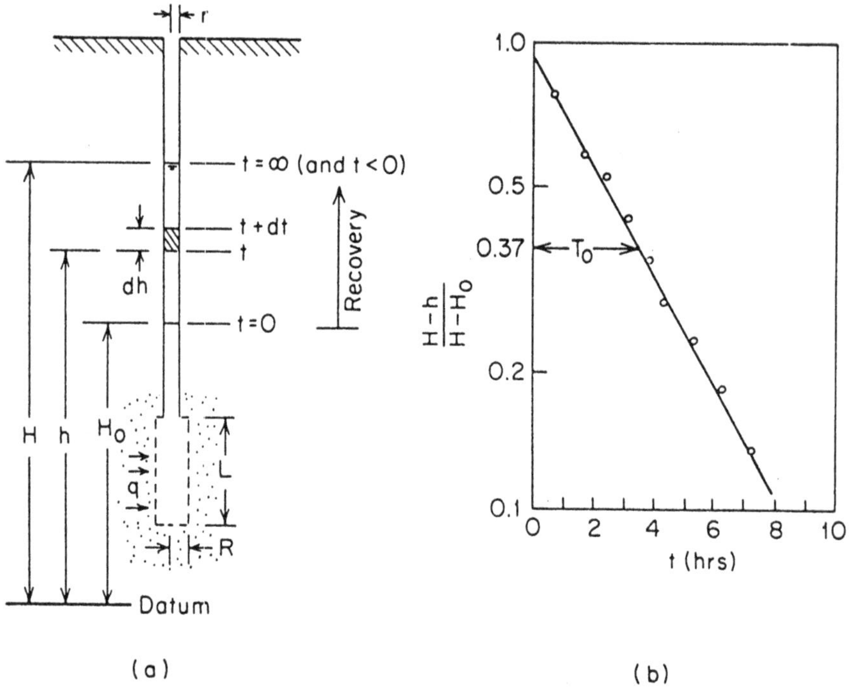

Figure 10-17.—Hvorslev piezometer test: (a) geometry; (b) method of analysis.

Analysis of this test requires graphing the well head changes versus time. Well head changes are given by $(H-h)/(H-H_0)$ (figure 10-17(a)) and should be plotted against time of each reading. Plotting this field data on a semi-log graph (with $[H-h]/[H-H_0]$ on the y axis as a log scale) should approximate a straight line (figure 10-17b). At the point where $(H-h)/(H-H_0) = 0.37$, the corresponding time value on the x axis is equal to T_0, which is defined by Hvorslev as the basic time lag (see Freeze and Cherry [1979] for details on solution). Using the graphical solution for T_0, the dimensions for the test cavity, and the appropriate shape factor, F, a solution for hydraulic conductivity can be found by:

$$K = \frac{\pi r^2}{FT_0} \qquad\qquad 10\text{-}18$$

The U.S. Department of the Navy Naval Facilities *Engineering Command Design Manual 7.1* (1992) gives equations for calculating F for various hole and casing dimensions. These dimensions are shown on figure 10-18 and selecting the proper configuration for the shape factor calculation, *F*, is important.

(e) Bouwer (1989a and b).—This method assumes no aquifer storage and finite wellbore storage. The wells can be partially penetrating and partially screened. The method was originally developed for unconfined aquifers but can also be used for confined or stratified aquifers if the top of the screen or perforated section is located some distance below the upper confining layer. The analysis is based on the Thiem equation and determines hydraulic conductivity, *K*, of the aquifer around the well from the equation:

$$K = \frac{r_c^2 \ln(R_e/r_w)}{2L_e} \frac{1}{t} \ln(\frac{y_0}{y_t}) \qquad\qquad 10\text{-}19$$

where:

L_e = length of the screened, perforated, or otherwise open section of the well

y_0 = the vertical distance difference between water level inside the well and the static water table at time zero

y_t = y at time t

t = time of reading

r_w = radius of the well plus gravel or developed zone

r_c = inside radius of the well casing

R_e = effective radial distance, the distance over which y returns to the static level

L_w = length from the water table to the bottom of the well

See figure 10-19 for configuration.

Values of R_e, the effective radial distance, were determined using an electrical resistance analog network. The effective radial distance is influenced by well diameter, well screen length, the

CONDITION	DIAGRAM	SHAPE FACTOR, F	PERMEABILITY, K BY VARIABLE HEAD TEST	APPLICABILITY
			(FOR OBSERVATION WELL OF CONSTANT CROSS SECTION)	
(A) UNCASED HOLE.		$F = 16\pi DSR$	$K = \dfrac{R}{16 DS} \times \dfrac{(H_2 - H_1)}{(t_2 - t_1)}$ FOR $\dfrac{D}{R} < 50$	SIMPLEST METHOD FOR PERMEABILITY DETERMINATION. NOT APPLICABLE IN STRATIFIED SOILS. FOR VALUES OF S, SEE FIGURE 13.
(B) CASED HOLE, SOIL FLUSH WITH BOTTOM.	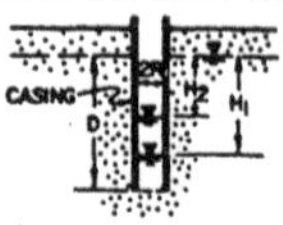	$F = \dfrac{11R}{2}$	$K = \dfrac{2\pi R}{11(t_2 - t_1)} \ln\left(\dfrac{H_1}{H_2}\right)$ FOR $6" \leq D \leq 60"$	USED FOR PERMEABILITY DETERMINATION AT SHALLOW DEPTHS BELOW THE WATER TABLE. MAY YIELD UNRELIABLE RESULTS IN FALLING HEAD TEST WITH SILTING OF BOTTOM OF HOLE.
(C) CASED HOLE, UNCASED OR PERFORATED EXTENSION OF LENGTH "L".	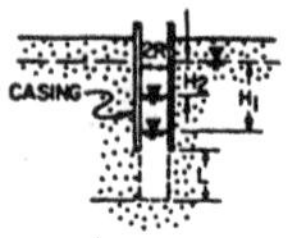	$F = \dfrac{2\pi L}{\ln\left(\frac{L}{R}\right)}$	$K = \dfrac{R^2}{2L(t_2 - t_1)} \ln\left(\frac{L}{R}\right) \ln\left(\dfrac{H_1}{H_2}\right)$ FOR $\dfrac{L}{R} > 8$	USED FOR PERMEABILITY DETERMINATIONS AT GREATER DEPTHS BELOW WATER TABLE.
(D) CASED HOLE, COLUMN OF SOIL INSIDE CASING TO HEIGHT "L".	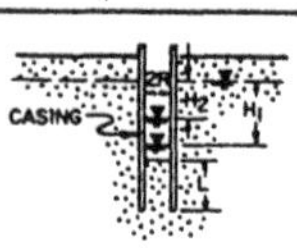	$F = \dfrac{11\pi R^2}{2\pi R + 11L}$	$K = \dfrac{2\pi R + 11L}{11(t_2 - t_1)} \ln\left(\dfrac{H_1}{H_2}\right)$	PRINCIPAL USE IS FOR PERMEABILITY IN VERTICAL DIRECTION IN ANISOTROPIC SOILS.
(E) CASED HOLE, OPENING FLUSH WITH UPPER BOUNDARY OF AQUIFER OF INFINITE DEPTH.	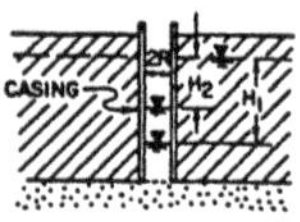	$F = 4R$	**(FOR OBSERVATION WELL OF CONSTANT CROSS SECTION)** $K = \dfrac{\pi R}{4(t_2 - t_1)} \ln\left(\dfrac{H_1}{H_2}\right)$	USED FOR PERMEABILITY DETERMINATION WHEN SURFACE IMPERVIOUS LAYER IS RELATIVELY THIN. MAY YIELD UNRELIABLE RESULTS IN FALLING HEAD TEST WITH SILTING OF BOTTOM OF HOLE.
(F) CASED HOLE, UNCASED OR PERFORATED EXTENSION INTO AQUIFER OF FINITE THICKNESS: (1) $\frac{L}{T} \leq 0.2$ (2) $0.2 < \frac{L}{T} < 0.85$ (3) $\frac{L_3}{T} = 1.00$ NOTE: R_0 EQUALS EFFECTIVE RADIUS TO SOURCE AT CONSTANT HEAD.		(1) $F = C_s R$ (2) $F = \dfrac{2\pi L_2}{\ln(L_2/R)}$ (3) $F = \dfrac{2\pi L_3}{\ln\left(\frac{R_0}{R}\right)}$	(1) $K = \dfrac{\pi R}{C_s(t_2 - t_1)} \ln\left(\dfrac{H_1}{H_2}\right)$ (2) $K = \dfrac{R^2 \ln\left(\frac{L_2}{R}\right)}{2L_2(t_2 - t_1)} \ln\left(\dfrac{H_1}{H_2}\right)$ FOR $\frac{L}{R} = >8$ (3) $K = \dfrac{R^2 \ln\left(\frac{R_0}{R}\right)}{2L_3(t_2 - t_1)} \ln\left(\dfrac{H_1}{H_2}\right)$	USED FOR PERMEABILITY DETERMINATIONS AT DEPTHS GREATER THAN ABOUT 5 FT. FOR VALUES OF C_s, SEE FIGURE 13. USED FOR PERMEABILITY DETERMINATIONS AT GREATER DEPTHS AND FOR FINE GRAINED SOILS USING POROUS INTAKE POINT OF PIEZOMETER. ASSUME VALUE OF $\frac{R_0}{R}$ = 200 FOR ESTIMATES UNLESS OBSERVATION WELLS ARE MADE TO DETERMINE ACTUAL VALUE OF R_0.

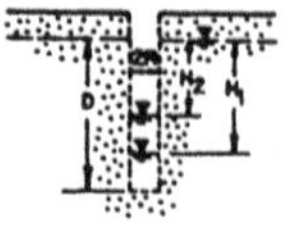

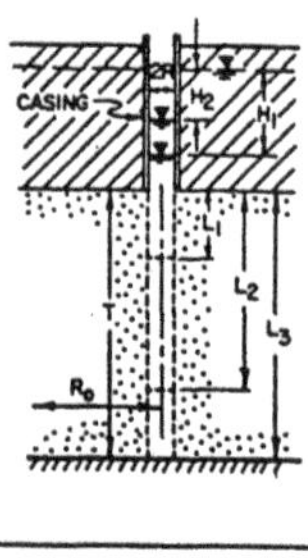

Rows A–D left label: OBSERVATION WELL OR PIEZOMETER IN SATURATED ISOTROPIC STRATUM OF INFINITE DEPTH

Rows E–F left label: OBSERVATION WELL OR PIEZOMETER IN AQUIFER WITH IMPERVIOUS UPPER LAYER

Figure 10-18.—Shape factors for computation of permeability from variable head tests.

depth of the well, and the thickness of the aquifer. Various values for r_w, L_e, L_w, and H were used in the analog network for analysis of their impacts on R_e.

The term $ln(R_e/r_w)$ is related to the geometry of the test zone and the amount of aquifer penetration of the well. Two separate solutions are required to address partially penetrating wells and

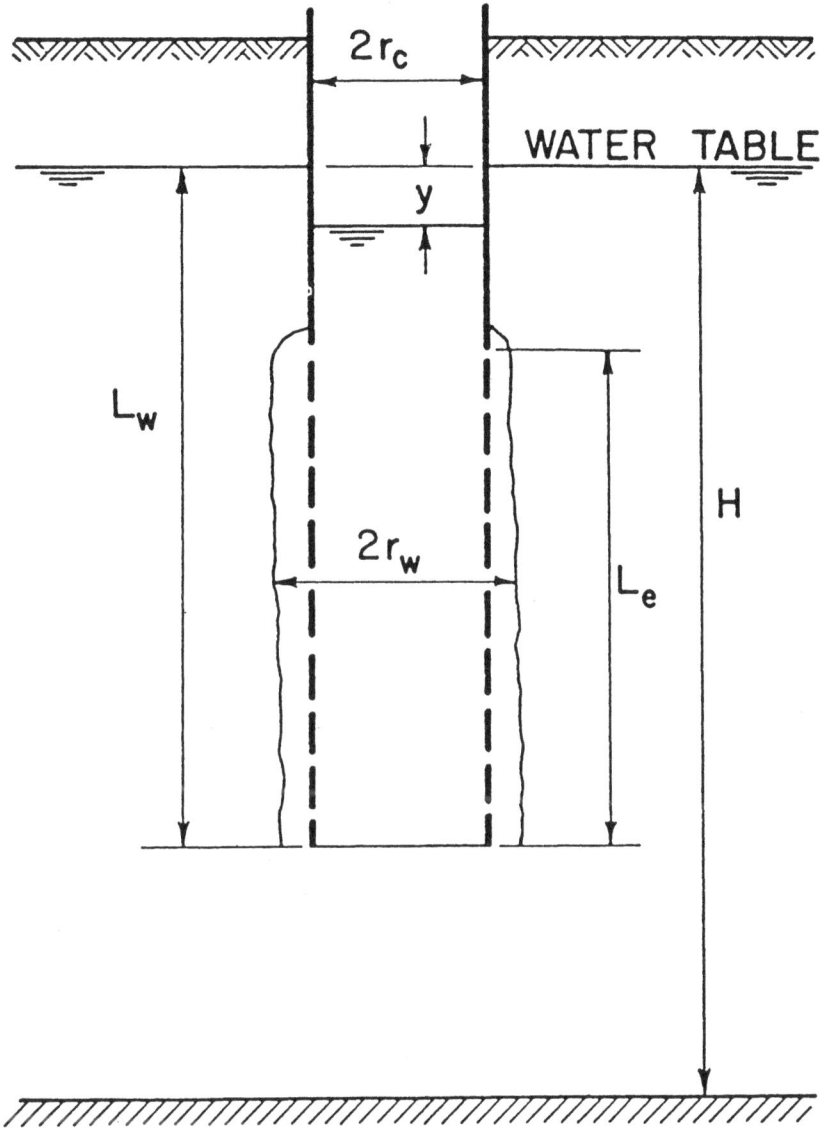

Figure 10-19.—Geometry and symbols for slug test on partially penetrating, screened well in unconfined aquifer with gravel pack and/or developed zone around screen (Bouwer, 1989).

fully penetrating wells. For the partially penetrating case, an empirical equation relating $ln(R_e/r_w)$ to the geometry of the test zone is:

$$\ln\frac{R_e}{r_w} = \left[\frac{1.1}{\ln(L_w/r_w)} + \frac{A + B\ln[(H-L_w)/r_w]}{L_e/r_w}\right]^{-1} \qquad \text{10-20}$$

In this equation, A and B are dimensionless coefficients that can be read on figure 10-20. It should be noted that an effective upper limit of $\ln[(H-L_w)/r_w]$ is 6. If the computed value of $\ln[(H-L_w)/r_w]$ is greater than 6, then 6 should be used in the equation for $\ln(R_e/r_w)$. When $H = L_w$, or the well is fully penetrating, then the value of C should be used from figure 10-20 in the equation:

$$\ln\frac{R_e}{r_w} = \left[\frac{1.1}{\ln(L_w/r_w)} + \frac{C}{L_e/r_w}\right]^{-1} \qquad \text{10-21}$$

Values of the field test data should be plotted as recovery, y, versus time for each data point reading. The values of y should be plotted on a y-axis log scale, and values for corresponding time should be plotted on the x axis. The points should approximate a straight line, which indicates good test data. Areas of the data that plot a curved line (usually at the beginning of the test or near the end of the test) should not be used in the computation.

Numerical computer analysis and solutions to the Bouwer and Rice method are available from various vendors. Using the computer solution may be much easier; however, it is important to recognize the limitations of the solutions and the appropriate cases where the tests apply.

(f) Cooper, Bredehoeft, and Papadopulos (1967).—This method of analysis is used for slug tests on a confined aquifer. The assumptions for solution are:

- Aquifer is confined

- Aquifer has infinite areal extent

- Aquifer is homogeneous, isotropic, and of uniform thickness

- Aquifer potentiometric surface is initially horizontal

- Volume of water, V, is injected or withdrawn instantaneously

- Pumping well is fully penetrating

- Flow to pumping well is horizontal

- Flow is unsteady

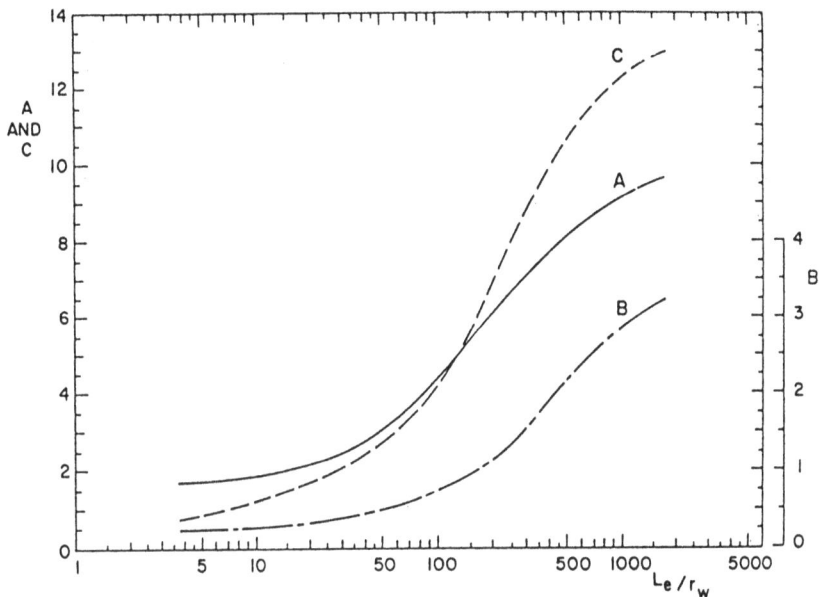

Figure 10-20.—Dimensionless parameters A, B, and C as a function of L_e/r_w (F for calculation of $ln(R_e/r_w)$. From: Bouwer, Herman, 1989, The Bouwer and Rice Slug Test - An Update, Ground Water, vol. 27, No. 3, May-June, p. 305.

- Water is released instantaneously from storage with decline of hydraulic head

- Diameter of well is very small so that storage in the well can be neglected.

From the field test data, a plot of the ratio of H/H_0 (head in well at time t/head in well at $t = 0$) versus time at each head reading is constructed on a semi-log graph. The time value is plotted on the horizontal logarithmic axis and H/H_0 is plotted on the vertical arithmetic axis. The procedure then is to match this curve to a set of type curves (figure 10-21), and select a point on the field curve that fits the type curve at $Tt/r^2 = 1.0$. The time value, t, is then selected from the field curve graph, and the equation $Tt/r^2 = 1.0$ is solved for T. The hydraulic conductivity is then T divided by the aquifer thickness. The storage coefficient is solved from the particular type curve that is matched by using the curve value, a, and substituting into the equation:

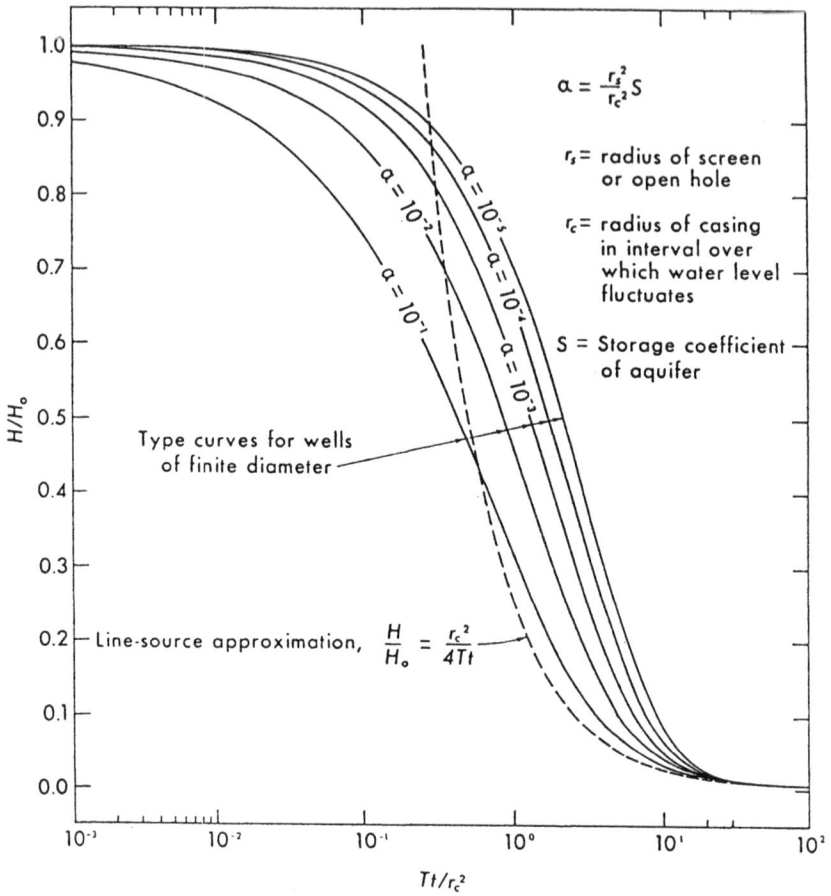

Figure 10-21.—Type curves for instantaneous charge in well of finite diameter (Cooper et al., 1967).

$$S = \frac{ar_c^2}{r_s^2}$$ 10-22

where:

a = a value from 10^{-1} to 10^{-10}
r_c^2 = radius of well casing
r_s^2 = radius of well screen

The shape curve varies only slightly between orders of magnitude and, therefore, determining S by this method can be difficult. Computer software programs are also available for this analysis.

(g) Barker and Black (1983).—This method for determining hydraulic conductivity in fissured rocks considers two cases: (1) where the rock matrix is relatively impermeable and (2) where the diffusivity of the rock matrix is large or the slab thickness is small. A schematic of the slug test is shown on figure 10-22. The initial piezometric surface is taken to be horizontal, and the head in the borehole is assumed to be instantaneously increased by an amount H_o at time $t = 0$. The ratio of H_t/H_o depends on three dimensionless parameters, α, β, γ, in addition to time; many combinations of these dimensionless parameters produce almost identical curves. However, in case (1) the parameter β is small, and the Cooper, Bredehoeft and Papadopulos (1967) analysis can be used. In case (2), γ is small, the rock matrix is in near-equilibrium with the fissures, and the aquifer behaves as would a homogeneous aquifer with a storage coefficient equal to the sum of the storage coefficient of the fissure and the matrix. The storage coefficient of the matrix usually dominates in such cases.

10-6. Auger-Hole Test for Hydraulic Conductivity.—

(a) Introduction.—The auger-hole test measures the average horizontal hydraulic conductivity of the soil profile from the static water table to the bottom of the hole. This test can be run in the presence of a barrier either at or below the bottom of the hole. It is most useful in fine-textured soils at shallow depths.

This section describes the equipment, procedures, and calculations used in performing this test. The development of the analytical details of the auger-hole test are given in a paper by Maasland and Haskew (1958).

(b) Equipment.—Equipment requirements for the auger-hole test are flexible, but the following items have been used successfully:

- An 80-millimeter (nominal 3-inch) diameter auger with three 1.5-meter (5-foot) extension handles and a 110-millimeter (nominal 4-inch) diameter auger.—An 80-millimeter-diameter auger is used initially for the auger-hole test. In the finer textured soils, the pressure required

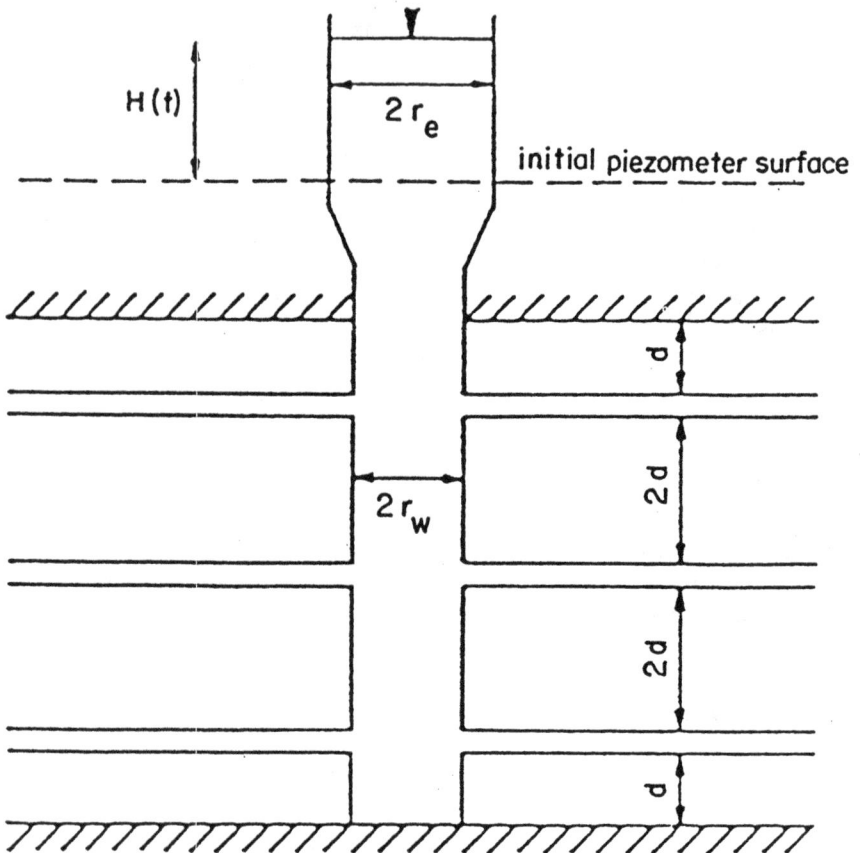

Figure 10-22.—Schematic representation of a slug test in a
fissured aquifer (Barker and Black, 1983).

for the initial augering causes a thin, dense seal to form on
the sides of the hole. This seal is hard to remove even with
a hole scratcher.

However, reaming the 80-millimeter hole with a 110-milli-
meter-diameter auger applies less pressure to the sides of
the hole, and the resulting seal is very thin and easier to
remove. The removal of this thin seal is essential to obtain
reliable data from the test. Three 1.5-meter extension
handles for the augers are usually sufficient for most test
holes.

The Durango- and Orchard-type augers are suitable for most soils, but the Dutch-type auger is preferable for some of the high clay and cohesive soils. Samples from the Durango-type auger are less disturbed than those from the other two types, thus permitting a more reliable evaluation of soil structure. Figure 10-23 shows photographs of the different types of soil augers generally used in drainage investigations.

- Equipment used to record changes in water-table elevation.—Two types of equipment have been used to record the recovery of the water table. The first type consists of a data logger with a preprogrammed logarithmic sampling schedule connected to a pressure transducer. The second type consists of a recorder board, recording tape, and float apparatus. The data logger setup can record recovery data points beginning at time zero, which is impossible to do using the float and recorder board. This capability allows the test to be conducted in materials with higher hydraulic conductivity rates than can be done with a float apparatus.

Water-table recovery data collected on a data logger can be downloaded directly to a computer. A spreadsheet can then be set up to compute test results.

- Recorder board, recording tape, and float apparatus.—This equipment is preferable to manual measuring equipment such as anelectric sounder because it is less expensive, easier to construct, simpler tooperate, and provides a permanent record. The board commonly used is 50 millimeters (2 inches) thick by 100 millimeters (4 inches) wide by 250 millimeters (10 inches) long. A notch 65 millimeters (2-1/2 inches) long and wide enough to hold a nylon roller is made 25 millimeters (1 inch) from one end and 15 millimeters (1/2-inch) from a side. A nylon roller, which can be taken from a regular chair caster, is installed in the notch and fastened in place.A pointer is fastened directly over the roller to act as a reference point during the test. A 50-millimeter (2-inch) diameter recess is drilled near the roller to hold the stopwatch and is located so that the operator can observe the stopwatch and mark on the recording tape without looking up from the stopwatch. A

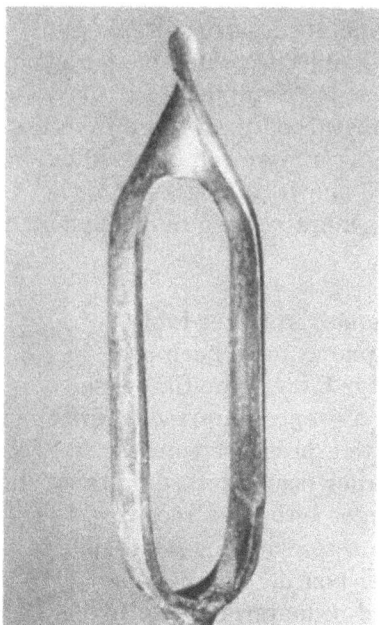

Dutch or open

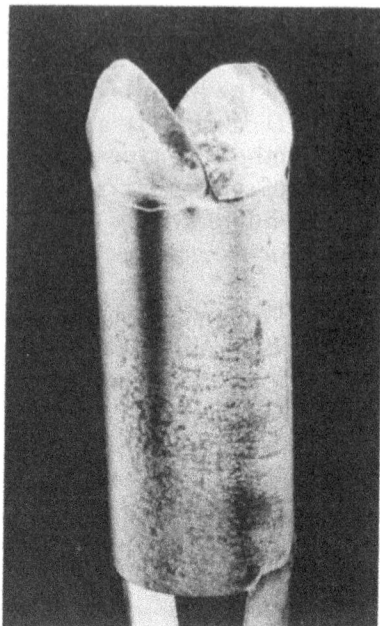

Orchard.

Durango.

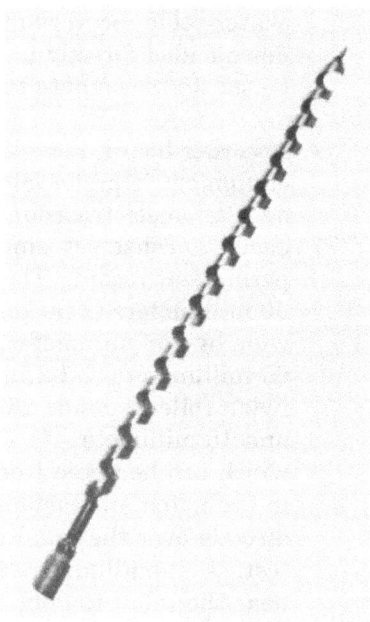

Ship or helical.

Figure 10-23.—Types of hand soil augers.

threaded metal plate for attaching a tripod is attached to the underside of the board on the opposite end from the roller and stopwatch.

The float should be less than 75 millimeters (3 inches) in diameter and weighted at the bottom. It should also be sufficiently buoyant and counterbalanced to prevent any lag in the rise of the float as the water-table rises in the hole. A counterweight that weighs slightly less than the float is used to keep the float string tight. The float should have sloping shoulders so it will be less likely to catch on pebbles or roots on the sides of the open hole or on the joints and perforations in the casing.

Recorder tapes are made from 1.5-meter (5-foot) graph paper strips cut 20 millimeters (3/4 inch) wide and backed with strapping tape. Paper staples are fastened at both ends so the strip can be connected to the float and counterweight. Figure 10–24 shows a schematic of the equipment setup for the auger-hole test.

- Tripod.—Any rigidly constructed tripod can be used. Planetable tripods furnish a rigid support and allow fast setting up and leveling of the recording board.

- Measuring rod or tape.—A measuring rod can be made, or a tape with a weight on the bottom can be used.

- Hole scratcher.—A hole scratcher can be made in a number of ways. The easiest method uses a wooden cylinder, 85 millimeters (3-1/2 inches) in diameter by 75 millimeters (3 inches) long, with small nails protruding as necessary for the auger being used. The heads of the nails, after they have been driven into the cylinder, are cut off to create sharp edges which will break the seal around the periphery of the hole. A 13-millimeter (1/2-inch) coupler attached to the wooden cylinder allows the scratcher to use the same extension handles as the augers. A more efficient hole scratcher can be made from an 85-millimeter (3-1/2-inch) outside-diameter black iron pipe cut 125 millimeters (5 inches) long. A 13-millimeter coupling is then welded to an 85-millimeter diameter by 7-millimeter (1/4-inch) thick plate which, in turn, is welded to one end of the pipe. Holes 3 millimeters (1/8 inch) in diameter are then drilled into the pipe in a staggered pattern. Concrete nails are then

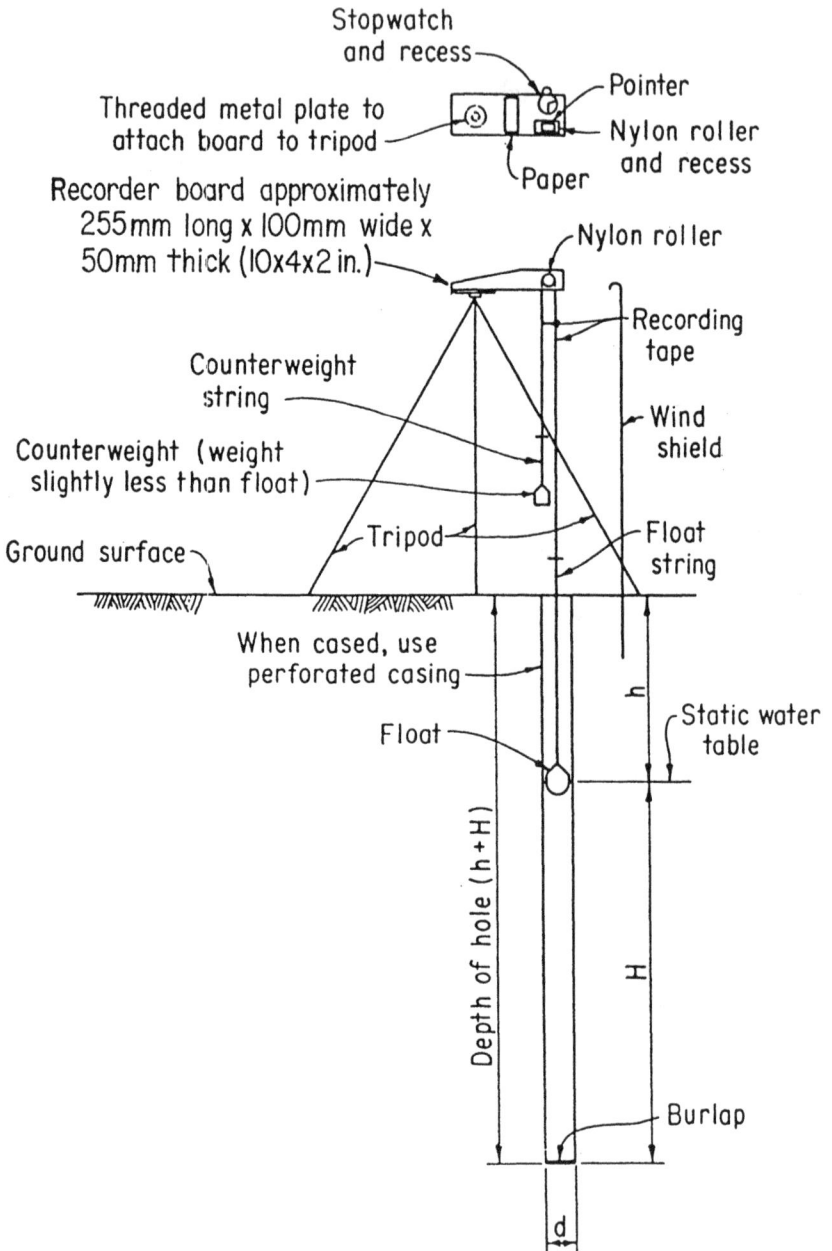

Figure 10-24.—Equipment setup for the auger-hole
or piezometer test.

inserted through each drilled hole from the inside of the pipe. The length of the nails used depends on the diameter of the auger to be used. A wooden block, 80 millimeters (3-1/4 inch) in diameter and 125 millimeters (5 inches) long, is then placed inside the pipe to hold the nails in place. The block can be held in position by drilling a few holes at the pipe ends for holding screws. As different auger-hole diameters are required, longer or shorter nails can be placed in the scratcher. A typical hole scratcher is shown on figure 10–25.

- Bailer or pump.—A bailer can be made from a 1-meter length of 85-millimeter (nominal 3-1/2-inch) diameter, thin-walled conduit with a rubber or metal foot valve at one end and a handle at the other end. Bailers longer than 1 meter are difficult to insert and remove from the auger hole. The hole in the foot valve should be large enough to allow water to enter as rapidly as possible. The bailer should be weighted at the bottom to increase its ability to submerge. Present-day requirements for water quality sampling have made many types of commercial bailers available. They are manufactured from a variety of materials which range from teflon to stainless steel. A lightweight stirrup pump, similar to the one shown on figure 10-25, capable of pumping about 1.5 liters per second (20 gallons per minute), is preferable to the bailer.

- Stopwatch.—Any standard stopwatch or digital watch with seconds registered is satisfactory when using the float apparatus. All readings should be made from a single reference time, which is the beginning of bailing, and all time should be accounted for during the test.

- Inside calipers.—An ordinary pair of inside calipers can be used to determine the diameter of the hole. To prevent the points of the caliper legs from gouging the walls of the auger hole, small flat plates should be welded to the legs. An extension rod screwed into the top of the calipers is used to measure the hole diameter at various depths. The average hole diameter is used in the calculations. The diameter is difficult to measure below the water table with ordinary inside calipers because the water surface reflects light and prevents visual determination of the contact of the calipers with the sides of the hole. For this reason, the

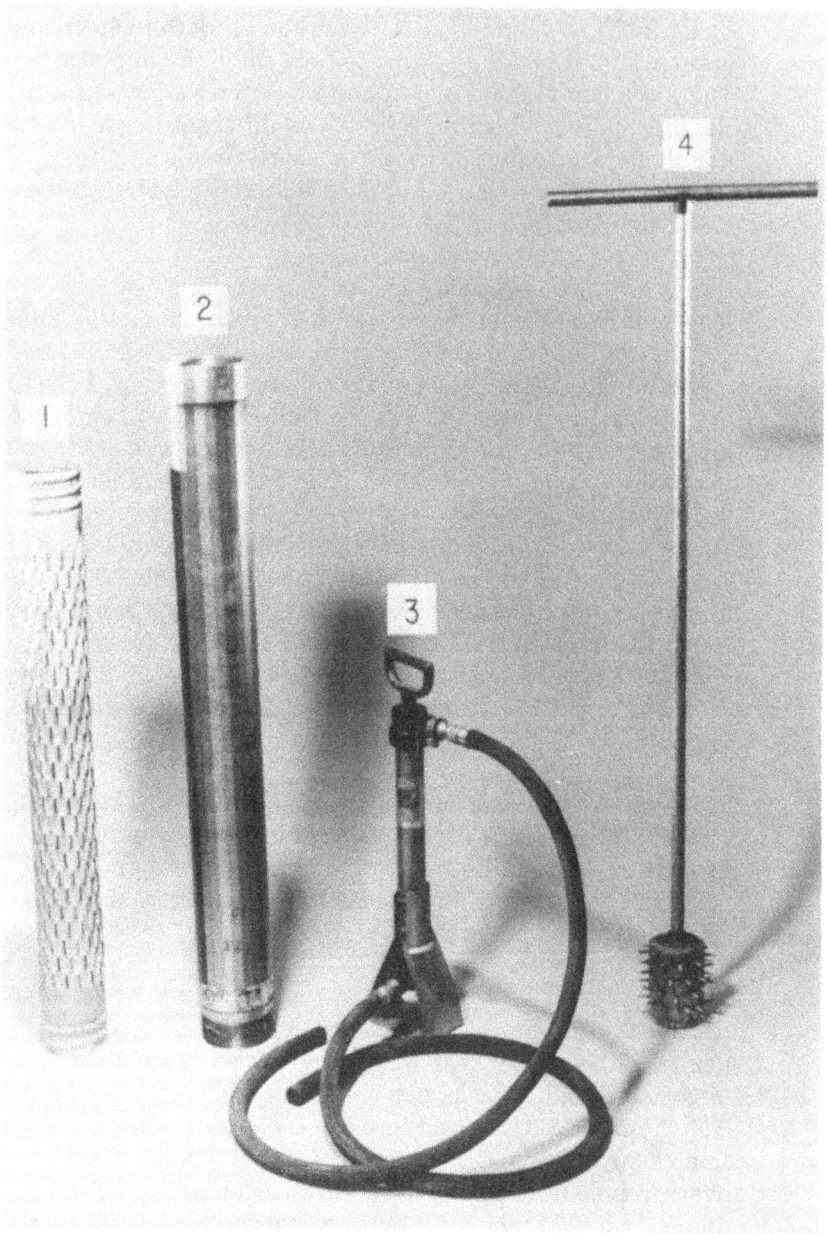

Figure 10-25.—Equipment for auger-hole test. Item (1) perforated
casing, (2) wire-wound well screen, (3) stirrup pump, and
(4) hole scratcher.

average hole diameter can be determined by the measurements made about 0.3 meter (1 foot) below the ground surface and just above the water table.

- Burlap.—Burlap or a similar permeable material will prevent soils from entering at the bottom of the hole. Each hole requires a piece measuring about 0.6 m^2 (2 ft^2).

- Perforated casing or wire-wound well screen.—This protection is necessary for auger holes in unstable soils. The casing or screen should have the same or a slightly larger outside diameter than the hand auger. As the screen or casing is pushed into the ground, the casing and the periphery of the hole make definite contact. Commercial well screen with at least a 10-percent perforated area is the most desirable; however, if this material is not available, a thin-walled downspout casing with 4- to 5-percent perforations is satisfactory. In most agricultural soils, about 200 5- by 25-millimeter hacksaw perforations per meter will give 4- to 5-percent perforations. Commercially available slotted polyvinylchloride casing has also proven adequate for conducting auger-hole tests. Figure 10-25 shows a typical perforated casing and wire-wound well screen.

- Mirror or strong flashlight.—Either one of these items can be used to examine the sides of the auger hole and facilitate measurements with the calipers.

- Windshield.—When wind protection is required, a windshield consisting of a 1- by 1-meter sheet of plywood has been used satisfactorily.

(c) Procedure.—The most efficient team for performing the auger-hole field test for hydraulic conductivity consists of two people. One operates the recorder board, puts the float in the hole, and operates the stopwatch, and the other operates the bailer or pump. After the water level in the hole has stabilized, an experienced team can perform the entire test in 10 to 15 minutes in most soils.

At sites where detailed soil profile data do not exist, a pilot hole will have to be drilled and logged, and test zones must be selected.

The hole should be augered vertically and as straight as possible to the required depth. If the soil is homogeneous throughout the

profile, the hole can be excavated to the total depth to be tested.
When the soil is heterogeneous, tests should be made for each
change in texture, structure, and color. If the material is highly
permeable throughout the profile to be tested, it is best to stop the
hole about 0.6 or 1.0 meter (2 to 3 feet) below the water table so
that one bailing will draw the water down to about the bottom of
the hole. Upon completion of the augering, the sides of the hole
should be scratched to break up any sealing effect caused by the
auger. Scratching is not necessary in the coarser textured soils.
Burlap is then forced to the bottom of the hole and tamped lightly
to prevent any soils from entering the bottom. The sealing effect
can be overcome by allowing the water table to rise to the static
water level, and then gently pumping or bailing the water out to
develop the best flow characteristic. Afterward, time must be
allowed for the water table to reach static level before running the
test. Prior to starting the test, the depth to the static water table
from the ground surface, the total depth of the hole, and the
distance from the static water table to the bottom of the hole
should be measured carefully.

Figure 10-26 shows a sample data and computation sheet for the
test. To begin the test, the tripod with the recorder board,
recording tapes, and float apparatus is placed near the hole so the
float can be centered over the hole and moved freely into it. The
float is then lowered into the hole until it floats on the static water
table level. After a short time period, to allow the water to return
to static level, a zero mark is made on the tape, and the counterweight is positioned so the full change of water-table level can be
recorded. This positioning may require that the counterweight
hang inside the casing. The float is then removed, and the water
is bailed or pumped from the hole as quickly as possible to
minimize the amount of water which returns before the readings
are started. For best results, sufficient water should be bailed or
pumped from the hole so all readings can be completed before the
water-level rises to half its original height, or $0.5\,H$. One or two
passes with the bailer are usually sufficient for most agricultural
soils. As the last bail is withdrawn from the hole, or the pump
starts drawing air, the float should be placed in the hole as quickly
as possible. When the water-level rises rapidly, the float can be
left in the hole and below the bailer or foot valve, which will
minimize the amount of water returning into the hole before the
first reading can be made. The stopwatch is started at the moment
the first bailer is withdrawn, or when pumping begins, and should
run continuously until completion of the test.

HOLE NO. _____ E - 4 _____ LOCATION SAMPLE FARM

OBSERVER _D.M.S._____ DATE AUGUST 6, 1992

HOLE: CASED [X]   UNCASED [ ]

HOLE DIAMETER 102 MILLIMETERS (4 inches)

LOG   DESCRIPTION

GROUND SURFACE   GROUND SURFACE

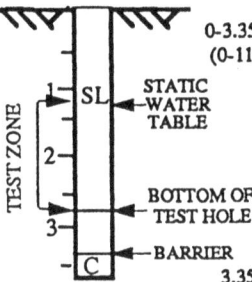

0-3.35 m: Light brown sandy loam(SL),
(0-11ft)   friable, nonsticky, granular.
           Wet at 1.52 m. (5ft)
           Slightly compacted below
           1.83 m (6ft). Appears to have
           good hydraulic conductivity.

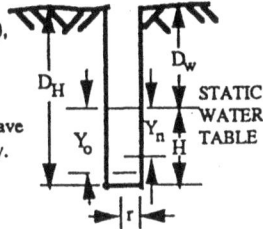

STATIC WATER TABLE

BOTTOM OF TEST HOLE

BARRIER

TEST ZONE

SL

C

3.35-3.66 m: Blue gray clay (C), Sticky,
(11-12ft)   structureless. Appears to
              be impermeable.

$r = 0.051$ m (0.167 ft)

$D_H = 2.74$ m (9.0 ft)

$D_w = 1.46$ m (4.8 ft)

$H = 1.28$ m (4.2 ft)

$Y_o = 0.96$ m (3.15 ft)

$0.8 Y_o = 0.77$ m (2.52 ft)

TIME SECONDS	Δt	Y_n meters (feet)		ΔY meters (feet)	
0	--	0		--	
13	13	0.960	(3.15)	Y_o	
23	10	0.927	(3.04)	0.033	(0.11)
33	10	0.893	(2.93)	0.034	(0.11)
43	10	0.860	(2.82)	0.033	(0.11)
53	10	0.823	(2.70)	0.037	(0.12)
63	10	0.789	(2.59)	0.034	(0.11)
73	10	0.759	(2.49)	0.030	(0.10)
83	10	0.732	(2.40)	0.027	(0.09)
93	10	0.704	(2.31)	0.028	(0.09)

$0.8 Y_o$ at 73

$\overline{Y}_n = \dfrac{0.960 + 0.759}{2} = 0.860$ meter (2.82 feet)

$\dfrac{\overline{Y}_n}{r} = \dfrac{0.860}{0.051} = 16.86$

$\Delta Y = 0.0335$ meter (0.11 ft)

$C = 390$ (from chart)

$\Delta t = 10$ seconds

$K = C \dfrac{\Delta Y}{\Delta t} = 1.31$ meters (4.3 feet) per day

$\dfrac{H}{r} = \dfrac{1.28}{0.051} = 25.10$

or 5.45 centimeters (2.151 inches) per hour

Figure 10-26.—Data and computation sheet on auger-hole test
for hydraulic conductivity.

When using the recorder board and float mechanism, using equal time intervals is convenient, starting from the initial tick mark on the recorder tape. As equal time intervals are read on the stopwatch, the operator marks the tape opposite the pointer. Measurements are continued until recovery of water in the hole equals about 0.2 of the depth initially bailed out or, stated another way, until a reading on the measuring tape of Y_o has been reached (Y_o is the distance the water in the hole was lowered by bailing). Upon completion of the test, the final time is recorded at the last tick mark on the recorder tape. Any irregularities in the record can be quickly observed on the recorder tape, and if readings are highly irregular, the test should be rerun after the static water table has been re-established. Only the period covering the regularly spaced tick marks below $0.8Y_o$ is used in the computations. One irregular spacing usually occurs at the beginning of the test while the float is steadying. As the water rises above $0.8Y_o$, the marks will no longer be equally spaced, but will become closer together with each successive reading. The beginning ofthe shorter spacings usually will occur around $0.8Y_o$, but two or three extra readings are recommended to show that the spacings are definitely getting closer together.

The use of a pressure transducer and a data logger eliminates or greatly reduces many of the problems related to recording water-table recovery discussed in the above paragraphs. With this equipment, the pressure transducer is placed near the bottom of the hole and calibrated to the static water level. The data logger is started just prior to removing the bailer from the hole. Running the data logger until 50-percent recovery has occurred will provide adequate data for computation of the hydraulic conductivity rate.

(d) *Calculations.*—Upon completion of the auger-hole field test for hydraulic conductivity, the time intervals and the corresponding distances between tick marks on the recorder tape are transferred to the computation sheet. Sample computations are shown on figure 10-26. The initial Y_n for time zero can be computed or extrapolated from a Y_n versus time curve if the time from start of pumping to the first tick mark is less than 10 seconds.

Determining the initial Y_n is necessary only when the time interval between the starting time and the first measurement is longer than about 5 seconds and the water-level recovery rate is very fast. Extrapolating the data to determine Y_o, or the initial Y_n, is not always reliable. Every effort should be made to keep the time interval between the start of pumping and the first tick mark as short as possible. This short-time interval is particularly important in sand and gravel with rapid recovery rates.

Care should be taken in selecting consistent, consecutive time intervals and water-table rises to be used in determining the average distance from static water table to the water surface in the hole during the test period, $\overline{Y}_n$; the average incremental rise during incremental time intervals, ΔY; and the average incremental time interval between ticks, Δt. Water-table recovery data collected by a data logger using a properly programmed logarithmic sampling schedule will provide data points beginning at time zero. This early time data greatly reduces, if not eliminates, the concerns discussed in the preceding paragraphs. Because it is difficult to start the data logger at the exact time water-table recovery begins, the early time data should be plotted to determine the point when computations should begin. The C value needed in the computations shown on figure 10-26 is determined from the graphs of figures 10-27 or 10-28, which are intended for use where the barrier is considered to be at infinity or at zero distance below the bottom of the hole. The C values plotted against the dimensionless parameter $\overline{Y}n/r$ simplify the determination of C for a wide range of values of H/r and $\overline{Y}n/r$. For the usual case where no barrier is present, or the barrier is equal to or greater than H below the bottom of the hole, figure 10-27 should be used to determine C. If the hole has been terminated on a slowly permeable zone, figure 10-28 should be used. If the hole penetrates into a slowly permeable zone below a permeable zone, figure 10-28 should be used with H as the distance from the level of the static water table to the slowly permeable layer instead of to the bottom of the hole, as is the usual case. The hydraulic conductivity can then be determined by multiplying the C factor by $\Delta Y/\Delta t$. The resulting hydraulic conductivity has units of meters per day (feet per day) or centimeters per second (inches per hour).

(e) Limitations.—The auger-hole test furnishes reliable hydraulic conductivity data for most conditions; however, the results are entirely unreliable when the hole penetrates into a zone under piezometric pressure. Small sand lenses occurring between less permeable layers make the test more difficult to perform and may yield unreliable data. Water flowing into the hole through the lenses falls on the float apparatus and causes erratic readings. The auger-hole test also cannot be used when the water table is at or above the ground surface because surface water or water

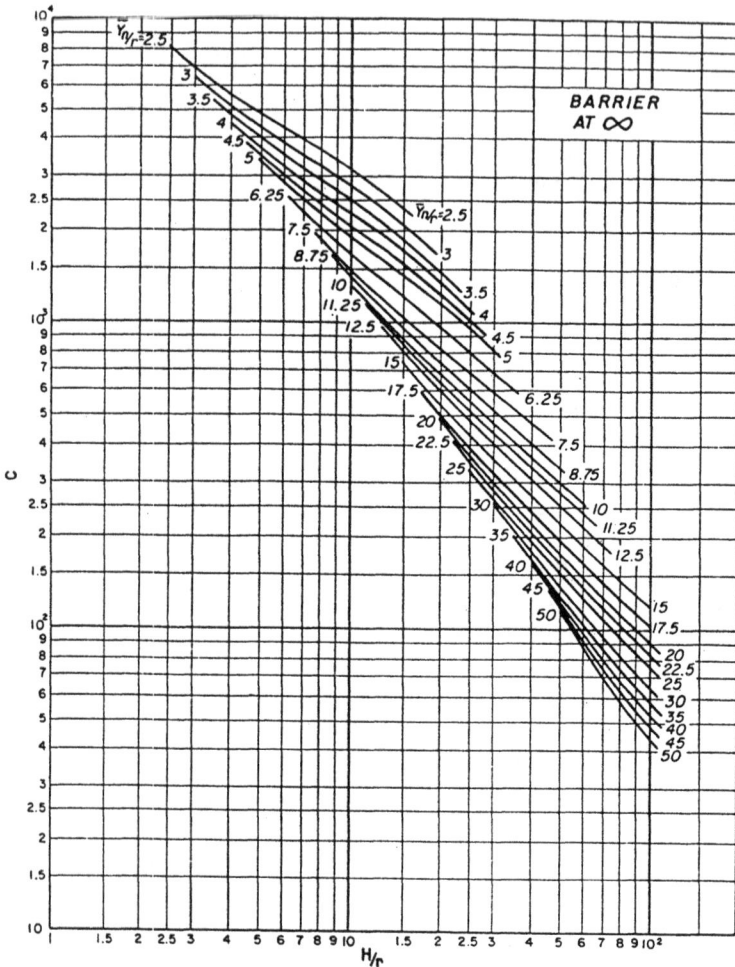

Figure 10-27.—Values of C when barrier is below bottom
of hole during auger-hole test (Massland and Haskew, 1958).

running through permeable surface layers will cause erroneous
readings. A depth of more than 5 meters (20 feet) to water table,
although not a limitation as far as obtaining valid data is
concerned, makes obtaining reliable data extremely difficult.

Comparatively high hydraulic conductivity rates, in the
magnitude of 6 meters per day (10 inches per hour) or more, make
the auger-hole test difficult to perform because the bailer cannot
remove the water as fast as it enters. A pump will remove the

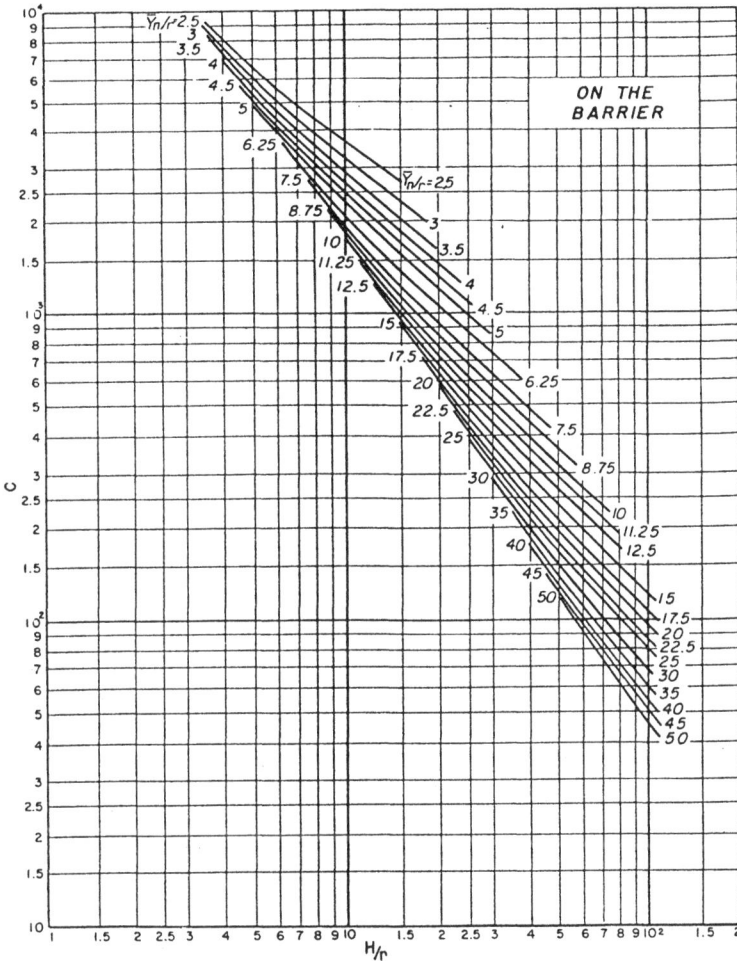

Figure 10-28.—Values of C when barrier is at bottom of hole
during auger-hole test (Massland and Haskew, 1958).

water from the hole rapidly, but in very permeable soils only one
or two readings can be obtained before recovery exceeds 0.2 of the
initial drawdown. A hydraulic conductivity can be calculated from
only one or two readings, but the results could be erroneous. The
use of a data logger to collect water-table recovery data will solve
this problem, which occurs when using float-activated equipment.
Tests have been successfully run in alluvial materials having
hydraulic conductivity rates of over 30 meters per day (50 inches
per hour) using a data logger.

At the other extreme, auger-hole tests in soils with hydraulic conductivity rates in the range of 0.0006 to 0.006 meter per day (0.001 to 0.01 inch per hour) usually give such erratic readings that accurate values cannot be obtained. However, the results can be important in determination of drainage requirements even though exact values are not obtained. However, the knowledge that hydraulic conductivities are very high or very low can be useful from a practical standpoint.

The difficulty usually encountered in augering or digging a hole of uniform size through rocky or coarse-gravel material can prevent the performance of an auger-hole test. Casing can sometimes be used to stabilize the walls of the hole if a test is needed in these materials.

(f) Step Tests in Layered Soils.—Step tests are used to determine the hydraulic conductivity of layered soils. Step tests are simply a series of auger-hole tests in or near the same hole location but at different depths. The hole is initially augered to within 75 to 100 millimeters (3 or 4 inches) of the bottom of the first texture change below the water table, and then the first auger-hole test is run and the hydraulic conductivity is computed. The hole is then augered to within 75 to 100 millimeters of the bottom of the next texture change, the second test is run, and the average hydraulic conductivity for both layers can then be determined. The procedure continues until the last layer to be tested has been reached. The hydraulic conductivity value calculated for each step will be the average value from the water table to the depth of the hole.

The hydraulic conductivity for the individual texture is found from the formula:

$$K_n x = \frac{K_n D_n - K_{n-1} D_{n-1}}{d_n} \qquad \text{10-23}$$

where:

$K_{n'x}$ = hydraulic conductivity to be determined
K_n = hydraulic conductivity obtained in the nth step of test
K_{n-1} = hydraulic conductivity obtained in the (n–1) step
d_n = thickness of the nth stratum $(D_n - D_{n-1})$
D_n = total thickness of the nth step from the static water level

D_{n-1} = total thickness from the static water level for the (n–1) step

n = number of the test

x = step number

Test errors may produce negative results, and the test should be rerun. If the results are still negative after a rerun, the piezometer test described in section 10-7 should be used. A sample calculation sheet for the step test is shown on figure 10-29.

10-7. Piezometer Test for Hydraulic Conductivity.—

(a) Introduction.—The piezometer test measures the horizontal hydraulic conductivity of individual soil layers below a water table. This test is preferred over the auger-hole test when the soil layers to be tested are less than 0.5 meter (18 inches) thick and when individual layers below the water table are to be tested. In ground-water investigations, an important application of this test is to provide data for determining which layer below the free water table functions as the effective barrier layer. This test also provides reliable hydraulic conductivity data for any soil layer below the water table.

(b) Equipment.—Suggested equipment required for the piezometer test is:

- Casing of 25- to 50-millimeter (1- to 2-inch) inside diameter is recommended, consisting of a thin-walled electrical conduit for depths to 4 meters and black iron pipe with smooth inside walls for depths greater than 4 meters.

- Ship auger which fits inside the casing.

- Pipe-driving hammer, consisting of a piece of 50-millimeter (2-inch) iron pipe which fits over the casing with a 5-kilogram (10-pound) weight fixed to the pipe. A small sledge hammer can be used in place of the 5-kilogram weight. The casing may also be pushed with a small drill rig.

- Hand-operated pitcher pump with hose and foot valve, or a bailer which will fit inside the casing.

The hydraulic conductivity for a specific layer is given by:

$$K_{n,x} = \frac{K_n D_n - k_{n-1} D_{n-1}}{d_n}$$

$$d_n = D_n - D_{n-1}$$

FIELD TEST DATA

STEP NO.	K_n, cm/h	d_n, meters	D_n, meters
1	$K_1 = 5.8$	$d_1 = 1.76 - 1.31 = 0.45$	$D_1 = 1.76 - 1.31 = 0.45$
2	$K_2 = 3.8$	$d_2 = 2.19 - 1.76 = 0.43$	$D_2 = 2.19 - 1.31 = 0.88$
3	$K_3 = 3.0$	$d_3 = 2.59 - 2.19 = 0.40$	$D_3 = 2.59 - 1.31 = 1.28$
4	$K_4 = 2.0$	$d_4 = 3.38 - 2.59 = 0.79$	$D_4 = 3.38 - 1.31 = 2.07$

CALCULATIONS FOR HYDRAULIC CONDUCTIVITY OF SPECIFIC LAYERS

$$K_{1,1} = K_{SL} = \frac{K_1 D_1}{d_1} = \frac{(5.8)(0.45)}{0.45} = 5.8 \text{ centimeters per hour (2.3 inches per hour)}$$

$$K_{1,2} = K_L = \frac{K_2 D_2 - K_1 D_1}{d_2} = \frac{(3.8)(0.88) - (5.8)(0.45)}{0.43} = 1.71 \text{ centimeters per hour} \quad (0.64 \text{ inch per hour})$$

$$K_{1,3} = K_{SCL} = \frac{K_3 D_3 - K_2 D_2}{d_3} = \frac{(3.0)(1.28) - (3.8)(0.88)}{0.40} = 1.24 \text{ centimeters per hour} \quad (0.53 \text{ inch per hour})$$

$$K_{1,4} = K_{CL} = \frac{K_4 D_4 - K_3 D_3}{d_4} = \frac{(2.0)(2.07) - (3.0)(1.28)}{0.79} = 0.38 \text{ centimeters per hour} \quad (0.15 \text{ inch per hour})$$

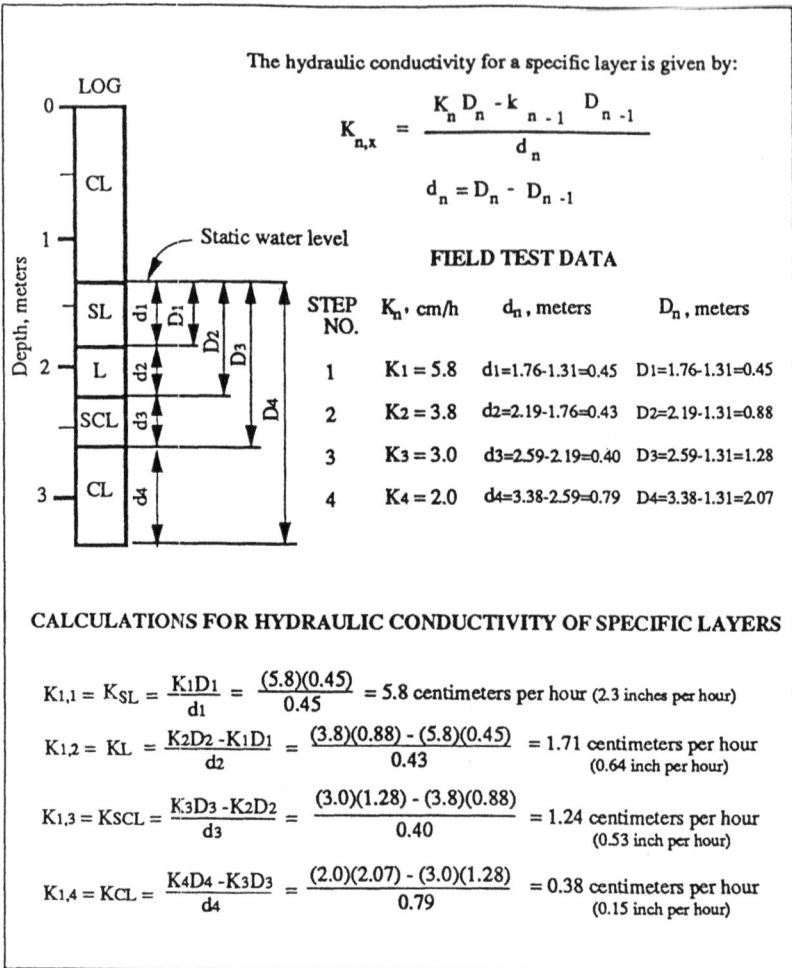

Figure 10-29.—Data and computation sheet on step test
for hydraulic conductivity.

- Recorder board, recording tapes, and float apparatus or an electrical sounder. The float resembles the float made for the auger-hole test but is of smaller size to fit into the smaller diameter casing. The counterweight must be adjusted accordingly. A transducer and data logger may be used as in the auger-hole test.

- Computation sheets, clipboard, stopwatch, measuring tape or rod, windshield, and casing puller.

- Bottle or vegetable brush for cleaning soil film from inside of test pipe. The brush should be fitted with a coupler that attaches to the auger handle.

(c) Procedure.—A two-man team is desirable in performing the piezometer field test for hydraulic conductivity. The test layer should be at least 300 millimeters (12 inches) thick so that a 100-millimeter (4-inch) length of encased hole, or cavity, can be placed in the middle of it. This placement is especially important if a marked difference in the texture, structure, or density of the layers exists above and below the test layer. After the test layer has been selected, the topsoil is removed from the ground surface, and a hole is augered to within 0.5 meter (2 feet) of the test layer. Some operators prefer to auger 150 to 300 millimeters (6 to 12 inches), then drive the casing and repeat this process for the entire depth of the hole. However, this method is slow, and experience shows its use is generally not warranted. Other operators jet the casing to within 0.5 to 0.75 meter (2 to 3 feet) of the test layer and then auger and drive the casing the remaining distance. This procedure requires additional equipment that usually cannot be moved into a waterlogged field. The augering and driving procedure is always used for the last 0.5 meter (2 feet) to assure a good seal and also to minimize soil disturbance. The casing is stopped at the depth selected for the top of the 100-millimeter- (4-inch-) long cavity, and the cavity is then augered below the casing. After some recovery has occurred, the pipe should be cleaned with a bottle brush to remove the soil film that the float may cling to.

The size and shape of the cavity are important in the test, so care should be taken to assure that the cavity has the predetermined length and diameter. If the soil in the test layer is so unstable that the cavity will not remain open during the test, screens should be made that can be pushed down inside the casing. For a 25-millimeter (1-inch) inside-diameter casing and a 100-millimeter (4-inch) cavity, the screen should be 125 millimeters (5 inches) long and have a 24-millimeter (15/16-inch) outside diameter. A rigid point should be welded on the bottom of the screen to facilitate pushing it down inside the casing. A pole about 20 millimeters (3/4-inch) in diameter can be used to push the screen to the bottom of the cavity. A small bent nail or hook placed on the opposite end of the pole will allow the screen to be reclaimed at the end of the test by hooking the nail into the screen

and pulling it out. The cavity is cleaned by gently pumping or
bailing water and sediment out of the hole until the discharge is
clear.

After the water table has returned to equilibrium, the recorder
board and float apparatus are set up, and the float is dropped
down the casing. Figure 10-24 shows the equipment setup. When
the float comes to rest, the pointer is set at zero on the recorder
tape, the float is removed from the hole, and the water is pumped
or bailed out. A small foot valve for the suction line can be made
similar to larger commercial types, or a bailer similar to that used
in the auger-hole test can be made. After pumping or bailing the
water, the float is immediately dropped down the casing. When
the float starts to rise, a tick mark is made on the recorder tape
and at the same time the stopwatch is started. Select a convenient
time interval between observations and make corresponding tick
marks on the recorder tape. Removal of all of the water from the
piezometer is not essential because measurements can be obtained
and used anywhere between the static water-table level and the
initial bailed-out level. Obtaining three or four readings during
the first half of the water rise will give consistent results.

(d) Calculations.—After completion of the piezometer test, the
hydraulic conductivity is calculated from the equation developed by
Kirkham (1945):

$$K = \frac{3{,}600\pi \left(\dfrac{D}{2}\right)^2 \ln\left(\dfrac{Y_1}{Y_2}\right)}{A\,(t_2 - t_1)} \qquad\qquad 10\text{-}24$$

where:

K = hydraulic conductivity in centimeters per hour
(inches per hour)

Y_1 and Y_2 = distance from static water level to water level at
times t_1 and t_2 in centimeters (inches)

D = diameter of casing in centimeters (inches)

$t_2 - t_1$ = time for water level to change from Y_1 to Y_2 (seconds)

A = a constant for a given flow geometry in centimeters
(inches)

A sample calculation using this equation is shown on
figure 10-30. The constant A may be taken from the
curves shown on figures 10-30 or 10-31.

Location: Hole C-2 -- Sample Farm
Observer: A.P.B.   Date: October 9, 1974

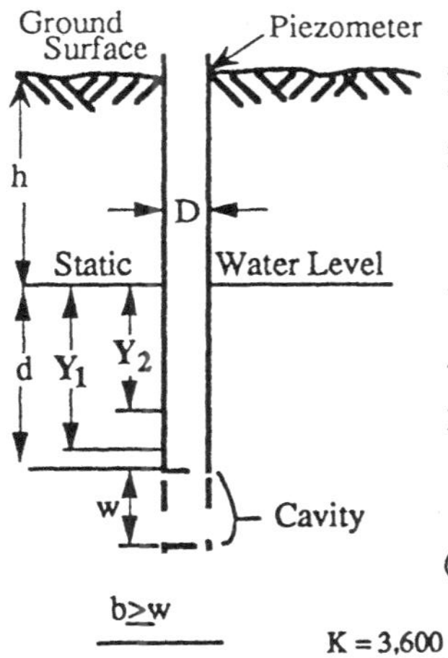

h = 218.44 centimeters (86.00 inches) Ground Surface to static water level
D = 2.54 centimeters (1.00 inch) Inside diameter of piezometer and cavity
d = 237.74 centimeters (93.60 inches) Static water level to bottom of piezometer
w = 10.16 centimeters (4 inches) Length of cavity
A = 33.27 centimeters (13.1 inches) Constant for a given flow geometry taken from curve.
K = Hydraulic conductivity, centimeters per hour (inches per hour)
b = depth to texture change
Y_1, Y_2 = Distance from static water level to change from in centimeters (inches).
$(t_2 - t_1)$ = Time for water to change from Y_1 to Y_2 (seconds)

$$K = \frac{3{,}600\,\pi\,(D/2)^2 \log_e (Y_1/Y_2)}{A(t_2 - t_1)}, \text{ centimeters per hour (inches per hour)}$$

Time (seconds)		Y, centimeters (inches)		A, cent. (inches)	t_2-t_1	Y_1/Y_2	$\log_e Y_1/Y_2$	$3{,}600\,\pi\,(D/2)^2$ cm¹ sec/hr (in² sec/hr)	K cm/hr (in/hr)
Initial (t_1)	Final (t_2)	Initial(Y_1)	Final(Y_2)						
0	30	218.44 (86.00)	197.87 (77.90)	33.27 (13.1)	30	1.104	0.099	18241.47 (2827.44)	1.80 (0.71)
30	60	197.87 (77.90)	178.44 (70.25)	33.27 (13.1)	30	1.109	0.103	18247.47 (2827.44)	1.88 (0.74)
60	90	178.44 (70.25)	160.02 (63.00)	33.27 (13.1)	30	1.115	0.109	18241.47 (2827.44)	1.99 (0.78)
90	120	160.02 (63.00)	145.47 (57.27)	33.27 (13.1)	30	1.100	0.095	18241.47 (2827.44)	1.74 (0.68)
120	150	145.47 (57.27)	131.17 (51.64)	33.27 (13.1)	30	1.109	0.103	18241.47 (2827.44)	1.88 (0.74)

Average for 5 readings = 1.86 (0.73)

EXAMPLE
D = 2.54 cm (1 in)
w = 10.16 cm (4 in)
w/D = 10.16/2.54 = 4
A/D = 13.1
A = 33.27 cm (13.1 in)

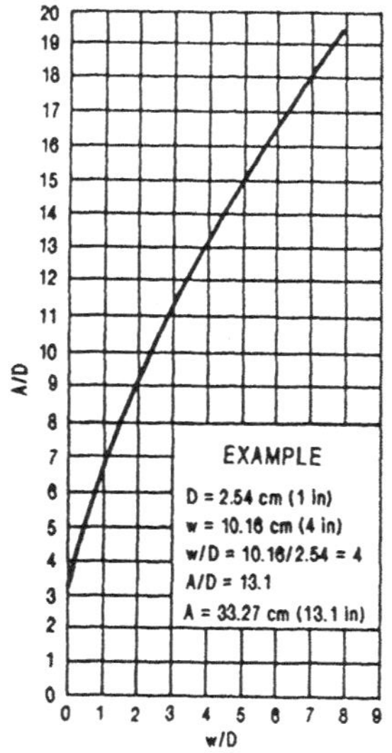

A as a function of D and w.
Redrawn from LUTHIN & KIRKHAM (1949).
Revised by USBR (Mantei, 1972).

Figure 10-30.—Data and computation sheet on piezometer test for hydraulic conductivity.

VALUE OF Δ/H, CONSTANT

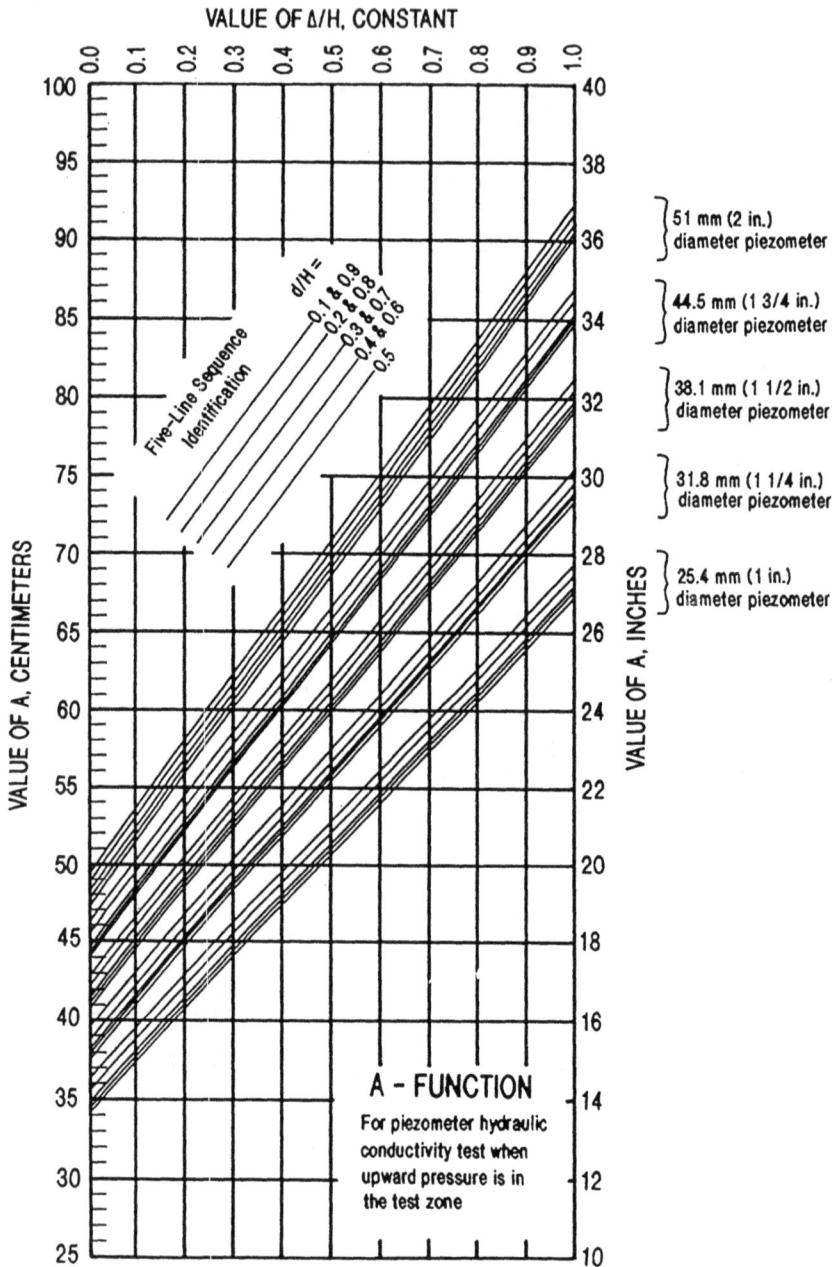

Figure 10-31.—Chart for determining A-function on piezometer test for hydraulic conductivity when there is upward pressure in the test zone.

The curve on figure 10-30 is valid when d and b are both large compared to w (d = distance from the static water level to bottom of piezometer; b = distance below bottom of cavity to top of the next zone; and w = length of cavity). According to Luthin and Kirkham (1949), when $b = 0$ and d is much greater than w, the curve will give an A factor for $w = 4$ and $d = 1$, which will be about 25 percent too large.

The chart on figure 10-31 is used for determining A when piezometric pressures exist in the test zone. When pressures are present, additional piezometers must be installed. The tip of the second piezometer should be placed just below the contact between layers in a layered soil (figure 10-32). In deep uniform soils, the second piezometer tip should be placed an arbitrary distance below the test cavity.

After installing the piezometer, the following measurements should be made:

- Distance, H, in meters (feet), between piezometer tips

- Difference, Δ, in meters (feet), between water levels in the piezometer at static conditions

- Distance, d', in meters (feet), between center of the lower piezometer cavity and the contact between soil layers in layered soils

The A value from figure 10-31 is used in equation 10-24 to determine the hydraulic conductivity.

(e) Limitations.—Installation and sealing difficulties encountered in gravel or coarse sand material comprise one of the principal limitations of the piezometer test for hydraulic conductivity. Also, when the casing bottoms in coarse gravel, a satisfactory cavity cannot be obtained. The practical limit of hole depth is about 6 meters (20 feet), both for installation and water removal with a stirrup pump. Duplicate tests in soils of very low hydraulic conductivity (0.0025 to 0.025 cm/h) are always in the low range, but can vary as much as 100 percent. However, this much variation has little consequence in this low range. Test layers less than about 250 to 300 millimeters (10 to 12 inches) thick and lying between more permeable materials will not give reliable results

Piezometer No.

1   **2**

$\triangle_1 = 2.2$ m

Piezometer
water surface

Ground surface

$\triangle_2 = 1.4$ m

0

LOAM

1

SILT
LOAM

Water table

2

$$K = \frac{3,600\,\pi\,(D/2)^2 \log_e(Y_1/Y_2)}{A(t_2 - t_1)}$$

3   FINE
    SANDY
    LOAM

Notes:
  d' = Distance from top of test
    layer to center of test
    cavity.
  H = Distance from water table
    to center of test cavity.

4

DEPTH, METERS

SILTY
CLAY
LOAM

5

$H_1 = 6.1$ m

$H_2 = 5.1$ m

**SAMPLE CALCULATION FOR PIEZOMETER
TEST WITH UPWARD PRESSURE IN TEST
ZONE**

Test in the silty clay - - Find A - Function using
piezometers 1 and 2

Diameter of piezometer is 3.8 centimeters.

$H = H_1 - H_2 = 6.1 - 5.1 = 1$ meter (3.3 feet)

$\triangle = \triangle_1 - \triangle_2 = 2.2 - 1.4 = 0.8$ meter (2.6 feet)

$\triangle/H = 0.8/1.0 = 0.8$

6   SILTY
    CLAY

d'

d' = distance from ground surface to center of test cavity in
piezometer
    No. 2 minus the distance from ground to top of silty
clay layer
    = 6.3 - 6.0 = 0.3 meter (1.0 foot)
$d'/H = 0.3/1 = 0.3$

7   SAND &
    GRAVEL

A = 71.6 centimeters (from A- function chart)
    (28.2 inches)
Use recovery data from piezometer No. 2 to
determine K value for the silty clay layer.

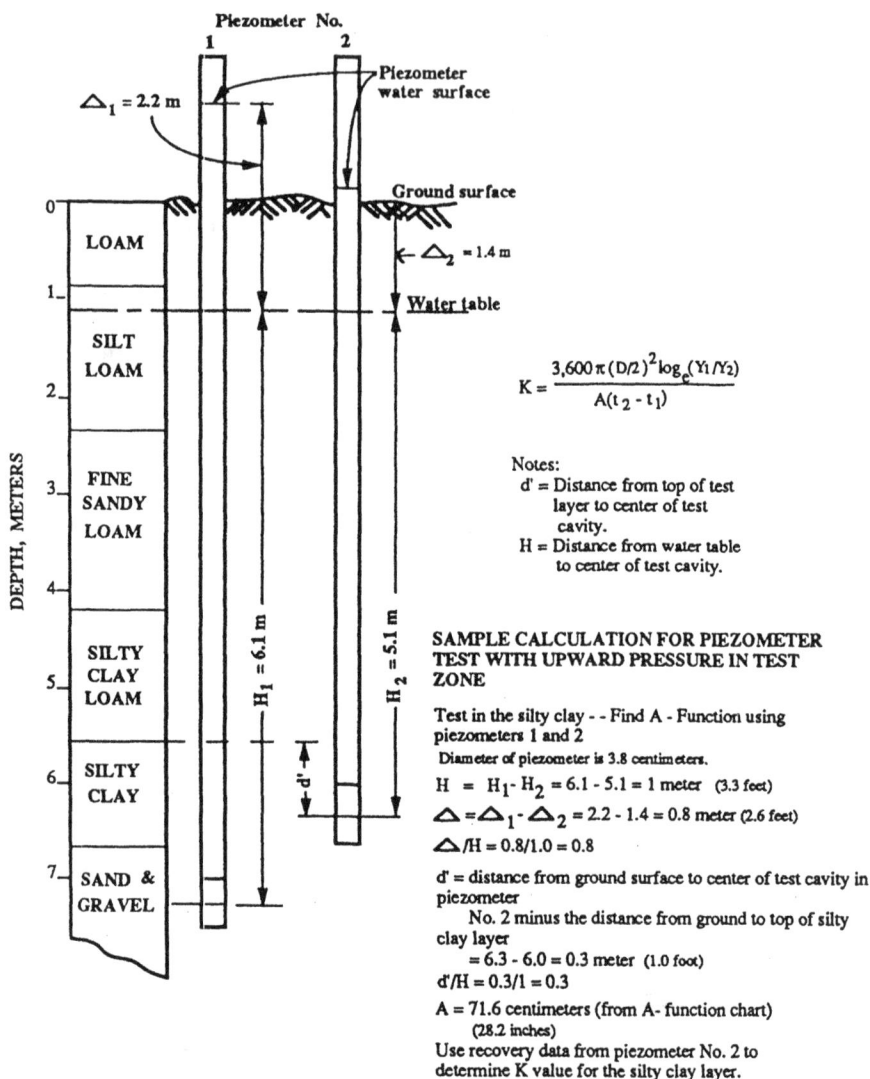

Figure 10-32.—Sample calculation for piezometer
test with upward pressure in the test zone.

because of the influence of the more permeable materials. The size
of the casing is a matter of preference, as long as it is
25 millimeters (1 inch) or more in diameter. Field experience has
shown that a 38-millimeter (1-1/2-inch) inside-diameter piezometer

provides adequate open area for float operation. Pipe diameters greater than 50 millimeters (2 inches) are difficult to install properly.

10-8. Pomona Well Point Method.—This method resembles the piezometer test discussed in the preceding paragraphs, except that this method measures discharge for a fixed drawdown rather than the water-table recovery rate. These differences allow data collection in unstable materials where an open cavity is difficult to maintain. This test method can also be used in materials where the water recovery rate is very rapid.

The setup may be identical to the piezometer test or it may employ a driven well point.

After installation is complete and the well has been developed, the test is conducted by pumping at a rate to maintain a fixed drawdown. The discharge is measured for 1 out of every 5 minutes until a steady rate is obtained. When the system reaches equilibrium, the discharge rate is measured. The hydraulic conductivity rate is determined by:

$$K = Q/Ah \qquad\qquad 10\text{-}25$$

where:

K = hydraulic conductivity
Q = discharge rate
A = a constant for a given flow geometry (figures 10-30 and 10-31)
h = head difference

Layered soils can easily be investigated, and the soil need not support a cavity if a screened well point is used. Even when the cavity is unsupported, as in the piezometer setup, substantially less hydrostatic pressure is exerted on the cavity than in the piezometer test. The primary limitations are the time required to conduct the test and the impracticality of measuring low permeabilities.

10-9. Single Well Drawdown Test for Hydraulic Conductivity.—Coarse sands and gravel usually make the auger-hole (pump-out) and piezometer tests difficult to run. An alternative pump-out test can be made to obtain a rough estimate of hydraulic conductivities in these materials. The test is a small-scale version of a regular pump test for large wells.

Equipment for the test is the same as that used for the auger-hole test except the recorder board and tripod are not used. A gasoline-driven pump with a valved discharge should be used. A calibrated bucket and a stopwatch should be used to determine flow rate.

Hole preparation is much the same as for the auger-hole test; however, hand augering is usually too difficult. Once the hole is prepared and the static water level is measured, water is pumped from the hole at a constant rate. After some time, the water level in the hole will reach a steady-state level. Steady state can be assumed to exist when the water level in the hole drops less than 30 millimeters (0.1 foot) in 2 hours. When steady-state conditions exist, the flow rate and depth of water in the hole are recorded. These data, along with the distance from the static water level to the bottom of the hole, are used in one of the equations shown on figure 10-33. Use the equation that most nearly approaches the test conditions.

This method should be used only in highly permeable sands and gravel to obtain an estimate of hydraulic conductivity when the auger-hole or piezometer tests fail to give satisfactory results.

10-10. Shallow Well Pump-In Test for Hydraulic Conductivity.—

(a) Introduction.—The shallow well pump-in test for hydraulic conductivity, also known as the well perimeter test, is used when the water table is located below the zone to be tested. Essentially, this test consists of measuring the volume of water flowing laterally from a well in which a constant head of water is maintained. The lateral hydraulic conductivity determined by this test is a composite rate for the full depth of the tested hole.

(b) Equipment.—Equipment requirements for the shallow well pump-in test include the following items previously described for

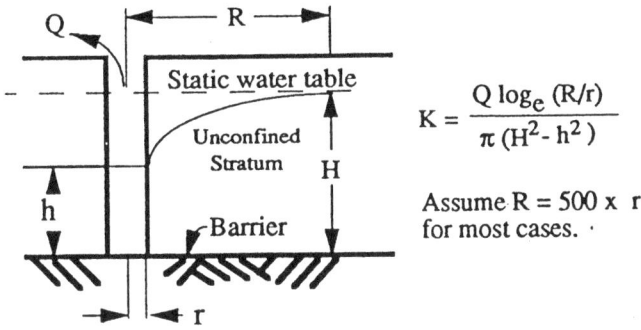

$$K = \frac{Q \log_e (R/r)}{\pi (H^2 - h^2)}$$

Assume R = 500 x r for most cases. ·

(a) Pumping from a uniform unconfined stratum, water table in stratum being pumped.

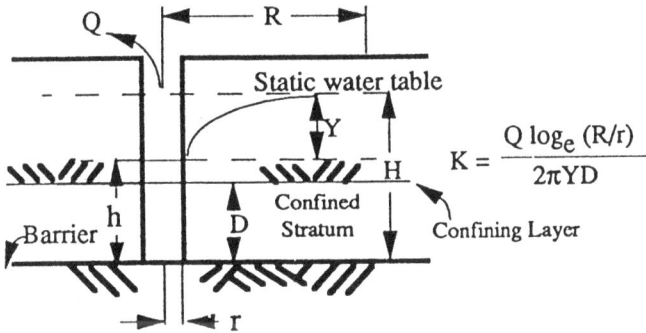

$$K = \frac{Q \log_e (R/r)}{2\pi YD}$$

(b) Pumping from a confined stratum, water table above stratum being pumped.

K = Hydraulic conductivity, $m^3/m^2/day$ $(ft^3/ft^2/day)$
Q = Flow rate at steady state conditions, m^3/day (ft^3/day)
Y = Drawdown from static water surface = H-h, m (ft)
H = Height of static water table above bottom of hole, m (ft)
h = Depth of water in hole at steady state pumping
   conditions, m (ft)
D= Flow thickness of strata between bottom of the hole
   and overlying (confining) stratum, m (ft)
R = Distance from centerline of well to point of zero
   drawdowm, m (ft)
r = Effective radius of well, m (ft)

Figure 10-33.—Determination of hydraulic conductivity by pumping from a uniform or confined stratum.

the auger-hole test in section 10-5: 75- and 100-millimeter (3- and 4-inch nominal) diameter soil augers, hole scratcher, perforated casing, burlap, and wristwatch with a second hand. Additional equipment items are:

- Water-supply tank truck of at least 1,200-liter (350-gallon) capacity with gasoline-powered water pump.

- Calibrated head tank, 200-liter (50-gallon) minimum. This tank should have fittings so that two or more tanks can be connected when required.

- An 8-meter (25-foot) long 25- to 50-millimeter (1- to 2-inch) diameter, heavy-walled hose for rapid filling of head tank from supply tank.

- Wooden platform to keep head tank off the ground and to prevent rusting.

- A 25-millimeter (1-inch) diameter pipe, 1 meter long, to be driven into the ground and wired to head tank to keep tank in position.

- Constant-level float valve (carburetor), which must fit inside the casing.

- A rod threaded to fit the threads on top of the carburetor, used to regulate the depth that the float valve is lowered into the hole.

- Sufficient 10- or 12.5-millimeter (3/8- or 1/2-inch) inside diameter flexible rubber tubing to connect tank to carburetor.

- Plexiglas cover, 300- by 300-millimeter (12- by 12-inch) by 3 millimeters (1/8 inch) thick, with hole in center for carburetor rod, and two other holes, one for rubber tubing and one for measuring water level and temperature of water in the hole.

- Filter tank and filter material.

- Steel fencepost with post driver, four required per site. About 25 meters (80 feet) of fencing wire (needed only when site must be fenced).

- Thermometer which can be lowered into hole, Celsius scale preferred.

- A 3-meter (10-foot) steel tape, clipboard, computation sheet, and a 400-millimeter (16-inch) tiling spade.

Figure 10-34 shows a schematic of the equipment setup for this test.

The constant-level float valve (carburetor) suggested for use in this test and in the ring perimeter test, described later, can be constructed out of various materials and can be made in different shapes. The only requirements are that it must fit inside a 100-millimeter (4-inch) diameter hole, have adequate capacity, cause minimum aeration of water, and control the water level within ±15 millimeters. Material to construct a carburetor that has proven satisfactory consists of the following:

- 1/2 meter (20 inches) of 20- by 3-millimeter (3/4- by 1/8-inch) metal strap

- One large tractor carburetor, needle valve, a needle valve seat at least 3 millimeters (1/8 inch) in diameter and a float made of styrofoam

- Two 20- by 6-millimeter (3/4- by 1/4-inch) bushings

- One 20-millimeter (3/4-inch) coupling

A photograph of a typical carburetor is shown on figure 10-35.

(c) Procedure.—A two-man team can efficiently install the equipment and conduct the shallow well pump-in test. The hole for the test should first be hand augered with a 75-millimeter (3-inch nominal) diameter auger and then reamed with the 100-millimeter (4-inch nominal) diameter auger. A complete log, including texture, structure, mottling, and color, should be obtained for use in interpreting and projecting results. The hole should be carefully scratched after completion to the desired depth to break up any compaction caused by the 100-millimeter auger and to remove any loose material on the sides. In unstable soils, a thin-walled perforated casing should be installed, with perforations extending from the bottom of the hole up to the predetermined controlled water level. A commercial well screen or slotted-PVC

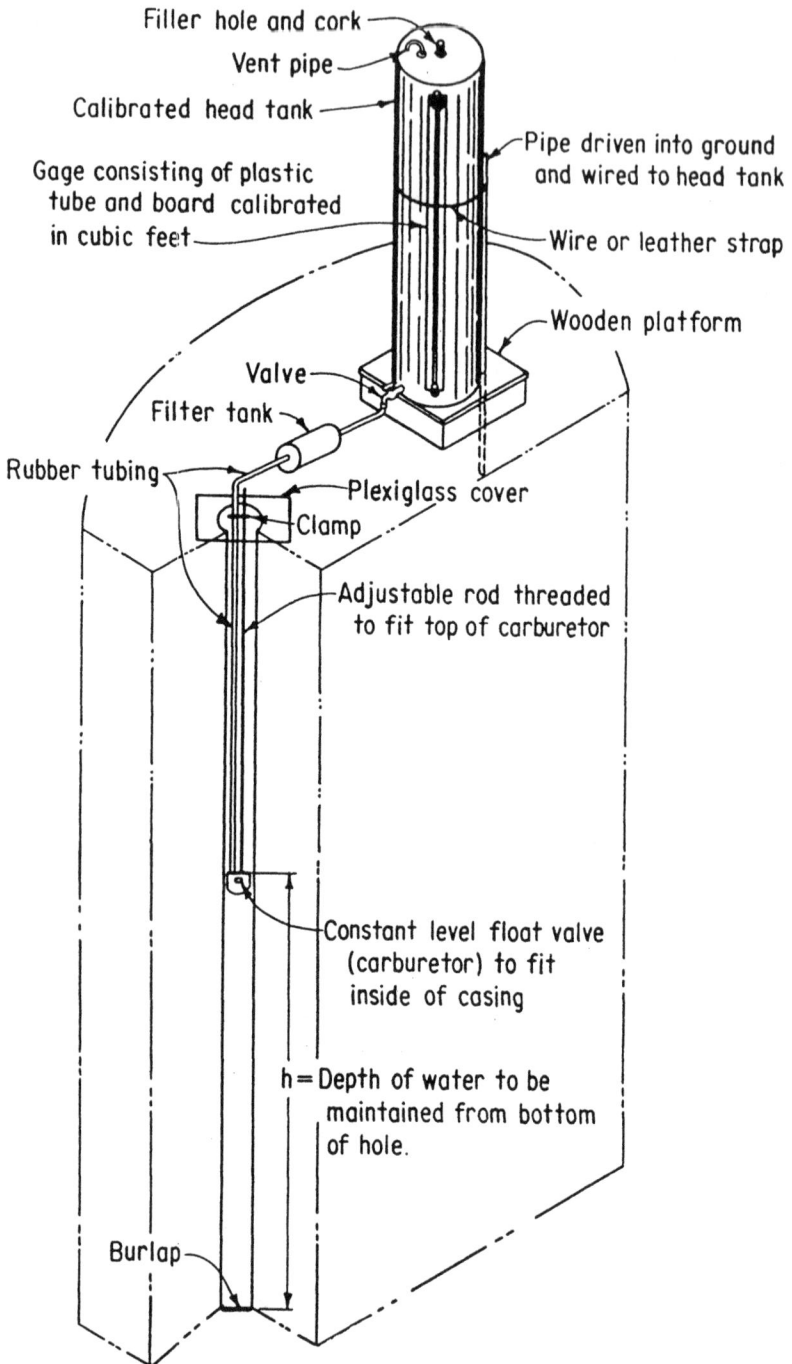

Figure 10-34.—Equipment setup for a shallow well pump-in test.

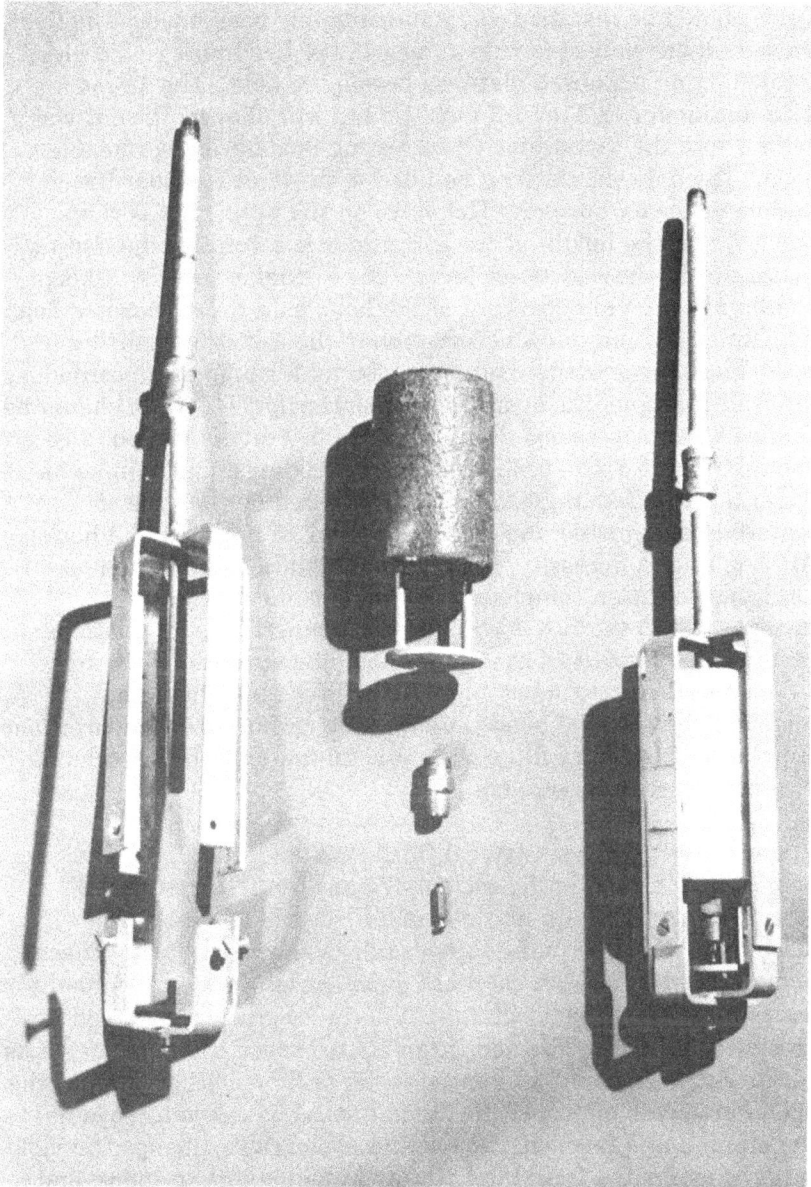

Figure 10-35.—Typical constant-level float valve used in hydraulic conductivity tests. Fully assembled float valve is shown on the right.

casing should be used, but when not available, a 100-millimeter
(4-inch nominal) diameter, thin-walled casing with about
180 uniformly spaced, hand-cut perforations per meter,
3 millimeters wide by 25 millimeters long (1/8 inch wide by 1 inch
long), will be satisfactory for most soils. The constant-level float
valve should be installed and approximately positioned. The float
valve is then connected with tubing to the head tank, which is
located on an anchored platform beside the hole. The 10- or
12.5-millimeter (3/8- or 1/2-inch) tubing will allow sufficient water
to flow into the carburetor when testing moderately permeable
soils. The hole should then be filled with water to about the
bottom of the carburetor. The valve on the head tank is then
opened, and the height of the carburetor is carefully adjusted to
maintain the desired water level. The Plexiglas cover will keep
small animals and debris out of the hole, hold the carburetor float
adjusting rod, and allow observation of the carburetor during the
test. The time and the reading on the tank gauge are recorded
after all equipment is operating satisfactorily. The tank should be
refilled when necessary. Each time the test site is visited, a record
should be kept of the time, tank gauge readings, and volume of
water added. Reading times are determined by the type of
material being tested and will range from 15 minutes to 2 hours.
Although not a necessity, the use of automatic recorders is
desirable so that a complete record may be kept of water
movement into the hole. When water temperature fluctuations
exceed 2 °C, viscosity corrections should be applied. If the test
water contains suspended material, a filter tank should be
installed between the head tank and the carburetor. Polyurethane
foam is a satisfactory filter material. In-line milk filter socks have
also been used successfully.

Figure 10-36 shows a typical filter tank and material. The
nomographs shown on figures 10-37a and 10-37b are used to
estimate the minimum and maximum volume of water to be
discharged during a pump-in hydraulic conductivity test. These
nomographs provide an excellent guide to determine the amount of
water that should be discharged into the hole before the readings
become unreliable. The nomographs are especially useful in sands
because the minimum amount of water will be discharged into the
hole in a very short time. Readings should be taken as soon as
the minimum is reached. To use the nomograph, the specific yield
must be estimated from the hydraulic conductivity, texture, and
structure of the soil. Knowing the depth of water maintained from
the bottom of the hole, h, and the radius of the hole, r, the
minimum and maximum amounts of water needed to meet the

Figure 10-36.—Typical filter tank and filter material.

conditions set up in the mathematical model can be determined. When the minimum amount has been discharged into the soil, the hydraulic conductivity should be computed following each reading. The test can be terminated when a relatively constant hydraulic conductivity value has been reached and the total volume discharged into the soil is not greater than the maximum value taken from the nomograph.

(d) Calculations.—A sample computation sheet for the shallow well pump-in test is shown on figure 10-38. Figures 10-39a, 10-39b, 10-40a, and 10-40b show equations and nomograph used in the computations. The use of these figures depends upon the depth of water maintained from the bottom of the hole, h, and the depth of the water table or depth to an impervious strata from the surface of water maintained, T_u. The h value can be determined accurately, but the depth to an impervious or restrictive zone, T_u, requires a deep pilot hole near the test site. Any zone which appears, from visual inspection, to have a much lower hydraulic conductivity than the zone above should be considered as a restrictive zone for determining T_u. A water table should also be considered a barrier when estimating T_u. If an in-place hydraulic conductivity test in this zone indicates the zone is not restrictive, the hydraulic conductivity can be recomputed using a larger T_u value and the appropriate equation or nomograph.

Height of water
Radius of well
h/r

Height of water

h, m

Minimum volume
Maximum volume
V, m³
Min | Max

Estimated
specific yield
S

Texture	Structure	Hydraulic conductivity m/day
Coarse sand Gravel	Single grain	>12.0
Medium sand	Single grain	6.0-12.0
Loamy sand Fine sand	Medium crumb Single grain	3.0-6.0
Fine sandy loam Sandy loam	Coarse subangular blocky & granular Fine crumb	1.5-3.0
Light clay loam Silt Silt loam Very fine sandy loam Loam	Med. prismatic & subangular blocky	0.5-1.5
Clay Silty clay Sandy clay Silty clay loam Clay loam Silt loam Silt Sandy clay loam	Fine & medium prismatic, angular blocky, & platy	0.12-0.5
Clay Clay loam Silty clay Sandy clay loam	Very fine or fine prismatic, angular blocky, & platy	0.06-0.12
Clay Heavy clay loam	Massive Very fine or fine columnar	<0.06

Key to solving Nomograph

h/r   h   V min. max   S

Answer

Formula:

$$V_{min} = 2.09 \, S \left[h \sqrt{\log_e(\frac{h}{r} + \sqrt{(\frac{h}{r})^2 + 1}) - 1} \right]^3$$

$$V_{max} = 15 \, S \left[h \sqrt{\log_e(\frac{h}{r} + \sqrt{(\frac{h}{r})^2 + 1}) - 1} \right]^3$$

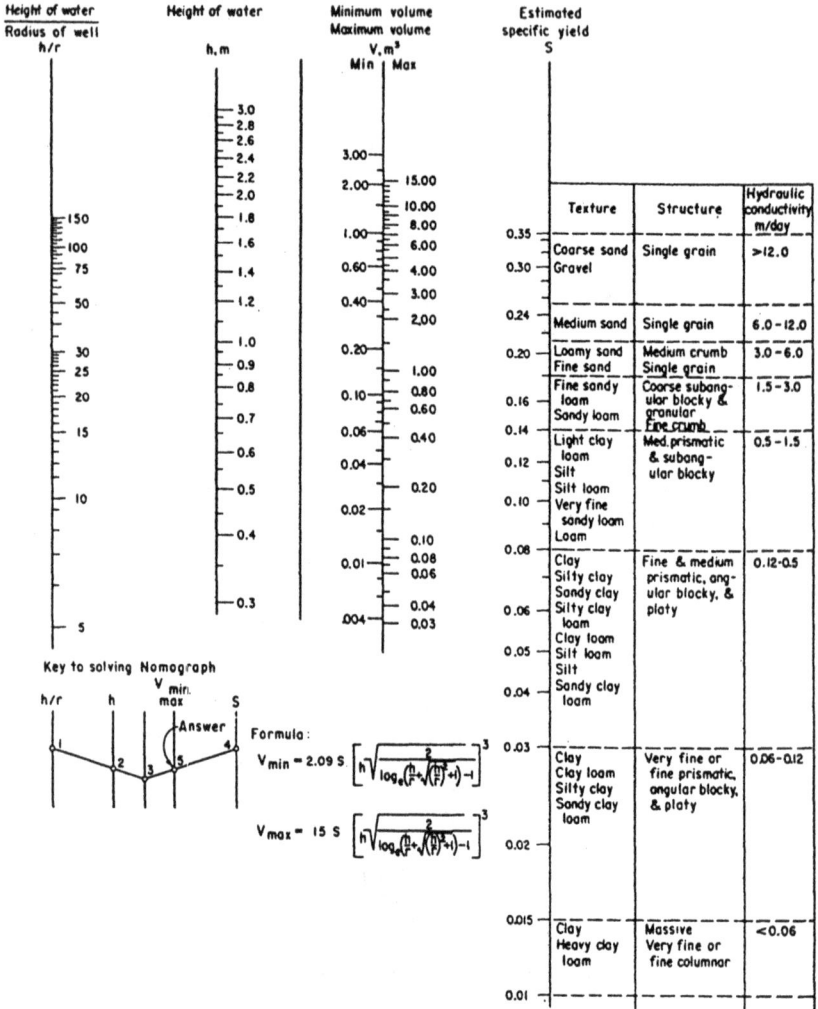

Figure 10-37a.—Nomograph for estimating the minimum and maximum volume of water to be discharged during a pump-in hydraulic conductivity test (metric units).

Height of water / Radius of well — h/r

Height of water — h, ft

Minimum volume / Maximum volume — V, ft^3 — Min | Max

Estimated specific yield — S

Texture	Structure	Hydraulic conductivity in/h
Coarse sand Gravel	Single grain	>20.0
Medium sand	Single grain	10.0–20.0
Loamy sand Fine sand	Medium crumb Single grain	5.0–10.0
Fine sandy loam Sandy loam	Coarse subangular blocky & granular Fine crumb	2.5–5.0
Light clay loam Silt Silt loam Very fine sandy loam Loam	Med. prismatic & subangular blocky	0.8–2.5
Clay Silty clay Sandy clay Silty clay loam Clay loam Silt loam Silt Sandy clay loam	Fine & medium prismatic, angular blocky, & platy	0.2–0.8
Clay Clay loam Silty clay Sandy clay loam	Very fine or fine prismatic, angular blocky, & platy	0.10–0.2
Clay Heavy clay loam	Massive Very fine or fine columnar	<0.10

Key to solving Nomograph

h/r h V_{min}/V_{max} S

Answer

Formula:

$$V_{min} = 2.09\,S\left[h\sqrt{\dfrac{2}{\log_e\left(\frac{h}{r}+\sqrt{\left(\frac{h}{r}\right)^2+1}\right)-1}}\right]^3$$

$$V_{max} = 15\,S\left[h\sqrt{\dfrac{2}{\log_e\left(\frac{h}{r}+\sqrt{\left(\frac{h}{r}\right)^2+1}\right)-1}}\right]^3$$

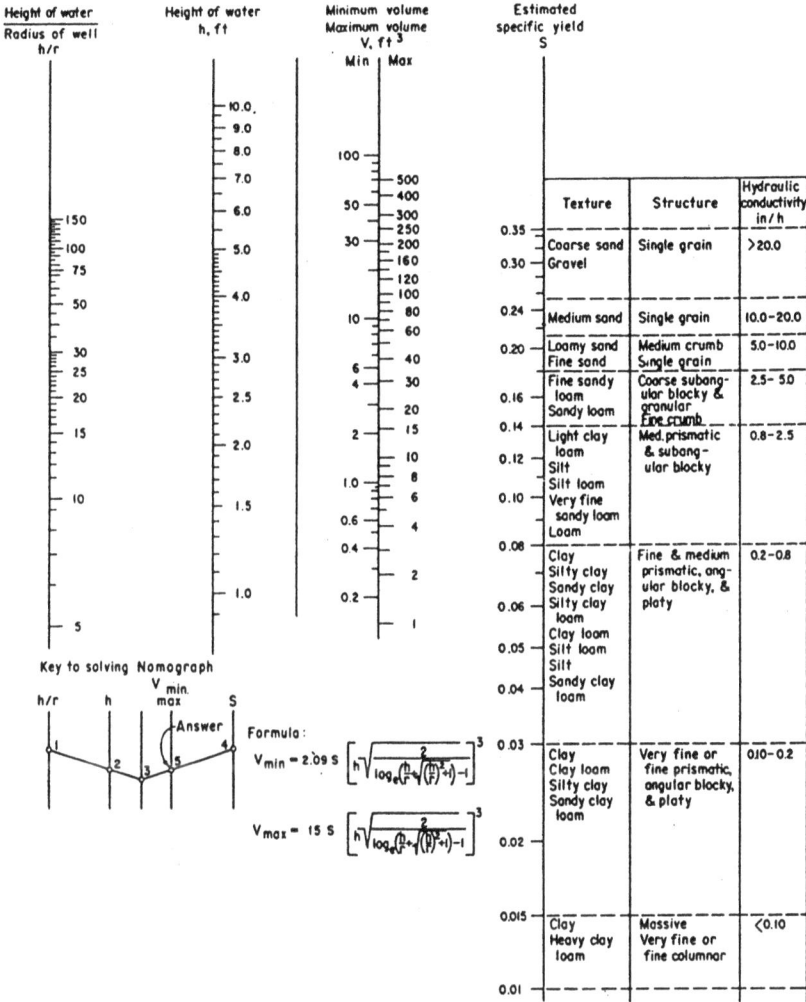

Figure 10-37b.—Nomograph for estimating the minimum and maximum volume of water to be discharged during a pump-in hydraulic conductivity test (U.S. customary units).

Locaton: __Hole C-3--Sample Farm__
Obserfer: __A.P.B.__ Date: __October 8, 1974__

D = 1,83 meters (6.0 feet)
r = 0.051 meters (0.167feet)

Water table or impervious strata = 2.13 meters (7.0 feet) below ground surface

T_u = 1.37 meters (4.5 feet) Depth of water table or impervious strata from surface of water maintained

h = 1.07 meters (3.5 feet) Depth of water maintained from bottom of hole

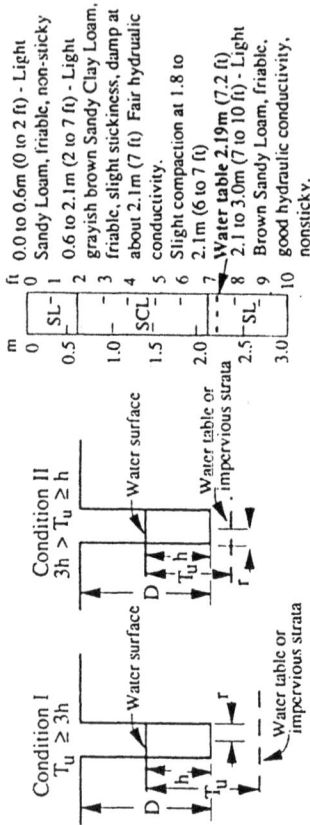

Condition I
$T_u \geq 3h$

Condition II
$3h > T_u \geq h$

m	ft	
0	0	0.0 to 0.6m (0 to 2 ft) - Light Sandy Loam, friable, non-sticky
0.5 SL	1 / 2	0.6 to 2.1m (2 to 7 ft) - Light grayish brown Sandy Clay Loam, friable, slight stickiness, damp at about 2.1m (7 ft) Fair hydraulic conductivity.
1.0 SCL	3 / 4	
1.5	5	Slight compaction at 1.8 to 2.1m (6 to 7 ft)
2.0	6 / 7	Water table 2.19m (7.2 ft)
2.5 SL	8 / 9	2.1 to 3.0m (7 to 10 ft) - Light Brown Sandy Loam, friable. good hydraulic conductivity, nonsticky.
3.0	10	

Initial		Final		Time	Tank Reading Initial		Final		Q		Temp. of water, C	Viscosity of water, Centipoise	Adjusted Q.		K.	
Date	Time	Date	Time	min	m³	ft³	m³	ft³	m³/min	ft³/min			m³/min	ft³/min	m/d	in /hr
10-8	0800	10-8	1100	180	0	0	0.173	6.12	0.000963	0.034	16	1.1111	0.000708	0.025	0.73	1.20
10-8	1100	10-8	1400	180	0	0	0.169	5.97	0.000939	0.033	18	1.0559	0.000481	0.017	0.49	0.80
10-8	1400	10-8	1800	240	0	0	0.170	6.00	0.000708	0.025		Note: Connected two barrels for greater capacity.				
10-8	1800	10-9	0530	690	0	0	0.351	12.41	0.000509	0.018						
10-9	0530	10-9	1130	360	0	0	0.193	6.82	0.000536	0.019	16	1.1111	0.000536	0.019	0.52	0.85
10-9	1130	10-9	1800	390	0	0	0.217	7.65	0.000556	0.020	19	1.0299	0.000515	0.018	0.50	0.82
10-9	1800	10-10	0530	690	0	0	0.343	12.10	0.000497	0.018	13	1.2028	0.000538	0.019	0.52	0.85
10-10	0530	10-10	1130	360	0	0	0.188	6.63	0.000522	0.018	15	1.1404	0.000535	0.019	0.52	0.85

Adjusted to average tank water temperature. -- see Figure 3 - 20 for method.
Remarks: No trouble with apparatus, assumed test satisfactory and results reliable.
Calculation: h/r = 1.07/0.051 = 20.96 hT_u = 1.07/1.37 = 0.78
Q (average after stabilization) = 0.000536 cubic meter (0.019 cubic feet) per minute
3h (or 3 x 1.07m) > T_u (1.37m) > h(1.07m), so use Condition II.
From nomograph (Fig. 3 - 18a&b): K = 0.52 meter per day (0.85 in per hour)

Figure 10-38.—Data and computation sheet on shallow well pump-in test for hydraulic conductivity.

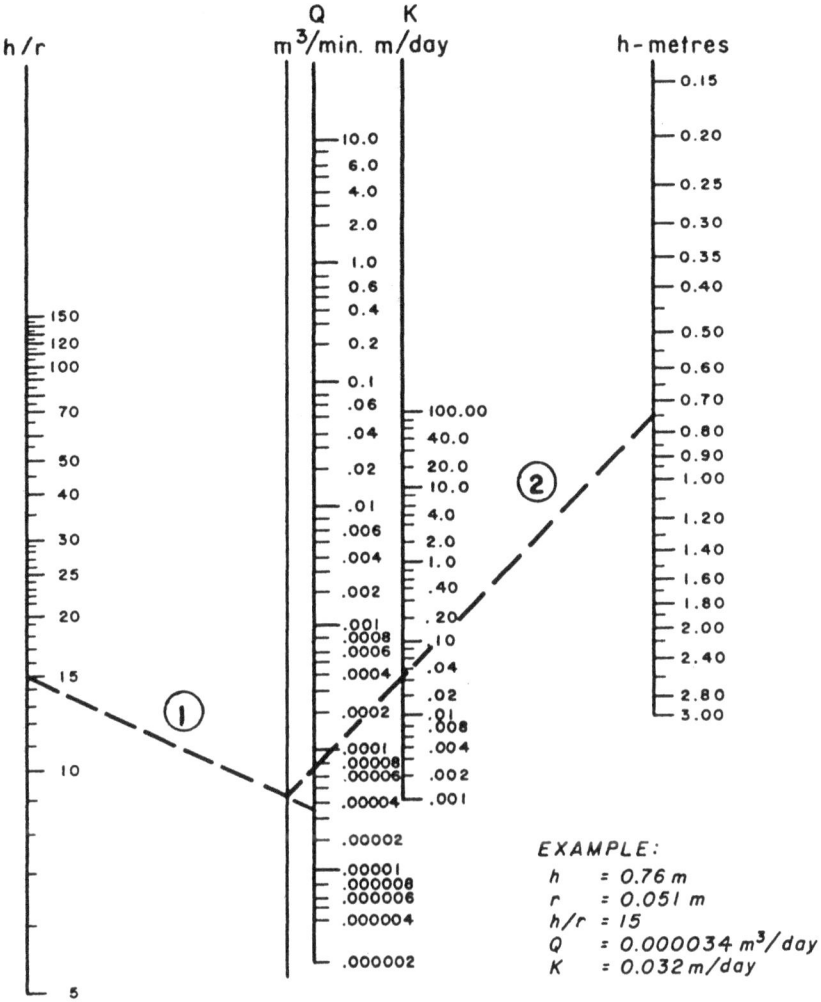

Figure 10-39a.—Nomograph for determining hydraulic conductivity from shallow well pump-in test data for condition I (metric units).

$$\frac{h}{r}$$ $$\begin{array}{c}Q\\ft^3/min\end{array}$$ $$\begin{array}{c}K\\in/h\end{array}$$ $$\begin{array}{c}h\\feet\end{array}$$

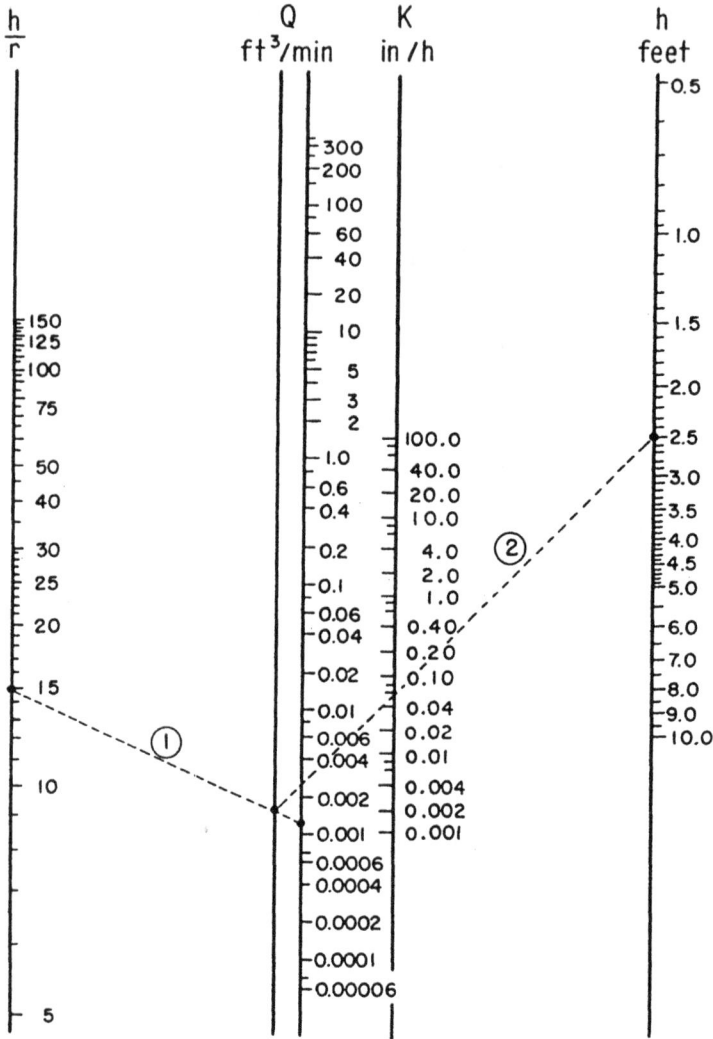

Example:

  h = 2.5 ft

  r = 0.167 ft

  h/r = 15

  Q = 0.0012 ft³/min

  K = 0.06 in/h

CONDITION I

$$T_u \geq 3h$$

$$K = \frac{720\left[\log_e\left(\frac{h}{r} + \sqrt{\left(\frac{h}{r}\right)^2 + 1}\,\right) - 1\right]Q}{2\pi h^2}$$

Figure 10-39b.—Nomograph for determining hydraulic conductivity from shallow well pump-in test data for condition I (U.S. customary units).

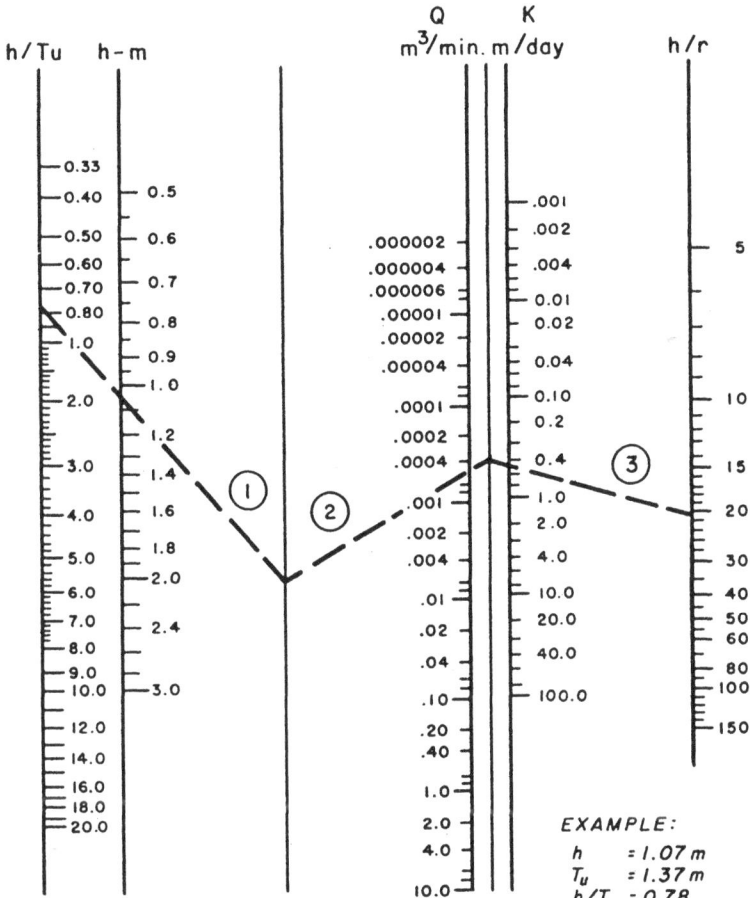

Figure 10-40a.—Nomograph for determining hydraulic conductivity
from shallow well pump-in test data for condition II
(metric units).

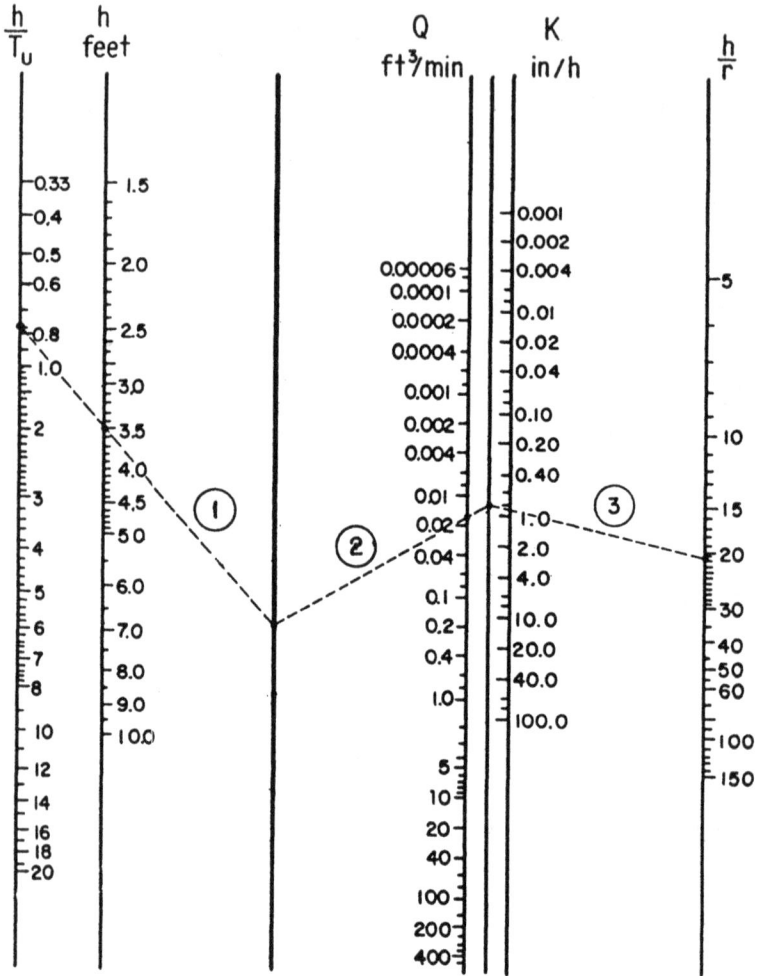

$\dfrac{h}{T_U}$ $\dfrac{h}{feet}$ Q ft^3/min K in/h $\dfrac{h}{r}$

Example:

h = 3.5 ft

T_U = 4.5 ft

$h/T_U = 0.78$

$Q = 0.019$ ft^3/min

r = 0.167 ft

$h/r = 20.96$

K = 0 85 in/h

CONDITION II

$3h \geq T_U \geq h$

$$K = 720\left[\dfrac{3\,log_e \frac{h}{r}}{\pi h(h+2T_U)}\right]Q$$

Figure 10-40b.—Nomograph for determining hydraulic conductivity from shallow well pump-in test data for condition II (U.S. customary units).

(e) Limitations.—The time required to set up the equipment and complete the test constitutes the principal limitation of this test. Also, a relatively large amount of water is required, especially if the material has a hydraulic conductivity over 4 to 6 cm/h. In soils high in sodium, the water used should contain 1,500 to 2,000 mg/L of salts, preferably calcium. Rocky material or coarse gravel may prevent augering the hole to accurate dimensions. Also, comparisons of electric analog test results with values from the auger-hole test show that the *h/r* ratio must be equal to or greater than 10.

Water moving outward from the hole sometimes causes the fines near the surface to form a seal before a constant hydraulic conductivity rate has been reached. If a constant rate cannot be obtained by the time the estimated maximum flow has occurred, the fines can be flushed back into the hole by removing the equipment and bailing all water out of the hole or by gently surging the hole with a solid surge block and then pumping the water out. This procedure is not always successful, but should be tried before abandoning the test site. Use of a filter on the supply line will generally prevent this problem.

10-11. Test for Determining Infiltration Rate.—Although most ground-water investigations concentrate on testing the hydraulic conductivity of aquifers and aquitards, the infiltration rates of surface soils must sometimes be considered in order to know the recharge potential of a study area. Several procedures have been developed for measuring infiltration rates through the vadose zone and vertical hydraulic conductivities in saturated soils. Because they are seldom used in ground-water investigations, they are not presented here. Those wishing to investigate these parameters are referred to the *Drainage Manual* (Bureau of Reclamation, 1993) where several tests are described in detail.

10-12. Bibliography.—

Ahrens, T.P., and A.C. Barlow, 1951, "Permeability Tests Using Drill Holes and Wells," including comments regarding equipment, etc., Bureau of Reclamation Geology Report No. G-97.

Barker, J.A., and J.H. Black, 1983, "Slug Test in Fissured Aquifers," Water Resources Research, vol. 10, No. 6, pp. 1558-1564.

Black, J.H., March 1978, "The Use of the Slug Test in Groundwater Investigations," Water Services, pp. 174-178.

Black, J.H., 1985, "The Interpretation of Slug Tests in Fissured Rocks," *Quarterly Journal of Geology*, London, vol. 18, pp. 161-171.

Bouwer, H., 1989a, "The Bouwer and Rice Slug Test - An Update," Ground Water, vol. 27, No. 3, May-June, pp. 304-309.

Bouwer, H., 1989b, "The Bouwer and Rice Slug Test - An Update," (discussion), Ground Water, vol. 27, No. 5, September-October, pp. 715-717.

Bureau of Reclamation, 1993, *Drainage Manual*, p. 118.

Chirlin, G.R., 1989, "A Critique of the Hvorslev Method for Slug Test Analysis: The Fully Penetrating Well," Ground Water Monitoring Review, pp. 130-138.

Cooper, H.H., Jr., J.D. Bredehoeft, and I.S. Papadopulos, 1967, "Response of a Finite Diameter Well to an Instantaneous Charge of Water," Water Resource Research, vol. 3, No. 1, pp. 263-269.

Dawson, K.J., and J.D. Istok 1991, "Aquifer Testing," Lewis Publishers.

Freeze, R.A., and J.A. Cherry, 1979, "Groundwater," Prentice-Hall, Inc., Englewood Cliffs, New Jersey.

Ferris J.G., D.B. Knowles, R.H. Brown, and R.W. Stallman, 1962, "Theory of Aquifer Tests," U.S. Geological Survey Water-Supply Paper 1536-E, pp. 104-105.

Jarvis, D.H., May 5, 1949, "Theory of Falling Head Perimeter in Unsaturated Material" memorandum to C.W. Jones, Bureau of Reclamation.

Hvorslev, M.J., 1951, "Time Lag and Soil Permeability in Groundwater Observation," Bulletin 36, U.S. Army Corps of Engineers, Waterways Experiment Station, Vicksburg, Mississippi.

Kirkham, D., 1945, "Proposed Method for Field Measurement of Permeability of Soil Below a Water Table," Soil Science Society of America Proceedings, vol. 10, pp. 58-69.

Luthin, J.N., and D. Kirkham, 1949, "A Piezometer Method for Measuring Permeability of Soil In Situ Below a Water Table," Soil Science, vol. 68, pp. 349-358.

Maasland M., and H.C. Haskew, May 1958, "The Auger Hold Method of Measuring the Hydraulic Conductivity of Soils and its Application to Tile Drainage Problems," paper presented at the International Commission on Irrigation.

Marsily, Ghislain de, 1986 and 1991, "Quantitative Hydrogeology," Academic Press Inc., D.M., "Practical Handbook of Ground-Water Monitoring," Lewis Publishers.

U.S. Department of the Navy, Navy Facilities Engineering Command, 1992, "Soil Mechanics," NAVFAC Design Manual 7.1.

Zangar, C.N., April 1953, "Theory and Problems of Water Percolation," Engineering Monograph No. 8, Bureau of Reclamation.

Witherspoon, October 1977, "Transient Flow in Tight Fractures," paper presented at Invitational Well-Testing Symposium, Division of Geothermal Energy, U.S. Department of Energy, Berkeley, California.

Chapter XI

WELL DESIGN, COMPONENTS, AND SPECIFICATIONS

11-1. General.—Chapter VI discussed relationships and conditions in ideal, isotropic, homogeneous aquifers of uniform thickness and infinite areal extent and the response of such aquifers to the discharge of ideal, fully penetrating wells. However, ideal aquifer conditions are seldom encountered, and wells must be designed with practical aspects in mind to economically use an aquifer. The more important of these practical aspects are discussed in this chapter.

A generally accepted, all-inclusive standard for water well design is not available. The American Water Works Association Publication, AWWA-A100 (1990), describes a number of commonly used design standards. The material is too extensive to describe all the different standards or even a majority of those in common use in this publication. Water well design involves the consideration of many factors, including: desired yield and pump size; casing diameters, wall thicknesses, and lengths; screen materials, diameters, and slot sizes; percentages of open area and length; the type and characteristics of the aquifer or aquifers to be developed; sanitation; control of corrosion and encrustation; local well drilling designs and practices; and State and local statutes and regulations. These factors might be assumed to preclude standardization; however, many of them are basic to all wells and these common factors make possible a degree of design standardization. Figures 11-1 through 11-8 illustrate this degree of standardization for drilled wells involving different types of construction.

The basic information desired before undertaking the design of a well is:

- Thickness, character, and sequence of materials above the water-table or at the top of a confined aquifer

- Thickness, character, and sequence of the aquifer(s), nature of the permeability (interstitial or resulting from secondary voids), and the degree of confinement of the aquifer

- Size and gradation of aquifer materials

PULL BACK CONSTRUCTION

Figure 11-3.—Straight wall, cable tool drilled well for pull back construction.

TELESCOPING CONSTRUCTION

Figure 11-2.—Gravel-packed, rotary drilled well for telescoping construction.

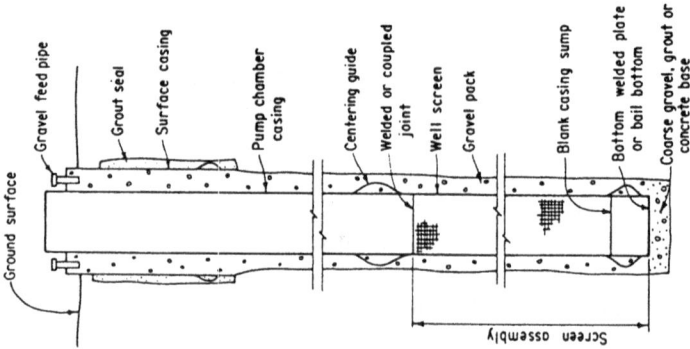

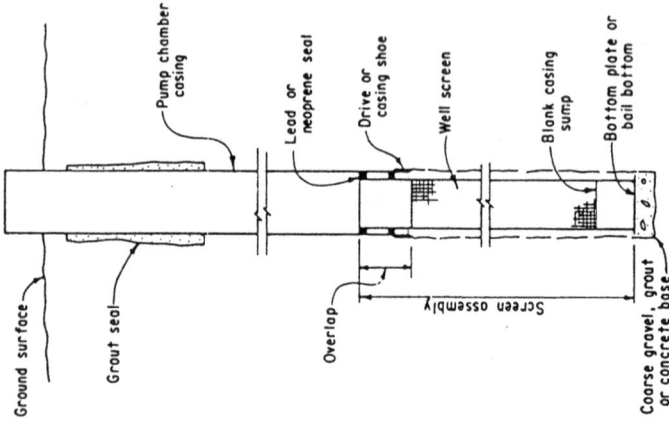

SINGLE STRING CONSTRUCTION

Figure 11-1.—Gravel-packed, rotary drilled well for single string construction.

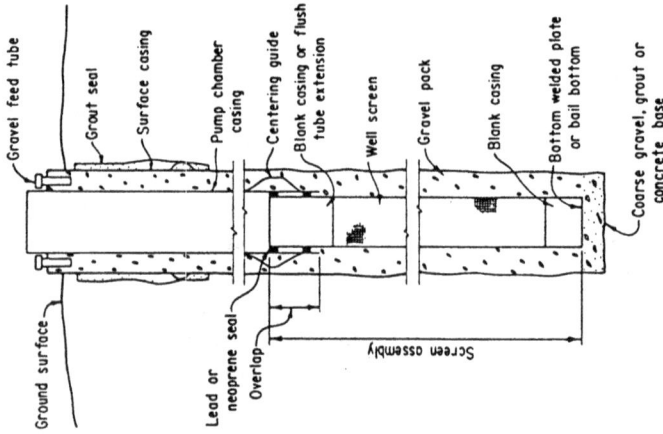

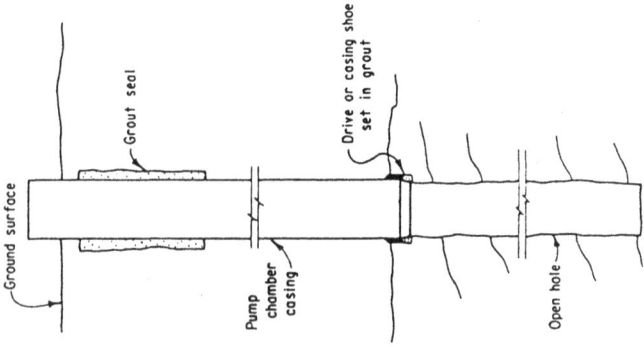

WELL IN CONSOLIDATED
MATERIAL (ROCK)

Figure 11-6.—Straight wall, cable tool drilled well in consolidated material (rock).

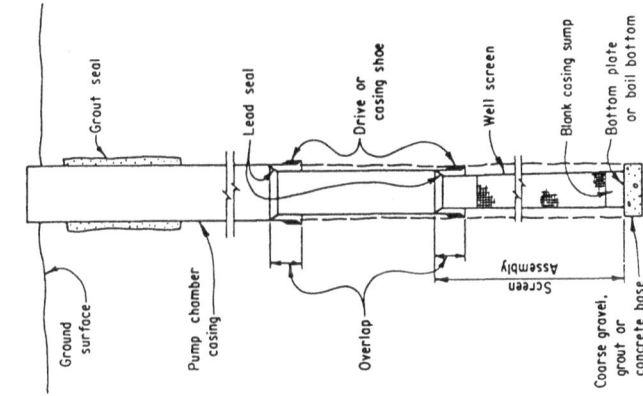

MULTIPLE TELESCOPING CONSTRUCTION

Figure 11-5.—Straight wall, cable tool drilled well for multiple telescoping construction.

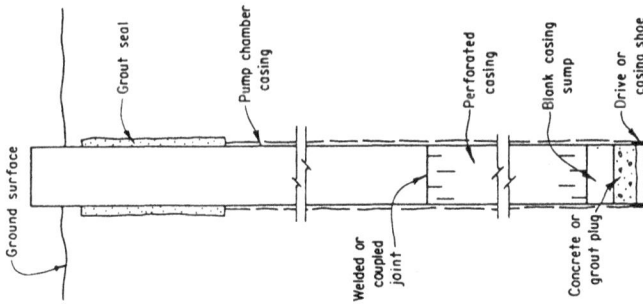

SINGLE STRING CONSTRUCTION

Figure 11-4.—Straight wall, cable tool drilled well for single string construction.

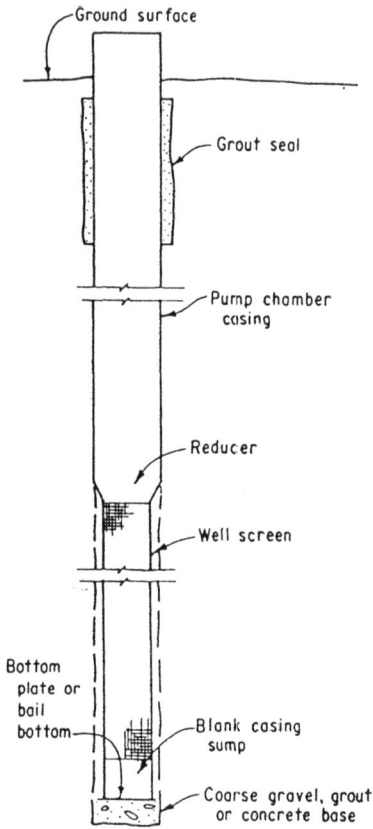

SINGLE STRING CONSTRUCTION
USING A REDUCER

Figure 11-7.—Straight wall,
rotary drilled well for single
string construction using a
reducer.

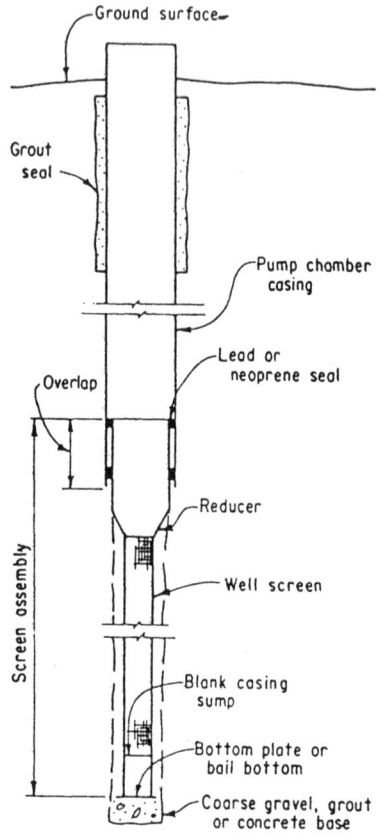

TELESCOPING CONSTRUCTION
USING A REDUCER

Figure 11-8.—Straight wall,
rotary drilled well for telescoping construction using
a reducer.

- Transmissivity and storativity of the aquifer

- Water-level conditions and trends

- Quality of water

- Design and construction features of wells previously constructed in the area

- Operation and maintenance history of previously constructed wells

- Purpose and desired yield of the proposed well

Unfortunately, all the information desired is seldom available, even in a developed area. In an undeveloped area, all that may be known initially is the approximate location of the proposed well and the desired yield.

For major wells (yields greater than 400 to 500 liters per minute [100 to 125 gal/min]), the designer should obtain all available information on existing wells, and aquifer transmissivity characteristics as described in chapters V and IX should be determined. These data should be supplemented by a geological investigation and the drilling of a pilot hole. A properly drilled and sampled pilot hole will furnish an accurate lithologic log, aquifer samples for mechanical analyses, information on static water-levels, water samples for analyses, and the type of aquifer present. The site, if found to be inadequate for any reason, can be abandoned without the major cost of drilling a producing well. If the site is found to be satisfactory, a design can be prepared and specifications written. Pilot hole data permit the preparation of firm design and specifications features which minimize the unknown factors and the risks during construction. Consequently, the contractor can foresee many problems, assemble necessary equipment and materials, and more accurately schedule the construction program. Contract savings resulting from data gained in drilling a pilot hole may far exceed the cost of the pilot hole.

For minor wells, the well drilling costs may be about the same as for a pilot hole; thus, the latter may not be economical. The recommended procedure for the design of such wells is to make a preliminary design based on the desired yield, readily available information, and evaluation of existing geological knowledge of the area. Additions or changes in the preliminary design, such as in pump chamber depth and screen slot size and setting, can be made on the basis of information obtained during the drilling.

The designations used for various components of a well are not standardized. Various terms are used for similar components in different publications and in different parts of the country. In the following sections, the terminology favored by the Bureau of Reclamation is used, but other terms commonly encountered are also noted.

11-2. Corrosion.—Metal screens, casings, and pumps used in wells may be subjected to deterioration and eventual destruction by reaction with the environment in which they are placed. This destructive process is called corrosion. For purposes of this discussion, corrosion of well components can be considered either chemical or galvanic. Chemical corrosion results from chemical reaction of the metals with components of the soil or water in which they are placed. Chemical corrosion usually results in the corroded metal going into solution and being carried away from the point of attack. Galvanic corrosion arises from the action of electrolytic cells formed between dissimilar metals or surface conditions in the well, which results in attack and corrosion of metals at anodic points and frequently the deposition of the corrosion products at the cathodes. Organism-induced corrosion is initiated, accelerated, and aggravated by certain algae, fungi, and bacteria. These organisms do not attack the metal directly but cause environmental changes which, in conjunction with the byproducts of their metabolism, cause chemical and galvanic corrosion (Kuhlman, 1959; Moehrl, 1961; Pennington, 1965).

Corrosion, to some degree, is inevitable in any well installation. Accurate prediction of the type of corrosion and the intensity of attack to be expected is difficult to impossible. Where additional wells are to be constructed in locations where there are existing wells, discussion with well owners, well drillers, and well service and supply companies will usually establish the presence or absence of prevailing aggressive corrosion conditions in an area and possibly the cause. In areas where no experience record exists, the designer has little guidance. Analysis and interpretation of water samples generally will indicate whether the water will cause chemical corrosion or be conducive to galvanic corrosion but may give little information regarding the intensity or severity of the effects (Kuhlman, 1959; Moehrl, 1961).

As a general guide, chemical corrosion can be anticipated when CO_2, HCO_3, O_2, H_2S, HC, C_2, H_2SO_4, or salts of these substances are present in excess of 5 to 10 milligrams per liter (mg/L), the pH of the water is below 7, and the Ryznar Index is above 9 (Ryznar, 1944). If the Ryznar Index is above 9, the pH below 7.8, and the total dissolved solids content above about 300 mg/L, galvanic corrosion may occur. With galvanic corrosion, the products of the attack on the anodic areas are frequently deposited on the cathodic areas, causing blocking of screen slots and water channels in pumps.

Bacterial corrosion is more difficult to predict or anticipate. Whether the bacteria are present everywhere in an aquifer or whether they are introduced into a well during drilling or servicing is not known, but substantial evidence exists that the latter does occur. Some so-called iron bacteria not only foster corrosion but also block screens and give a disagreeable taste and odor to the water. Other types of bacteria are not so readily recognized and can be identified only by special sampling, incubation, and study techniques, which are expensive and time consuming. Determinations, therefore, are not usually made unless experience in an area shows that such investigations are necessary. Because so little is known of the origin or control of bacterial corrosion, good practice requires sterilization of every well on completion of drilling and at the time the permanent pump is installed (Grange and Lund, 1969; Moehrl, 1961).

In the following galvanic series tabulation, metals and alloys are arranged in order of increasing resistance to corrosion. Galvanic corrosion between two adjacent metals or alloys in the tabulation is low. The farther two metals are apart in the list, the greater the potential which will develop between them and the more aggravated the corrosion attack will be on the anodic material. Because surface area is also important if two dissimilar materials must be used, the more cathodic material should have as small a surface area as possible, and the anodic material should have as large a surface area as possible to diffuse and spread out the attack.

When corrosive environments are known to be present, consideration should be given to the use of corrosive-resistant metals in the screen and in various parts of the pump.

Except under unusual conditions, using less costly metals and increasing the wall thickness of the casing and column pipe is generally more economical than using the more costly metals. The increased wall thickness provides more metal to resist corrosion and thus increases the life of the casing and column pipe (Pennington, 1965).

When the well screen is installed by telescoping, it may be possible to remove and replace the screen. In this situation, the choice of metal for the screen becomes a problem of balancing the cost of the more expensive metal against the replacement cost of the less expensive metal. When the well screen is attached directly to the casing, a dielectric coupling should be provided between the two components if they are of dissimilar metals to provide corrosion protection to both.

The use of corrosive-resistant metals in the manufacture of different parts of the pump to be installed in the well must also be considered. Here the problem is additionally complicated because not only are the individual parts attacked by chemical corrosion, but also dissimilar metals in contact with each other are conducive to galvanic corrosion. When corrosion is known to be present, the metals of the pump parts must be chosen with regard to their location in the tabulation as related to the intensity of corrosion expected.

Galvanic Series
(Arranged in order of increasing resistance to corrosion)

Anodic-corroded end	Cathodic-protective end
Magnesium	Inconel (active)
Magnesium alloys	Brass or copper
Zinc	Bronze
Aluminum 2S	Monel
Cadmium	Silver solder
Aluminum 17ST	Nickel (passive)
Steel, iron, or cast iron	Inconel (passive)
Chromium-iron (active)	Chromium-iron (passive)
Ni-resist	Chromium-nickel-iron (passive)
Chromium-nickel-iron (active)	Silver
Lead, tin, or lead-tin solders	Gold or platinum
Nickel (active)	

11-3. Encrustation.—Encrustation may be considered the opposite of corrosion insofar as it is characterized by the accumulation of minerals deposited primarily in and about openings in the screen and in the voids of the formations surrounding the well. This accumulation hinders water from entering the well, eventually the efficiency of the well declines, the drawdown increases, and the discharge decreases. The rate of mineral deposition depends upon the character and quality of the water involved and apparently increases with increased drawdown and the entrance velocity of the water. In some wells, a screen may be completely blocked within a few months or a year, whereas in others, the deposition is so low that the effects are not recognizable over many years. As far as can be determined, the

materials of which the well and screen are constructed have little
influence upon the rate or character of the deposition unless
galvanic corrosion is associated with the encrustation.

A total dissolved solids value in excess of 150 mg/L, high calcium
and iron bicarbonate content, a Ryznar Index below 7, and a pH
above 7.5 are indicative of possible encrustation problems (Ryznar,
1944).

The most common forms of mineral encrustation encountered in
wells are summarized as follows:

- Precipitation of iron and calcium carbonates in the form of a
  hard, brittle, cement-like material adhering to the screen and
  frequently cementing the gravel pack or aquifer particles for
  some distance from the screen.

- An accumulation of iron and manganese hydroxides or
  hydrated oxides on the screen or in the formation immediately
  adjacent to the screen. The hydroxides are insoluble, jelly-like
  masses unless oxygen is present, in which case, they are
  oxidized into oxides or hydrated oxides having a black, brown,
  or reddish granular appearance.

- Occasionally, silts and clays in suspension are deposited about
  the screen and reduce the entrance of water. These deposits
  are a relatively common occurrence in some types of
  gravel-pack construction, particularly when the pack is too
  coarse or development of the well has been inadequate.

- In some parts of the Western United States where lignite beds
  are prevalent, decomposition of the lignite results in the
  formation of a slimy black or brown viscous material about the
  screen and in the adjacent aquifer.

Although mineral encrustation cannot be avoided entirely in an
area where conditions are favorable for its formation, some
practices can retard encrustation. The simplest practice is to
ensure maximum inlet area in the screen and consequently
minimum inlet velocity of the water. In addition to having
maximum inlet area, the screen should have uniform openings over
the entire surface; these openings should be clean cut without
burrs or rough edges which encourage deposition.

The second important factor is proper development of the well to
ensure that as many fines as possible are withdrawn from the
aquifer in the immediate vicinity of the screen. If used, a gravel
pack should have a permeability sufficiently higher than the
aquifer so that a reduction in velocity occurs as the water passes
through the pack. This practice ensures minimum velocity and
maximum open area in the formation and the gravel pack.

**11-4. Surface Casing (Soil Casing, Conductor Casing,
Outer Casing).--**

(a) Description and Purpose.—Steel surface casing may or may
not be used, depending upon local conditions, established drilling
practices, and Government regulations. The casing may be
temporary and may be removed when completing the well, or it
may be a permanent part of the structure. Surface casing is
installed, where possible, from near the ground surface through
unstable, unconsolidated, or fractured materials a short distance
into a firm, stable (or massive), and relatively impermeable
material. The purpose of surface casing is to:

- Simplify and facilitate drilling a well by supporting unstable
  materials so they will not cave and fall into the hole

- Minimize washing and erosion of the side of the hole by
  drilling fluids and tools

- Reduce loss of drilling fluids

- Facilitate installing or pulling back other casing

- Facilitate placing a sanitary seal

- Serve as a reservoir for gravel pack and under some
  circumstances to provide a degree of protection against
  upward caving around the well

- Seal against artesian leakage, erosion, cratering, and eventual
  loss of control of flow

- Provide a structural base for the well from which other casing,
  screen, etc., may be suspended

- Protect other above-ground components of the well from
  damage

- Facilitate drilling a straight and plumb hole below the casing

- Provide for re-entry into the well to rehabilitate or replace portions of the casing or other components such as screens

Table 11-1 gives the recommended minimum diameters of surface casing for various pump chamber casing diameters. Temporary surface casing normally is pulled from the hole on completion of the well.

Concurrent with pulling the temporary casing, the annular space about the permanent casing is normally filled with grout as described in section 15-3.

When permanent surface casing is installed, the first operation in construction of the well is to drill an oversized hole and to install, center, and grout the surface casing. Table 11-2 shows the minimum hole diameter for different casing sizes. After the surface casing has been installed and the grout has set, the well is drilled deeper through the bottom of the surface casing. The hole drilled is usually about 50 millimeters (2 inches) smaller in diameter than the outside diameter of the surface casing.

In gravel-packed wells of the depth and diameter of most Bureau of Reclamation wells, the gravel pack is usually placed through nominal 50- to 100-millimeter (2- to 4-inch) coupled tremie pipes. To permit insertion of these pipes, the annular space between the wall of the hole and pump chamber casing must be adequate to permit passage of the pipe couplings.

Schedule 10 or 20 weight pipe is usually adequate for permanent surface casing to depths shown in table 11-3. For setting depths greater than those shown in table 11-3, heavier casing should be used.

In areas where unstable material extends to depths beyond normal surface casing placement, the casing may have to be supported at the ground surface. I-beams welded to the casing may provide adequate support, as shown on figure 11-9.

(b) Design Particulars for Surface Casing.—The wall thickness, depth of setting, and weight of temporary surface casing is usually left to the discretion of the contractor whose responsibility includes its installation and removal. However, the minimum diameter

Table 11-1.—Minimum pump chamber and permanent surface casing diameters

Well yield		Nominal pump chamber casing diameter		Surface casing diameter			
				Naturally developed[2]		Gravel-packed wells	
L/min	gal/min	mm	in[1]	mm	in	mm	in
Up to 340	90	150	[1a]6	200-250	8-10	450	18
190-570	50-150	200	[1b]8	250-300	10-12	500	20
380-1,890	100-500	250	[1b]10	300-350	12-14	550	22
1,135-5,680	300-1,500	300	[1b]12	400-450	16-18	600	24
1,890-4,570	500-2,000	400	[1b]16	400-450	16-18	650	26
5,680-11,355	1,500-3,000	400	[1b]16	450-500	18-20	700	28
4,570-18,925	2,000-5,000	500	[1c]20	500-550	20-22	750	30
1,1355-18,925	3,000-5,000	600	[1c]24	600-650	24-26	850	34
15,140-30,280	4,000-8,000	700	[1c]28	650-700	26-28	900	36

[1] Based on use of deep well turbine pumps with the following nominal rotation rates:
   [1a] 3,600 revolutions per minute (r/min).
   [1b] 1,800 r/min.
   [1c] 1,200 r/min.
[2] Larger values should be used if nominal 3/4- to 1-inch pipe will be inserted into the annular space for grouting the pump chamber casing as the temporary surface casing is withdrawn.

Table 11-2.—Minimum hole diameter required for adequate grout thickness around different sizes of casing

Nominal pump chamber casing diameter		Hole diameter for casing with couplings[1]		Hole diameter for casing with welded joints	
mm	in	mm	in	mm	in
150	6	260	10-3/8	240	9-5/8
200	8	315	12-7/8	290	11-5/8
250	10	—		345	13-3/4
300	12	—		395	15-3/4
350	14	—		425	17
400	16	—		475	19
450	18	—		525	21
500	20	—		575	23
550	22	—		625	25
600	24	—		675	27
650	26	—		725	29
700	28	—		775	31
750	30	—		825	33
800	32	—		875	35
850	34	—		925	37
900	36	—		975	39

[1] Coupled pipes are not normally available or used.

must be large enough not only to give a minimum 38-millimeter (1-1/2-inch) thick grout seal about the pump housing casing but also to permit the installation of the tremie pipe. Table 11-4 shows suggested minimum temporary surface casing diameters to use with various nominal pipe sizes of pump chamber casing and nominal 25 millimeter (1-inch) diameter tremie pipe.

Table 11-3.—Permanent surface casing - wall thickness, diameter, and maximum depth of setting

Nominal diameter		ASA schedule no. or class	Wall thickness		Weight - plain end		Maximum depth of setting	
mm	in		mm	in	kgs per m	lbs per ft	m	ft
200	8	20	6	0.250	34.48	23.36	126	420
250	10	20	6	0.250	41.40	28.04	70	235
300	12	20	6	0.250	49.28	33.38	42	140
350	14	10	6	0.250	54.20	36.71	32	105
400	16	10	6	0.250	61.75	42.05	21	70
450	18	10	6	0.250	69.98	47.39	15	50
500	20	20	10	0.375	116.04	78.60	36	120
550	22	20	10	0.375	127.20	86.61	28	95
600	24	20	10	0.375	139.70	94.62	21	70
650	26	Std.	10	0.375	151.51	102.63	16	55
700	28	20	12	0.500	216.80	146.85	32	105
750	30	20	12	0.500	232.58	157.53	26	85
800	32	20	12	0.500	248.33	168.21	21	70
850	34	20	12	0.500	264.11	178.89	18	60
900	36	20	12	0.500	289.73	189.57	15	50

Permanent surface casing is usually installed with a minimum stickup of 0.3 meter (1 foot) above the original ground surface. However, the design of surface facilities may require more or less stickup.

When drilling with rotary rigs using water or other drilling fluid, a hole is normally cut in the surface casing to permit fluid flow between the slush pit and the well. Grouting is, therefore, usually stopped when the grout approaches this hole. On completion of the well, a patch of the same weight material as the casing should be welded over the hole. When the foundation is poured, the unfilled space about the top of the casing is then filled with concrete. Minimum depth of setting, diameter wall thickness, and weight of permanent surface casing should be given in the specifications. In some instances, to facilitate operation, contractors may elect to install, at their own expense, casing exceeding the minimum requirements.

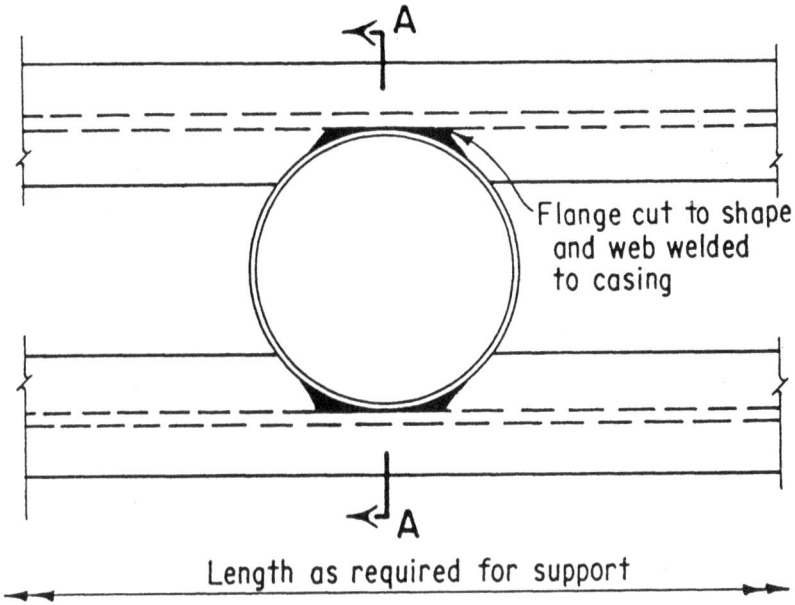

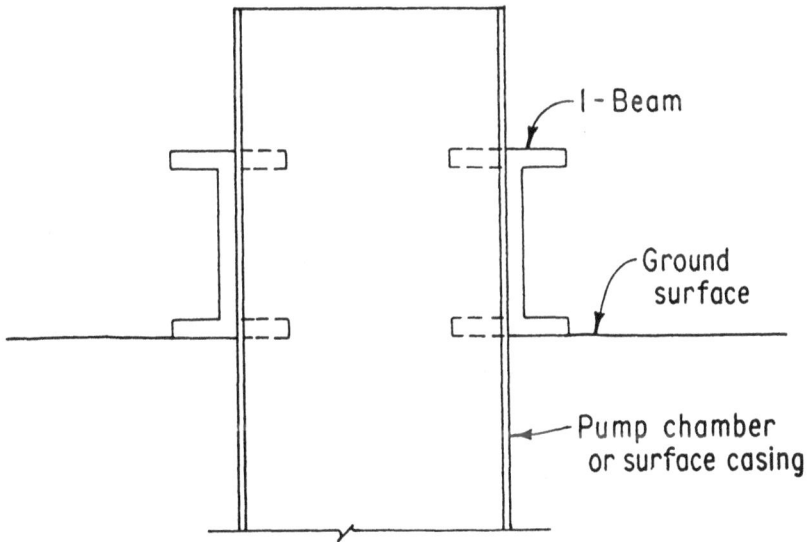

Flange cut to shape
and web welded
to casing

Length as required for support

I-Beam

Ground
surface

Pump chamber
or surface casing

SECTION A-A

Figure 11-9.—Support of a casing by I-beams.

Table 11-4.—Minimum temporary surface casing diameters to permit grout around pump housing casing with 25-mm (1-in) grout pipes

Nominal pump chamber casing diameter		Temporary surface casing diameter (welded joints on pump chamber casing)	
mm	in	mm[1]	in
150	6	250	10
200	8	300	12
250	10	250	14
300	12	400	16
350	14	450	18
400	16	500	20
450	18	550	22
500	20	600	24
550	22	650	26
600	24	700	28

[1] Add 50 mm (2 in) to the diameter to accommodate pump chamber casing with coupled joints.

11-5. Pump Chamber Casing (Pump Housing Casing, Working Casing, Inner Casing, Flow Pipe Protective Casing).—

(a) Description and Purpose.—Pump chamber casing is an essential part of every well. In single string construction of uniform diameter, it comprises all casing above the screen. In other types of construction, it is the casing within which the pump bowls will be set.

The pump chamber casing furnishes a direct connection between the surface and the aquifer, and when permanent surface casing is not used, it seals out undesirable surface or shallow ground water and supports the side of the hole. Materials used for casing are commonly steel, plastic, or fiberglass. Steel or plastic is usually required by regulation.

(b) Design Particulars for Pump Chamber Casing.—Pipe conforming to API Standard 5L (American Petroleum Institute, 1963), either Grade A or B, is commonly used for pump chamber casing. Pipe conforming to ASA Standard for Wrought Steel Pipe, ASA B36.10-59 Schedules 30 and 40, or to standard weight pipe in regard to weight and wall thickness, are also used for well casing. A wall thickness of less than 6.35 millimeters (0.250 inch) is not recommended because of corrosion possibilities. Where hard driving, deep setting in unconsolidated materials, or aggressive corrosion attack is anticipated, heavier casing should be used.

Lighter weight pipe may be used for temporary installations in shallow wells and when pipe is set in an open hole in stable consolidated rock. Tables 11-5a and 11-5b gives data on commonly used pipe of various diameters and weights and suggested maximum setting depths.

Couplings may be available on casing up to 305 millimeters (12 inches) in diameter, but most casing joints are welded. Plain ends beveled for welding and threaded ends should conform to API Standard 5L, Sections 7 and 8 (American Petroleum Institute, 1963). Other threads and couplings can be used, but the API Standards are recommended because of their desirable design characteristics. Threads on all pipe and couplings should be undamaged and brushed clean and doped before joining. They should be joined without cross-threading and tightened to the maximum possible to ensure watertight joints. Welds should be multiple pass, fully penetrating, continuous running welds, and the weld should be brushed with a steel brush and peened to remove slag and stresses between each pass. When stainless steel or other alloys are used, welding rods and procedures recommended by the manufacturer should be used. Welding, braising, or sweating of joints between dissimilar metals or alloys such as stainless and low carbon steel or brass should be avoided. Such materials should be insulated from each other by a dielectric coupling to avoid galvanic corrosion.

In areas of the Southwestern United States, two-ply casing composed of 6- to 12-gauge sheet steel in diameters between 200 and 750 millimeters (8 and 30 inches) is commonly used in wells drilled with mud scows (see tables 11-6a and 11-6b). This casing is manufactured in 1.5-meter (5-foot) lengths, each of which telescopes one-half of its length into the next section. The casing is forced down the hole by hydraulic jacks, and the outer overlap is welded to the telescoping casing. The casing is usually perforated in place. Numerous successful wells have been constructed with such casing. Thiscasing is not generally accepted in other parts of the country, however, nor is it recommended for use in wells of a permanent nature.

Aggressive corrosion, heavy encrustation, or both, often suggest use of special alloys, nonmetallic pipes, or special coatings of low carbon steel pipes for pump housing and other casing. Coatings are not recommended. Installing a casing without breaking the coating on the outside of the casing or installing a pump without breaking the coating on the inside of the casing is practically impossible. Where the coating is broken, a point of aggressive, concentrated corrosion attack may result.

Table 11-5(a).—Data on standard and line pipe commonly used for water well casing

Normal size (mm)	Outside diameter (mm)	Outside diameter, couplings (mm)	Schedule or class [1]	Wall thickness (mm)	Weight per meter-plain end (kg)	Inside diameter (mm)	Suggested maximum setting[2] (m)
100	112	130	—	5.56	1.40	103.17	357
			40	6.02	1.49	102.26	318
150	166	185	—	6.35	2.35	155.57	212
			40(S)	7.11	2.62	154.05	255
200	216	241	20	6.35	3.09	206.37	126
			30	7.04	3.42	205.00	158
			40(S)	8.18	3.95	202.72	208
250	269	294	20	6.35	3.88	260.35	70
			30	7.04	4.73	257.45	123
			40(S)	9.27	5.60	254.51	174
300	319	350	20	6.35	4.62	311.15	42
			30	8.38	6.05	307.09	96
			(S)	9.52	6.85	304.80	130
			40*	10.31	7.41	303.22	154
350	350	375	10	6.35	5.08	342.90	32
			20	7.92	6.32	339.75	58
			30(S)	9.52	7.55	336.55	105
			40	11.12	8.76	333.35	148
400	400	425	10	6.35	5.82	393.70	21
			20	7.92	7.24	390.55	42
			30(S)	9.52	8.65	387.35	72
			40	12.7	11.45	381.00	148
450	450	475	10	6.35	6.55	444.50	15
			20	7.92	8.16	441.35	30
			S	9.52	9.76	438.15	51
			30	11.12	11.35	434.95	81
			40	14.27	14.49	428.65	148
500	500	525	10	6.35	7.29	495.30	10
			20(S)	9.52	10.87	488.95	38
			30	12.7	14.40	482.60	88
			40*	15.09	17.02	477.57	134
550	550	—	10	6.35	8.03	546.10	9
			20(S)	9.52	11.98	539.75	28
			30	12.70	15.58	533.40	66
600	600	—	10	6.35	8.77	596.90	6
			20(S)	9.52	13.09	590.55	21
			30	14.27	17.47	581.05	72
			40	17.47	23.67	574.65	123
650	650	—	10	7.92	11.86	644.55	9
			S	9.52	14.19	641.35	16
			20	12.70	18.83	635.00	40
700	700	—	10*	7.92	12.78	695.35	8
			(S)	9.52	15.27	692.15	14
			20	12.70	20.31	685.80	32
			30	15.88	25.27	679.45	63

Note: See footnotes at end of table.

Table 11-5(a).—Data.on standard and line pipe commonly used for water well casing-continued

Normal size (mm)	Outside diameter (mm)	Outside diameter, couplings (mm)	Schedule or class [1]	Wall thickness (mm)	Weight per meter-plain end (kg)	Inside diameter (mm)	Suggested maximum setting[2] (m)
750	750	—	10*	7.92	13.70	746.15	6
			(S)	9.52	16.41	742.75	10
			20	12.70	21.79	736.60	26
			30	15.88	27.12	730.25	51
800	800	—	10*	7.92	14.63	796.85	6
			(S)	9.52	17.52	793.75	9
			20	12.70	23.24	838.20	21
			30	15.88	28.96	831.85	42
850	850	—	10*	7.92	15.55	847.75	4
			(S)	9.52	18.62	844.55	8
			20	12.70	24.74	838.20	18
			30	15.88	30.81	831.85	34
900	900	—	10*	7.92	16.47	898.55	3
			(S)	9.52	19.73	895.35	6
			20	12.70	26.22	889.00	15
			30	15.88	32.66	882.65	30

[1] ASA Standard B36.10 schedule numbers (S) indicate standard weight pipe; * indicates a non-API standard.
[2] Maximum settings were estimated for the worst possible conditions in unconsolidated formation. A design factor of approximately 1.5 was used for steel with yield strength less than 2,812 kg/cm². A 50-percent increase in depth of setting beyond those given is considered safe under favorable conditions.

Plastic pipe such as polyvinyl chloride (PVC), polyethylene, acrylonitrile-butadiene-styrene (ABS), or rubber modified plastics offer many advantages such as light weight, ease of installation, corrosion resistance, and low price. Pipe of diameters up to 150 millimeters (6 inches) has been used successfully in water wells in firm, consolidated rock to depths in excess of 240 meters (800 feet). However, these materials lack the tensile, yield, and impact strengths; elasticity; and the ease of joining of low carbon steel pipe. Plastic pipe is manufactured of suitable wall thicknesses and in diameters up to about 250 to 300 millimeters (10 to 12 inches) for a setting depth of about 50 meters (150 feet) in unconsolidated formations.

Fiberglass reinforced plastic pipe up to 250 millimeters (10 inches) in diameter with 4.5- to 5-millimeter (0.180- to 0.200-inch) wall thickness has been used extensively in water wells in some areas to depths of about 100 meters (300 feet). However,

Table 11-5(b).—Data on standard and line pipe commonly used for water well casing

Normal size (in)	Outside diameter (in)	Outside diameter, couplings (in)	Schedule or class [1]	Wall thickness (in)	Weight per foot- plain end (lbs)	Inside diameter (in)	Suggested maximum setting[2] (m)
4	4.5	5.2	—	0.219	10.10	4.062	1,190
			40	0.237	10.79	4.026	1,060
6	6.6	7.4	—	0.250	17.02	6.135	705
			40(S)	0.280	18.97	6.065	850
8	8.6	9.6	20	0.250	22.36	8.125	420
			30	0.277	24.70	8.071	525
			40(S)	0.322	28.55	7.981	695
10	10.8	11.8	20	0.250	28.04	10.250	235
			30	0.307	34.24	10.136	410
			40(S)	0.365	40.48	10.020	580
12	12.8	14.0	20	0.250	33.38	12.250	140
			30	0.330	43.77	12.090	320
			(S)	0.375	49.56	12.00	435
			40*	0.406	53.56	11.938	515
14	14.0	15.0	10	0.250	36.71	13.500	105
			20	0.312	45.68	13.376	195
			30(S)	0.375	54.57	13.250	350
			40	0.438	63.37	13.124	495
16	16.0	17.0	10	0.250	42.05	15.500	70
			20	0.312	52.36	15.376	140
			30(S)	0.375	62.58	15.250	240
			40	0.500	82.77	15.000	495
18	18.0	19.0	10	0.250	47.39	17.500	50
			20	0.312	59.03	17.376	100
			(S)	0.375	70.59	17.250	170
			30	0.438	82.06	17.124	270
			40	0.562	104.76	16.876	495
20	20.0	21.0	10	0.250	52.73	19.500	35
			20(S)	0.375	78.60	19.250	125
			30	0.500	104.13	19.000	295
			40*	0.594	123.06	18.802	445
22	22.0	—	10	0.250	58.07	21.500	30
			20(S)	0.375	86.61	21.250	96
			30	0.500	114.81	21.00	220
24	24.0	—	10	0.250	63.41	23.500	20
			20(S)	0.375	94.62	23.250	70
			30	0.562	140.80	22.876	240
			40	0.688	171.17	22.624	410
26	26.0	—	10	0.312	85.73	25.376	30
			(S)	0.375	102.63	25.250	55
			20	0.500	136.17	25.000	135
28	28.0	—	10*	0.312	92.41	27.376	25
			(S)	0.375	110.41	27.250	45
			20	0.500	146.85	27.000	105
			30	0.625	182.73	26.750	210

Note: See footnotes at end of table.

Table 11-5(b).—Data on standard and line pipe commonly used for water well casing-continued

Normal size (in)	Outside diameter (in)	Outside diameter, couplings (in)	Schedule or class [1]	Wall thickness (in)	Weight per foot- plain end (lbs)	Inside diameter (in)	Suggested maximum setting[2] (ft)
30	30.0	—	10*	0.312	99.08	29.376	20
			(S)	0.375	118.65	29.250	35
			20	0.500	157.53	29.000	85
			30	0.625	196.08	28.750	170
32	32.0	—	10*	0.312	105.76	31.376	20
			(S)	.375	126.66	31.250	30
			20	0.500	168.21	33.000	70
			30	0.625	209.43	32.750	140
34	34.0	—	10*	0.312	112.43	33.376	15
			(S)	0.375	134.67	33.250	25
			20	0.500	178.79	33.000	60
			30	0.625	222.78	32.750	115
36	36.0	—	10*	0.312	119.11	35.376	10
			(S)	0.375	142.68	35.250	20
			20	0.500	189.57	35.000	50
			30	0.625	236.13	34.750	100

[1] ASA Standard B36.10 schedule numbers (S) indicate standard weight pipe; * indicates a non-API standard.
[2] Maximum settings were estimated for the worst possible conditions in unconsolidated formation. A design factor of approximately 1.5 was used for steel with yield strength less than 40,000 lb/in². A 50-percent increase in depth of setting beyond those given is considered safe under favorable conditions.

the conditions are exceptional, and collapse under normal development procedures has been reported to be a problem. The pipe offers all the advantages of plastic pipe plus greater collapse strength but is not as strong as steel.

Stainless steel and various copper alloys are satisfactory well casing from nearly all standpoints. However, they are expensive and justifiable only in permanent wells in very corrosive environments.

The collapse resistance of pipe increases with the wall thickness, the elasticity, and the yield strength of the material and decreases with an increase in pipe diameter. From all standpoints except corrosion resistance, low carbon steel pipe is the most satisfactory material. Pennington (1965) determined that the corrosion resistance follows an exponential curve such that doubling the wall thickness extends the life of the pipe about four times.

Table 11-6(a).—Recommended maximum depth in meters of setting for California stovepipe and similar sheet steel and steel-plate fabricated casing

Diameter, (millimeter)	Gauge[1]						Thickness (millimeters)			
	12		10		8	6	5	6	8	10
	D[2]	S[3]	D	S	D	D				
200	102	38	225	78	X[4]	X	X	X	X	X
250	45	18	117	40	X	X	96	216	X	X
300	30	10	68	22	117	X	54	130	262	X
350	18	6	42	14	75	X	34	81	159	X
400	12	4	27	9	50	82	22	54	108	189
450	9		20	6	34	57	16	38	78	134
500	6		14		26	42	10	27	54	96
550			10		18	32	X	X	X	X
600			8		14	24	6	15	30	56
650			6		10	18	X	X	X	X
750			3		8	12	3	8	15	28

Table 11-6(b).—Recommended maximum depth in feet of setting for California stovepipe and similar sheet steel and steel-plate fabricated casing

Diameter (inches)	Gauge[1]						Thickness (inches)			
	12		10		8	6	3/16	1/4	5/16	3/8
	D[2]	S[3]	D	S	D	D				
8	340	125	750	260	X[4]	X	X	X	X	X
10	150	60	390	135	X	X	320	750	X	X
12	100	35	225	75	390	X	180	435	875	X
14	60	20	140	45	250	X	115	270	530	X
16	40	15	90	30	165	275	75	180	360	630
18	30		65	20	115	190	55	125	260	445
20	20		45		85	140	35	90	180	320
22			35		60	105	X	X	X	X
24			25		45	80	20	50	100	185
26			20		35	60	X	X	X	X
30			10		25	40	10	25	50	95

[1] U.S. standard gauge.
[2] D = telescoping.
[3] S = single thickness.
[4] X = not commonly made in these sizes.

In view of these advantages, the low carbon steel casing with increased wall thickness is, in most cases, the most satisfactory and economical material to use in areas of moderate corrosion. However, these criteria do not apply to perforated casing.

The pump chamber casing should have a nominal diameter at least 50 millimeters (2 inches) larger than the nominal diameter of the pump bowls of the required capacity. Table 11-1 shows recommended minimum diameters for various desired yields using deep-well turbine pumps of standard manufacture. The diameter should be increased to 100 millimeters (4 inches) for deep settings of the larger pumps.

The top of the pump housing casing should be set at least 0.3 meter (1 foot) above the proposed top elevation of the pump foundation. Any excess may be cut off when the permanent pump is installed.

Depth of setting of a pump, hence depth of the pump chamber casing, is determined by estimating projected pumping levels (section 15-9) and considering the following factors:

- Present static water-level

- Minimum static water-level of record in area

- Long-time water-level trends in the area

- Probable drawdown at desired yield

- Possible interference by other wells or boundary conditions

- Required pump submergence

- Ten pipe diameters between the suction cone of the pump and the top of any reduction in pipe size (desirable but not essential)

- The presence of any telescoping overlap

In some areas, one or more of these factors may not be in harmony and compromise will be necessary. The controlling factors in permanent well installations are usually sanitation, stability, and an estimated minimum usable well life of 25 years.

Pump housing casing should always be installed plumb and straight. Deviation from the vertical should not exceed two-thirds of the inside diameter of the casing per 30 meters (100 feet) of depth. A 12-meter (40-foot) long dolly 12.5 millimeters (1/2 inch) smaller than the inside diameter, should pass through the casing without binding (section 16-7). If used, surface casing should be installed plumb and straight and the pump housing casing should be centered in the casing. If the pump chamber casing is longer than about 15 meters (50 feet) and the annular space is greater than 25 millimeters (1 inch), the casing should have centering guides at the bottom and at 12-meter (40-foot) intervals up to the surface, where it should be firmly anchored until the foundation is placed and set. Where hard driving is required in setting casing, such as with a cable tool rig, several reductions in size of the pump chamber casing may be necessary so that the desired diameter can be set at the required depth by telescoping. With the exception of telescoping nominal 12-inch into nominal 14-inch pipe, all standard weight and thinner walled pipes with couplings and drive shoes will telescope into the next larger size. Twelve-inch pipe with welded joints and no drive shoe will telescope into some 14-inch pipe.

Pump chamber casing should be grouted in except when it is run inside a grouted-in surface casing. This grouting may be done as the temporary surface casing is withdrawn if such casing is used, or an oversized hole may be drilled with rotary equipment.

In areas of thick, unstable materials, a pump chamber casing may have to be supported from the surface in a manner similar to surface casing (section 11-2(a) and figure 11-9).

Pump chamber casing is sometimes perforated, or screened sections are included in it, above the pump bowls. This practice should be avoided if possible in permanent installations. If the drawdown increases to depths below the screened section, cascading results in entrained air in the discharge, which may cause pump cavitation and other adverse effects. Also, periodic exposure of a screen to the atmosphere is conducive not only to corrosion but also to growth of organisms which may plug the screen.

11-6. Screen Assembly.—

(a) Description and Purpose.—The purposes of a screen are to:

* Stabilize the sides of the hole,
* Keep sand out of the well,
* Facilitate flow into and within the well, and
* Permit development of the aquifer adjacent to the screen.

The screen proper may range from pipe perforated in place to carefully fabricated, cage-type, wire-wound screen with accurately sized slot openings. The screen assembly may consist of only a screen or perforated section, or of a screen, associated blank casing, bottom seals, etc.

Formerly, most drilled water wells were constructed with cable tools and finished by perforating the casing in place with a Mills knife or other similar tool. However, this practice is rapidly declining. The size of the openings resulting from in-place perforating cannot be accurately controlled, and the sizes range from 3 to 12 millimeters (1/8 inch to 1/2 inch) wide and from 25 to 50 millimeters (1inch to 2 inches) long. The perforations are large and irregular and have rough, ragged edges that encourage corrosion attack and encrustation. On heavier wall casing, the perforators sometimes do not make an opening but only dimple the pipe. Maximum percentage of open area is about 3 to 4 percent. Good development is almost impossible, and wells so perforated are usually sand pumpers unless the aquifer is comprised of relatively clean, coarse sand and gravels.

Perforated casing made by sawing, machining, or torch cutting the slots in the casing is commonly used in wells. Slot openings range from about 0.25 to 6.25 millimeters (0.010 to 0.250 inch), and the maximum percentage of open area is about 12 percent for the larger slots. Sawed and machine-cut slots are satisfactory if properly sized and if entrance velocity limits can be met. In some instances, entrance velocity limits can be met by enlarging the diameter of the well and screen or increasing the depth of the well and the length of the screen. Torch-cut slots usually have rough edges and slag remnants adhering to them. The finest possible slot is about 3 millimeters (0.125 inch). Wells screened with perforated pipe of any kind are usually more difficult to develop than wells that use continuous slot or louvre screens, and if slots are not accurately sized to the aquifer, the wells may be sand pumpers.

Numerous screens are manufactured with punched or stamped perforations. Slots range from 1.5 to 6 millimeters (0.060 to 0.250 inch) and they often have rough and ragged edges. Maximum percentage of open area is about 20 percent. Some of these screens are made of 8-gauge and lighter stock and are not suitable for settings at depths much greater than 30 to 45 meters (100 to 150 feet), depending upon diameter.

A number of louvre-type screens are manufactured of 7-gauge or heavier materials and are available in six- or eight-slot sizes ranging from about 0.75 to 3.75 millimeters (0.030 to 0.150 inch). The slots are usually accurately sized and wire brushed to remove roughness or irregularities. Open areas range from about 3 percent for the smaller slot sizes to about 20 percent for the larger.

Some types of screen are made by winding wire around a perforated pipe base with or without spacers between the wires and the pipe. Almost any slot size can be readily obtained and open area in those made with spacers compares favorably with that of louvre-type screen of the same slot size. Such screens are satisfactory in clean, rather coarse materials with few or no fines. However, where aquifers contain a large percentage of fines, the channels between the spacers between the pipe and wire wrapping may become clogged and seriously reduce the entrance area. The pipe base is often made by winding stainless steel wire on a low carbon steel pipe. In corrosive waters, this combination may result in rapid corrosion of the pipe base. Such screens should always be made of a single metal or alloy. The only real advantage of pipe-base screens is their superior tensile and collapse strengths.

Cage-type wire-wound screens consist of a continuous winding of round or specially shaped wire on a cage of vertical rods. The wire is attached to the rods by welding or using dovetailed connections. Almost any slot size is readily available, usually in increments of about 0.125 millimeter (0.005 inch) from 0.150 to 6.25 millimeters (0.006 to 0.250 inch). Screens are made in both telescoping and pipe sizes. The former will just telescope through a casing of the same nominal size of the screen, whereas the latter has the same inside diameter as the casing and may be joined to it by welding or coupling. Cage-type wire-wound screens are the most efficient available. Open areas are the largest obtainable and slot sizes can be closely matched to aquifer gradations. Although such screens are more expensive initially than other types of screen, they are usually more economical, especially when used in thin but highly productive aquifers.

Most screens are made in lengths ranging from 1.5 to 6 meters
(5 to 20 feet) which can be joined by welding or couplings to give
almost any length of screen and combination of slot sizes desired.
Couplings and welding materials should be composed of the same
materials as the screen.

Screens are available in diameters ranging from 1-1/4 to
60 inches. Screen diameters should be selected on the basis of
the desired yield from the well and thickness of the aquifer.
Table 11-7 gives the recommended minimum diameters for various
well capacities. These diameters may be increased to obtain
acceptable entrance velocities if necessary, and smaller diameters
are sometimes specified in the interest of economy. With smaller
diameters, the initial cost is lower, but well efficiency is also
lowered. Smaller diameters are not recommended for permanent
installations. However, equal efficiency as well as certain other
advantages may result when installing 9 meters (30 feet) or more
of screen by increasing the diameter toward the top of the well
provided that a satisfactory average entrance velocity results
(e.g., the screen might consist of 3-meter [10-foot] lengths of 8-,
10-, and 12-inch diameters connected by reducers, with the topmost
screen of the minimum diameter recommended in table 11-7 for
the desired yield).

In uniform aquifers, a continuous length of screen is normally installed. In thick, nonuniform aquifers, however, usual practice is
to set the screen opposite only the best aquifer materials and set
blank pipe opposite the poorer materials between the screen
sections. The blank sections between screens should be pipe size
where pipe size screens are used or flush tubing where telescoping
size screens are used. Blank pipe or flush tube extensions may
extend from the top of the screen up to the pump chamber casing,
to which it may be attached by welding, a coupling, a reducer, or
the extensions may telescope for 1.5 meters (5 ft) or more into the
casing. Where the extension is relatively short, it is sometimes
referred to as a flush tube extension or overlap pipe; otherwise, it
is called a riser pipe. Installation of a bottom sump consisting of
1.5 to 3 meters (5 to 10 ft) of blank casing or a flush tube extension
below the bottom of the lowest screened section in the well is
recommended. Use of the bottom sump in well design is not
common, but it offers several advantages. During development,
materials drawn into the well settle in the sump and do not
encroach on the screen. In addition, sand enters all wells in
unconsolidated materials when they are pumped. The sump
provides a storage space for sand, which settles to the bottom and

Table 11-7.—Recommended minimum screen
assembly diameters

Discharge		Minimum nominal screen assembly diameter	
L/min	gal/min	mm	in
Up to 190	50	50	2
190 to 475	50 to 125	100	4
475 to 1,330	125 to 350	150	6
1,330 to 3,040	350 to 800	200	8
3,040 to 5,320	800 to 1,400	250	10
5,320 to 9,500	1,400 to 2,500	300	12
9,500 to 13,300	2,500 to 3,500	350	14
13,300 to 19,000	3,500 to 5,000	400	16
19,000 to 26,600	5,000 to 7,000	450	18
26,600 to 34,200	7,000 to 9,000	500	20

prolongs the effective operation of the total screen. The sump also
provides a suitable place for the attachment of centering guides at
the bottom of the screen assembly.

Screen assemblies set in unconsolidated materials should have a
bottom seal. This seal may consist of steel plate welded to the
bottom, a bail bottom which is welded or coupled to the bottom of
the assembly and has a bail on the upper surface to facilitate
installation, any of a variety of float shoes, bail down shoes, self-
sealing jets, and other special fixtures, or a concrete plug. The
bottom seal not only precludes heaving of materials up into the
well under certain circumstances but also provides a bearing area
for support of the screen assembly.

Screens may be made of many different metals and metal alloys,
plastics, fiberglass-reinforced plastic (fiberglass), and coated base
metals. The least expensive and most commonly available screens
are fabricated of low carbon steel. Screens made from the
nonferrous metals and alloys, plastics, fiberglass, and exotic

materials are used in areas of aggressive corrosion and encrustation to prolong well life and efficiency or where permanence and continuous service are essential.

A well screen is particularly susceptible to corrosion attack and encrustation by mineral deposits. The perforations expose more surface area to a reactive environment than a pipe of similar size. In addition, water flowing to and through a screen brings a constantly renewed supply of reactive materials into contact with it and at the same time removes protective coatings or corrosive products which otherwise would offer some protection against further attack. Minimum hydrostatic pressure and maximum water velocity occur at the well face, which may result in release of carbon dioxide and other dissolved gases, some of which are aggressive corrosion agents. These and related factors also upset the chemical balance of the water so that carbonates of calcium, magnesium, iron, and other minerals may deposit on the screen and in the formation adjacent to it, blocking the slots and reducing the open area. Encrustation is commonly remedied by acidizing the well. However, even when inhibited acids are used, such treatment results in some corrosion of the screen.

Concrete, vitrified clay, and several other materials have been and in some parts of the world still are used for well casing and screen. However, they are rarely used in modern society and so will not be discussed here.

Plastic and fiberglass screens are practically immune to corrosion attack. Though unavoidable, encrustation is reported to be less troublesome and can be removed without damage to the screen. However, these screens commonly have relatively low percentages of open area. Nonreinforced plastics are subject to creep under sustained load with resultant changes in slot sizes. The collapse resistance of plastic screens in unconsolidated materials is questionable, particularly in wells deeper than about 45 meters (150 feet), unless wall thicknesses are properly sized to resist stresses. Increasing the wall thickness increases the cost to where stainless steel or other similar alloys may be cost competitive and more satisfactory.

Coated steel or other base metal screens are not recommended. To be effective, the coating would have to be applied after the perforations were made. The coating would change the slot sizes an indeterminate amount. Moreover, a coated screen cannot be installed in a well without some of the coating being scratched or otherwise damaged. These spots, or holidays, then become points for aggressive, concentrated corrosion attack.

When all factors are considered, the most satisfactory screen materials, except under unusual conditions, are steel, stainless steel, or some of the metal alloys.

(b) Design Particulars of Screen Assemblies.—Achieving a satisfactory screen design generally is not possible until a pilot hole or the well has been drilled, logged, sampled, mechanical analyses have been made of the formation samples, and chemical analyses have been made of water samples. In addition, if other wells are present in the area, the depth, design, and history of such wells should be examined for data on corrosion and encrustation experience, sand pumping, and drawdown.

Casing in a typical well normally consists of low carbon steel pipe of adequate wall thickness, not only to support the hole but also to give a satisfactory life span consistent with the corrosion potential. Screens are a different matter, however, because of their construction. Corrosion resistance cannot readily be accomplished by increasing the weight of the screen as in casing because of the critical role of slot width. Instead, corrosion resistance for screens is increased by using a suitable corrosion-resistant material.

Table 11-8 shows the more commonly used metallic screen materials in order of increasing costs. As mentioned previously, plastic and fiberglass screens have high corrosion resistance, but their use should be limited to relatively shallow, small diameter wells of low capacity or in unusual applications where other materials are unsuitable.

Where exceptionally deep settings or unstable ground is encountered which would require extra column strength and collapse resistance, some of the wire-wound screens are available in extra strength designs. These screens have a reduced percentage of open area compared to the standard designs but are still superior to other types. Where perforated casing is used, extra strength can be gained by increasing the wall thickness of the casing

The hydraulic advantages of various patterns and types of slots or perforations are discussed in section 3-8. Although these factors

Table 11-8.—Well screen material

Material[1]	Acid resistance	Corrosion resistance in normal ground water
Low carbon steel	Poor	Poor [2,3]
Toncan and Armco iron	Poor	Fair [2,3]
Admiralty red brass	Good	Good [3]
Silicon red brass	Good	Good [3]
304 stainless steel	Good	Very good
Everdure bronze	Very good	Very good [4]
Monel metal	Very good	Very good [4]
Super nickel	Very good	Very good [4]

[1] Other materials are available for use under special situations such as installations in aquifers containing high-temperature corrosive brines.

[2] Not recommended for permanent installations where incrustation is a serious problem.

[3] Not recommended for permanent installations where sulfate-reducing or similar bacteria are present or where water contains more than 60 p/m SO_4.

[4] Recommended only in areas where corrosion is very aggressive.

are significant, the most important characteristics of a screen are the slot size and amount of open area. Slot size is determined from the mechanical analyses of the formation samples obtained from the pilot hole or well.

If the uniformity coefficient of a sample for a naturally developed well is 5 or less, a slot size should be selected which will retain from 40 to 50 percent of the aquifer. If representation of the sample is questionable, or if corrosion may be a problem, a slot size which will retain 40 to 45 percent of the aquifer should be used. If the sample is representative and corrosion is not a problem but encrustation is anticipated, a slot size is selected which will retain from 45 to 50 percent of the aquifer.

If the uniformity coefficient of a sample is greater than 5, a slot size which will retain from 30 to 50 percent of the aquifer should be selected. Thirty-to 40-percent retention is used if the sample is

representative, corrosion is not a problem, or if encrustation is
anticipated. Forty to 50 percent is used if there is doubt that the
sample is representative or if corrosion is a problem. The upper
limit of retention in each case is the extreme, and if screens are
not available in a standard slot size, the next smaller standard
size should be selected. Slot sizes are commonly described in
0.001 inch (0.025 millimeter); thus, a No. 60 slot has a 0.060-inch
(1.5 millimeter) slot width. Figure 11-10 shows slot sizes for some
representative screens.

Most aquifers are not homogeneous and uniform but consist of
layers containing granular materials of different gradation, size,
and uniformity coefficient. Consequently, a single slot size is
seldom suitable for use throughout a well. A suggested method of
treating this usual occurrence in large capacity wells is to arrange
a table based on the log and mechanical analyses of the samples in
descending order as they were encountered in the hole. Blank
casing is set opposite zones consisting of clay or silt or in which
more than 20 percent of the grains pass the 100-mesh screen. The
selected slot size is then entered on the table adjacent to the
interval represented by each usable aquifer sample. If coarser
materials in beds 1.5 meters (5 feet) or less in thickness are
interlayered with finer aquifer material, attempting to screen them
separately is seldom worthwhile, so screen suitable for the finest is
used throughout.

Aquifers pack and settle during development of a well and a fine
sand may migrate downward to a point opposite a screen which is
too coarse. Therefore, where finer material overlies coarser
material, the finer screen should be extended downward into the
coarser material at least 10 percent of the thickness of the coarser
material.

The diameter of the pump chamber is determined by the size of
the pump required to discharge the desired yield. The diameter of
the screen, however, is determined by the desired yield. Table 11-7
indicates the recommended minimum diameters of screen for a
range of discharges. In Bureau of Reclamation wells, pump
chamber diameter is usually 50 or moremillimeters (2 or more in)
larger than the screen diameter, and pipe size screen assemblies
are generally used because they will telescope through the larger
diameter pump chamber casing.

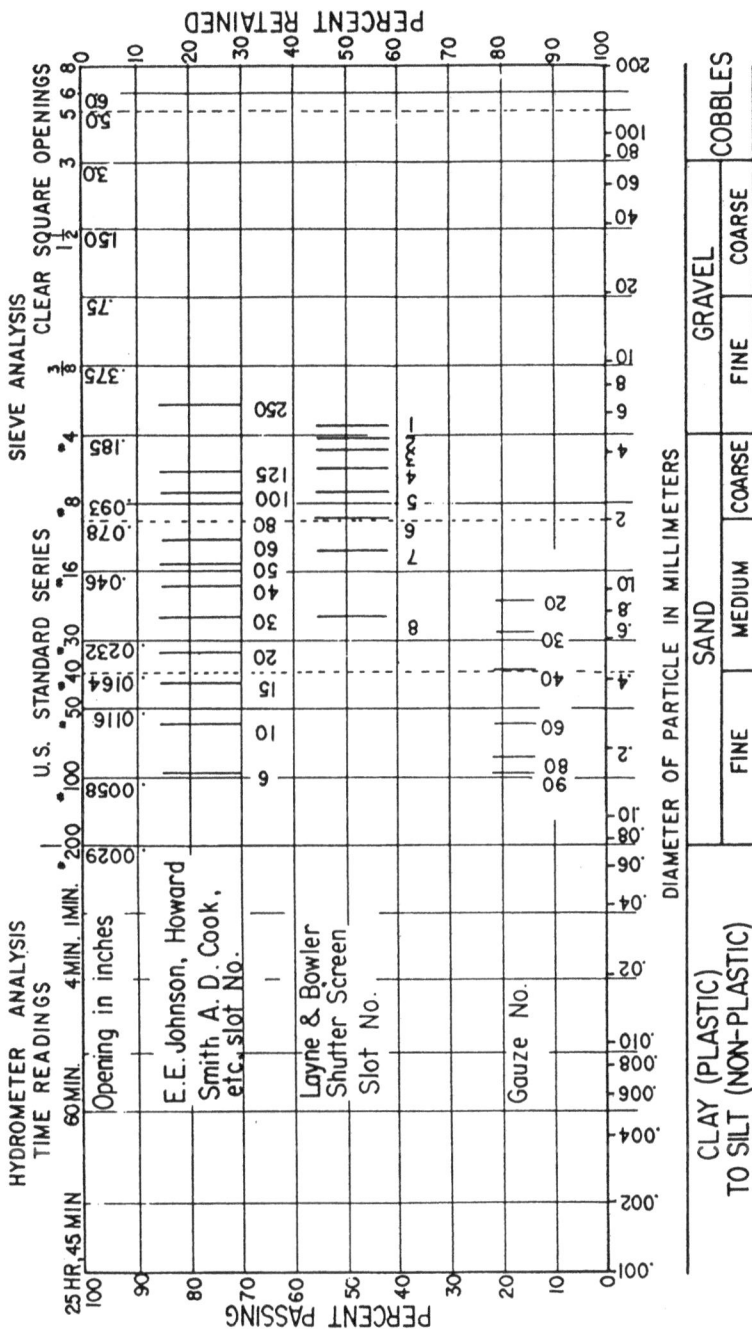

Figure 11-10.—Slot sizes for representative well screens.

When the minimum diameter and slot sizes of the screen have been determined, the average entrance velocity can be estimated by dividing the desired yield in cubic meters (cubic feet) per second by the total open area of the screen in square meters (feet).

Table 11-9 gives minimum open areas of some representative screens of various slot sizes and diameter in square meters per meter (square feet per foot) of length. Estimates based on interpolation are usually adequate for combinations between those shown in the table.

The values given in table 11-9 were computed from the manufacturer's published data. The square meters (feet) and percentage open area for louvre or shutter screens, perforated screens, and slotted pipe are believed to be accurate within a few percent. On wire-wound screen, the open area may be more than 30 percent less than shown because of the manufacturer's method of stating open area and lack of data on vertical wire sizes used. In selecting diameters and open areas for cage-type wire-wound screen, a smaller diameter or slot size may actually have more open area because of changes in wire size to maintain strength. Some savings in cost may be realized in some instances by selecting a smaller diameter screen or slot size and at the same time obtaining an equal or greater open area.

The average entrance velocity through the screens, neglecting blockage by aquifer or gravel pack material, should be 0.03 meter (0.1 foot) per second or less. If this velocity is exceeded by more than about 0.015 meter (0.005 foot) per second, the screen diameter should be increased or the screen lengthened if possible to give the desired maximum entrance velocity. The greater the percentage ofscreen open area, the shorter the length of screen required to obtain an acceptable entrance velocity. Consideration should also be given to the effect of percentage of open hole as discussed in section 6-7. In confined aquifers, full penetration and maximum percentage of open hole are recommended where aquifer depth and thickness make such construction economically feasible. Where aquifers are deep and thick, judgment should be used to determine the most economical combination of penetration and open hole. In unconfined aquifers, full penetration and a 35- to 50-percent open hole at the bottom of the well are recommended, depending upon the thickness, stratigraphy, productivity of the aquifer, and the economy of such construction. Where the aquifer is deep and thick, judgment is again necessary to determine the most

Table 11-9.—Minimum open areas of screens in square feet
per linear foot and the percentage of open area [1]

(Cage-type wire-wound screen—telescoping sizes [from Johnson Division, UOP Inc.])

Screen size, inches	Slot sizes, thousandth of an inch						
	10	20	40	60	80	100	150
4	0.139	0.243	0.389	0.493	0.555	0.514	0.611
	14.1	24.7	39.6	50.2	56.5	52.3	62.2
4.5	0.160	0.278	0.444	0.555	0.632	0.583	0.694
	14.3	25.0	39.9	49.9	56.8	52.4	62.3
5	0.180	0.312	0.493	0.618	0.701	0.752	0.777
	14.4	25.0	39.6	49.6	56.3	52.4	62.4
5.625	0.194	0.347	0.548	0.687	0.784	0.722	0.861
	14.1	25.2	39.8	49.9	57.0	52.5	62.6
6	0.208	0.368	0.451	0.590	0.694	0.777	0.916
	14.1	24.9	30.6	40.0	47.1	52.7	62.1
8	0.194	0.354	0.604	0.784	0.916	1.027	1.110
	9.9	18.0	30.7	39.9	46.6	52.2	56.5
10	0.249	0.451.	0.763	0.992	1.166	1.312	1.409
	10.0	18.1	30.6	39.8	46.8	52.7	56.6
12	0.291	0.534	0.902	0.999	1.200	1.367	1.666
	9.8	18.1	30.6	33.9	40.7	46.3	56.5
14	0.264	0.493	0.847	1.110	1.339	1.513	1.853
	8.2	15.3	26.4	34.5	41.7	47.1	57.7
16	0.298	0.555	0.964	1.263	1.527	1.735	2.110
	7.8	14.6	25.3	33.2	40.2	45.6	55.5
16 sp	0.305	0.576	0.992	1.298	1.568	1.776	2.165
	7.9	15.0	25.8	33.8	40.9	46.3	56.5
18	0.340	0.638	1.110	1.450	1.749	1.985	2.415
	7.9	14.9	26.0	33.9	40.9	46.5	56.6
18 sp	0.347	0.645	1.117	1.464	1.763	1.999	2.443
	8.0	14.9	25.8	33.8	40.7	46.2	56.6
20	0.263	0.479	0.881	1.221	1.499	1.721	2.193
	5.5	10.0	18.4	25.5	31.3	36.0	45.8
24	0.319	0.596	1.097	1.513	1.874	2.138	2.727
	5.3	10.0	18.5	25.5	31.6	36.0	46.0

[1] Top value shown is ft²/ft and bottom value is percent.

Table 11-9.—Minimum open areas of screens in square feet per linear foot and the percentage of open area[1]-continued

(Cage-type wire-wound screen—telescoping sizes [from Cook Well Strainer Company])

Screen size, inches	Wire size, inches	Slot size, thousandth of an inch										
		10	20	25	30	40	50	60	70	80	100	125
4	0.09	0.097	0.178	0.213	0.245	0.302	0.350	0.392	0.430	0.461	0.516	—
		10.0	18.2	21.7	25.0	30.7	35.7	40.0	43.8	47.0	52.6	—
6		0.136	0.247	0.295	0.339	0.418	0.485	0.542	0.593	0.637	0.722	—
		10.0	18.2	21.7	25.0	30.8	35.7	40.0	43.8	47.0	52.6	—
8		0.196	0.356	—	0.490	0.604	0.700	0.784	0.858	0.923	1.032	—
		10.0	18.2	—	24.9	30.8	35.7	39.9	43.8	47.0	52.6	—
8	.1467	0.125	0.236	0	0.334	0.421	0.500	0.571	0.625	0.694	0.797	0.904
		6.4	12.0	0	17.0	21.5	25.5	29.9	32.4	35.4	40.6	46.1
10	.09	0.248	0.447	0.553	0.621	0.765	0.869	0.993	1.101	1.168	1.307	1.440
		10.0	18.7	21.7	25.0	30.8	35.7	39.9	43.6	47.0	52.6	58.0
10	.1467	0.159	0.299	—	0.423	0.533	0.633	0.723	0.805	0.879	1.010	1.146
		6.4	12.0	—	17.0	21.5	25.5	29.1	32.4	35.4	40.6	46.1
10	.1875	0.126	0.241	0.294	0.345	0.440	0.528	0.607	0.682	0.749	0.872	1.003
		5.0	9.6	11.7	13.7	17.5	21.0	24.2	27.1	29.8	34.7	40.0
12	.09	0.300	0.546	—	0.751	0.926	1.115	1.208	1.316	1.414	1.582	1.754
		10.0	18.1	—	24.9	30.8	35.7	39.9	43.7	47.0	52.6	58.0
12	.1467	0.192	0.362	—	0.512	0.646	0.767	0.876	0.974	1.064	1.222	1.387
		6.4	12.0	—	17.0	21.4	25.5	29.1	32.4	35.3	40.5	46.1
12	.1875	0.153	0.292	0.356	0.417	0.532	0.638	0.733	0.824	0.905	1.053	1.213
		5.0	9.7	11.7	13.7	17.5	21.0	24.2	27.1	29.8	34.7	40.0
14	.1875	0.180	0.345	—	0.492	0.630	0.755	0.874	0.962	1.071	1.248	1.435
		5.0	9.7	—	11.9	13.7	17.5	21.0	24.2	29.8	34.7	40.0
16	1.469	0.245	0.460	—	0.652	0.821	0.975	1.117	1.239	1.353	1.555	1.764
		6.4	12.0	—	17.0	21.4	25.5	29.1	32.4	35.3	40.5	46.1
16	.1875	0.191	0.367	—	0.447	0.523	0.669	0.803	0.925	1.138	1.325	1.527
		5.0	9.7	—	11.7	13.7	17.5	21.0	24.2	29.8	34.7	40.0
18	.1875	0.219	0.416	0.510	0.596	0.762	0.913	1.050	—	1.296	1.505	1.735
		5.0	9.7	11.7	13.7	17.5	21.0	24.2	—	29.8	34.7	40.0
20	.1875	0.246	0.468	0.571	0.669	0.854	1.027	1.177	—	1.458	1.690	1.945
		5.0	9.7	11.7	13.7	17.5	21.0	24.2	—	29.8	34.7	40.0
24	.1875	0.298	0.567	0.692	0.811	1.034	1.240	1.426	1.599	1.760	2.047	2.180
		5.0	9.7	11.7	13.7	17.5	21.0	24.2	27.1	29.9	34.7	40.0

[1] Top value shown is ft³/ft and bottom value is percent.

Table 11-9.—Minimum open areas of screens in square feet per linear foot and the percentage of open area[1]-continued

(Cage-type wire-wound screen—telescoping sizes [from Howard Smith Company])

Screen size, inches	Slot size, thousandth of an inch								
	8	10	12	14	16	20	30	40	50
4	0.104	0.125	0.145	0.167	0.187	0.229	0.305	0.382	0.437
	9.0	11.2	13.0	15.0	16.8	20.6	27.4	34.3	39.2
6	0.118	0.146	0.174	0.194	0.222	0.271	0.368	0.451	0.437
	8.0	9.9	11.8	13.2	15.1	18.4	25.0	30.6	29.6
8	0.160	0.194	0.229	0.264	0.298	0.354	0.493	0.604	0.700
	8.1	9.8	11.6	13.4	15.2	18.0	25.1	30.7	35.6
10	0.201	0.250	0.291	0.333	0.375	0.451	0.625	0.763	0.888
	8.1	10.1	11.7	13.3	15.1	18.1	25.1	30.6	35.6
12	0.243	0.291	0.347	0.396	0.444	0.534	0.736	0.902	1.047
	8.3	9.9	11.8	13.4	15.1	18.2	25.0	30.7	35.6
14	0.215	0.263	0.312	0.360	0.403	0.492	0.687	0.854	1.006
	6.3	7.8	9.2	10.6	11.9	14.6	20.3	25.3	29.8
16	0.194	0.243	0.284	0.333	0.375	0.479	0.680	0.853	1.006
	5.2	6.5	7.6	8.9	10.0	12.8	18.2	22.9	26.9

[1] Top value shown is ft²/ft and bottom value is percent.

Table 11-9.—Minimum open areas of screens in square feet
per linear foot and the percentage of open area[1]-continued

(Cage-type wire-wound screen—pipe size screen [from Johnson Division, UOP Inc.])

Screen size, inches	Slot size, thousandth of an inch						
	10	20	40	60	80	100	150
4	0.174	0.305	0.472	0.597	0.680	0.639	0.756
	14.3	25.2	38.9	49.2	56.1	52.7	62.4
6	0.174	0.319	0.534	0.694	0.812	0.916	0.986
	5.7	18.4	30.8	40.0	46.8	52.8	56.8
8	0.222	0.410	0.694	0.902	1.055	1.187	1.277
	9.8	18.1	30.7	39.9	46.7	52.5	56.5
10	0.285	0.514	0.868	0.958	1.152	1.305	1.596
	10.1	18.2	30.8	34.0	40.9	46.3	56.7
12	0.264	0.500	0.868	1.131	1.367	1.548	1.888
	7.8	14.9	25.9	33.8	40.9	46.3	56.5
14	0.298	0.555	0.965	1.263	1.527	1.735	2.110
	8.1	15.1	26.3	34.4	41.6	47.3	47.5
16	0.340	0.639	1.110	1.450	1.749	1.985	2.415
	8.1	15.2	26.4	34.6	41.7	47.3	57.6

[1] Top value shown is ft²/ft and bottom value is percent.

Table 11-9.—Minimum open areas of screens in square feet
per linear foot and the percentage of open area[1]-continued

(Cage-type wire-wound screen—pipe size screen [from Johnson Division, UOP Inc.])

Screen size, inches	Slot size, thousandth of an inch								
	30	40	50	60	70	80	90	100	125
8	0.21	0.27	0.32	0.37	0.41	0.45	0.49	0.52	0.60
10	.26	.33	.39	.45	.50	.55	.59	.63	.72
12	.31	.39	.46	.53	.59	.65	.70	.75	.80
14	.35	.45	.53	.61	.68	.75	.81	.87	.99
16	.40	.51	.60	.69	.78	.85	.92	.98	1.12
Approximate percent open area	9.2	11.7	13.9	16.0	17.9	19.5	21.2	22.8	26.0

[1] Top value shown in ft²/ft and bottom value is percent.

Table 11-9.—Minimum open areas of screens in square feet per linear foot and the percentage of open area[1]-continued

(Cage-type wire-wound screen—pipe size screen [from Johnson Division, UOP Inc.])

Screen size, inches	Wire size, inches	Slot size, thousandth of an inch									
		10	20	25	30	40	50	60	80	100	125
6	0.09	0.171	0.312	0.372	0.429	0.528	0.612	0.686	0.800	0.902	—
		10.0	18.2	21.7	25.0	30.8	35.7	40.0	46.7	52.6	—
6	.1467	0.111	0.208	0.252	0.295	0.372	0.442	0.504	0.614	0.705	0.799
		6.4	12.0	14.5	17.0	21.4	25.4	29.0	35.4	40.5	46.0
6	.1875	0.191	0.117	0.214	0.249	0.319	0.384	0.477	0.547	0.635	0.732
		5.0	9.7	11.7	13.7	17.5	21.0	24.2	29.9	34.7	40.0
8	.1875	0.118	0.225	0.274	0.321	0.410	0.492	0.567	0.707	0.814	0.938
		5.1	9.6	11.7	13.7	17.5	21.0	24.2	29.9	34.7	40.0
10	.1875	0.145	0.276	0.336	0.394	0.503	0.693	0.694	0.857	0.997	1.140
		5.0	9.7	11.7	13.7	17.5	21.0	24.2	29.9	34.7	40.0
12	.1875	0.171	0.326	0.398	0.467	0.595	0.709	0.821	1.013	1.179	1.356
		5.0	9.7	11.7	13.7	17.5	21.0	24.2	29.8	34.7	40.0
18	.1875	0.250	0.477	0.582	0.683	0.871	1.043	1.200	1.482	1.724	1.984
		5.0	9.7	11.7	13.7	17.5	21.0	24.2	29.8	34.7	40.0
20	.1875	0.277	0.528	0.644	0.755	0.963	1.154	1.327	1.637	1.905	2.194
		5.0	9.7	11.7	13.7	17.5	21.0	24.2	29.8	34.7	—

[1] Top value shown is ft²/ft and bottom value is percent.

Table 11-9.—Minimum open areas of screens in square feet per linear foot and the percentage of open area[1]-continued

(Cage-type wire-wound screen—pipe size screen [from Howard Smith Company])

Screen size, inches	Slot size, thousandth of an inch								
	8	10	12	14	16	20	30	40	50
4	0.104	0.125	0.146	0.167	0.187	0.229	0.305	0.382	0.437
	8.8	10.6	12.4	14.2	15.9	19.4	25.9	32.4	35.6
6	0.139	0.174	0.201	0.236	0.264	0.312	0.430	0.534	0.618
	8.0	10.0	11.6	13.6	15.2	18.0	24.8	30.8	35.6
8	0.187	0.229	0.264	0.305	0.340	0.408	0.562	0.694	0.805
	8.3	10.1	11.7	13.5	15.0	18.1	24.8	30.7	35.6
10	0.229	0.278	0.333	0.382	0.423	0.514	0.701	0.861	0.979
	8.1	9.9	11.8	13.6	15.0	18.2	24.8	30.6	35.6
12	0.222	0.271	0.319	0.368	0.416	0.500	0.701	0.874	1.020
	6.6	8.1	9.5	11.0	12.4	14.9	20.9	26.1	30.5
14	0.243	0.298	0.353	0.403	0.451	0.548	0.770	0.958	1.124
	6.6	8.1	9.6	11.0	12.3	14.9	21.0	26.1	30.7
16	0.278	0.340	0.402	0.465	0.521	0.632	0.881	1.096	1.284
	6.6	8.1	9.6	11.1	12.5	15.1	21.0	26.2	30.7

[1] Top value shown is ft²/ft and bottom value is percent.

Table 11-9.—Minimum open areas of screens in square feet per linear foot and the percentage of open area[1]-continued

(Wire-wound on pipe base [from Howard Smith Company])

Screen size, inches	Pipe perforations	Slot size, thousandth of an inch								
		8	10	12	14	16	20	30	40	50
4	0.208	0.083	0.115	0.135	0.155	0.174	0.210	0.291	—	—
	17.6	6.4	9.0	10.6	12.1	13.6	16.4	22.8	—	—
6	0.310	0.103	0.126	0.150	0.172	0.194	0.236	0.333	0.418	0.496
	17.9	5.6	6.8	8.1	9.3	10.5	12.8	18.0	22.6	26.9
8	0.375	0.133	0.163	0.193	0.222	0.251	0.305	0.430	0.541	0.638
	16.6	5.6	6.9	8.1	9.3	10.5	12.8	18.1	22.7	26.9
10	0.408	0.160	0.202	0.239	0.275	0.311	0.379	0.532	0.670	0.793
	14.5	5.5	6.9	8.1	9.4	10.6	12.9	18.1	22.8	27.0
12	0.491	0.193	0.237	0.281	0.323	0.366	0.445	0.626	0.780	0.931
	14.7	5.6	6.8	8.1	9.3	10.6	12.9	18.1	22.7	26.9
14	0.525	0.212	0.260	0.308	0.355	0.401	0.487	0.685	0.862	1.021
	14.3	4.9	6.9	8.1	9.4	15.3	12.9	18.0	22.7	27.0
16	0.624	0.211	0.296	0.341	0.404	0.457	0.555	0.799	0.980	1.162
	14.9	4.9	6.9	7.9	9.4	10.6	12.8	18.1	22.7	27.0
18	0.691	0.269	0.332	0.393	0.452	0.511	0.622	0.874	1.083	1.301
	14.6	5.6	6.9	8.1	9.4	10.6	12.9	18.1	22.4	27.0
20	0.708	0.300	0.369	0.436	0.502	0.566	0.690	0.970	1.221	1.416
	13.5	5.6	6.9	8.2	9.4	10.6	12.9	18.1	22.8	26.5

[1] Top value shown is ft²/ft and bottom value is percent.

Table 11-9.—Minimum open areas of screens in square feet per linear foot and the percentage of open area[1]-continued

(Louvre or shutter-type screen—standard [3/16- to 1/4-inch wall] [from Roscoe Moss Company])

Screen size, inches	Slot size, inch					
	1/16	3/32	1/8	5/32	3/16	1/4
6	0.017	0.025	0.039	0.042	0.050	0.068
	0.9	1.4	1.9	2.4	2.8	3.9
8	0.025	0.038	0.050	0.063	0.076	0.101
	1.1	1.6	2.2	2.7	3.3	4.4
10	0.027	0.040	0.055	0.069	0.083	0.111
	0.9	1.4	1.9	2.4	2.9	3.9
12	0.036	0.055	0.073	0.092	0.111	0.147
	1.0	1.6	2.1	2.7	3.3	4.3
14	0.036	0.055	0.073	0.092	0.111	0.147
	0.9	1.4	1.9	2.4	2.9	3.8
16	0.046	0.069	0.092	0.115	0.138	0.183
	1.0	1.5	2.1	2.6	3.1	4.1
18	0.046	0.069	0.092	0.115	0.138	0.183
	0.9	1.4	1.8	2.3	2.8	3.7
20	0.055	0.083	0.111	0.138	0.165	0.222
	1.0	1.5	2.0	2.5	3.0	4.1

[1] Top value shown is ft²/ft and value is percent.

Table 11-9.—Minimum open areas of screens in square feet per linear foot and the percentage of open area[1]-continued

(Full flow screen [3/16- to 1/4-inch wall] [from Roscoe Moss Company])

Screen size, inches	Slot size, inch					
	1/16	3/32	1/8	5/32	3/16	1/4
6	0.050	0.076	0.101	0.127	0.151	0.202
	2.8	4.3	5.8	7.3	8.7	11.6
8	0.067	0.101	0.135	0.169	0.202	0.235
	2.9	4.4	5.9	7.4	8.9	10.3
10	0.106	0.165	0.222	0.282	0.346	0.472
	3.7	5.8	7.8	10.0	12.2	16.7
12	0.132	0.206	0.278	0.353	0.432	0.589
	3.9	6.1	8.3	10.5	12.9	17.6
14	0.132	0.206	0.278	0.353	0.432	0.589
	3.4	5.4	7.3	9.2	11.3	15.4
16	0.158	0.247	0.333	0.424	0.519	0.707
	3.6	5.6	7.6	9.7	11.8	16.1
18	0.184	0.289	0.389	0.494	0.605	0.825
	3.7	5.9	7.9	10.1	12.4	16.9
20	0.212	0.331	0.444	0.564	0.691	0.944
	3.9	6.1	8.2	10.4	12.7	17.4

[1] Top value shown is ft²/ft and value is percent.

Table 11.9.—Minimum open areas of screens in square feet per linear foot and the percentage of open area[1]-continued

(134 shutter screen—3 gauge [0.25-inch wall] [from the Layne and Bowler Company])

Screen size, inches	Slot size, thousandth of an inch				
	30	55	80	105	130
4	0.039	0.072	0.104	0.138	0.168
	3.3	6.1	8.8	11.7	14.2
6	0.074	0.133	0.196	0.256	0.318
	4.3	7.8	11.5	15.1	18.7
8	0.094	0.172	0.251	0.329	0.410
	4.2	7.7	11.2	14.7	18.3
10	0.126	0.231	0.336	0.441	0.545
	4.5	8.4	12.2	16.1	19.8
12	0.147	0.270	0.392	0.514	0.637
	4.5	8.3	12.0	15.7	19.5
16	0.200	0.364	0.532	0.696	0.863
	4.8	8.7	12.7	16.6	20.6
20	0.220	0.403	0.586	0.770	0.952
	4.7	8.5	12.4	16.3	20.2
24	0.283	0.518	0.754	0.990	1.225
	4.5	8.2	11.9	15.7	19.5

[1] Top value shown is ft²/ft and value is percent.

Table 11-9.—Minimum open areas of screens in square feet per linear
foot and the percentage of open area[1]-continued

(Punched screens—gravel guard well screen, 0.25-inch wall [from Doerr Metal Products])

Screen size, inches	Slot size, inch			
	1/32	1/16	1/8	3/16
8	0.054	0.120	0.263	0.410
	2.5	5.7	12.5	19.5
10	0.069	0.153	0.335	0.522
	2.6	5.8	12.8	19.9
12	0.084	0.185	0.407	0.634
	2.7	5.9	12.9	20.1
14	0.098	0.218	0.478	0.746
	2.7	5.9	13.0	20.2
16	0.111	0.245	0.538	0.839
	2.7	5.9	12.8	20.0
18	0.126	0.278	0.610	0.951
	2.7	5.9	12.9	20.2
24	0.160	0.352	0.773	1.21
	2.5	5.6	12.3	19.2

[1] Top value shown is ft²/ft and value is percent.

Table 11-9.—Minimum open areas of screens in square feet per linear
foot and the percentage of open area [1]-continued

(Slotted pipe[2]—horizontally slotted casing)

Pipe size, inches	Slot size, inch			
	1/8	5/32	3/16	1/4
10	0.061	0.076	0.090	0.120
	2.1	2.7	3.2	4.3
12	0.074	0.092	0.109	0.145
	2.2	2.8	3.3	4.3
14	0.085	0.106	0.127	0.167
	2.3	2.9	3.5	4.6
16	0.098	0.122	0.145	0.192
	2.3	2.9	3.5	4.6
18	0.109	0.136	0.163	0.216
	2.3	2.9	3.5	4.6
20	0.115	0.144	0.173	0.228
	2.2	2.8	3.3	4.3

[1] Top value shown is ft²/ft and value is percent.
[2] Slots are 1.5 inches long on 6-3/8-inch centers on a plane around the pipe or on 1-1/4-inch centers vertically with each horizontal row staggered.

Table 11-9.—Minimum open areas of screens in square feet per linear foot and the percentage of open area[1]-continued

(Oil field milled slotted casing[2])

Pipe size, inches	Slot size, thousandth of an inch					
	100	120	140	180	200	250
6	0.017	0.020	0.023	0.030	0.033	0.042
	0.98	1.6	1.8	2.3	2.5	3.2
8	0.022	0.027	0.031	0.040	0.044	0.056
	0.97	1.2	1.4	1.8	1.9	2.5
10	0.028	0.033	0.039	0.050	0.056	0.069
	1.0	1.2	1.4	1.8	2.0	2.4
12	0.033	0.040	0.047	0.060	0.067	0.083
	0.99	1.2	1.4	1.8	2.0	2.5
14	0.039	0.047	0.054	0.070	0.078	0.097
	1.1	1.3	1.5	1.9	2.1	2.6
16	0.044	0.053	0.062	0.080	0.089	0.111
	1.1	1.3	1.5	1.9	2.1	2.7
18	0.050	0.060	0.070	0.090	0.100	0.125
	1.0	1.2	1.4	1.8	2.1	2.5
20	0.056	0.067	0.078	0.100	0.111	0.139
	1.0	1.2	1.4	1.8	2.1	2.5

[1] Top value shown is ft²/ft and value is percent.
[2] Vertical 2-inch-long slots spaced at two diameter centers on staggered horizontal rows around the pipe. Vertical spacing or horizontal rows on 3-inch centers.

economical combination. Basically, the screen should always be placed at the bottom of the well, and the screen length should not be less than 35 percent of the estimated thickness of the aquifer penetrated by the well.

Screen components such as blank sections of pipe or flush tube extensions, bottom plates, float-down or jetting shoes, centering guides, and other accessories should be fabricated from the same material as the screen; otherwise, plastic or concrete components should be used. Dissimilar metals should never be incorporated in the screen assembly. When the screen assembly is fabricated from nonferrous metals, the assembly should be separated from the low carbon steel pump housing casing by neoprene, plastic, cement, other nonmetallic materials or couplings.

In straight wall wells where the diameter of the hole results in an annular space about the screen greater than about 50 millimeters (2 inches), a formation stabilizer should be used (section 11-11(b)).

A common misconception holds that straightness and plumbness are unimportant in the screen assembly because the pump is not set in the assembly. The hole should be straight enough, however, to permit installing the screen without having to force or drive the screen down. If installed crooked or too much out of plumb, the screen is subject to bending stresses that may cause slot enlargement or collapse. Plumbness of the screen therefore should meet the same criteria as the casing in not deviating from the vertical more than two-thirds of the inside diameter of the screen per 30 meters (100 feet) and the axis of the screen assembly should coincide with that of the pump housing casing or riser pipe in the vicinity of their junction.

11-7. Drive Shoes.—When casing is driven into place, particularly in cobbly or bouldery materials, the bottom of the casing should be reinforced with a hardened steel ring or drive shoe which is screwed or welded to the casing. The drive shoe should have an outside diameter and beveled cutting edge that are about the same as that of couplings for the diameter of pipe being used. The cutting edge shaves irregularities off the side of the hole as the casing is driven and will split or force large rocks into the side of the hole, thus preventing the bottom of the casing from collapsing. Commercially available drive shoes come in two patterns, regular and Texas. The Texas pattern is longer and somewhat more rugged than the regular pattern and is used where driving is particularly difficult. The selection of the type of shoe to be used is usually left to the discretion of the contractor.

11-8. Reducers and Overlaps.—

(a) Description and Purpose.—When drilling with cable tools, a point is reached where the casing can be driven no farther because of skin friction and other factors. When this point is reached, a smaller casing is telescoped into the installed casing and drilling is continued using a smaller bit. On completion of the drilling, the smaller casing may be cut off at a point some distance above the bottom of the larger casing. On some deep holes, six or more such reductions may be required. The starter casing should always be of suitable diameter so that the diameter of the pump chamber casing at the depth of pump setting will be adequate.

In other designs, particularly when the hole is drilled uncased as with most rotary rigs, the pump chamber casing may be directly attached to the screen assembly riser pipe or extension by a coupling or reducer. The entire casing and screen assembly is then lowered into the well as a continuous string of pipe, and additional lengths are added at the surface as the string is lowered. If the string is allowed to rest on the bottom, the weight of the entire string is carried by the screen, which is the weakest section. Thus, the possibility of buckling or collapse of the screen is increased. Because of this hazard, single string construction should provide for maintaining the casing and screen string in tension until the well is developed and permanently anchored at the surface. This construction technique is particularly important in wells exceeding about a 30-meter (100-foot) depth. Designs in which the screen assembly is telescoped into place offer some advantage because the screen does not support the entire column weight. If necessary, the screen can be withdrawn and replaced, an operation which is impossible with a solidly connected line of pipe (see section 11-2(a)).

(b) Design Particulars of Reducers and Overlaps.—A commonly used length of overlap between casing and screen assembly is 1.5 to 3 meters (5 to 10 feet). In extremely deep wells or where the possibility of settlement of the screen assembly is present, more overlap may be necessary (Ahrens, 1970; Reinke and Kill, 1970; Driscoll, 1986). An overlap should always be sealed as described in section 11-7.

Reducers used between the casing and the screen assembly are, in many instances, fabricated by the contractor or local shops, both of which have a tendency to use flat conical sections. From the standpoint of hydraulic efficiency and strength, the upper straight end and conical section of reducers should be fabricated of the same weight or wall thickness and material as the larger pipe to which they will be attached. The conical section of the reducer should have a length at least 10 times the difference in diameter of the two pipes which it will connect.

11-9. Seals.—

(a) Description and Purpose.—The grout seal commonly placed around the permanent surface casing or the pump chamber casing is primarily a sanitary seal which should be of sufficient thickness, depth, and imperviousness to prevent any surface-water or poor quality ground water from entering the well. Native clay, bentonite, and other materials are also used as grout and may be satisfactory from the sealing standpoint. However, a good cement-based grout mixed with a proper amount of bentonite or aluminum powder will produce a better seal as well as serve other useful functions. Such a mixture protects the casing against corrosion attack, and if the casing is removed by corrosion, the grout serves as a concrete casing. If correctly installed, grout forms a bond between the casing and the soil, stabilizes the soil about the well, and sometimes acts as a keystone to limit the extent of upward caving. From the standpoint of a sanitary seal, grouting the full length of casing is probably unnecessary; but in view of the other functions the grout performs, the practice is recommended. Once the placement equipment is installed, additional amounts of grout are relatively inexpensive. Care must be used when grouting plastic casing with neat-cement grout. The heat of hydration can cause the plastic casing to weaken. Typically, the grout should not be more than 50 millimeters (2 inches) thick to avoid this condition (Driscoll, 1986).

Where casings or screen assemblies are telescoped down the hole, a seal should be placed at the top of each telescoped section. Two types of seals or packers are commonly used. A commercial, swaged lead seal fits on the top of the smaller casing and is swaged out with a special tool against the inside of the larger casing. A correctly installed swaged lead seal is practically leakproof so far as permitting the entrance of water from outside the casing. In addition, the seal prevents sand and gravel from being carried up the annular space and into the well when the well is pumped. A swaged lead seal also permits casing and screen assemblies to be pulled out of a well if necessary. The installation of a swaged lead seal will not be permitted for drinking water wells.

Neoprene rubber seals are vulcanized around the smaller casing at the factory. The outside diameter of the seals, which has flexible lips, is slightly larger than the inside diameter of the larger casing. If the inside of the casing is wet when the smaller casing is telescoped in place, the seals slide down readily. When in place, they form a tight seal which keeps out undesirable water and stops upward movement of sand or gravel in the annular space. In addition, seals act as insulation, separating dissimilar materials which otherwise would cause galvanic corrosion. They permit easier removal of casing or screen assemblies, if necessary, than other seals. Where insulation as well as sealing is desired, usual practice is to use two or more of the seals spaced about 0.45 to 0.90 meter (1-1/2 to 3 feet) apart to ensure separation of the dissimilar metals in the overlap.

Where the difference in diameter of the overlapping casings is sufficient to permit insertion of a 15-millimeter (1/2-inch) or larger pipe into the annular space, a neat cement-bentonite grout seal may be placed. The grout acts also as an insulation. It has sufficient strength to resist the flow of water but is readily broken if the lower casing needs to be pulled.

Designs may be encountered in which a grout seal is placed in the annular space between a surface casing and the pump chamber casing and the surface casing is not grouted in. The theory is that caving of unstable materials will create a positive seal about the surface casing. This theory cannot be depended upon, so permanent surface casing should always be grouted in. However, the top of the annular space between the surface casing and the pump chamber casing should always be tightly sealed with an expanding packer, a concrete plug, or a steel ring welded to the wall of the pump chamber casing and the top of the surface casing (Ahrens, 1970; Reinke and Kill, 1970; Driscoll, 1986).

(b) Design Particulars for Seals.—The grout seal around surface or pump chamber casing should have a minimum thickness of 40 millimeters (1-1/2 inches) about the pipe or about the couplings, if they are used. The annular space should be flushed with water before commencing the grouting operations. The grout should be introduced at the bottom of the space to be grouted and should be placed in a continuous operation. If cement grout is used, it should be entirely placed before the occurrence of initial set. AWWA-A1OO-66, Standard for Deep Wells, Section Al-8.4, outlines various acceptable methods used for placing grout. For most water wells, however, placement through a tremie pipe is acceptable.

11-10. Gravel or Concrete Base.—When a well is bottomed in fine sands, plastic clay, or other soft or unstable material, a recommended practice is to overdrill the well 0.9 to 1.2 meters (3 to 4 feet). This interval should be filled with coarse gravel or concrete to provide a firm base for the casing and screen.

11-11. Centering Guides.—Where casing or screen assemblies over 12 meters (40 feet) long are installed in holes having nominal diameters 50 millimeters (2 inches) or larger than the outer diameter of the casing, centering guides should be installed. The guides hold the casing in the center of the hole as well as offer support against bending and buckling because of axial and unbalanced horizontal loads. Centering guides are essential for centering casing and screens for grouting and gravel packing. The guides should be placed at the bottom and at about 12- to 15-meter (40- to 50-foot) intervals up the hole. In gravel packed wells, care should be exercised to keep the centering guides on approximately straight lines from top to bottom so as not to interfere with the insertion of tremie pipes. Centering guides should not be welded directly to the screen proper if avoidable. Preferably, a short section of blank casing to which the centering guides can be welded should be inserted in the screen at approximately the desired intervals. Centering guides may be of wood, plastic, strap steel, or alloy. Wood guides may not be permitted in drinking water wells. Metallic guides should always be of the same alloy as the casing or screen assembly to which they are attached. The guides are set in a plane around the casing at 90- or 120-degree intervals. Figures 11-11 and 11-12 show designs of acceptable centering guides (Ahrens, 1970; Driscoll, 1986).

11-12. Tremie Pipes.—In gravel packed wells, the pack material is generally installed through one or two temporary tremie pipes which are withdrawn in stages as the pack is placed. Tremie pipes consist of nominal 50- to 100-millimeter (2- to 4-inch) coupled steel pipe. The diameter depends on the pack, grain size, clearance, and other factors. The design of the well must provide for adequate annular space to permit passage of the tremie pipes including couplings. Many different designs are used for

Wall of casing

Top fillet welded on sides to casing

150 mm (6 in)

6 X 30 mm ($\frac{1}{4}$ x 1$\frac{1}{4}$ in.) Strap steel

Length of guide equal to 2D to 4D depending on hole diameter

Centering guide 3 or 4 guides per round

Wall of hole

Casing

Annular space

D

90° or 120°

PLAN

Bottom not attached to casing if later pulling of casing is anticipated

150 mm (6 in.)

DETAIL

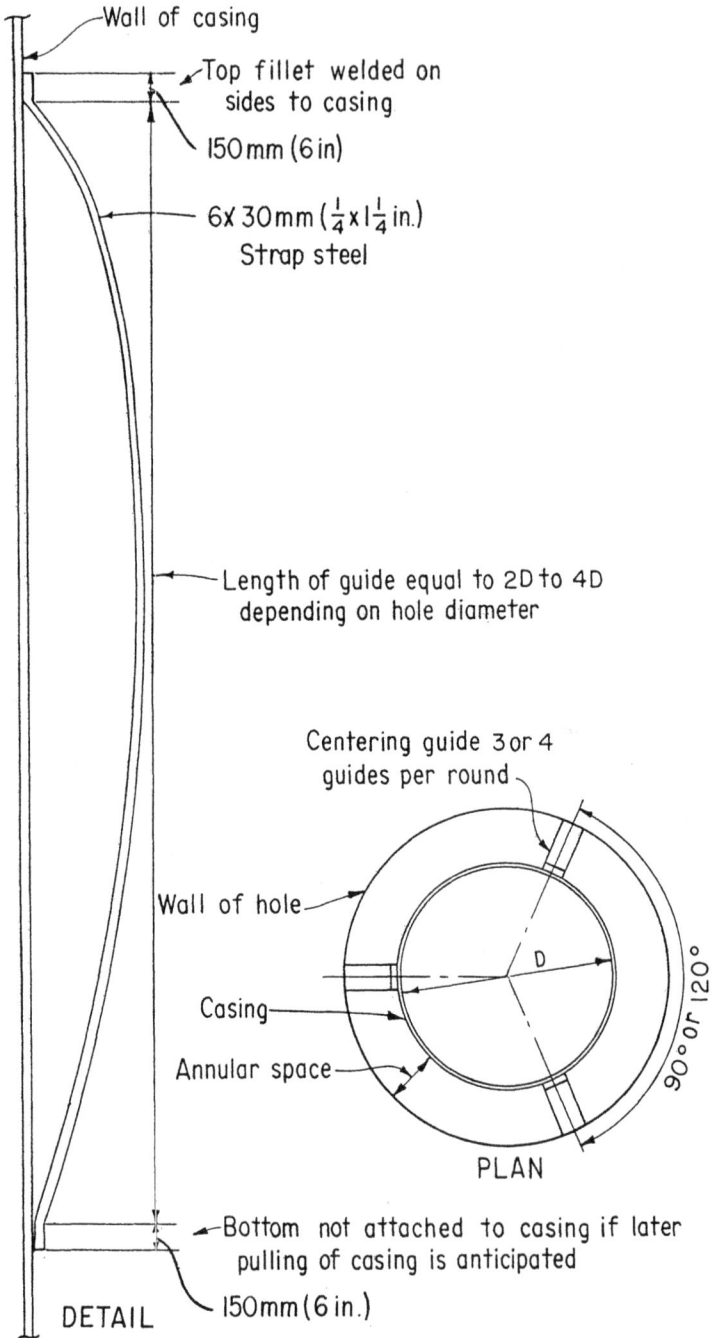

Figure 11-11.—Steel strap centering guide.

Well casing

300 mm (12 in.) or more

150 mm (6 in.)

90° or 120°

Casing

Steel band

Centering guide 50mm (2 in.) or thicker lumber

PLAN

19 mm ($\frac{3}{4}$ in.) or wider steel band

DETAIL

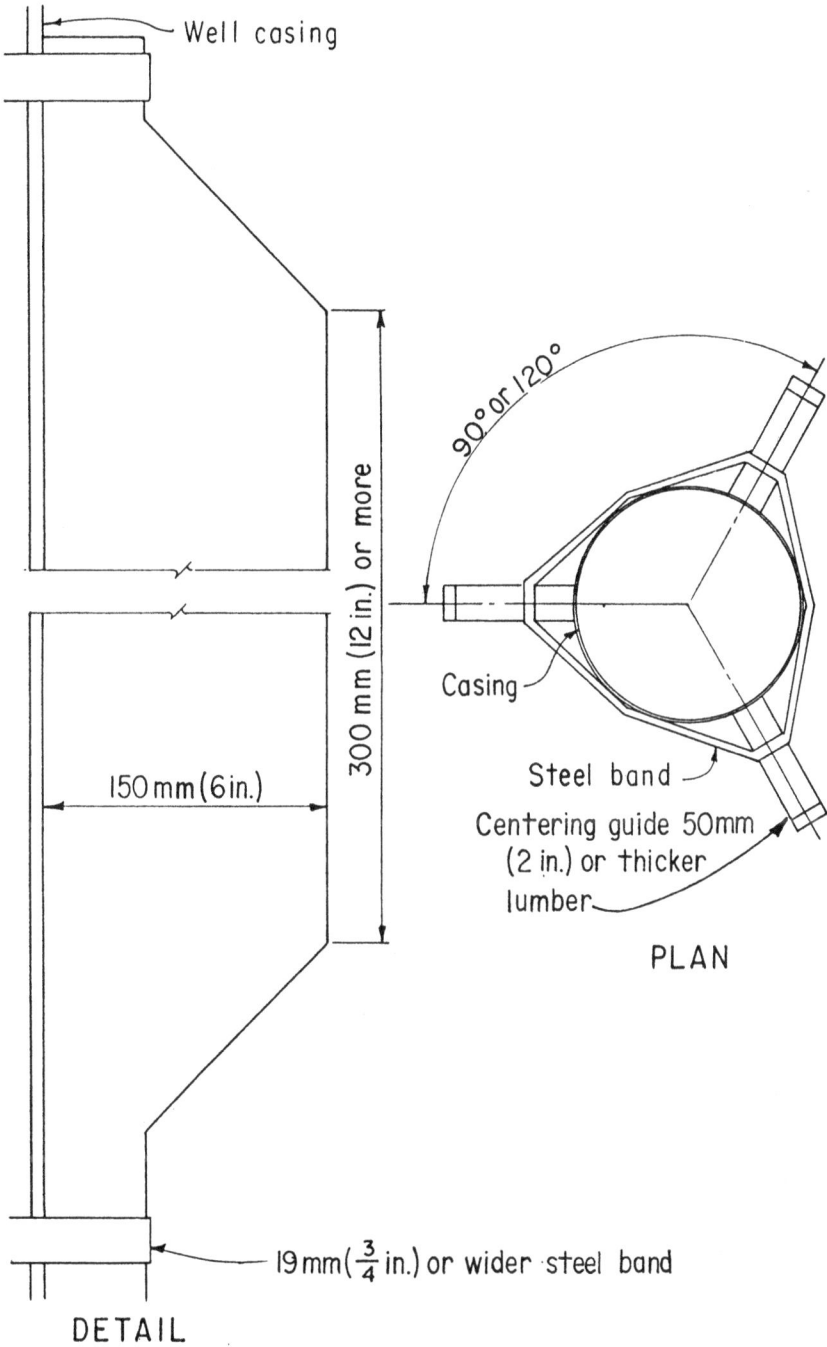

Figure 11-12.—Wood block centering guide.

accommodating gravel packs in wells. Most designs provide for the gravel pack to be extended at least 6 meters (20 feet) above the top of the uppermost screen. In some designs, the annular space above the pack is filled with grout and no provision is made for the addition of gravel pack if required.

Prudent design calls for both adequate and protected storage for the gravel pack by continuing the pack for at least 3 meters (10 feet) above the top of the screen. Also, a means of permanent replenishment of the pack from the surface should be provided. Such features minimize the possibility of direct aquifer contact with the screen in the event of excessive pack settlement. Storage can be provided between the pump chamber and surface casings. Replenishment can be accomplished through permanent tremie pipes installed when the well is completed. One or two permanent tremie pipes are installed between the surface and pump chamber casing and extending to the desired depth. If required, a concrete seal is then placed around them. The top of permanent tremie pipes should always be threaded and sealed with a screw cap (Ahrens, 1970; Driscoll, 1986).

Sizes of temporary tremie pipe may be left to the discretion of the contractor, but permanent tremie pipe should be of adequate size and weight for permanent service.

11-13. Gravel Packs and Formation Stabilizers.—

(a) Description and Purpose.—Where casing and screen is set in an oversized hole where the annular space is larger than 50 milli-meters (2 inches) but gravel pack construction is not intended, a formation stabilizer should be placed in the annular space. The stabilizer does not need to be carefully selected in regard to gradation as long as the smaller grains are larger than the screen slot size and the largest are 9 millimeters (3/8 inch) or less. The purpose of the formation stabilizer is to support the pipe againstunbalanced forces which might arise during development of the well and to facilitate development of the well (Driscoll, 1986).

The principal functions of a gravel pack are to:

• Stabilize the aquifer and minimize sand pumping

• Permit use of the largest possible screen slot with resultant maximum open area

- Provide an annular zone of high permeability, thus increasing the effective radius of the well and the yield

A gravel pack normally should not be used unless aquifer conditions make pack use unusually advantageous. Use of a pack usually increases the cost and difficulty of constructing a well.

Gravel packs should be designed to have a small coefficient of uniformity, and grain sizes should be carefully selected to match the aquifer material. Screens should then be selected to pass not more than 5 percent of the pack material. Maximum grain size of a pack should be 9 millimeters (3/8 inch) if placed through a nominal 100-millimeter (4-inch) tremie pipe. Minimum design thickness depends on the ability to place the pack. A 12-millimeter (1/2-inch) thick pack is theoretically adequate. Maximum design thickness should be 200 millimeters (8 inches) because of the difficulty of development through a thicker pack (Ahrens, 1970; Campbell and Lehr, 1973; Kruse, 1960; Reinke and Kill, 1970; Driscoll, 1986; Walton, 1970).

Conditions which especially favor the use of a pack include:

- Presence of a fine, uniform sand aquifer

- Presence of a layered aquifer with alternating sand and clay layers

- A requirement for maximum yield from a marginal aquifer

- Presence of a friable sandstone or similar aquifer

(b) Placement Procedures and Design of Gravel Packs.—If the design permits, the screen assembly should be supported at the surface and kept under tension while the gravel pack is being placed.

Gravel pack should be placed in a manner which ensures complete filling of the annular space and minimizes bridging and segregation. In wells drilled to depths of up to 150 meters (500 feet), gravel pack is best placed through two tremie pipes placed 180 degrees apart and extending initially to within about 1.5 meters (5 feet) from the bottom of the hole. The inside diameter of the tremie pipes should be at least 12 times the diameter of the coarsest pack material if placed by gravity and

10 times if pumped. As the pack is being placed, the tremie pipes are raised in such a manner that the free fall of pack material below the bottom of the pipe does not exceed 1.5 meters (5 feet). The placement of gravel should be continued at a uniform rate until completed. Gravel may be poured or shoveled into the tremie pipes dry or may be washed or pumped in. If washed in, a constant flow of gravel and clean, sediment-free water is fed into each tremie pipe. If pumped in, the gravel-water ratio should be 0.75 cubic meter (1 cubic yard) of gravel to 6,800 to 11,400 liters (1,800 to 3,000 gallons) of water. The gravel-water ratio will depend upon the stability of the hole walls, the grain size of the gravel pack, and the type and size of pump available that will permit a uniform, constant pumping rate without causing caving of the wall of the hole.

In rotary drilled holes, the fluid in the hole should be circulated and the viscosity reduced by dilution with water until the Marsh Funnel velocity is less than 30 seconds before gravel is introduced into the well. A desirable practice, if well design and available equipment permit, is to pump fluid from the bottom of the well as the gravel pack is pumped in. To avoid collapse of the hole after the drilling fluid has been thinned, the pumping rate should be adjusted to maintain the water-level in the well above the static water-level. In some cases, however, the aquifer may be too permeable to permit building up an adequate head above the static water-level.

Other methods have been devised to place gravel pack in deep rotary drilled wells where use of tremie pipes would be impracticable. Such equipment as the crossover tool permits pumping of gravel through the drill pipe and into the annulus.

Numerous formulas based on standard mechanical analyses of grain size have been developed for the selection of gravel pack gradations. None are entirely satisfactory, but those described in this discussion are usually adequate. A number of terms commonly used in the literature must be carefully examined in regard to their meaning. In most ground-water literature, the grain size terms, D_{10}, D_{60}, D_{100}, etc., refer to the percent retained sizes. Bureau of Reclamation practice is to refer to them as the percent passing or the percent smaller than (e.g., the uniformity coefficient, C_u, in Bureau of Reclamation terminology, is the D_{60}/D_{10} size ratio; in most other literature, it is referred to as the D_{40}/D_{90} size ratio).

The terminology used in most ground-water literature is used in the following discussion, and all references to an MA plot refer to the percent retained values on the right side of Bureau of Reclamation form 7-1415 (figure 7-10). The criteria for (1) and (2) below generally have been taken from Kruse 1960, but have been modified to conform to Bureau of Reclamation field experience. Criteria for (3) have been taken from Driscoll (1986).

(1) Where the uniformity coefficient, C_u, of aquifer material is less than 2.5:

    (a) Gravel pack material with C_u between 1 and 2.5 is preferable, and the 50-percent size should be a maximum of 6 times the 50-percent size of the aquifer material.

    (b) If uniform pack material is not readily available, use of gravel pack material with a C_u between 2.5 and 5 is acceptable. Select the gravel pack to have a 50-percent size not greater than 9 times the 50-percent size of the aquifer.

    (c) Normally, the screen slots should not pass more than 10 percent of the pack material.

(2) Where the uniformity coefficient, C_u, of aquifer material is between 2.5 and 5:

    (a) Gravel pack material with C_u between 1 and 2.5 is preferable, and the 50-percent size of pack material should be not more than 9 times the 50-percent size of the formation.

    (b) An acceptable but less desirable criterion is to use gravel pack material with C_u between 2.5 and 5, and the 50-percent size of the pack material should be not more than 12 times the 50-percent size of the formation.

    (c) The screen slots should not pass more than 10 percent of the pack material unless conditions permit.

Conditions 1a and 2a are the most efficient packs and most readily developed, but pack material with a low uniformity coefficient is sometimes not readily available and may be costly.

(3) Where the uniformity coefficient, C_u, of the formation is greater than 5:

  (a) Multiply the 70-percent retained size of the formation by 6 and 9 and locate the points on the graph.

  (b) Through these points, draw two parallel lines representing materials having uniformity coefficients of 2.5 or less.

  (c) Prepare specifications for gravel pack material falling between the two lines.

  (d) Select a screen slot size which will retain 90 percent or more of the pack material.

Regardless of the criteria used in selecting the gravel pack, the gravel should be washed, screened, rounded where possible, abrasive-resistant, dense, and of siliceous materials with less than 5-percent flat grains. The pack should contain not more than 5-percent earthy or soft materials such as clay, shale, or anhydrite, or readily soluble materials such as limestone or gypsum. Regulations usually require the pack material and any water used to be disinfected before being introduced in the well.

A mechanical analysis should be run on three random samples of the pack material. Each sample should be taken from a different part of the shipment to ensure conformance to the gravel pack gradation requirements. The gravel is acceptable if 95 percent passes the coarsest designated screen, plus or minus 8 percent of that designated is held on smaller screens, and not more than 10 percent passes the finest designated screen.

Bureau of Reclamation form 7-1415 (figure 3-10), although based on the U.S. Standard Series (fourth root of two ratio), does not include all the screen sizes available. Uniformity coefficients of materials plotting between two of the adjacent solid lines on form 7-1415 will be around 1.55 to 1.60 and between two alternate solid lines between about 2 and 2.8. At times, closer approximations may be desirable in selecting slot sizes and similar values. Table 11-10 lists the approximate volume of the annulus between various sizes of holes and casings.

Table 11-10.—Volume of annulus between casing or screen and hole for grout and gravel pack

Casing outside diameter		Hole diameter		Annulus volume		Casing outside diameter		Hole diameter		Annulus volume	
mm	in	mm	in	m³/m	ft³/ft	mm	in	mm	in	m³/m	ft³/ft
60	2-3/8	100	4	0.005	0.06	350	14	450	18	0.06	0.70
		150	6	0.01	0.17			500	20	0.10	1.11
		200	8	0.03	0.32			550	22	0.15	1.57
114	4-1/2	200	8	0.02	0.24			600	24	0.19	2.07
		250	10	0.04	0.43			650	26	0.24	2.62
		300	12	0.06	0.67			700	28	0.30	3.21
165	6-5/8	250	10	0.03	0.31	400	16	500	20	0.07	0.79
		300	12	0.05	0.55			550	22	0.12	1.24
		350	14	0.08	0.83			600	24	0.16	1.75
		400	16	0.11	1.16			650	26	0.21	2.29
		450	18	0.14	1.53			700	28	0.27	2.88
215	8-5/8	300	12	0.04	0.38			750	30	0.33	3.51
		350	14	0.06	0.66	450	18	550	22	0.08	0.87
		400	16	0.09	0.99			600	24	0.13	1.37
		450	18	0.13	1.36			650	26	0.18	1.92
		500	20	0.16	1.78			700	28	0.23	2.51
268	10-3/4	350	14	0.04	0.44			750	30	0.29	3.14
		400	16	0.07	0.77			800	32	0.35	3.82
		450	18	0.11	1.14	500	20	600	24	0.09	0.96
		500	20	0.14	1.55			650	26	0.14	1.50
		550	22	0.19	2.01			700	28	0.19	2.09
		600	24	0.23	2.51			750	30	0.25	2.73
320	12-3/4	400	16	0.05	0.51			800	32	0.32	3.40
		450	18	0.08	0.88			850	34	0.38	4.12
		500	20	0.12	1.29	550	22	650	26	0.10	1.05
		550	22	0.16	1.75			700	28	0.15	1.64
		600	24	0.21	2.25			750	30	0.21	2.27
		650	26	0.26	2.80	600	24	700	28	0.11	1.13
								750	30	0.16	1.77
								800	32	0.23	2.44

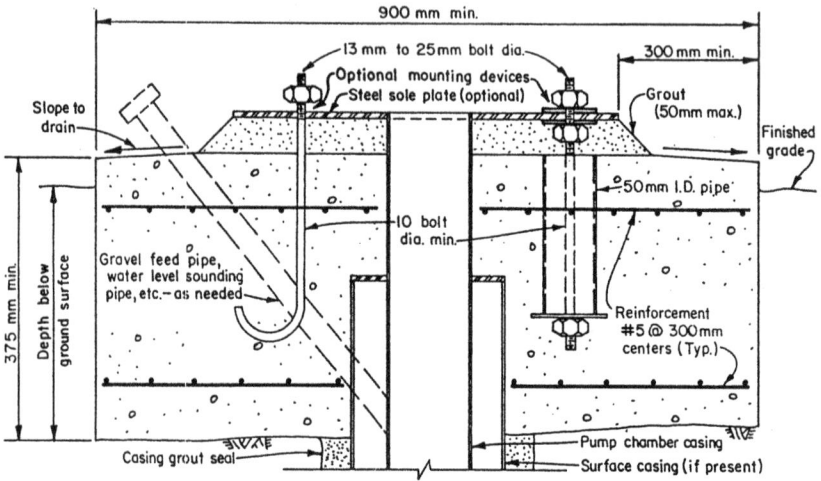

Figure 11-13(a).—Schematic section of a typical concrete pump foundation (millimeters).

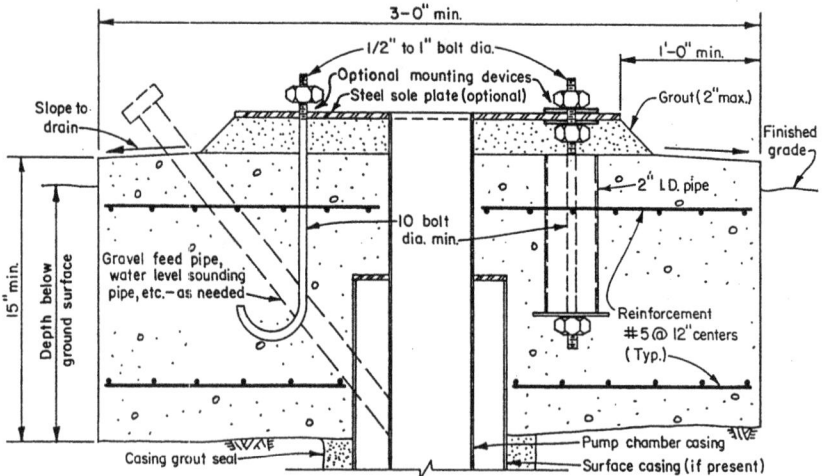

Figure 11-13(b).—Schematic section of a typical concrete pump foundation (inches).

11-14. Pump Foundations.—Surface-mounted pumps must be supported on foundations capable of resisting all thrust and torque loads. Supporting pumps by mounting them directly onto well casings is not recommended. Foundations generally should be constructed of minimum 25,900-kilopascal (3,750-pounds-per-square-inch) concrete placed on solid ground. A typical schematic section of a pump foundation is shown on figure 11-13. The sleeve shown on the right mounting bolt option is meant to provide flexibility of the bolt because the holes in the sole plate seldom line up exactly with the bolts. The sleeve can also be used with the "J" bolt option shown on the left.

Steel reinforcement of the concrete foundation is recommended. The pump should be secured to the foundation with anchor bolts. Setting plans indicating head base dimensions are available from pump manufacturers. Where necessary, a pump foundation pedestal should be constructed to raise the pump base above the elevation of any probable flood waters or surface runoff which might inundate the area. If a foundation pad is constructed about the pedestal, it should be an integral part of the pedestal, and its surface should slope gently away from the pedestal so water will not accumulate around it. Pump pits are restricted in most States, especially for domestic or municipal water supplies, and are not recommended for Bureau of Reclamation installations. If protection of the pump is required, a surface pumphouse is preferred to a pump pit.

11-15. Special Well Types.—

(a) Drainage Wells.—Drainage wells are usually conventional ground-water wells designed for the special purpose of relieving and controlling a high water-table. For drainage purposes, wells are designed to prevent the water-table from encroaching within a certain depth below the land surface and are deliberately located to interfere with each other to accomplish this purpose.

For an area to be susceptible to drainage by wells, a suitable aquifer must be present and the soil lying between the root zone and the aquifer must have adequate vertical permeability to permit deep percolation.

Little basic difference exists in the design of a well for drainage and the design of a well for the production of water. However, a distinct difference exists in the criteria used in the design. In drainage wells, the basic purpose is to lower and maintain the water-level to a given depth within a given period of time. A given volume of water must be removed to accomplish this purpose. Three factors—volume, drawdown, and time—are the parameters that, along with the aquifer characteristics, will give the most economical installation from the standpoint of initial and operational costs.

(b) *Inverted Wells.*—A special type of well, called an inverted, recharge, or injection well, is used to return surplus or unwanted surface-water to an aquifer. Such wells usually operate by gravity and have been used, where geological conditions are favorable, to dispose of irrigation waste. Other applications include injection of fresh water to build a barrier against intrusion of saltwater, disposal of industrial wastes and treated sewage effluent, and recharge of ground water.

The design and construction methods used for inverted wells that operate by gravity are similar to methods used for pumping wells. Where injection pressures are used, however, special design features may be necessary to control such pressures. Inverted wells may be designed with backwashing and flushing features to permit cleaning the wells if they become plugged by silt or other foreign matter being carried into them with the recharge water. When large volumes of surface-water must be recharged through inverted wells, settling ponds and filters usually are used to reduce the sediment load. Water quality standards may require the recharge water to be chlorinated or otherwise treated. Federal, State, and local regulations should be checked early in the planning process.

(c) *Pressure Relief Wells.*—Another special purpose well, called a pressure relief well, is used to reduce and control excessive artesian head. As the name implies, the purpose of this type of well is to reduce the pressure in artesian aquifers, thereby reducing the upward leakage of ground water through the overlying materials or reducing hydraulic uplift. Such wells have been used to drain agricultural lands and reduce pressure under engineering structures such as dams and powerplants and unstable earth masses such as landslides.

(d) *Collector Wells.*—In many areas, aquifers are too thin, contain poor quality water, or for other reasons cannot furnish water in the desired quantity or quality to a standard well. Under such circumstances, a collector well may be a solution. A collector well commonly consists of a concrete caisson 2 to 5 meters (6 to 15 feet) in diameter which is sunk to an adequate depth to directly intercept a thin aquifer or to permit horizontal screens to be extended radially into such an aquifer (Campbell and Lehr, 1973; Walton, 1970).

Drawdown resulting from such a well is spread over a relatively large area and is less than that which would result from a single well pumping the same volume. Also, temperature and quality of the water may be subject to a degree of control. The collecting-type well requires intensive local exploration and testing to determine conditions and data for design purposes. Construction, which may take 10 months to a year, requires special skill, knowledge, and equipment, and is usually expensive. However, many installations have been constructed and operated economically where conditions were such that other types of development were impracticable.

11-16. Purposes of Well Specifications and Available Standards.—Specification of water wells is a critical element in the procurement process and one which must be undertaken with great care. Clear and explicit requirements result in:

- Full and open competition leading to more favorable bids

- A sound baseline upon which contractors can develop their bids

- Sound standards by which to evaluate the contractors' performance

- Wells with the desired yield, discharge, and drawdown characteristics

The desired results can be accomplished if the specifications clearly define the requirements, including time, in language that lends itself to only one reasonable interpretation, and clearly state the means to determine quality and conformance to contract requirements prior to acceptance. As a minimum, each set of specifications must include a description of the finished product, any performance standards that the product must meet, the time allowed for construction, and basis for payment. Beyond these basic elements, each water well specification will be tailored to the needs to be filled and the source aquifer.

Two basic formats for contracts may be used for well construction. One is where a pilot hole has been drilled, a log and mechanical analysis of samples are available, and a firm design can be made. From this information, the amount and nature of drilling, materials, and related items can be specified. Lump-sum bids can be requested for definite work and unit price bids requested for the remainder of the work. Where such specifications can be prepared, bids are usually lower and more realistic because the contractor knows the drilling condition that will be encountered, the equipment needed, and the amount of materials to order. The amount of risk involved in well drilling, which is usually reflected in the bids, may be measurably reduced by providing the bidders with definitive information.

The second format is where the contractor must either drill and log a pilot hole and design the well from that information, or drill the well and base the design of the screen, casing, and pump components on that information. In this case the bids will usually reflect the uncertainties that exist as well as anticipated standby time while decisions are made and approved.

Each set of specifications must meet or exceed all applicable Federal, State, and local codes and standards. A few years ago, good judgment and a little care would "get you through." Now, a multitude of regulations have been designed to protect both quantity and quality of ground water, and the list grows almost daily. These regulations are intended to protect the natural resources from depletion and pollution. Table 11-11 provides current information as of this writing, but the designer should independently verify the information case by case. The American Water Well Association and Environmental Protection Agency are also primary sources of information.

The following paragraphs will provide some guidance in the preparation of specifications, but the designer must use independent judgment to arrive at the desired conclusion. A sample specification is included in the "Well Construction" module of the Bureau of Reclamation's *Comprehensive Construction Training Program*.

(a) *Materials*.—Well specifications should provide that the completed well will be constructed of material which will be compatible with the environment and which will give an adequate well life. The qualities of common well construction materials, except for water and concrete, were discussed earlier in this

Table 11-11.—Summary of State ground-water monitoring well construction requirements

State	Licensing/ registration required	Construction standards/ guidelines	Permit/ notice required	Reports required	Agency contact
Alabama	No	Yes	No	Yes	Ground Water Section Dept. of Environmental Management 1751 Federal Drive Montgomery AL 36130 (205) 271-7832
Alaska	No	No	No	No	Ground Water Section Dept. of Environmental Conservation PO Box 0 Juneau AK 99811-1800 (907) 465-2653
Arizona	Yes	Yes	Yes	Yes	Dept. of Water Resources 15 S. 15th Ave. Phoenix AZ 85007 (602) 542-1581
Arkansas	Yes	Yes	No	Yes	Arkansas Water Well Commission One Capitol Mall, Ste. 2C Little Rock AR 72201 (501) 682-1025
California	Yes	Yes	Yes	Yes	Dept. of Water Resources PO Box 942836 Sacramento CA 94236-0001 (916) 327-1641
Colorado	Yes	Yes	Yes	Yes	Div. of Water Resources 1313 Sherman St. Denver CO 80203 (303) 866-3581
Connecticut	No	Yes	Yes	Yes	Dept. of Environmental Protection Hazardous Waste Section 165 Capitol Ave. Hartford CT 06106 (203) 566-1848
Delaware	Yes	Yes	Yes	Yes	Dept. of Natural Resources and Environmental Control Div. of Water Resources PO Box 1401 Dover DE 19903 (302) 736-3665
Florida	Yes	Yes	Yes	Yes	Dept. of Environmental Regulation Twin Towers Office Bldg. 2600 Blairstone Rd. Tallahassee FL 32399-2400
Georgia	No	Yes	No	Yes	Environmental Protection Div. Georgia Geologic Survey 19 Martin Luther King Jr. Dr., S.W. Atlanta GA 30334 (404) 656-3214

Table 11-11.—Summary of State ground-water monitoring well construction requirements

State	Licensing/ registration required	Construction standards/ guidelines	Permit/ notice required	Reports required	Agency contact
Hawaii	Yes	No	Yes	Yes	Dept. of Commerce and Consumer Affairs 1010 Richard St. Honolulu HI 96813 (808) 548-7637
Idaho	Yes	Yes	Yes	Yes	Dept. of Water Resources Ground Water Protection Section 351 N. Orchard Boise ID 83704 (208) 327-7900
Illinois	Yes	No	Yes	Yes	Dept. of Public Health 535 W. Jefferson Springfield IL 62761 (217) 785-4306
Indiana	Yes	Yes	No	Yes	Dept. of Natural Resources Division of Water 2475 Director's Row Indianapolis IN 46241 (317) 232-4176
Iowa	Yes	No	Yes	Yes	Dept. of Natural Resources Wallace Bldg. Des Moines IA 50309 (515) 281-8693
Kansas	Yes	Yes	Yes	Yes	Dept. of Health and Env. Bureau of Water Protection Forbes Field Topeka KS 66620 (913) 296-1500
Kentucky	Yes	No	Yes	Yes	Dept. of Surface Mining Reclamation and Enforcement Capital Plaza Tower 18 Reilly Rd. Frankfort KY 40601
Louisiana	Yes	Yes	No	Yes	Dept. of Transportation and Development Office of Public Works PO Box 94245 Baton Rouge LA 70804-9245 (504) 379-1434
Maine	No	Yes	Yes	Yes	Dept. of Environmental Protection Statehouse Station 17 Augusta ME 04333 (207) 289-2651
Maryland	Yes	Yes	Yes	Yes	Board of Water Well Drillers Dept. of Env. 2500 Broening Hwy. Baltimore MD 21224 (301) 631-3168

Table 11-11.—Summary of State ground-water monitoring well construction requirements

State	Licensing/ registration required	Construction standards/ guidelines	Permit/ notice required	Reports required	Agency contact
Massachusetts	Yes	Yes	No	Yes	Div. of Water Resources 100 Cambridge St. Boston MA 02202 (617) 727-3267
Michigan	No	Yes	Yes	Yes	Dept. of Public Health Ground Water Quality Control Section PO Box 30035 Lansing MI 48909 (517) 335-8300
Minnesota	Yes	Yes	No	Yes	Dept. of Health Div. of Environmental Health 717 Delaware St., S.E. Minneapolis MN 55440 (612) 623-5339
Mississippi	Yes	Yes	No	Yes	Bureau of Land and Water Resources Dept. of Natural Resources PO Box 10631 Jackson MS 39209 (601) 961-5200
Missouri	No	No	No	No	Dept. of Natural Resources Div. of Geology and Land Survey PO Box 250 Rolia MO 65401 (314) 364-1752
Montana	Yes	Yes	No	Yes	Board of Water Well Contractors 1520 E. Sixth Ave. Helena MT 59620 (406) 444-6643
Nebraska	Yes	Yes	No	Yes	Dept. of Health Div. of Environmental Health/Housing Surveillance Box 95007, State Office Bldg. Lincoln NE 68509 (402) 471-2541
Nevada	Yes	Yes	Yes	Yes	Div. of Water Resources 201 S. Fall St. Carson City NV 89710 (702) 885-4380
New Hampshire	Yes	No	No	Yes	NH Water Well Board PO Box 208 Concord NH 03301 (603) 271-3406
New Jersey	Yes	Yes	Yes	Yes	Dept. of Environmental Protection Div. of Water Resources Bureau of Water Allocation CN-029 Trenton NJ 08625 (609) 984-6831

Table 11-11.—Summary of State ground-water monitoring well construction requirements

State	Licensing/ registration required	Construction standards/ guidelines	Permit/ notice required	Reports required	Agency contact
New Mexico	Yes	Yes	Yes	Yes	State Engineer's Office Water Rights Div. Bataan Memorial Bldg. Santa Fe NM 87503 (505) 827-6120
New York	No	Yes	No	No	Dept. of Environmental Conservation Bureau of Municipal Waste 50 Wolf Rd. Albany NY 12233 (518) 457-2051
North Carolina	Yes	Yes	Yes	Yes	Div. of Environmental Management Groundwater Section PO Box 27687 Raleigh NC 27611 (919) 733-5083
North Dakota	Yes	Yes	No	Yes	State Board of Water Well Contractors 900 E. Blvd. Bismarck ND 58505-0187 (701) 224-2754
Ohio	No	Yes	Yes	Yes	Ohio EPA Div. of Ground Water 1800 WaterMark Columbus OH 43266
Oklahoma	Yes	Yes	No	Yes	Water Resources Board Ground Water Division PO Box 53585 1000 N.E. 10th Oklahoma City OK 73152 (405) 271-2516
Oregon	Yes	Yes	Yes	Yes	Water Resources Dept. 3850 Portland Rd., N.E. Salem OR 97310 (503) 378-8456
Pennsylvania	Yes	Yes	Yes	Yes	Dept. of Environmental Resources Bureau of Topo. and Geo. Survey PO Box 2357 Harrisburg PA 17120 (717) 787-5828
Rhode Island	No	Yes	No	Yes	Dept. of Environmental Management Air and Hazardous Materials Div. 75 Davis St. Providence RI 02908 (401) 277-2797
South Carolina	Yes	Yes	Yes	Yes	Board of Certification of Env. System Operators 2221 Devine St., Ste. 320 Columbia SC 29205 (803) 734-9140

Table 11-11.—Summary of State ground-water monitoring well construction requirements

State	Licensing/ registration required	Construction standards/ guidelines	Permit/ notice required	Reports required	Agency contact
South Dakota	Yes	Yes	No	Yes	Dept. of Water and Natural Resources Water Rights Div. Joe Foss Bldg. Pierre SD 57501 (605) 773-3352
Tennessee	No	No	No	No	Ground Water Protection Div. Dept. of Health and Environment TERRA Bldg., 5th Fl. 150 9th Ave., N. Nashville TN 37219-5404 (615) 741-0690
Texas	Yes	Yes	No	Yes	Water Well Drillers Board PO Box 13087 Austin TX 78711 (512) 463-7999
Utah	Yes	Yes	Yes	Yes	Div. of Water Rights 1636 W. North Temple Salt Lake City UT 84116 (801) 538-7242
Vermont	No	Yes	No	Yes	Dept. of Water Resources and Environmental Engineering Water Quality Div. 103 S. Main St. Waterbury VT 05676 (802) 244-5638
Virginia	No	Yes	No	No	Div. of Technical Services Dept. of Health 1100 Monroe Bldg. 101 N. 14th St. Richmond VA 23219 (804) 786-1750
Washington	Yes	Yes	No	Yes	Dept. of Ecology Mail Stop PV-11 Olympia WA 98504 (206) 459-6045
West Virginia	Yes	Yes	Yes	No	Office of Env. Health (804) 558-2981
Wisconsin	No	Yes	No	Yes	Bureau of Solid and Hazardous Waste Management Box 7921 Madison WI 53707
Wyoming	No	Yes	Yes	Yes	State Engineers Office Herschler Bldg. 122 W. 25th Cheyenne WY 82002 (307) 777-7354

chapter. What is important here is to ensure that the specifications adequately convey to the prospective bidders exactly which materials will be acceptable and why. Local contractors often are accustomed to designing the wells they construct and they may not be aware of the special requirements of the job. For instance, if stainless steel screen is required for longevity, a contractor who does not normally use stainless steel may not realize that a black iron casing will be sacrificially destroyed by corrosion if connected directly to the screen.

Water used to prepare sealing mixtures should generally be of drinking water quality, compatible with the type of sealing material used, free of petroleum and petroleum products, and free of suspended matter. In some cases, water considered nonpotable, with a maximum of 2,000 mg/L chloride and 1,500 mg/L sulfate, can be used for cement-based sealing mixtures. The quality of water to be used for sealing mixtures shall be determined where unknown.

Cement used in sealing mixtures shall meet the requirements of American Society for Testing and Materials C150, *Standard Specification for Portland Cement,* including the latest revision thereof. Types of Portland cement available under ASTM C150 for general construction are:

Type I -
General purpose. Similar to American Petroleum Institute Class A.

Type II-
Moderate resistance to sulfate. Lower heat of hydration than Type I. Similar to API Class B.

Type III -
High early strength. Reduced curing time but higher heat of hydration than Type I. Similar to API Class C.

Type IV -
Extended setting time. Lower heat of hydration than Types I and III.

Type V -
High sulfate resistance.

Special cement setting accelerators and retardants and other additives may be used in some cases. Special field additives for Portland cement mixtures shall meet the requirements of ASTM C494, *Standard Specification for Chemical Admixtures for Concrete,* and latest revision thereof.

(b) *Methods of Construction.*—The method of construction is usually best left to the contractor. Each contractor has certain methods which have worked well; requiring a change will almost certainly raise the bid. It may also invoke protests, thereby causing delays and possibly award to a lesser qualified bidder. Unless definable reasons exist for using a certain method (i.e., possible contamination of an aquifer or the need for a specific logging procedure), the added costs usually cannot be justified. When a method of construction is specified, it must be clear as to what is expected, it must be compatible with any common procedures which will be used on other work items, and reasons should be stated for the restrictions. At times, specified methods may be unavoidable or even desirable, but they should be used with discretion.

(c) *Time of Performance.*—The time of performance should be as flexible as possible and still meet the needs of the project. Well construction is a high risk process because of the infinite variability of geologic formations. Weather conditions may also be a factor. If the bidders feel that adequate time is available to cover hidden contingencies, the bids will usually be more favorable than if time is unduly limited. Time limits are necessary and appropriate for any contract, but for best results they must not be arbitrarily short.

(d) *Payment Process.*—The method of measuring progress for payment items must be clearly stated and fair to both parties. Some items can be easily measured. The length of casing or screen used is very apparent. Some items are ambiguous, such as the cost of standby time. The specifications should be very clear as to how each item will be paid. If it is clear, the bidders can account for it in any way they wish; if not, claims, delays, and added costs will result.

11-17. Bibliography.—

Ahrens, T.P., 1970, "Basic Considerations of Well Design, "*Water Well Journal*, vol. 24, No. 4, 5, 6, and 8.

American Petroleum Institute, March 1963, "Specification for Line Pipe," API Std. 5L, American Petroleum Institute Division of Production, Dallas, Texas.

American Society of Agricultural Engineers, December 1964, *Designing and Constructing Water Wells for Irrigation*, St. Joseph, Missouri.

American Water Works Association, 1990, *AWWA Standard for Water Wells*, AWWA-A100-90, New York.

Associated Drilling Contractors of California, 1960, *Recommended Standards for Preparation of Water Well Construction Specifications*, Sacramento.

Buyalski, Clark P., 1986, Gravel Pack Thickness for Ground-Water Wells Report No. 1, Bureau of Reclamation, REC-ERC-86-7.

Campbell, M.D. and J.H. Lehr, 1973, *Water Well Technology*, McGraw-Hill, New York.

Driscoll, F.G. (principal author and editor), 1986, *Groundwater and Wells*, 2nd edition, Johnson Division, St. Paul, Minnesota.

Grange, J.W., and E. Lund, May 1969, "Quick Culturing and Control of Iron Bacteria," Journal of the American Water Works Association, vol. 6, No. 5, pp. 242-245.

Kuhlman, F.W., November 1959, "Corrosion of Iron in Aqueous Media," Canadian Mining and Metallurgical Bulletin No. 52, pp. 713-729.

Kruse, E.G., March 1960, "Selection of Gravel Packs for Wells in Unconsolidated Aquifers," Colorado State University, Agricultural Experiment Station, Technical Bulletin No. 66, Fort Collins.

Moehrl, K.E., February 1961, "Corrosion Attack in Water Wells," *Corrosion*, vol. 17, No. 2, pp. 26-27.

Oregon Drilling Association, 1968, *Manual of Water Well Construction Practices for the State of Oregon*, 2nd edition, Salem, Oregon.

Pennington, W.A., March 9, 1965, "Corrosion of Some Ferrous Metals in Soil with Emphasis on Mild Steel and on Gray and Ductile Cast Irons," Bureau of Reclamation, Chemical Engineering Branch Report No. ChE-26.

Reinke, J.W. and D.L. Kill, 1970, "Modern Design Techniques for Efficient High Capacity Irrigation Wells," Paper No. 70-732, presented at American Society of Agricultural Engineers Winter Meeting.

Ryznar, J.W., April 1944, "A New Index for Determining the Amount of Calcium Carbonate Scale Formed by Water," Journal of the American Water Works Association.

State of California, 1968, "Water Well Standards—State of California," California Department of Water Resources Bulletin 74, Sacramento.

State of California, July 1968, "Idaho Minimum Well Construction Standards," Idaho Department of Reclamation, Boise.

State of Colorado, 1967, "Rules and Regulations and Water Well Drilling Pump Installation Contractors' Law," Colorado Division of Water Resources, Denver.

State of Maryland, 1969, "Water Resources Regulation 2.3, Well Drillers and Well Construction," Maryland Water Resources Commission and Department of Water Resources, Annapolis.

State of Michigan, 1966, "Ground-Water Quality Control," Michigan Department of Public Health, Act 294PA 1965 and Rules, Lansing.

University of Nebraska, August 1957, "Nebraska Minimum Standards for Gravel Packed Irrigation Wells," University of Nebraska Extension Service, College of Agriculture, E.C. 57-702, Lincoln.

State of Nevada, 1969, "Rules and Regulations for Drilling Wells and Other Related Materials," Nevada Department of Conservation and Natural Resources, Division of Water Resources, Carson City.

State of New Mexico, 1966, "Rules and Regulations Governing Drilling of Wells and Appropriation and Use of Ground-Water in New Mexico," New Mexico State Engineer, Santa Fe.

State of Ohio, October 1946, "Ohio Water Well Construction Code," Ohio Water Resources Board, Columbus.

State of Oregon, 1972, "Rules and Regulations Prescribing General Standards for the Construction and Maintenance of Water Wells in Oregon," Oregon State Engineer, (Preliminary Draft Subject to Revision), Salem.

State of Pennsylvania, 1961, "Construction Standards Individual Water Supplies," Pennsylvania Department of Health, Harrisburg.

State of Wisconsin, 1972, "Well Construction and Pump Installation," Wisconsin Administrative Code, Chapter NRI12, Wisconsin Department of Natural Resources, Madison.

State of Wyoming, 1971, "Regulations and Instructions, Water Well Minimum Construction Standards," Wyoming State Engineer, Cheyenne.

U.S. Department of Health, Education, and Welfare, 1965, "Recommended State Legislation and Regulations, Urban Water Supply, Water Well Construction and Individual Sewage," Public Health Service, Washington, DC.

Vertical Turbine Pump Association, 1962, *Turbine Pump Facts*, Pasadena, California.

Walton, W.C., 1970, *Ground Water Resources Evaluation*, McGraw-Hill, New York.

Chapter XII

WATER WELL DRILLING AND DEVELOPMENT

12-1. Introduction.—Most wells are drilled by mechanically powered equipment normally referred to as drill rigs. This chapter is intended to acquaint the reader with the major types of drill rigs and the capabilities and limitations of such rigs. The well drilling methodology used should fit the subsurface conditions, as well as the desired diameter and depth of the well. Well development, which stimulates the completed well and increases its production, is also discussed. It involves surging the water up and down and bailing the well to remove drilling muds and stabilize the gravel pack and aquifer. Finally, the well is sterilized to prevent corrosion and inhibit organic organisms. A more detailed description of well sterilization is included.

12-2. Drilling and Sampling with Cable Tool Rigs and Variations.—

(a) Drilling Methods.—The cable tool method of drilling, often referred to as the standard method, churn drill, percussion method, or facetiously referred to as the yo-yo, is one of the oldest, most versatile, and simplest drilling devices (Gordon, 1958).

The cable tool drills by lifting and dropping a string of tools suspended on a cable. A bit at the bottom of the tool string strikes the bottom of the hole, crushing, breaking, and mixing the cuttings. A string of tools in ascending order consists of a bit, a drill stem, jars, and a swivel socket, which is attached to the cable. Cuttings are removed from the hole with a bailer or a scow.

In stable rock, an open hole can be drilled, but in unconsolidated or raveling formations, casing must be driven down the hole during the drilling. Above the water table or in otherwise dry formations, water is added to the hole to form a slurry of the cuttings so they may be readily removed by a bailer. The bottom of the casing is usually fitted with a drive shoe to protect the casing during driving.

In some unconsolidated formations, casing can be sunk by merely bailing and driving so that samples are relatively unbroken and representative.

As casing is driven in unconsolidated formations, the vibration causes the sides of the hole to collapse against the casing and compact. Frictional forces increase until it is no longer possible to drive the casing. When this occurs, a smaller diameter casing is telescoped inside of the casing already in place, and drilling is continued using a smaller diameter bit. On deep holes, as many as four or five reductions in casing may be required.

Cable tool rigs are generally limited to drilling maximum hole diameters of 600 to 750 millimeters (24 to 30 inches) and to depths of less than 600 meters (2,000 feet).

The cable tool rig is probably the most versatile of all rigs in its ability to drill satisfactorily under a wide range of conditions. Its major drawback, compared to some other rigs, is its slower rate of progress and depth limitations.

The initial cost of a cable tool rig complete with tools is one-half to two-thirds that of a rotary rig of equivalent capacity. The rigs are usually compact, require less accessory equipment than other types, and are more readily moved in rugged terrain. The simplicity of design, ruggedness, and ease of maintenance and repair of the rigs and tools are particularly advantageous in isolated areas. They generally require less skilled operators and a smaller crew than other rigs of similar capacity. The low horsepower requirements are reflected in lower fuel consumption, an important aspect where fuel costs are high or sources of fuel are remote.

Although slower than other rigs in drilling some formations, cable tool rigs can usually drill through boulders and fractured, fissured, broken, or cavernous rocks which often are beyond the capabilities of other rigs. In addition, much less water is required for drilling than with most other commonly used rigs, an important consideration in arid and semiarid zones. Also, sampling and formation logging are simpler and more accurate with the cable tool rig. The cuttings bailed from each drilled interval usually represent about a 1.5-meter (5-foot) zone. When casing is used, there is little chance of contamination of the sample. A skilled driller can usually recognize a change in formation by the response of the rig to the changed drilling condition and can then take samples at a shorter interval.

The samples are not greatly contaminated by drilling mud and clay; shale and silt fractions are less likely to be lost by dispersion in the drilling fluid. Cuttings of unconsolidated formations are usually not finely pulverized, and cuttings are usually of sufficient size to permit ready identification and description. A more reliable method of sampling involves the use of a drive barrel sampler driven by drilling jars. This method provides a means of obtaining representative to undisturbed samples from moderate to great depths. When promising aquifer materials are encountered, they are readily tested for yield and quality of water by bailing or, if of sufficient importance, by pumping.

The disadvantages of a relatively slow rate of progress and the economical and physical limitations on depth and diameter have been mentioned previously. A further disadvantage of the cable tool rig is the necessity of driving casing coincident with drilling in unconsolidated materials. This requirement precludes the use of electric logs, which are desirable in many instances. Gamma logs may be taken inside a casing although the practice is not recommended. The driving of casing necessitates a heavier wall pipe than would otherwise be required in some installations. Also, screens often must be set by pullback or bail-down methods. The pullback method in deep or large diameter wells is sometimes extremely difficult, and the bail-down method may cause alignment problems and cannot be used in hard bedrock.

Mud scow drilling (Driscoll, 1986; National Water Well Association, 1971; and Campbell and Lehr, 1973) uses a heavy scow in place of a bit for drilling large diameter holes in gravel and finer materials. The scow is a heavy pipe, 3 meters (10 feet) or more in length, fitted at the bottom with a heavy shoe similar to a drive shoe and generally fitted with a heavy steel knife blade welded across the diameter of the shoe. The scow is often fitted with a flapper valve to create a "suction" on the upstroke to move formation material into suspension for easy bailing. Typically, bailing is performed with the scow. When drilling with a mud scow, the casing used is usually California double-walled stovepipe (see section 10-3) in 1- to 1.5-meter (3- to 5-foot) lengths which are jacked down as drilling proceeds. The method is used mostly in the Southwestern part of the United States.

The cable tool rig is readily adapted to drilling 50- to 100-millimeter (2- to 4-inch) diameter holes with jet or hollow-rod tools in soft formations such as clay or sand. Jet drilling (Speedstar Division, 1967; Bennison, 1947; Gibson and Singer, 1969; National Water Well Association, 1971; Campbell and Lehr, 1973) is basically a percussion method combined with a pressure pump.

The drill pipe is lifted and dropped, which chops up material at the bottom of the hole. The water helps to jet the broken material loose and carries the cuttings up the hole where they are discharged into the pit. The method is useful in installing observation holes and small capacity water wells.

Both the hollow-rod and jetting methods have the disadvantages of the direct circulation rotary rig discussed below in regard to sampling of formations and water and the measurement of static water levels.

(b) Sampling.—The accessories and equipment for cable tool rigs are fairly uniform and standard, although the sampling procedures of different drillers are variable. To ensure obtaining good samples meeting a minimum standard, Bureau of Reclamation specifications usually require that they be taken at each 1- to 1.5-meter (3- to 5-foot) interval or at each change in material, whichever is less. To ensure reliable sampling, Bureau of Reclamation requires samples to be deposited in a sample box. One such box contains four separate sample compartments. Each sample batch is mixed and quartered until a 2-liter (2-quart) representative sample remains. This sample is placed in two separate l-liter (1-quart) containers, each marked with the well designation, the date, and the drilled interval it represents. After the sample is obtained, the compartment is thoroughly cleaned and flushed before another sample is placed in it. A typical sample box is shown on figure 12-1.

12-3. Drilling and Sampling with Direct Circulation Rotary Rigs and Variations.—

(a) Drilling Methods.—The rotary rig drills by turning a fishtail, toothed cone, or similar bit at the bottom of a string of drill pipe. The typical string consists of a bit which scrapes, grinds, fractures, or otherwise breaks the formation drilled; a drill collar of heavy-walled pipe which adds weight to the bit and helps to maintain a straight hole; and a drill pipe which extends to a kelly (shaft) near the surface which imparts rotation. As the bit is turned, drilling fluid (mud) is pumped down the pipe to lubricate and cool the bit, jet material from the bottom of the hole, and to clean the hole by transporting the cuttings to the surface in the annular space between the hole wall and the drill pipe. The drilling fluid also forms a thin layer of mud on the wall of the hole which reduces seepage losses and, together with the hydrostatic head exerted by the mud column, holds the hole open (Speedstar Division, 1967;

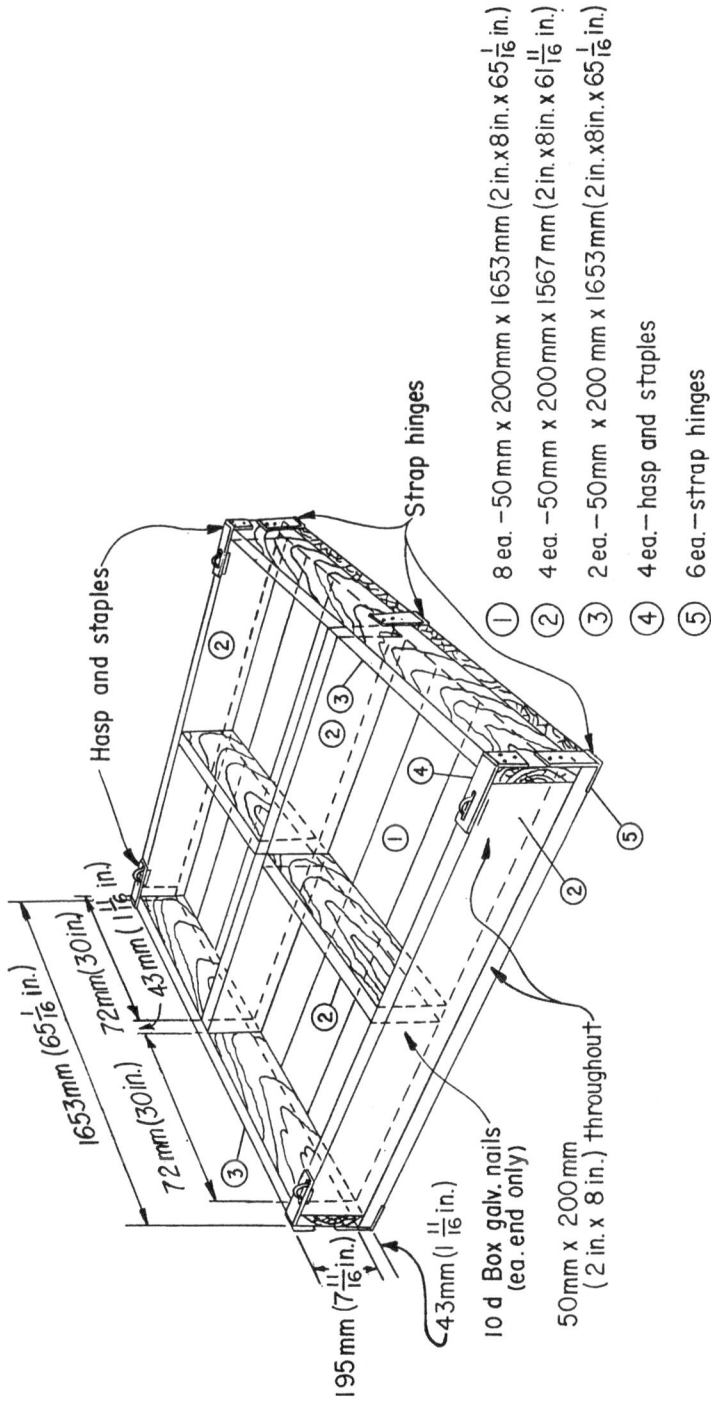

195mm (7 11/16 in.)

43mm (1 11/16 in.)

1653mm (65 1/16 in.)

72mm (30 in.)

72mm (30 in.)

43mm (1 11/16 in.)

Hasp and staples

Strap hinges

10 d Box galv. nails (ea. end only)

50mm x 200mm (2 in. x 8 in.) throughout

① 8 ea. −50mm x 200mm x 1653mm (2 in. x 8 in. x 65 1/16 in.)

② 4 ea. −50mm x 200mm x 1567mm (2 in. x 8 in. x 61 11/16 in.)

③ 2 ea. −50mm x 200mm x 1653mm (2 in. x 8 in. x 65 1/16 in.)

④ 4 ea. −hasp and staples

⑤ 6 ea. −strap hinges

Figure 12-1.—Compartmented cuttings sample box for use with a cable tool drill.

U.S. Department of the Army, 1965; Driscoll, 1986; Bennison, 1947; Gibson and Singer, 1969; National Water Well Association, 1971; and Campbell and Lehr, 1973).

The selection of the correct mud and the maintenance of mud weight, viscosity, jelling strength, and a low percentage of suspended solids, together with a suitable uphole velocity, contribute to rapid, trouble-free drilling. Numerous drilling fluids are used, but for water wells, a suspension of bentonite or similar clays in water is most commonly used.

The cost of a rotary rig is considerably greater than that of a cable tool rig of equal capacity. Operation of the rotary rig requires much more training and skill than a cable tool rig and requires a larger crew. Maintenance and repair are more complex, and more water is required for drilling when using drilling fluids. Also, if the permeability of the aquifer is much greater than about 15 meters per day (50 feet per day), mud may invade the aquifer and jell at some distance from the wall of the hole. Although chemicals are available to break down the mud, full development of a water well in which mud has invaded the aquifer is frequently impossible.

Clayey materials are frequently mixed and incorporated in the drilling fluid and may not be recognized. Cuttings are usually fine, and because of a variable rate of travel in the mud stream, become mixed and separated and are not always representative. Because of the mud-filled hole and dispersal of cuttings in the return flow, possible aquifer materials may be overlooked in the drilling. Static water levels, water samples, and pumping tests are not readily obtained from aquifers without special equipment. An electric log is usually desirable in conjunction with the driller's formation log when interpreting the results of rotary drilling for well design purposes.

Despite the above disadvantages, the direct circulation rotary rig offers relatively rapid drilling in most formations, greater depth capacities, and an open hole which simplifies installation of casing, screen, and grout and permits the use of most geophysical well-logging equipment.

Although plain water is often used as a drilling fluid, the fluid usually consists of a suspension of native clay, bentonite, or organic thickeners in water. Native clays are seldom a satisfactory mud base. Bentonite is far more effective and efficient, particularly when the better grades are used. The desirable properties obtainable with bentonite are high viscosity, jelling strength, and a relatively low solids content in the mud. Organic bases have little or no jell strength but excellent viscous properties. They degrade with time instead of jelling and permit more rapid and thorough development of a well. For water wells, mud weight is usually 1 to 1.1 kilograms per liter (9 to 9-1/2 pounds per gallons), and the viscosity is 32 to 36 seconds from a marsh funnel. The sand content of the drilling fluid, where it is picked up by the mud pump, should not exceed about 2 percent. Drilling fluid should be tested at about 4-hour intervals for weight and viscosity. If it cannot be treated to obtain the desired properties, the old mud should be discarded and a new batch should be mixed. The use of bentonite or biodegradable drilling additives may be prohibited by State law.

Where a pilot hole has been drilled and suitable aquifers found, the pilot hole is normally reamed to the desired diameter.

(b) Sampling.—The method of sampling required by Bureau of Reclamation is to drill 1 to 1.5 meters (3 to 5 feet), raise the bit from the bottom of the hole, and continue circulation until all cuttings from the sample interval are cleared from the hole and caught in a sample catcher. Drilling is then resumed for another 1 to 1.5 meters (3 to 5 feet). The sample is then removed from the catcher and the sample catcher is thoroughly cleaned prior to drilling the next interval. The sample is mixed and quartered until about a 2-liter (2-quart) representative sample remains. This sample is placed in a pail or drum to which about 20 liters (5 gallons) of clear water are added, stirred, and permitted to settle for about 20 minutes. The muddy water is then decanted and the sample is placed in two 1-liter (1-qt) containers, each of which is clearly marked to show the well designation, the depth interval represented, and the date the sample was taken. Sampling by rotary methods using clay-based muds is not recommended as a basis for the design of wells in granular materials. An exception is drive sampling. Figure 12-2 shows a typical sample catcher. A convenient arrangement which saves considerable time is to use two sample catchers in parallel with a diversion gate at the end of the ditch, which permits the return flow to be diverted into either catcher. The cleaning of one catcher is then possible while the other is being used.

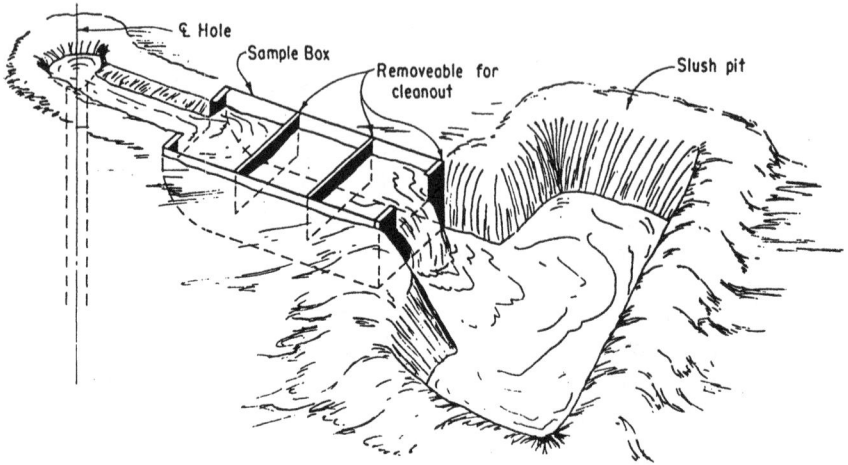

Figure 12-2.—Cuttings sample catcher for use
with a direct circulation rotary drill.

12-4. Air Rotary Drilling.—Air rotary drilling (Speedstar
Division, 1967; National Water Well Association, 1971; Campbell
and Lehr, 1973) was developed primarily in response to the need
for a rapid drilling technique in hard rock in arid areas. The rig,
bits, etc. are essentially the same as for direct circulation rotary
drilling except the fluid channels in the bit are of uniform diameter
rather than jets, and the mud pump is replaced by an air
compressor. Air is circulated down the drill string to cool the bit
and to blow the cuttings to the surface.

When initially developed, the air rotary method was used for
relatively small diameter holes in hard rock. Larger holes have
become possible through use of foams and oilier air additives, and
diameters up to about 200 millimeters (8 inches) have been drilled
successfully.

Sampling of cuttings is not adequate or practical for well design
because some delay occurs in bringing cuttings to the surface.
During this delay, considerable mixing of cuttings from various
depths occurs. The rigs are mostly applicable to hard rock terrains
where water is encountered in fractures or similar openings and
wells are completed as open holes.

Shortly after the development of air rotary drilling, the down-the-hole hammer bit was developed. This technique consists of a maximum 200-millimeter (8-inches) diameter bit working on the principle of the jackhammer, which replaces the conventional bit at the bottom of the drill string. This arrangement efficiently combines some of the advantages of the cable tool and rotary rig. The air used to activate the bit either blows the cuttings to the surface or lifts them by the principles of the airlift pump when drilling below the water table. The bit is particularly applicable to rapid drilling of hard rock.

Both rotary air and down-the-hole drilling in saturated materials are limited in depth by the available air pressure, which must be greater than that exerted by the column of water in the hole if the rig is to function.

12-5. Drilling and Sampling with Reverse Circulation Rotary Drills.—

(a) Drilling Methods.—The reverse circulation rotary rig operates essentially the same as a direct circulation rotary rig except that the water is pumped up through the drill pipe rather than down through it. Large capacity centrifugal or jet pumps similar to those used on gravel dredges are used. The discharge is directed into a large pit in which the cuttings settle out. The water then runs through a ditch and into the hole so that the water level in the hole is maintained at the ground surface (Speedstar Division, 1967; U.S. Department of the Army, 1965; Driscoll, 1986; National Water Well Association, 1971; Campbell and Lehr, 1973).

Velocity of water down the hole cannot exceed about a meter per minute to avoid erosion of the side of the hole at the restricted annulus around each flange joint. Consequently, the minimum hole diameter is about 400 millimeters (16 inches). The drag bits range in diameter from about 0.4 to 1.8 meters (16 to 72 inches). When boulders or cobbles too large to pass through the drill pipe are encountered, the bit is pulled from the hole and the larger rocks are removed with an orangepeel bucket. Recently, compound bits consisting of combinations of cones similar to the bits used in direct circulation rotary drilling and the use of drill collars have made drilling in rock more feasible.

The water velocity up the drill pipe is usually in excess of 120 meters per minute (400 feet per minute) and separation of cuttings is at a minimum. Samples caught are representative of the formation being drilled within depths of about 75 millimeters (a few inches) at most.

A differential head of 2.5 to 4 meters (8 to 13 feet) is required to maintain a stable hole. If the static water level is less than this range, an arrangement to increase the head must be devised.

The reverse circulation rig is probably the most rapid drilling equipment available for unconsolidated formations, but it requires a large volume of water which must be constantly replenished because drilling mud is seldom used.

Where the water table is in excess of 6 meters (20 feet) below ground surface, surface casing should be installed and grouted to minimize loss of water. The column of water in the hole acts to keep the hole open in a manner similar to the drilling fluid used in a direct circulation rotary rig. Because of the large hole diameters, reverse circulation drilled holes are usually gravel packed. The minimum hole diameter is 300 millimeters (12 inches).

The normally equipped reverse circulation rig can drill to a depth of about 135 meters (450 feet) at sea level. Deeper drilling is usually not possible because friction losses in the drill pipe and the weight of the cuttings-charged column of water become too great for the suction lift of the pump. However, by introducing air into the lower third of the drill pipe and using airlift pumping rather than centrifugal pumping, wells have been drilled to more than 360 meters (1,200 feet) at elevations of over 1500 meters (5000 feet) using this method.

(b) Sampling.--A common method of obtaining samples with a reverse circulation rig is to catch them either in a bucket or with a screen at the end of the discharge pipe. However, such samples are never representative.

The large volume of water discharging at a high velocity tends to wash fines over the edge of the bucket or through the screen mesh. To overcome this problem, the sampler shown on figure 12-3 was developed by the Cope Drilling Company of Idaho Falls, Idaho. This sampler permits the catching of representative samples without loss of fines.

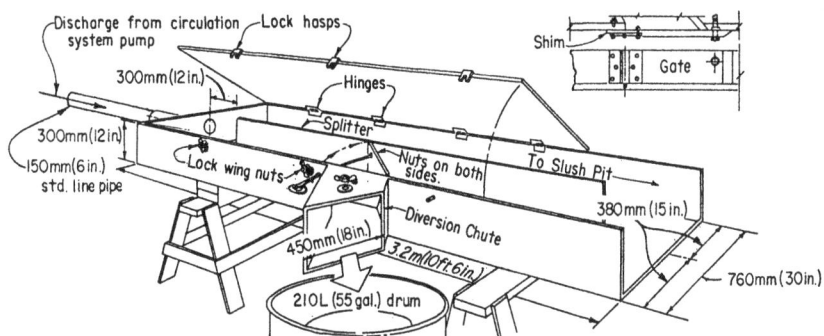

Original model was made of welded 5mm ($^3/_{16}$in) steel plate; however, if general measurements are followed, model could be made from wood. Discharge hose from Reverse Circulation Pump is connected to 150mm (6in) pipe so all materials go through sampler. Materials can be observed in open 600mm (24in) discharge without cover. Sampler is mounted on heavy sawhorses or similar supports with about a 150mm (6in) slope towards the slush pip into which material is discharged. To obtain sample, control gate is thrown open against splitter to divert sample through chute and into 210L (55gal) drum.

Figure 12-3.—Cope cuttings sample catcher for use
with a reverse circulation rotary drill.

Several drums are used so that samples may be caught at frequent intervals when drilling in a thick aquifer. Each sample is allowed to stand for about 10 minutes to permit the fines to settle to the bottom. The water is then decanted and the sample dumped on a clean plywood panel or similar surface, where it is mixed and quartered until a 2-liter (2-quart) representative sample remains. The sample is placed in two 1-liter (1-quart) containers, which are clearly marked with the well designation, the depth represented, and the date.

12-6. Other Drilling Methods.—Sonic or rotary-vibratory drilling is used as an alternative to direct or reverse rotary drilling. In sonic drilling, the drill head contains a mechanism to produce high frequency vibrations in the drill line. The drill bit is physically vibrating in addition to the rotation and downward pressure. This vibrating action causes unconsolidated materials to liquify. It also enhances drilling speeds in most consolidated materials by adding the force of high frequency vibration to normal rotary action. The drill head is vibrated between 40 and 120 cycles per second (Roussy, 1994) creating a resonance in the drill line that results in maximum vibration of the bit. The frequency of vibration is adjusted to match the length of the drill line so that resonance is achieved. The driller uses hydraulic drive pressure readings and actual drilling speed to judge when resonance is occurring. Sampling is the same as in all rotary drilling operations. Numerous other methods such as auger and chilled shot are used, but most of them have limited depth capacities, only special applications, small diameters, or are slow and costly.

12-7. Plumbness and Alignment Tests.—Each new well should be checked for conformance to the specifications regarding plumbness and alignment or straightness.

The measurements made are of the plumbness and straightness of the cased hole. Thus, an oversized hole may be out of line or plumb, but the casing may fall within the limits of the specifications. The casing should not be permitted to excessively encroach on the annulus and hinder placement of grout or gravel pack.

The usual standard for plumbness requires that the axis of the well casing not deviate from the vertical in excess of two-thirds the inside diameter of the casing per 30 meters (100 feet) of depth and that the deviation be reasonably consistent regarding direction. This requirement applies to both the casing and the screen (American Water Works Association, 1990).

The usual standard for alignment or straightness requires that a 12-meter (40-foot) long dolly can be passed freely through the pump housing casing without hanging. The dolly should be rigid and fitted with 0.3-meter (1-foot) wide rings which have a 12-millimeter (1/2-inch) smaller outside diameter than the inside diameter of the casing or screen being tested. The rings are placed at each end and in the center of the dolly (American Water Works Association, 1967).

The dolly is hung so it is centered at the top of the well with the supporting cable attached at the exact center of the dolly. The cable sheave or support should be adjusted and firmly fixed so that the cable is vertical between the support and top of the dolly. The dolly is then lowered in 1.5-meter (5-foot) increments, and the deviation of the cable from the center of the casing is measured for amount and direction at each 1.5-meter (5-foot) interval. During lowering, the cable should be watched to detect any deviation from the general direction of displacement or other sudden deflection.

The deviation of the well from the vertical at any depth can be computed by the equation:

$$X = \frac{D(H+h)}{h}$$

12-1

where:

X = well deviation at any given depth, millimeters (inches)

D = distance the cable departs from the center of the casing, millimeters (inches)

H = distance from the top of the casing to the top of the dolly, meters (feet)

h = distance from the suspension point of the cable to the top of the casing, meters (feet)

If the dolly passes freely through the casing and deviations from the vertical are within acceptable limits, the well is satisfactory. If trouble is encountered, it can be checked with a cage (see figure 12-4).

The cage should be at least 300 millimeters (1 foot) long and have a minimum outside diameter 12 millimeters (1/2 inch) smaller than the inside diameter of the casing. The cage is first set in the top of the casing and centered. Deviations of the casing from the vertical and the direction of deviation can be determined by measuring the distance and direction of movement of the cable from the center of the casing and applying equation 12-1. A special template, as shown on figure 12-5, can be used if desired to simplify measurement of the direction and amount of movement of the cable. The computed deviation can then be plotted on graph paper to determine conformance to the specifications or the location of any difficulty encountered.

A number of methods have been developed, including laser beam, gyroscope, single shot and multishot systems and magnetic systems to name a few. When properly used, each of these methods will produce accurate borehole surveys.

12-8. Well Development.—The primary purpose of well development or stimulation is to obtain maximum production efficiency from the well. Incidental benefits are stabilization of the structure, minimization of sand pumping, and reduction in the potential for future corrosion and encrustation conditions. Development also removes the mud cake from the face of the hole and breaks down the compacted annulus about the hole caused by drilling. Fines are removed from the pack and the aquifer, thus

Guide bolts here

Plumb line

First position

Second position

Exact center

**DETAILS OF
CAGE RING**

Washers

Bolt to
frame

Oversized holes
for adjustment

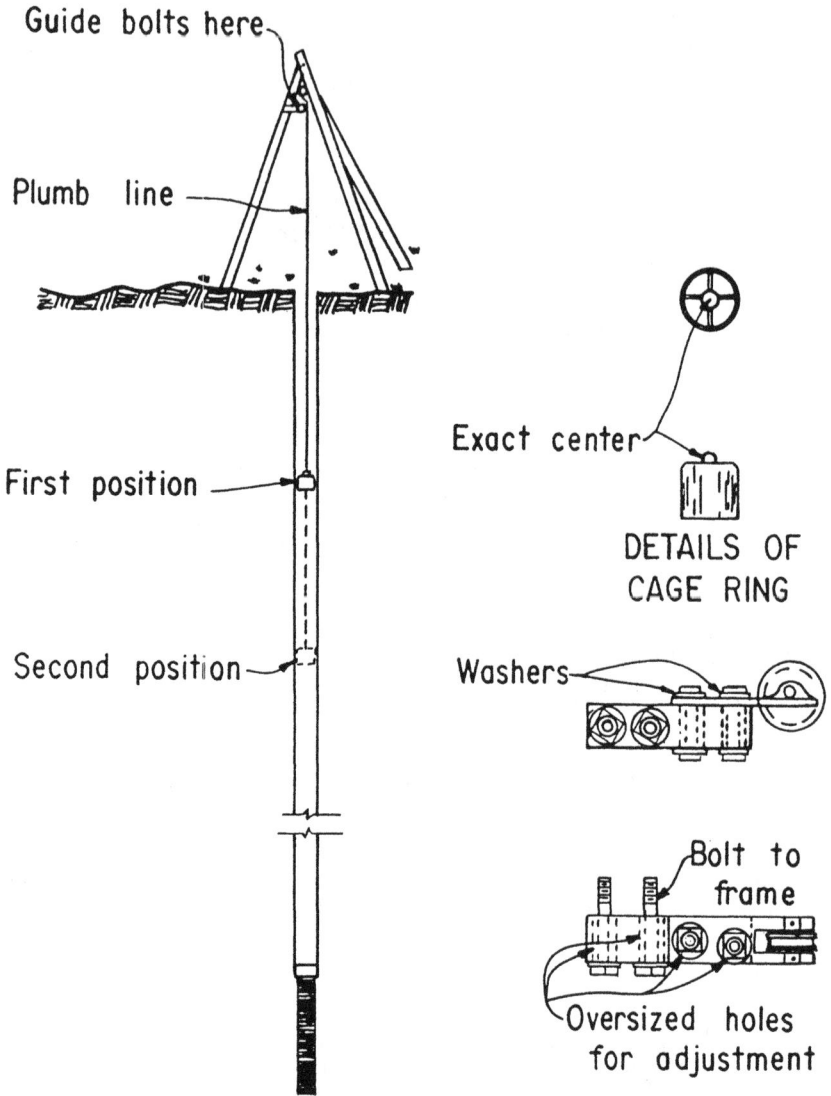

**TYPICAL ARRANGEMENT FOR
TESTING PLUMBNESS AND
ALIGNMENT OF A WELL**

**DETAILS OF
ADJUSTABLE
GUIDE**

Figure 12-4.—Cable suspended cage for checking
straightness and plumbness of wells.

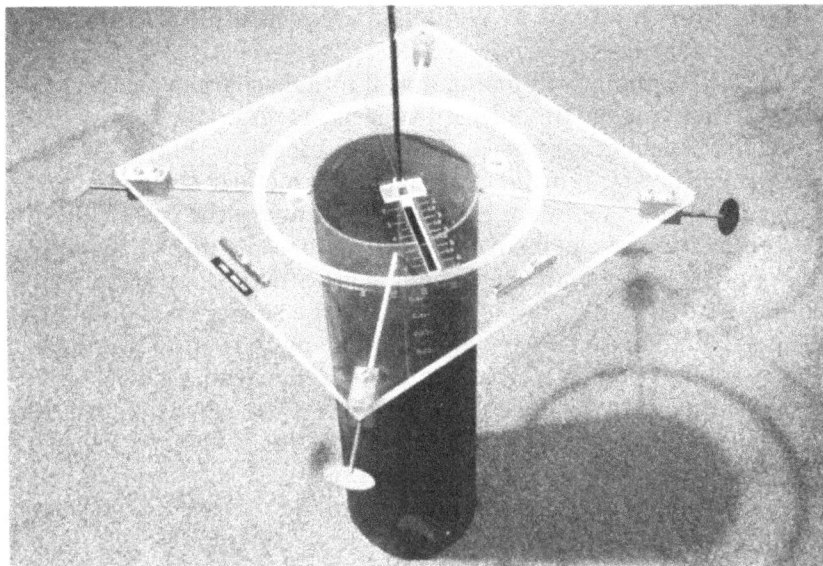

Figure 12-5.—Template for measuring deviation of a well.

increasing the porosity and the permeability of the pack and aquifer. Water is made to surge back and forth through the screen, pack, and aquifer and to flow into the well at higher velocities than during pumping at design rates. Material which is brought to stability under high development velocities and surging will remain stable under velocities present during normal pumping operations.

Proper and careful development will improve the performance of most wells. Well development is not expensive in view of the benefits derived and only under unusual circumstances or improper methods will it cause harm.

Quite frequently, when a screen is selected to reduce sand pumping in sandy gravel formations found in the Southwest, development will provide no measurable improvement in the specific capacity of the well.

Depending upon the circumstances, a number of methods and supplemental chemicals may be used in developing a well. Some of the common methods and the conditions for which they are used are described in the following sections.

12-9. Development of Wells in Unconsolidated Aquifers.—

(a) Overpumping.—Pumping a well at a discharge rate considerably higher than design capacity is often the only well development procedure used. However, except in thin, relatively uniform grained, permeable aquifers, this method alone is not recommended. The pump is normally set above the top of the screen; hence, development is primarily concentrated in the upper one-quarter or one-half the screen length. With the water moving in one direction only, stable bridging of the sand grains occurs so long as pumping continues. When pumping is stopped, the water in the column pipe drops back into the well, causing a reverse flow which destroys the bridging. When the well is again pumped, sand will enter the well until stable bridging is re-established. A well so developed may pump sand for several minutes each time the pump is started. This may continue for months or even years but may eventually clear up.

(b) Rawhiding (Pumping and Surging).—The arrangement for rawhiding is similar to that for overpumping. However, the pump must not be equipped with either a rachet or other device that could prevent reverse rotation of the pump, and no check valve is used. The well is pumped in steps (e.g., one-quarter, one-half, one, one and one-half, and two times the design capacity). At the beginning of each step, the well is pumped until the discharge is relatively sand free. The power is then shut off and the water in the column pipe is allowed to surge back into the well to break up bridging. The well may be surged one or more additional times by operating the pump until clear water is discharged at the surface and then stopping the pump. The rate of discharge is then increased and the same procedure is followed at each of the higher rates, with the final rate being at the maximum capacity of the pump or well. Rawhiding is definitely superior to simple overpumping, but when used alone will usually result in development of only the upper portion of the screened aquifer. Rawhiding is recommended as a finishing procedure following initial development by any of the methods described in the following subsections (c), (d), and (e) of this section.

During final development by rawhiding, the amount of sand discharged by the well is measured when pumping is resumed after each cycle of surging. The initial discharge on resumption of pumping is usually almost sand free. Within a few seconds or minutes, depending upon the rate of discharge and the depth of the well, the sand will increase to a maximum. This condition will usually persist for a short period, and then the amount of sand will begin to decrease until the discharge is practically sand free. At this time, the well should be surged again.

The approximate concentration of sand being discharged can be estimated by looking through the discharge stream. The sand will be concentrated at the bottom of the stream where it issues from a discharge pipe with free discharge. It will look like a dark gray or brown layer. If an orifice is attached to the end of the pipe, the sand will appear as a dark vein in the center of the jet. The orifice should always be removed to avoid sand cutting its edge during rawhiding.

The time of maximum concentration of sand can be judged closely by observing discharge flow at the beginning of each rate of discharge. A sample is taken when the sand discharge is maximum.

Sand traps are available which will permit relatively accurate determination of sand content of the discharge, but they are expensive, heavy pieces of equipment. An Imhoff cone is commonly used to catch samples (see figure 12-6). The cone should be held firmly with both hands, and the outside lip of the cone should be slipped into the bottom of the discharge stream to the center of the sand concentration. The cone fills in a fraction of a second, and the entire procedure must be done rapidly.

The cone is then set in a holder to permit the contents to settle for a few minutes, and then the sand content by volume is estimated.

The smallest division on a cone is 0.1 milliliter (0.006 in^3). About one-tenth of the smallest division on the scale is approximately 10 mg/L by volume or 20 mg/L by weight. For estimating purposes, multiply the volume by 2 to get weight. Acceptable sand content for various purposes is as follows:

- Municipal, domestic, and industrial supply - 0.01 mL or 20 mg/L by weight

- Sprinkler irrigation - 0.025 mL or 50 mg/L by weight

- Other irrigation (furrow, flooding, etc.) - 0.075 mL or 150 mg/L by weight

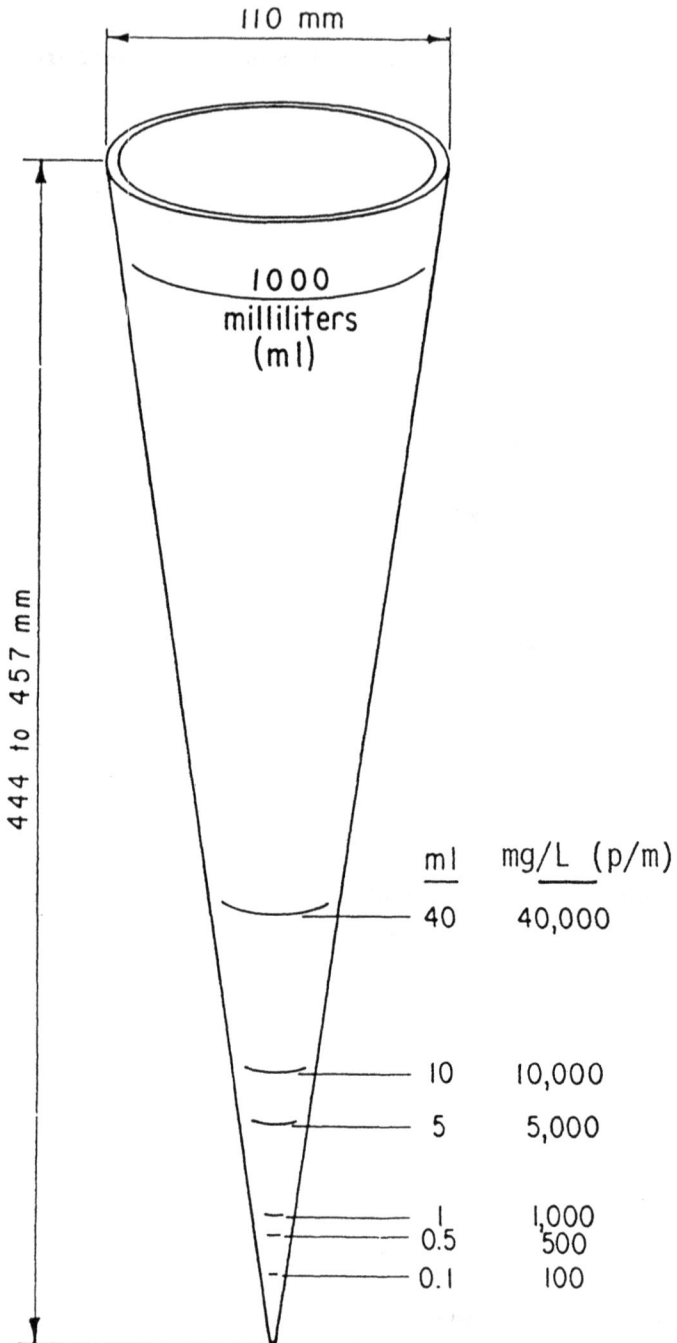

Figure 12-6.—Imhoff cone used in determination of sand content in pump discharge.

Two factors will yield high sand measurements resulting in artificially high calculations—first, the sand is measured during the period of highest concentration; second, the Imhoff cone is typically brought into the discharge stream from the bottom where concentrations of sand are highest. The error caused by these factors can be reduced by taking numerous cone measurements during discharge periods, beginning when sand is first noticed in the discharge and continuing until sand pumping has stabilized or discontinued. Cone measurements can be averaged or simply used to note how long sand pumping occurs. Because the sample is taken during the period of highest sand concentration in the discharge, the estimated sand content is probably high and on the safe side.

Rawhiding, pumping, and sampling should be continued at the maximum discharge rate until the desired sand content is reached.

Imhoff cones are made in two styles—one has a rounded bottom and the other has a more pointed bottom. The model with the pointed bottom is preferable for estimating small volumes of material. Most Imhoff cones are made of glass and the breakage is sometimes high, particularly when sampling high capacity wells. Plastic models which are less expensive, easier to clean, and less prone to breakage are available with a pointed bottom and a plug to facilitate easy emptying (U.S. Department of the Army, 1965; Driscoll, 1986; Gibson and Singer, 1969).

(c) Surge Block Development.—The surge block is one of the oldest and most effective methods of well development. Such blocks are particularly applicable for use with a cable tool rig, and often such a rig equipped with a surge block is used to develop a well drilled by other methods. Most surge blocks are made by well drilling contractors to fit their drill and the well under construction. The advantages of surging over pump development is that surging is a more vigorous development technique that settles the gravel pack and has a greater potential to remove the mudpack from the wall of the boring.

Solid, vented, and spring-loaded surge blocks are used. The solid and vented blocks consist of a body block 25 to 50 millimeters (1 to 2 inches) smaller in diameter than the well screen, and fitted with as many as four 6- to 12-millimeter (1/4- to 1/2-inch) thick disks of belting, rubber, or other tough material having a diameter the same as the inside diameter of the screen in which they will be used.

The solid surge block has a solid body, whereas the vented one has a number of holes drilled through the body parallel to the axis. The top of the body is fitted with rubber or similar flap valve which seals the holes on the upward stroke and permits water to move through them on the downstroke. Figure 12-7 shows a design for a vented surge block for use in a 200-millimeter (8-inch) diameter well screen. The same general design can be used for screens from 100 to 300 millimeters (4 to 12 inches) in diameter. Figure 12-8 shows a design for a larger diameter spring-loaded vented surge block. Similar designs can be used for solid surge blocks by eliminating the vents and flap valves.

The principle of surging is to draw fine grained material out of the formation and into the well on the upstroke and reduce or break sand bridging on the downstroke. As the block is moved up and down in the screen, the solid surge block imparts a surging action to the water which is about equal in both directions. The gentler downstroke of the vented surge block causes only sufficient backwash to break up any bridging which may occur, and the stronger upstroke pulls in the sand grains freed by the destruction of the bridging. The solid surge block is usually most effective in dirty sands containing large percentages of clay, silt, and organic matter; the vented surge block is best in cleaner sands.

The spring-loaded surge block may be vented or solid, is more effective than the other two blocks, and offers some advantage in avoiding sand or locking in the screen.

Before surging is begun, the well should be bailed clean and the surge block cable should be marked to identify the bottom of the well and top of the screen.

The surge block is attached to the bottom of a drill stem of adequate weight to ensure a fairly rapid downstroke under the action of gravity. Surging should be started above the screen to bring in the initial flow of sand, thus minimizing the hazard of sand locking the block in the screen.

Surging is then continued at the bottom of the screen using the longest stroke and slowest rate of which the rig is capable. The surge block is raised slowly through the screen as surging progresses until all the screen has been surged. The procedure is then repeated using a faster stroke. Several passes should be made until the maximum stroke rate is attained at which the line

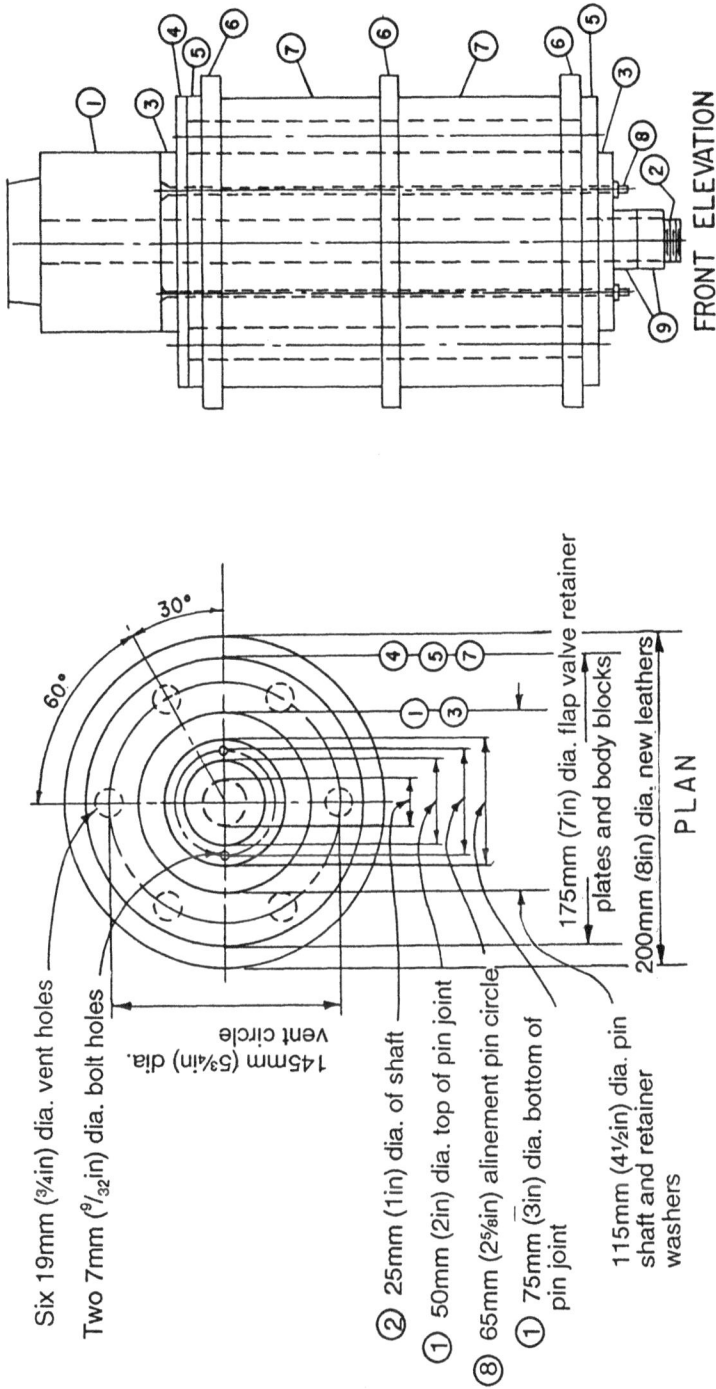

FRONT ELEVATION

PLAN

Six 19mm (¾in) dia. vent holes

Two 7mm (⁹/₃₂in) dia. bolt holes

145mm (5⅝in) dia. vent circle

② 25mm (1in) dia. of shaft

① 50mm (2in) dia. top of pin joint

⑧ 65mm (2⅝in) alinement pin circle

① 75mm (3in) dia. bottom of pin joint

115mm (4½in) dia. pin shaft and retainer washers

175mm (7in) dia. flap valve retainer plates and body blocks

200mm (8in) dia. new leathers

30°

60°

Figure 12-7.—Design of a vented surge block for a 200-millimeter (8-inch) well screen (sheet 1 of 2).

① API pin joint, collar tapped and threaded for a 25mm (1in) diameter shaft.

② One 25mm (1in) diameter shaft threaded the required length on each end.

③ Two 115mm (4½in) o.d. steel retainer washer 10mm (⅜in) thick with 25mm (1in) dia. center hole. Two 7mm (⁹/₃₂in) dia. alinement holes drilled 180º apart on a 65mm (2⅝in) dia. circle about the center. Alinement holes for tap washer countersunk for flathead bolts.

④ Flap valve, 6mm (¼in) thick flexible belting, 175mm (7in) o.d., 25mm (1in) dia. center hole, two 7mm (⁹/₃₂in) alinement holes 180º apart on a 65mm (2⅝in) dia. circle about the center.

⑤ Two steel retainer plates 10mm (⅜in) thick, 175mm (7in) o.d., 25mm (1in) center hole. Two 7mm (⁹/₃₂in) dia. alinement holes 180º apart drilled on a 65mm (2⅝in) dia. circle about the center. Six 20mm (¾in) dia. vent holes 60º apart drilled on a 145mm (5¾in) dia. circle about the center.

⑥ Three "leathers", 12mm (½in) thick belting, 200mm (8in) o.d., 25mm (1in) center hole. Two 7mm ((⁹/₃₂in) dia. alinement holes 180º apart drilled on a 65mm (2⅝in) dia. circle about the center. Six 20mm (¾in) dia. vent holes 60º apart drilled on a 145mm (5¾in) dia: circle about the center.

⑦ Body blocks of hardwood, marine plywood, etc. with 175mm (7in) o.d., 100mm (4in) thick, 25mm (1in) center hole, two 7mm (⁹/₃₂in) alinement holes 180º apart drilled on a 65mm (2⅝in) dia. circle about the center. Six 20mm (¾in) dia. vent holes 60º apart drilled on a 145mm (5¾in) dia. circle about the center.

⑧ Two flathead bolts, 6mm (¼in) dia., 275mm (11in) long with nuts.

⑨ Heavy nut and locknut to fit threads on 25mm (1in) shaft, or a castellated nut with cotter pin.

NOTE: For a solid surge block, the same materials and design may be used but omitting the flap valve ④ and the six 20mm (¾in) dis. vent holes specified in ⑦.

Figure 12-7.—Design of a vented surge block for a 200-millimeter (8-inch) well screen (sheet 2 of 2).

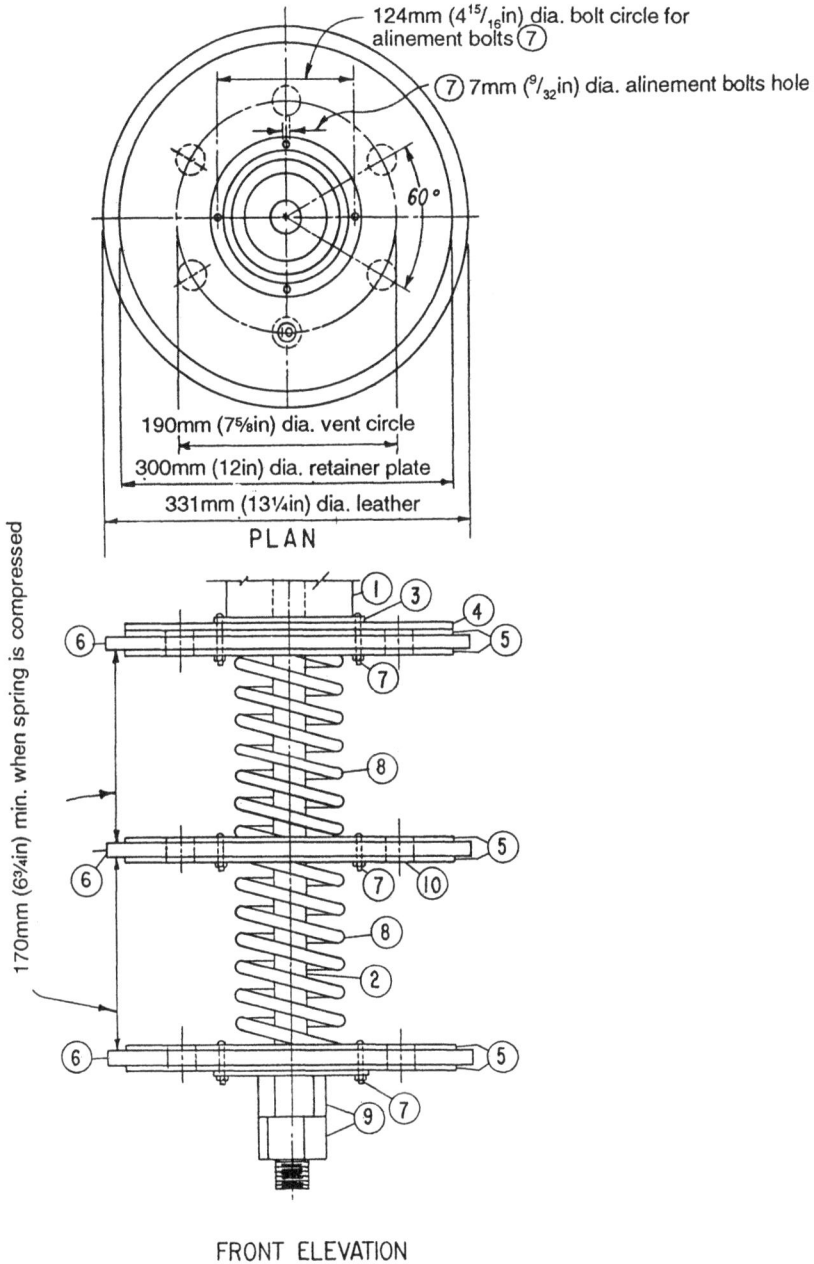

FRONT ELEVATION

Figure 12-8.—Design of a spring-loaded, vented surge block (sheet 1 of 2).

(1)   API Pin joint, collar tapped and threaded for shaft.

(2)   One 5mm ($^3/_{16}$in) or 12mm (½in) shaft of required length,
        threaded at both ends.

(3)   Two 142mm (5½in) dia. retaining washers, 10mm (⅜in) thick,
        center hole drilled to fit shaft. Four 7mm ($^9/_{32}$in) dia. alinement
        holes 90º apart drilled on a 124mm ($4^{15}/_{16}$in) dia. circle about
        center.

(4)   Flap valve, 6mm (¼in) to 12mm (½in) thick flexible belting or
        impregnated fabric, 300mm (12in) o.d., center hole drilled to
        fit shaft, Four 7mm ($^9/_{32}$in) alinement holes 90º apart drilled on
        a 124mm ($4^{15}/_{16}$in) dia. circle about center. (Desirable that at
        least two spare flop valves be available at the rig.)

(5)   Six retainer plates 10mm (⅜in) steel, 300mm (12in) o.d.,
        center hole drilled to fit shaft. Four 7mm ($^9/_{32}$in) alinement
        holes 90º apart drilled on a 124mm ($4^{15}/_{16}$in) dia. circle about
        center. Six 25mm (1in) dia. vent holes 60º apart drilled on a
        190mm (7⅝in) dia. circle about center.

(6)   Three "leathers,"6mm to 12mm (¼ to ½in) thick belting,
        331mm (13¼in) o.d., center hole drilled to fit shaft. Four 7mm
        ($^9/_{32}$in) alinement holes 90º apart drilled on a 124mm ($4^{15}/_{16}$in)
        dia. circle about center. Six 25mm (1in) dia. vent holes 60º
        apart drilled on a 190mm (7⅝in) dia. circle about center.

(7)   Alinement bolts. Four 70mm (2¾in), 56mm x 6mm (2¼in x
        ¼in), with lock washers and nuts.

(8)   Two coiled steel compression springs with squared ends,
        max. 100mm (4in) o.d., 12mm (½in) or heavier spring stock,
        installed under load. (Automobile coil springs have been
        satisfactory).

(9)   Heavy nut and lock nut threaded to fit shaft or castellated nut
        with cotter pin lock.

Figure 12-8.—Design of a spring-loaded, vented surge block (sheet 2 of 2).

can be maintained in tension on the downstroke. The drill stemshould not be permitted to fall out of plumb because of the hazard of falling against and possibly damaging the screen.

After each upward pass of surging, the block should be lowered to the bottom of the well to check the accumulation of material. When the material accumulated at the bottom begins to encroach upon the screen, the block should be pulled and the hole should be bailed clean. The rate of accumulation should be recorded to provide data on the progress of development.

Each time the surge block is removed from the well, the disks should be measured, and if they have worn so that their diameter is 20 millimeters (3/4 inch) smaller than the inside diameter of the screen, they should be replaced.

Surging is often done with a flap valve bailer. The action is similar to a vented surge block; if the difference between the diameter of the bail and inside diameter of the pipe is 25 millimeter (1 inch) or less, it is almost as effective as a surge block. If the difference in diameters is greater than 25 millimeters (1 in), use of a bailer for surging is relatively ineffective. Smaller diameter bailers are often built up with wrappings of burlap, clamp-on rings, etc., to fit the inside diameter of the pipe more closely.

The time required to properly surge a well depends upon the character of the aquifer material and its apparent response to development. As a guide, if the thickness of material accumulated at the bottom of the well is less than one-third of the inside diameter of the screen during 1/2 hour of surging on each 6 meters (20 feet) of screen, development with the surge block is probably adequate. On completion of surging, the well should be bailed clean to the bottom. Final development by rawhiding should precede testing of the well.

Development by a surge block should be started at the slowest possible rate and should be increased as the well develops. An overly vigorous initial development, particularly in clayey formations, can damage a well rather than improve it (U.S. Department of the Army, 1965; Driscoll, 1986; National Water Well Association, 1971; Campbell and Lehr, 1973).

Swabbing is a violent method which is not recommended for initial well development because it may pack fines tightly around a

screen. However, swabbing of wells which do not respond to
normal surging or chemical treatment is frequently effective in
breaking down a heavy wall cake or particularly tight annulus.
Normal development is usually possible once these are broken
down. Swabbing should be used only as a last resort and never
used in wells cased or screened with thin-walled pipe, California
stovepipe, plastic, or similar materials because these lighter
casings are likely to collapse. Swabbing is done using a surge
block which is lowered to the bottom of the screen and then pulled
up as rapidly as possible until suction is broken or the block comes
out of the casing.

(d) Development with Air.—Development of wells using compressed air is an effective method that requires considerable equipment and skill on the part of the operator. Figure 12-9 shows
typical installations for development of wells by air. Two methods,
backwashing and surging, are generally used (Driscoll, 1986):

- In the backwashing method, water is alternately pumped
  from the well by airlift and then forced through the screen
  and into the water-bearing formation by compressed air
  introduced through a tight seal at the top of the casing.
  The three-way valve (figure 12-9, arrangement A) is turned
  to deliver air down the air line, which pumps water out of
  the well through the discharge pipe.

  The bottom of the discharge pipe is usually set about
  1/2 meter (1 to 2 feet) above the top of the screen. When
  the discharged water becomes clear, the supply of air is cut
  off and the air cock is opened. The water in the well is
  allowed to return to the static level, which can be
  determined by listening to the escape of air through the air
  cock as the water rises in the casing.

  When air ceases to flow from the air cock, the water no
  longer rises in the well. The air cock is then closed and the
  three-way valve is turned to direct the air supply down the
  bypass into the well above the static water level. This air
  forces the water back through the aquifer, agitating it and
  breaking down bridges of sand grains. When the water
  surface in the well is lowered to the bottom of the drop pipe,
  the air escapes through the drop pipe and the water is not
  depressed to any greater depth, thus avoiding air logging of
  the formation. When water reaches the bottom of the drop
  pipe, air can be heard escaping from the discharge pipe and

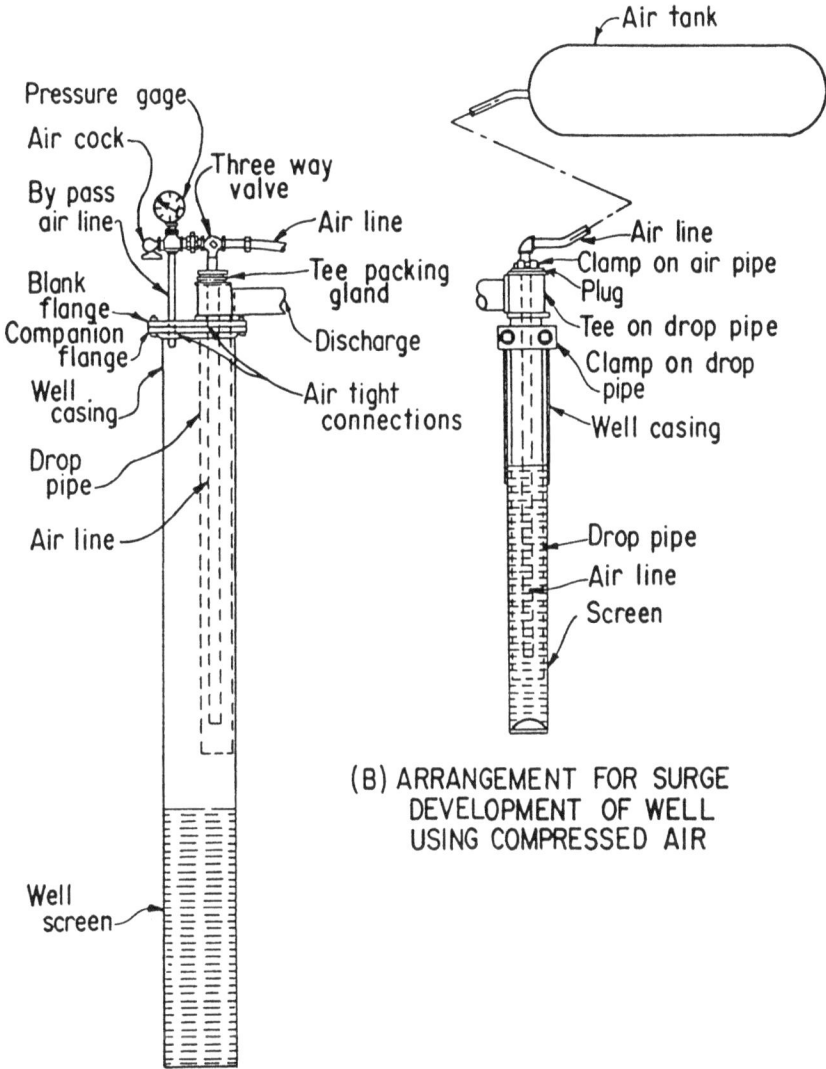

Pressure gage

Air cock

By pass air line

Three way valve

Air line

Tee packing gland

Blank flange

Companion flange

Discharge

Well casing

Air tight connections

Drop pipe

Air line

Well screen

Air tank

Air line

Air line

Clamp on air pipe

Plug

Tee on drop pipe

Clamp on drop pipe

Well casing

Drop pipe

Air line

Screen

(B) ARRANGEMENT FOR SURGE DEVELOPMENT OF WELL USING COMPRESSED AIR

(A) ARRANGEMENT FOR BACKWASH DEVELOPMENT OF WELL USING COMPRESSED AIR

Figure 12-9.—Typical installation for development of a well using compressed air.

the pressure gauge will show no increase in pressure with time. At this point, the supply of air is cut off and the air cock is opened to allow the water to return to its static level. Then the three-way valve is turned and the air supply is again directed down the air line to pump the well. This procedure is repeated until the well is thoroughly developed. The well should be sounded and, if necessary, bailed to remove any sand which may have accumulated on the bottom of the hole during the development operation. This method is not very effective except on short lengths of screen and open hole because development is limited to the upper part of longer screens or open hole.

• Figure 12-9, arrangement B, shows the arrangement and equipment necessary for development with air by the surging method. It should be noted that in this method, the drop pipe is placed inside the screen. For the best operation, the drop pipe should have a submergence of at least 60 percent (i.e., 60 percent of the total length of the drop pipe should be beneath the water surface when the well is pumping). However, a good operator can do an acceptable job of pumping with as little as 35-percent submergence. Development consists of a combination of surging and pumping. The sudden release of large volumes of air produces a strong surge within the well, and pumping is accomplished as with an ordinary airlift pump.

At the start of development, the drop pipe is lowered to within about 1/2 meter (2 feet) of the bottom of the screen, and the air line is lowered within it so that its lower end is about 1/3 meter (1 foot) above the bottom of the drop pipe. Air is turned on and the well is pumped until the water discharged is free of sand. The air tank is then pumped to maximum pressure while the air line is lowered 1/3 meter (a foot) below the bottom of the drop pipe. When the air tank is full, the quick-opening valve is thrown open, allowing the air in the tank to rush with great force into the well. This rushing air causes a brief but forceful surge of the water. The air line is immediately pulled back so that its lower end is again about 1/3 meter (1 foot) above the bottom of the drop pipe. Pumping by airlift is then resumed until the water is again free of sand, and the cycle is repeated until little or no sand is evident on pumping immediately after surging. The drop pipe is then raised about two screen diameters and the cycles of surging and pumping are repeated until the entire length of the screen has been developed. The drop pipe and air line should then be lowered to the bottom of the well and an effort should be made to pump out any sand which may have accumulated at the bottom of the screen. If this effort is not successful, it may be necessary to use a sand pump in the well to remove any accumulation of sand.

Under certain conditions, the foregoing procedures should be used with caution because of the potential hazard of air binding of the aquifer. A safer procedure is to maintain the air pipe about 1 meter (3 feet) up within the drop pipe and accomplish the surging by the falling column of water within the drop pipe.

In deep wells with water at a considerable depth, the use of compressed air may be limited by the volume and pressure capacities of available compressors. For example, a compressor with a 1,035-kPA (150 lb/in^2) pressure rating would be counterbalanced by a depth of water of 105 meters (350 feet). For the best results, a volume of free air at the ratio of 5-1/2:1 is required for pumping water. When surging and pumping with air, the discharge should be directed into a fairly large tank in which the sand discharged can be collected so that the degree of effectiveness and progress of development can be determined.

When developing by backwashing or surging with air, an arrangement should be made which will permit periodic measurement of the volume of sand accumulating in the bottom of the well. Should sand begin to encroach on the screen, development should be stopped and the well should be bailed or pumped clean before resuming the operation. On completion of development, the well should be bailed or pumped clean (Driscoll, 1986).

Figure 12-10 shows an effective arrangement for developing wells with long screen sections by surging with air. The double packer is set at the base of the eductor pipe, and all surging and pumping is done within the double packer. The double packer maximizes the effectiveness by confining the action to a specific length of the screen.

(e) Hydraulic Jetting.—Hydraulic jetting is most effective in open rock holes and in wells having cage-type wire-wound screen and some types of louvre screen. The jetting tool consists of a head with two or more 5- to 12-millimeter (3/16- to 1/2-inch) jet nozzles equally spaced in a plane about the circumference (see figure 12-11). The head is attached to the bottom of a string of 30-millimeter (1-1/4-inch) or larger pipe that is connected through a swivel and hole to a high pressure, high capacity pump. Water

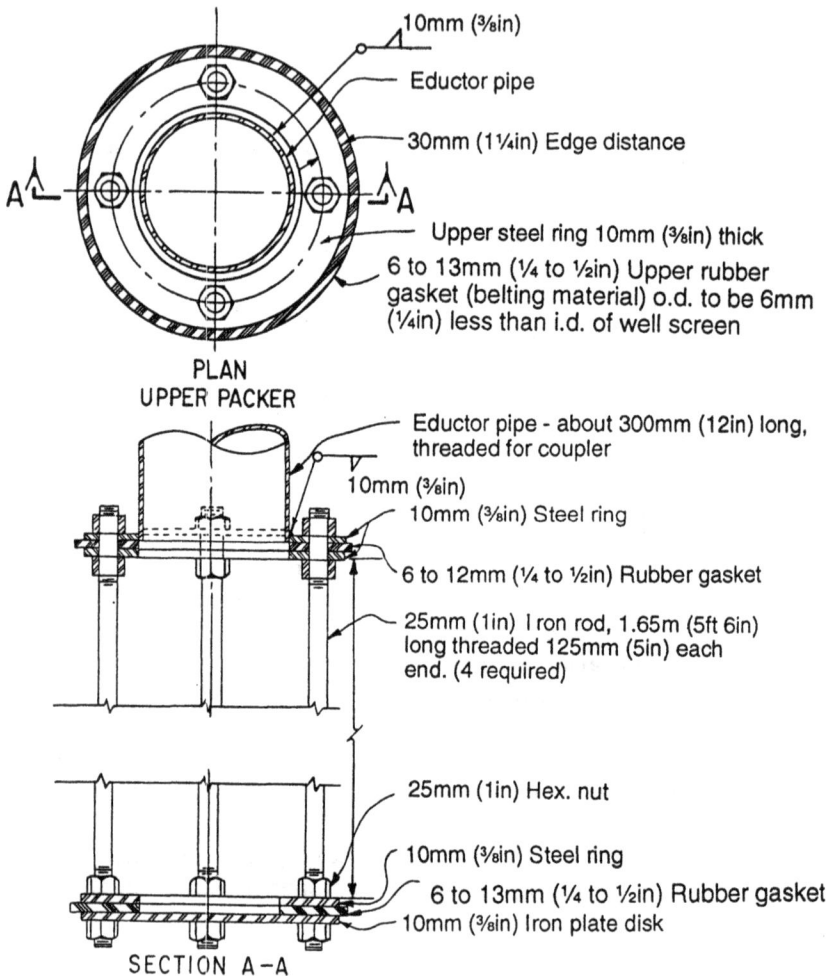

10mm (⅜in)

Eductor pipe

30mm (1¼in) Edge distance

Upper steel ring 10mm (⅜in) thick
6 to 13mm (¼ to ½in) Upper rubber gasket (belting material) o.d. to be 6mm (¼in) less than i.d. of well screen

A

A

PLAN
UPPER PACKER

Eductor pipe - about 300mm (12in) long, threaded for coupler

10mm (⅜in)
10mm (⅜in) Steel ring

6 to 12mm (¼ to ½in) Rubber gasket

25mm (1in) Iron rod, 1.65m (5ft 6in) long threaded 125mm (5in) each end. (4 required)

25mm (1in) Hex. nut

10mm (⅜in) Steel ring
6 to 13mm (¼ to ½in) Rubber gasket
10mm (⅜in) Iron plate disk

SECTION A–A

NOTE: This assembly is governed by the i.d. of the well screen. It should be made so that the 6 to 13mm (¼ to ½in) rubber gaskets (belting material) are about 6mm (¼in) diameter less than the i.d. of the well screen to be used.

Figure 12-10.—Double packer air development assembly.

① Concentric reducer, 25mmx6mm (1inx¼in) or 25mmx10mm (1inx⅜in) steel butt welding or similar, black iron (extra strong), two req'd. for 100mm (4in) coupling and 2 or 4 req'd for 150mm (6in) coupling.

② Line pipe T and C, 25mm (1in) grade A API 5L, black iron (extra strong), two req'd. for 100mm (4in) coupling and 2 or 4 req'd for 150mm (6in) coupling. Length should be such that small end of reducer① is flush with o.d. of guide ring⑥.

③ Standard coupling, 112mm (4.5in) long for 100mm (4in) coupling and 120mm (4.78in) long for 150mm (6in) coupling, black iron (extra strong)

④ Hexagon bushing (outside or inside), double tapped 100mmx50mm (4inx2in) for 100mm (4in) coupling, 150mmx75mm (6inx3in) for 150mm (6in) coupling, black iron (extra strong).

⑤ Bar, 50mmx6mm (2inx¼in) strap steel, 4 req'd.,length depends on i.d. of screen.

⑥ Guide ring, 50mmx6mm (2inx¼in) strap steel, o.d. of guide ring should be 25mm (1in) smaller that i.d. of screen.

⑦ Bar plug, cast iron.

⑧ Line pipe T and C, black iron (extra strong), length should be such as to permit easy removal when swivel is at max. height, sufficient pipe should be available to reach from casing collar to bottom of screen.
NOTE: Above material is for use with a 100 or 150mm (4 or 6in) nominal coupling.

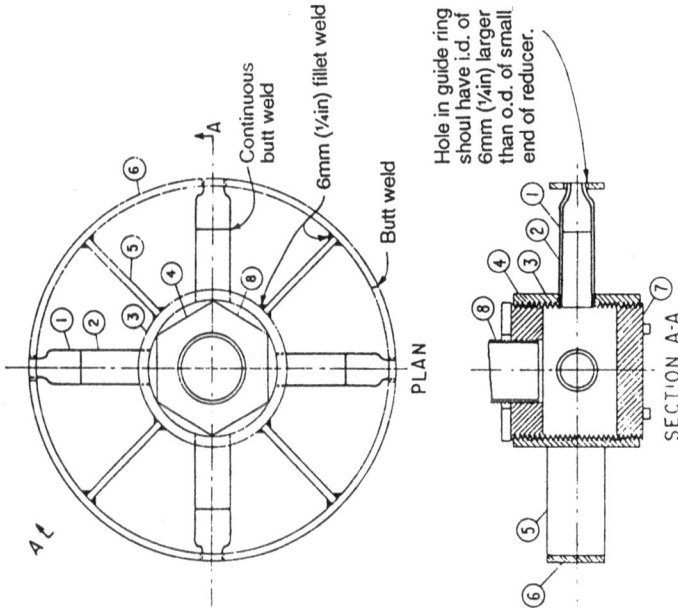

Figure 12-11.—Jetting tool for well development.

is pumped down the pipe in sufficient quantity and at sufficient pressure to give a nozzle velocity of 45 meters per second (150 feet per second) or more. Tables 12-1(a) abd 12-1(b) indicates the discharge rate for various exit velocities of different size nozzles and the required pressures in the jet head. Pumping pressures may have to be higher, depending on the number of nozzles, the required volume, and the sizes and arrangement of the plumbing.

For small diameter wells of 75 millimeters (3 in) and less, an efficient jetting tool can be made by attaching a coupling and plug to the bottom of the air line. Two or four 5-millimeter (3/16-inch) diameter holes are drilled through the coupling to serve as nozzles.

The jet head is turned at 1 revolution per minute or less but should not be maintained for more than 2 minutes at a given setting. The tool is successively raised a distance equal to about one-half the screen diameter after completion of jetting at each setting until the entire surface of the screen or the open hole has been developed. During jetting, the jet head should always be rotated. The sand-bearing water returning to the well is picked up by the jet stream, and if a jet impinges on one spot or a circumference for only a few minutes, it may erode a hole in the screen or even cut it.

The jetting tool has been effective in removing stubborn mud cakes from some holes and in opening up formations of dirty sand which have been plugged by too rapid and vigorous development by surging. The jet is particularly effective in developing gravel-packed wells.

If possible, water should be pumped from the well during development by jetting. Ideally, the pump discharge should exceed the jet discharge by 1.5 to 2 times. This practice removes the fines as they are washed into the screen and keeps ground water flowing into the well, thus avoiding the buildup of positive head in the well which acts to force the fines back into the formation. An airlift pump is usually used for this purpose.

The pump should discharge into a large tank which permits an appraisal of the effectiveness of the jetting on the basis of the materials collected in the bottom of the tank. The tank also permits recirculation of the water to the jet head, which is advantageous when chemical additives are used (Driscoll, 1986).

Table 12-1(a).—Approximate jet velocity and discharge per nozzle (SI metric)

Size of nozzle orifice (mm)	Effective pressures (kilopascals)							
	690		1,035		1,380		1,725	
	Velocity (m/s)	Charge (L/min)	Velocity (m/s)	Charge (L/min)	Velocity (m/s)	Charge (L/min)	Velocity (m/s)	Charge (L/min)
4.76	36	34	45	46	51	49	57	58
6.35	36	60	45	80	51	87	57	98
9.52	36	137	45	175	51	201	57	224
12.70	36	251	45	312	51	353	57	395

Table 12-1(b).—Approximate jet velocity and discharge per nozzle (U.S. customary)

Size of nozzle orifice (in)	Effective pressures (lb/in^2)							
	100		150		200		250	
	Velocity (ft/s)	Charge (gal/min)	Velocity (ft/s)	Charge (gal/min)	Velocity (ft/s)	Charge (gal/min)	Velocity (ft/s)	Charge (gal/min)
3/16	120	9	150	12	170	13	190	15
1/4	120	16	150	21	170	23	190	26
3/8	120	36	150	46	170	53	190	59
1/2	120	66	150	82	170	93	190	104

12-10. Development of Wells in Hard Rock.—Open hole
wells drilled in hard rock supposedly do not benefit from
development, but experience has shown this supposition to be in
error. In consolidated granular materials, a mud cake forms and
fines are forced into the walls of the hole by the drilling operation.
In fractured and jointed rocks where water yields depend upon the
interception by the well bore of water-filled cracks or solution
openings, such openings are frequently sealed by much the same
action as well as by mud invasion. Practically all the methods
used in developing screened wells can be used effectively in open-
hole hard-rock wells. Under some circumstances, however, some
additional practices may also be effective.

Wells in carbonate rock are often developed by the addition of
sulfamic acid or muriatic acid, which attack the carbonate rock and
enlarge existing openings and create new openings. When the acid
is spent, it is pumped to waste and the well is treated with
polyphosphates and surging or jetting. Under some circumstances,
shooting a well in limestone with dynamite or other explosives
using 20- to 50-kilogram (50- to 100-pound) charges of 60-percent
gelatin or equivalent every 1.5 meters (5 feet) to fracture the rock
has been effective. Hydrofracturing may also be effective. A
definite risk is associated with all of these procedures, and they
should be planned and carried out only under experienced and
informed direction using adequate equipment and safeguards.

Wells in sandstone drilled with cable tools or down-the-hole tools
should be developed using polyphosphates and vigorous surging.
Wells drilled with rotary rigs are sometimes under-reamed about
12 millimeters (1/2-inch) using plain water as a drilling fluid, after
which the well is bailed clean. The well is further developed using
polyphosphates and strong surging. In very competent, cemented
sandstones, shooting with 2- to 4-kilogram (5- to 10-pound) charges
of 50-percent gelatin at 1.5-meter (5-foot) intervals along the hole
to enlarge it, or shooting with a few heavier charges to break and
fracture the rock, are sometimes effective.

Regardless of the method of drilling, wells in basalt and
crystalline rocks should be developed using polyphosphates and
jetting, vigorous surging, or both. Spot shooting selected portions
of the hole with 20 to 50 kilograms (50 to 100 pounds) of
50-percent gelatin is sometimes effective in increasing yields.

Hydraulic fracturing has been of limited effectiveness in
increasing yields of sedimentary, crystalline, and volcanic rocks.

Inflatable packers on a pipe leading to the surface are used to isolate 1.5- to 3-meter (5- to 10-foot) lengths of the hole. The pipe and isolated section are filled with water and pump pressure is applied to fracture the rock. Continued pumping may result in another buildup in pressure and additional fracturing. Sand fracturing, a refinement of the method, consists of pumping selected sizes and types of sand into the fractures to prop them and keep them open. Some wells have showed an increase in yield of as much as 200 percent as a result of fracturing, but in all cases, the initial yield was small, from less than 3 to possibly 10 liters per minute (1 to possibly 3 gallons per minute) (Driscoll, 1986).

12-11. Chemicals Used in Well Development.—Numerous chemicals are used to aid in well development. The most commonly used chemicals are the polyphosphates. Polyphosphates act as deflocculants and dispersion agents for clays and other fine particles and help considerably in removing clays that occur naturally in the formation or that are introduced into the borehole through the use of drilling muds.

Two types of polyphosphates are generally used in developing wells. The crystalline polyphosphates consist of sodium tripolyphosphate (STP)($Na_5P_2O_{10}$), sodium acid pyrophosphate (SAPP)($Na_4P_2O_7$), and tetrasodium pyrophosphate (TSPP)($Na_1P_2O_7$). Sodium hexametaphosphate (SHMP)($NaPO_2$)$_6$ is also commonly used and is a glassy polyphosphate. The action of the polyphosphates is enhanced by the addition of chelating agents for removing heavy metals, and by the addition of wetting agents to increase the disaggregation of the clays. A chlorine solution, such as sodium hypochlorite, is frequently added to the polyphosphate mixture to control the bacterial growth that is often promoted by the presence of polyphosphates (Driscoll, 1986).

The polyphosphate mixture should be premixed before introduction into the borehole because the polyphosphates do not mix well in low temperature waters. The mixture should be left in the well long enough for the mixture to completely disaggregate and disperse the clays, usually overnight, before the well is developed.

Any one of the polyphosphates can be used, but the mixture usually used by Bureau of Reclamation consists of 7 kilograms (16 pounds) of STP, 1.8 kilograms (4 pounds) of sodium carbonate (a chelating agent), and 1 liter (1 quart) of 5.25-percent sodium hypochlorite for each 380 liters (100 gallons) of water in the screen and casing. A wetting agent, such as Pluronic F-68 or equivalent, is also added at the rate of 0.5 kilogram (1 pound) per 380 liters (100 gallons) of water when developing dirty formations.

When polyphosphates or commercially produced development compounds are not available, the common household phosphate-based detergents are a usable but expensive substitute. Most detergents have the disadvantage of containing a foaming agent, and excessive foaming may result when the well is pumped.

Polyphosphates and wetting agents should not be used in formations that contain thinly bedded clays and sands because the chemical action of the polyphosphates and wetting agents tends to make the clays near the borehole unstable and causes them to mix with the sand. This mixing reduces the hydraulic conductivity of the material near the borehole and results in clay being continually passed into the borehole with each pumping cycle.

12-12. Well Sterilization.—Many States and local political subdivisions require the sterilization of domestic and municipal water-supply wells to ensure the absence of pathogenic bacteria. All wells, regardless of the water use, should be sterilized on completion to prevent or retard the growth of corrosion or encrustation fostering organisms. Many of these organisms are not harmful, but they can accelerate and aggravate corrosion and encrustation problems and reduce the life of a well. Although sterilization may not always eliminate such problems, it is a worthwhile and relatively inexpensive precautionary measure.

12-13. Chlorination.—Sterilization is usually accomplished by introducing chlorine, or a compound yielding chlorine, into the water in the well and the immediate aquifer surrounding the well. Chlorine gas may be used, but the safest and usually most readily available materials to furnish chlorine for field operations are calcium hypochlorite ($Ca(ClO)_2$) or sodium hypochlorite ($NaClO$) and chlorinated lime. Calcium hypochlorite is available in granular or tablet form and contains about 70-percent available chlorine by weight. Sodium hypochlorite is available commercially in aqueous solutions ranging from about 3- to 15-percent chlorine. Commercially available chloride of lime is not a pure compound and does not have a definite formula but usually contains about 23-percent available chlorine.

Calcium hypochlorite is probably the least costly and most convenient material to use for sterilization. However, if the calcium in the ground water plus that added in the hypochlorite solution exceeds about 300 mg/L, a precipitate of calcium hydroxide may form which may reduce the permeability of the aquifer adjacent to the well. If the sterilization solution is mixed to give 1,000 mg/L available chlorine, about 280 mg/L calcium would be present which, when combined with the calcium already present in the natural water, could result in precipitation of calcium hydroxide. Because the percentage of calcium is much higher in chlorinated lime, the danger of such a similar occurrence is even greater. Consequently, sodium hypochlorite is recommended for use in well sterilization. Numerous household bleach solutions containing sodium hypochlorite are commonly available. These solutions usually contain about 3- to 5.25-percent available chlorine. Commercial solutions which may be purchased from chemical supply houses contain 15- to 20-percent chlorine. Thus, in using sodium hypochlorite, the chlorine content should always be determined if not shown on the container. Also, because sodium hypochlorite deteriorates with age, the freshness and concentration should be considered in its purchase and use.

The trade percentage may be converted to milligrams per liter of chlorine by the following equation:

$$\text{milligrams per liter} = (\text{trade percent})\,(10{,}000)$$

Thus, a 5.25-percent solution would be equivalent to approximately 52,500 mg/L chlorine.

For well sterilization for pathogens, usually 50 to 100 mg/L available chlorine and a contact time of from 30 minutes to 2 hours are specified. Many organisms, however, such as sulfate-reducing and filamentous iron bacteria, require 400 mg/L or more available chlorine and contact times up to 24 hours for an effective kill (Speedstar Division, 1967; U.S. Department of the Army, 1965; Driscoll, 1986).

Wells may contain oil and other organic materials which combine with and neutralize the effectiveness of the chlorine. In addition, an unknown amount of dilution takes place in the well. Therefore, to ensure an adequate concentration of chlorine in the well, the volume of water in the screen, casing, and gravel pack should be estimated and sufficient chlorine should be added to yield a chlorine concentration of about 1,000 mg/L.

The amount of various additives to use to obtain about
1,000 mg/L of chlorine in a well could be estimated as in the
following example:

Metric example:
   Known: well depth   130 meters
            casing size   0 to 90 meters; 440 millimeters
                             90 to 130 meters; 350 millimeters
            static water level   60 meters

Using table 12-2, find the volume of water in the well.

60 to 90 meters = 30 meters of 400-millimeter casing at 117.85 L/m
90 to 130 meters = 40 meters of 350-millimeter casing at 88.91 L/m

then:
$$(30 \text{ X } 117.85) + (40 \text{ x } 88.91) = 7091.9 \text{ or } 7,092 \text{ liters}$$

Using 70-percent calcium hypochlorite:

$$Wt \; (kg) = (liters \; of \; water) \; (1 \; kg/L) \left(\frac{concentration \; desired}{concentration \; of \; sterilant} \right)$$

$$Wt = (7092)(1)\left(\frac{0.001}{0.70} \right) = 10.1 \; kg$$

Using 23-percent chlorinated lime:

$$Wt = (7092 \,)(1) \left(\frac{0.001}{0.23} \right) = 30.8 \; kg$$

Using 5.25-percent sodium hypochlorite:

$$Vol = (7092)\left(\frac{0.001}{0.0525} \right) = 135 \; liters$$

English example:
   Known: well depth   425 feet
            casing size   0 to 300 feet; 16 inches
                           300 to 425 feet; 14 inches
            static water level   190 feet

Using table 12-2, find the volume of water in the well.

190 to 300 feet = 110 feet of 16-inch casing at 9.49 gal/ft
300 to 425 feet = 125 feet of 14-inch casing at 7.16 gal/ft

then:

(110 x 949) + (125 x 7.16) = 1,939 gal

Table 12-2.—Volume of water in well per meter (feet) of depth

Casing size (mm)	Volume (L/M)	Casing size (in)	Volume (gal/ft)
100	8.20	4	0.66
125	12.91	5	1.04
150	18.63	6	1.50
200	33.03	8	2.66
250	52.03	10	4.19
300	77.02	12	5.80
350	88.91	14	7.16
400	117.85	16	9.49
450	148.52	18	11.96
500	182.92	20	14.73
550	223.40	22	17.99
600	267.98	24	21.58

Using 70-percent calcium hypochlorite:

$$Wt(lb) = (gallons\ of\ water)(8.33\ lb/gal)\left(\frac{concentration\ desired}{concentration\ of\ sterilant}\right)$$

$$Wt = (1939)(8.33)\left(\frac{0.001}{0.70}\right) = 23\ lb$$

Using 23-percent chlorinated lime:

$$Wt = (1939\ )(8.33)\left(\frac{0.001}{0.23}\right) = 70\ lb$$

Using 5.25-percent sodium hypochlorite:

$$Vol = (1939)\left(\frac{0.001}{0.0525}\right) = 37\ gal$$

The solution should be thoroughly mixed in the well by surging with a bailer or other similar tool from the bottom of the well to the water surface or by surging with the pump. The solution should remain in the well for at least 6 hours, during which time the well should be surged at about 2-hour intervals.

In wells containing considerable oil or organic material in the water, or in which the aquifer contains considerable organic matter, the solution should be tested for residual chlorine after each cycle of surging and mixing. If the residual chlorine falls below the desired concentration, additional compounds should be added to bring it up to the desired concentration.

If the pump is in the well, the well should be pumped at the expiration of the contact time and the discharge should be diverted to flow back into the well to thoroughly flush the inside of the casing, the column pipe, and gravel pack, if present, for at least 30 minutes. The well should then be pumped to waste until there is little or no odor or taste of chlorine in the discharge (Campbell and Lehr, 1973).

To add calcium hypochlorite or chlorinated lime, the required amount is first determined from the volume of water in the well (see table 12-2). This amount of the compound is then placed in a bail or similar tool through which water can flow. The tool is then raised and lowered between the bottom of the well and the water level until the material is completely dissolved. Although this method is effective, it may require considerable time.

A more rapid method involves dissolving the chlorine compound in clear water using 8 liters of water per kilogram of compound or 1 gallon of water per pound. Depending upon temperature and the quality of the water, all the compound may not go into solution, but if some solid material remains, it can be broken up and stirred into suspension at the time the solution is poured into the well. It will readily dissolve in the well water.

If sodium hypochlorite is used, the solution may be poured into the well as received (U.S. Department of the Army, 1965).

Table 12-3 shows chlorine compounds and water required to give various concentrations of chlorine.

Sterilization of a well, except possibly for some pathogens, is seldom 100-percent effective. Organisms may be covered with encrustation or corrosion products or lodged in crevices not readily penetrated by the sterilizing solution. Although most such

Table 12-3.—Chlorine compounds and water required to give
various concentrations of chlorine

Chlorine	Units of water treated by one unit of 5.25% sodium hypochlorite		Units of water treated by one unit of 10% sodium hypochlorite		Units of water treated by one (lb) of dry calcium hypochlorite - 70% available	
mg/L	L	gal	L	gal	L	gal
10	19,950	5,250	38,000	10,000	70,000	8,400
20	9,975	2,625	19,000	5,000	35,000	4,200
30	6,650	1,750	12,540	3,300	23,335	2,800
40	4,997	1,315	9,500	2,500	17,500	2,100
50	3,990	1,050	7,600	2,000	14,000	1,680
60	3,325	875	6,270	1,650	11,670	1,400
70	2,850	750	5,396	1,420	10,000	1,200
80	2,508	660	4,750	1,250	8,750	1,050
90	2,223	585	4,218	1,110	7,780	935
100	1,995	525	3,800	1,000	7,000	840
1,000	200	52	380	100	700	84

organisms may be destroyed, the few remaining continue to multiply and, depending upon conditions, periodic sterilization may be required to control them.

In some instances, continuous chlorination may be required to control mineral encrustation in addition to organisms. When chlorination is necessary, chlorine gas should be discharged continuously through a suitable pipe to the bottom of the well. Where pathogenic contamination is present, chlorine should be injected into the pump discharge pipe and sufficient storage capacity should be provided to permit adequate contact time.

Sterilization of a new well is best deferred until installation of the permanent pump unless excessive delay is incurred. The required amount of sterilant should be added to the well just prior to installing the pump. After the pump is installed, but before it is bolted down permanently, the pump may be used to surge the well periodically to increase the effectiveness of the solution and flush the casing, pump column, and gravel pack, where present.

When sterilizing an existing well where both encrustation and organic fouling are present, it is generally required that acid is first added to remove encrustation, which will allow subsequent chlorine treatments to contact the organisms.

12-14. Other Sterilants.—A number of other sterilants are equal or superior to chlorine or chlorine compounds for controlling certain organisms. Such sterilants are usually more expensive and less readily available than chlorine and some are too toxic for use in potable water-supply wells. Some of those sterilants which are suitable in potable water supplies are:

- A mixture of a polyphosphate detergent and anthium dioxide (chlorine dioxide) for control of filamentous algae

- Cocomines and cocodiamines for sulfate-reducing bacteria

- Quaternary ammonium chloride compounds for general use

Most of these compounds are sold under proprietary names. The manufacturers should be consulted regarding recommended concentrations to use, contact time, and other factors.

Other sterilants not recommended for use in water-supply wells but which might be used in waste disposal or similar wells are:

- Copper sulfate
- Formaldehyde
- Some mercury compounds

12-15. Sterilization of Gravel Pack.—When wells are constructed using a gravel pack or formation stabilizer, sterilization of the pack at the time it is installed in the well is recommended. A fairly common and acceptable practice, in most cases, is to mix sterilant in each cubic meter (yd^3) of gravel as it is installed in the well.

Another method is to pour one of the chlorine compound solutions down the tremie pipes with the gravel. Recommended approximate amounts per cubic meter or cubic yard of gravel are as follows:

Metric:

- 84,000 mg/L calcium hypochlorite solution (0.45 kg per 3.8 liters of water), 3.7 liters per cubic meter

- 27,000 mg/L chlorinated lime solution (0.45 kg per 3.8 liters of water)

- 52,500 mg/L sodium hypochlorite solution (household bleach), 6.2 liters per cubic meter

English:

- 84,000 p/m calcium hypochlorite solution (1 lb per gallon of water), 3 quarts per cubic yard

- 27,000 p/m chlorinated lime solution (1 lb per gallon of water), 2.5 gallons per cubic yard

- 52,500 p/m sodium hypochlorite solution (household bleach), 5 quarts per cubic yard

When sterilization is completed, the well should be sealed and pumped to waste until the discharge has no odor or taste of chlorine. The waste solution may require treatment or other special disposition to minimize ecological effects.

12-16. Bibliography.—

"AWWA Standard for Deep Wells," 1967, AWWA A100-66, American Water Works Association, New York.

"AWWA Standard for Water Wells," 1990, AWWA A100-90, American Water Works Association, New York.

Bennison, E., 1947, *Ground-water, Its Development, Uses, and Conservation,* Edward E. Johnson Company, St. Paul, Minnesota.

Campbell, M.D., and J.H. Lehr, 1973, *Water Well Technology,* McGraw-Hill, New York.

Driscoll, F.G., 1986, *Groundwater and Wells,* 2d edition, St. Paul, Minnesota, 1089 p.

Gibson, U.P., and R.D. Singer, January 1969, *Small Wells Manual,* U.S. Agency for International Development, Washington, DC.

Gordon, R.W., 1958, *Water Well Drilling with Cable Tools,* Bucyrus-Erie Company, South Milwaukee, Wisconsin.

Roussy, R. 1994, Water Well Journal, p. 61.

National Water Well Association, 1971, *Water Well Drillers Beginning Training Manual,* Columbus, Ohio.

Speedstar Division Of Koehring, Co., 1967, *Well Drilling Manual*, Enid, Oklahoma, 72 p.

U.S. Department of the Army, 1965, *Well Drilling: Technical Manual*, Washington, DC, 5-297 pp.

Chapter XIII

INFILTRATION GALLERIES AND HORIZONTAL WELLS

13-1. Introduction.—The distinction between infiltration galleries and horizontal wells is indistinct and broad enough to allow considerable overlap. For the purposes of this manual, infiltration galleries are specifically intended to extract water from a surface-water body, usually a stream, to meet a water requirement by creating a gradient through the bed or bank of the surface-water body, toward the collection facility. A horizontal well is intended to extract water from a shallow aquifer to provide a water supply or some other purpose such as dewatering, stabilization, or skimming contaminants from the ground water.

13-2. Basic Components of an Infiltration Gallery.— Infiltration galleries usually are constructed to discharge into a sump at a point some distance above the bottom of the sump. The sump may be of almost any dimensions but commonly is a circular or square structure 1.2 to 2.5 meters (4 to 8 feet) in diameter or on a side. Depth should be adequate to permit the pump bowls to be set with adequate submergence and clearance.

The basic components of the sump and gallery design will depend upon type of water-supply needs (quality and quantity), local site conditions, and local construction standards. Normally, a solid collector pipe, tightly sealed to the infiltration gallery installation, conveys water to the inlet pipe of a sump. The actual infiltration gallery may be an agricultural horizontal drain (Bureau of Reclamation, 1993), horizontal infiltration tunnel, or suitable conduit into which water can infiltrate to be collected and transported to a central collection point.

13-3. Types of Infiltration Gallery Installations.—Lakes or perennial streams may not be suitable for simple intake structure pumping directly from the open water body. Where direct stream diversion of surface water is adversely affected by sediment-colloid content, an infiltration gallery becomes a practical economical alternative. Silt load, shoreline slope, presence of inert and physiologically harmless contaminants, rapid and unpredictable changes in water level, excessive wave action, and ice pressure are circumstances that make direct pumping difficult.

In any of these or similar cases, if minimum water depth is predictable, an infiltration gallery may provide an adequate and suitable water supply. The type of installation will depend upon the nature and permeability of the streambed or lakebed involved. In the case of a relatively impermeable bed, the gallery is placed beneath the channel or lakebed in a suitable excavation which is backfilled to completely surround and cover the gallery with selected gravel pack, which serves as an artificial aquifer.

The decision to place the gallery adjacent to or under the water surface depends on requirements of yield, quality, and site conditions for construction and maintenance. Normally, the subsurface geologic conditions of streambeds allow for economical near-surface installations within alluvial deposits (Bureau of Reclamation, 1992). Results of Bureau of Reclamation's Mariaville Test provided 0.04 to 0.05 m^3/sec (1.5 to 1.7 ft/sec) under gravity flow of sediment-free ground water per 305 meters (1,000 feet) of gallery from the Niobrara River alluvium.

Many ephemeral or intermittent streams are dry during a significant part of the year, but the channels are underlain by sand and gravel containing a significant and perennial underflow which may be captured more readily and completely by an infiltration gallery than by wells. These variable conditions dictate, to a large extent, the type of structure that will be required.

Instream infiltration galleries can be designed to divert substantial flows within environmental guidelines on wild and scenic rivers where concrete structures, steel grates, and noise from pumps are unacceptable.

13-4. Design of Infiltration Galleries.—Design considerations here will focus on infiltration galleries within shallow alluvial aquifers.

Design considerations for the actual sump should include: removal of unwanted entrapped air caused by cascading water conditions; pump(s) setting as to submergence and clearance, pump(s) bowl levels, and cycling criteria; maintenance accessibility of pumps and controllers; and suitable site location. The sump is usually cased with concrete or corrugated metal pipe, although wood shoring, brick, or concrete block have been used. The bottom is sealed with concrete or a metal plate. The top is finished with a reinforced concrete slab incorporating an inspection and clean out manhole and a hole through which the pump is installed (figure 13-1).

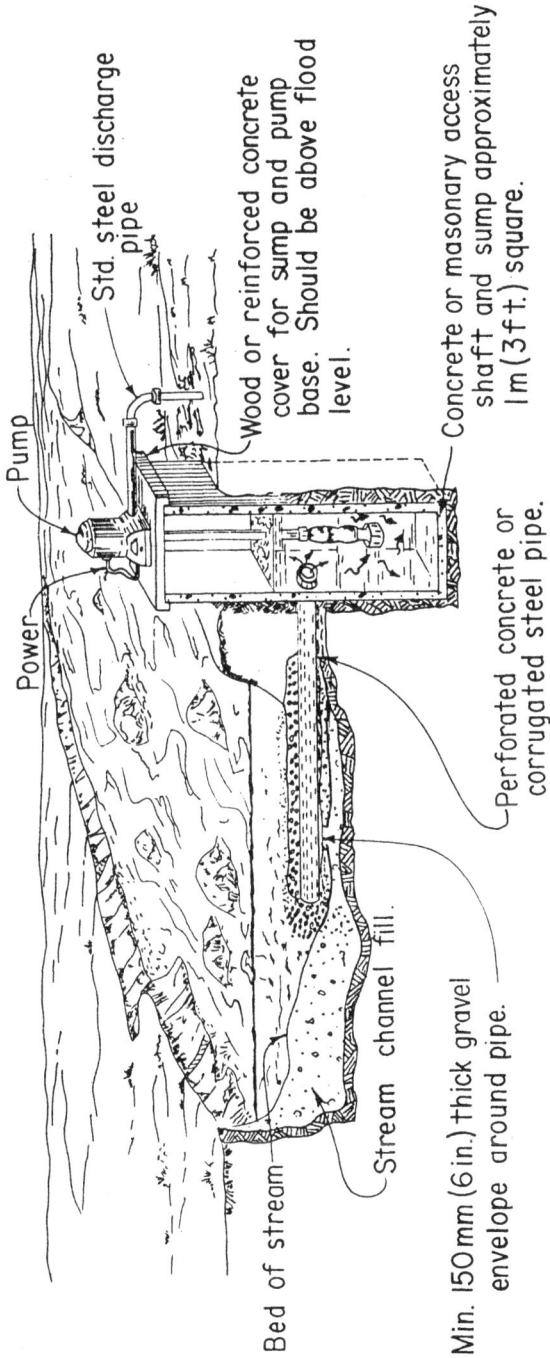

Std. steel discharge pipe

Pump

Power

Wood or reinforced concrete cover for sump and pump base. Should be above flood level.

Concrete or masonary access shaft and sump approximately 1m (3ft.) square.

Bed of stream

Perforated concrete or corrugated steel pipe.

Stream channel fill.

Min. 150mm (6in.) thick gravel envelope around pipe.

Figure 13-1.—Schematic section of an infiltration gallery constructed to intercept infiltration from a stream.

All points of possible leakage through the top are sealed with mastic, compression seals, or other waterstops. The top of the sump should be located at an elevation which will preclude its submergence by any floodflows or surface runoff.

The conductor pipe, usually fabricated of corrugated metal pipe, leads from the sump, to which it is tightly sealed, to the screen or manifold of the gallery. The pipe may be horizontal or may slope slightly downward toward the sump. Where more than one screen is used, the conductor pipe may connect to a manifold from which two or more screens, possibly of smaller diameter, extend into the permeable material. If only one screen is used, it should have the same diameter as that of the conductor pipe and may be normal to it or an extension of it. Depending upon conditions and purpose, the screen may be perforated corrugated culvert, cage-type wire-wound screen, slotted concrete, plastic, or other perforated materials.

Unless placed in very clean, permeable gravels, the screen should always be gravel packed. The same criteria are used in selecting the gravel pack and screen slot sizes as are used for gravel packed wells (section 11-11) or suitable agricultural drain envelope (Bureau of Reclamation, 1993). Other criteria are used in determining the thickness of the pack.

Under some circumstances, the gravel pack is protected by an overlying layer or a diversion wall of riprap.

If the screen is to be subjected to direct pumping pressure, provision should be made to permit back flushing of the gallery from a fairly large capacity storage tank or by direct pumping of clear stream or lake water into the sump.

The design of any infiltration gallery should provide for an average entrance velocity of 0.03 meter per second (0.1 foot per second) or less. The smaller the entrance velocity, within economical and physical limits, the better (Zangar, 1948).

Depth beneath the minimum water surface, whether surface or ground water, should be as great as physically possible and economically practical.

Permeability and storativity of a natural aquifer should be determined by pumpout tests, if possible. Where such tests are not feasible, permeability can sometimes be determined by permeability tests (see chapter X), and storativity can be approximated on a judgment basis from a number of representative samples of the material. Where galleries are placed in relatively impermeable streambed or lakebed material, the permeability of the gravel pack should be determined by laboratory tests on two or more representative samples.

Recommended minimum diameter of screen and conductor pipe is 450 millimeters (18 inches), although smaller sizes have been used successfully for domestic supplies in some areas. When installed in impermeable lakebeds or streambeds, or other places where a gravel pack is used, or where a relatively impermeable base underlies the gravel or sand, the invert of the screen should be located a distance equal to at least 1 diameter of the screen above the bottom of the excavation.

Because sealing by silt and the resultant decline in yield are common occurrences, infiltration galleries are commonly over-designed by increasing the screen diameter or length to compensate for decline in yield. The screens may also be fitted with a back-flushing mechanism using air, water, or both.

Estimates of yield are made with one of the following applicable equations:

- For a gallery surrounded by highly permeable gravel pack in slowly permeable material with minimum depth of water above streambed, channel, or lakebed.—Under this condition, the river or lake is assumed to have direct access to the gravel pack or backfill. The flow moves directly downward from the water body into the pack and then into the pipe (Zangar, 1948) (figure 3-2). The equation for determining the length of screen necessary to yield a given volume is:

$$L = \frac{Qd}{KHB} \qquad \text{13-1}$$

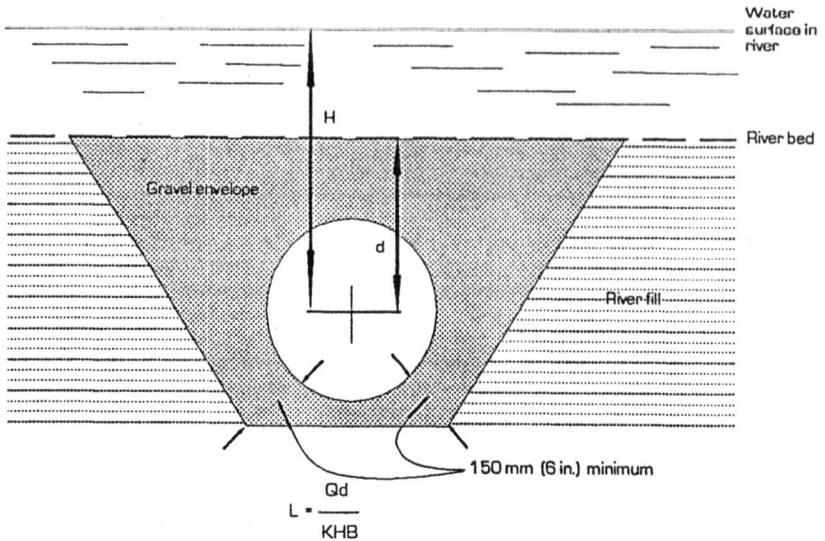

L = length of screen required, m (ft)
Q = desired discharge, m³/s (ft³/s)
d = distance from river bed to center of pipe , m (ft)
K = coefficient of permeability of gravel backfill, m/s (ft/s)
H = head acting on the center of the pipe (distance between minimum water
     surface elevation and the center of the pipe), m (ft)
B = average width of the trench backfilled with gravel

Figure 13-2.—Factors influencing flows to an infiltration gallery
screen placed in gravel backfill in a slowly permeable streambed.

where:

L = length of screen required, meters (feet)

Q = desired discharge, m³/s (ft³/s)

d = vertical distance between riverbed and center of screen, meters (feet)

K = coefficient of permeability of the gravel backfill, meters per second (feet per second)

H = head acting on the center of the pipe (vertical distance between minimum water surface elevation and the center of the pipe), meters (feet)

B = average width of the trench backfilled with gravel, meters (feet)

The axis of the gallery is commonly placed normal to the bank or shore, but it could be placed at any angle between parallel and normal so long as the minimum water-level requirements are met.

- For a gallery in permeable riverbed or lakebed with minimum depth of water above the bed (Zangar, 1948) (figure 13-3).—The equation for computing the length of screen required is:

$$L = \frac{Qln(\frac{2d}{r})}{2\pi KH}$$ 13-2

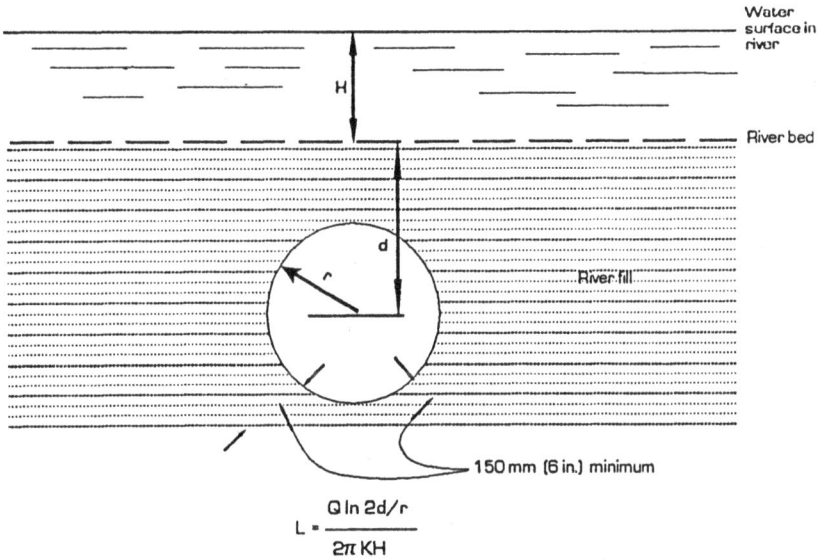

$$L = \frac{Q \ln 2d/r}{2\pi KH}$$

L = computed length of screen to yield desired discharge, m (ft)
Q = desired discharge, m³/s (ft³/s)
ln = natural logarithm
r = radius of pipe, m (ft)
K = coefficient of permeability of channel fill or lake bottom material, m/s (ft/s)
H = depth of water in the river at low flow, m (ft)
d = distance from river bed to center of pipe , m (ft)

Figure 13-3.—Factors influencing flows to an infiltration gallery screen in a permeable streambed.

where:

L = computed length of screen required to yield the desired rate discharge, meters (feet)

Q = desired rate of discharge, m³/s (ft³/s)

ln = natural log

d = vertical distance from bed of river to center of the pipe, meters (feet)

r = radius of the pipe, meters (feet)

K = permeability of the channel fill or lake bottom material meters per second (feet per second)

H = minimum depth of water above the lake or channel bed, meters (feet)

Depending upon the gradation of the sand and gravel, a gravel pack may or may not be required. When not required, the pipe is covered and the excavation backfilled with the materials removed from it. When the pack is required, the same criteria are used in selecting the gravel pack as were described for wells (section 11-11).

- For a gallery located in an ephemeral or intermittent stream channel filled with permeable material through which a perennial underflow is moving (Moody and Ribbens, 1965).—Under such conditions, the gallery is located normal to the axis of the channel and as far beneath the water table as is practical while still leaving a distance equal to at least one screen diameter thickness of permeable material beneath the invert of the screen. The yield per unit length of the gallery is estimated using equation 13-4, which has been rearranged from equation 13-3 derived by Moody and Ribbens (1965).

$$s(r,t) = \frac{q}{2K}\left\{\sqrt{\frac{4Kt}{\pi MS}}\ \exp\left(-\frac{r^2S}{4Tt}\right) + \frac{r}{M}\ erf\sqrt{\frac{r^2S}{4Tt}} - \frac{2}{\pi}\ln\left[\exp\left(\frac{\pi r}{2M}\right) - \exp\left(-\frac{\pi r}{2M}\right)\right]\right\} \qquad 13\text{-}3$$

$$q = \frac{2Ks}{\sqrt{\frac{4Kt}{\pi MS}}\ \exp\left(-\frac{r^2S}{4Tt}\right) + \frac{r}{M}\ erf\sqrt{\frac{r^2S}{4Tt}} - \frac{2}{\pi}\ln\left[\exp\left(\frac{\pi r}{2M}\right) - \exp\left(-\frac{\pi r}{2M}\right)\right]} \qquad 13\text{-}4$$

where:

s = distance between the water table and the top of the screen, meters (feet)

q = yield for a linear unit length of gallery, m^3/s (ft^3/s)

K = permeability of the aquifer, meters per second (feet per second)

t = time since pumping began on any pumping schedule, seconds

exp = exponential function, $exp\ x = e^x$

erf = error or probability function:

$$erf\ x = \frac{2}{\sqrt{\pi}} \int_0^x e^{-x^2} dx; \qquad\qquad 13\text{-}5$$

Tables for this function can be found in most standard mathematical tables publications.

M = undisturbed saturated thickness of the aquifer, meters (feet)

r = radius of the screen, feet

S = storativity of the aquifer, dimensionless

T = KM = transmissivity of the aquifer, m^2/s (ft^2/s)

π = 3.1416, e = 2.71828, and ln = natural log limits: $S{\leq}0.1M$

Because the equation gives the yield, q, per unit length, the required length for a desired Q is Q/q, where Q is the desired or required yield.

If s is larger than about 10 percent of M, Q will be smaller than estimated.

If the width and average depth of the saturated channel fill, the gradient of the top of the saturated underflow zone, and the permeability of the fill are known, the volume of underflow can be estimated from:

$$Q_t = KA\frac{h_1 - h_2}{L} \qquad\qquad 13\text{-}6$$

where:

Q_t = total underflow, m³/s (ft³/s)
K = permeability, meters per second (feet per second)
A = cross-sectional area, m² (ft²)
h_1-h_2 = elevations of the tops at the saturated section at any two points along a flowline approaching the gallery, meters (feet)
L = distance between h_1 and h_2, meters (feet)

The total estimated underflow would be the maximum recoverable, but seldom more than 60 to 75 percent of the total underflow can be intercepted.

The value of r has a small influence on the value of s and Q. Also, s and Q decrease with time until $Q = Q_t$ or 60 to 75 percent of Q, at which time the system stabilizes. When using the equation to design a gallery, three or four different combinations of s, t, and Q may be required to estimate a satisfactory relationship necessary to intercept the desired or available discharge.

13-5. Horizontal Wells.—Horizontal wells are used for a variety of purposes, including water supply, dewatering for construction, hazardous waste cleanup, and beach erosion control. The technology has advanced significantly in recent years to a point where horizontal wells are both economical to install and reliable as permanent installations. Horizontal wells provide the ability to develop relatively thin aquifers where vertical wells are inefficient. The wells can also be used in coastal areas, to skim fresh water from aquifers where saline water has intruded in the lower reaches.

Typically, water-supply installations use 150- to 200-millimeter (6- to 8-inch) corrugated, perforated plastic pipe, laid at depths of 4.5 to 6 meters (15 to 20 feet). Lengths from 120 to 250 meters (400 to 800 feet) are common. A pumping riser is provided at one end and a clean-out riser is provided at the opposite end (Bowman and Justice, 1992). Actual areal configuration of the well depends on geology and other site conditions. Rigid well screen and casing can be used for horizontal wells, but at the loss of the economy factor.

The hydrodynamics of horizontal wells are in the development stage as of this writing. However, an installation described by Bowman and Justice (1992) provides some insight to the magnitude of installations that are practical. The installation included 178 meters (593 feet) of 150-millimeter (6-inch) polyethylene pipe with a 200-millimeter (8-inch) pumping header and a 150-millimeter (6-inch) clean-out. The pipe was installed at a depth of about 4.9 meters (16 feet) in a highly permeable strata. A 20-horsepower submersible pump, capable of producing 1,600 liters per minute (420 gallons per minute) at 275 kPa (39 lbs/in^2) was installed. The actual production rate was 1,265 liters per minute (333 gallons per minute).

Where aquifer materials have good permeability but a relatively thin fresh water lens that is underlain by salt or brackish water, (figure 13-4) as is common near many coastal areas, the well is placed normal to the direction of flow of the ground water and at a depth which will preclude drawdowns great enough for saltwater to come up into the well.

Equation 13-4 is used to determine the optimum depth of the well, except M represents the thickness of the fresh water layer. However, s must be limited so that saltwater does not come up to the screen. The Ghyben-Herzberg principle is usually adequate to determine the maximum value of s (Walton, 1970):

$$h_s = \frac{d_s}{d_s - d_f} h_f \qquad \text{13-7}$$

where:

h_s = distance below mean sea level of the saltwater point head at the fresh water, saltwater interface

h_f = distance from top of the fresh water table to mean sea level or the point saltwater head

d_s = density of the saltwater

d_f = density of the fresh water

For seawater of average density of 1.027 and fresh water of average density of 1.000:

$$h_s = \frac{1.027}{1.027 - 1.000} h_f = 38.0 h_f \qquad \text{13-8}$$

The depth to the interface is about 38 times the distance between the water table and mean sea level or the saltwater piezometric surface at the point of measurement. For example, if h_f was

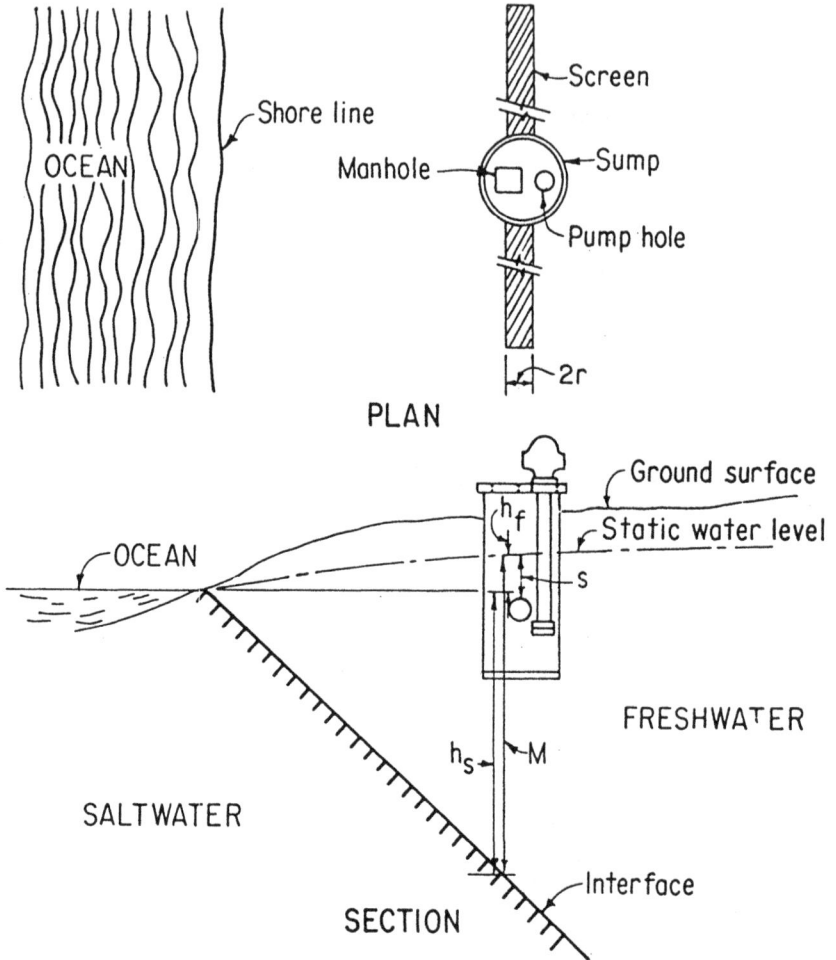

Figure 13-4.—Schematic section of an infiltration gallery
constructed to obtain fresh water near a seacoast.

0.9 meter (3 feet), the interface would be located about 35 meters
(114 feet) below the water table. If an infiltration gallery lowered
the water table 0.75 meter (2.5 feet), h_f would be 0.15 meter
(0.5 foot), and the saltwater could come up to only 5.7 meters
(19 feet) below the drawdown.

The saltwater, fresh water interface is not a sharp contact but a zone of brackish water between the fresh and saltwater bodies. Approximate saltwater point head, h_s, may be determined by installing a piezometer to below the brackish zone; slowly pump or bail water out of the casing until it is completely filled with saltwater, and permit the level to stabilize. The water-table elevation can be determined in a shallow observation well. The shallow fresh water-table elevation minus the elevation of the point head of saltwater equals h_f.

Advantages offered by horizontal wells include economical development of shallow, thin aquifers and the ability to selectively extract water from aquifers having stratified water quality. Saline water often intrudes on the lower layers of coastal aquifers while the shallower layers maintain fresh water through recharge from the surface. Pumping of the saline water can be avoided by completing horizontal wells in the shallow strata above the fresh water-saline water interface. In hazardous waste cleanup work, horizontal wells can be used to skim light nonaqueous phase liquids from the top of the water table.

Difficulties that may be experienced with horizontal wells include the possibility of collapsing the pipe, disturbance of large areas of land surface compared to vertical wells, and dewatering of the zone above the pipe. Pressure differences between the aquifer and the negative pressure inside the pipe caused by pumping can exceed the collapse strength of the pipe if flow into the pipe is restricted. The condition may develop as a result of filter clogging, particularly where a synthetic filter sock is employed on the collector pipe. The aquifer above the horizontal well may be dewatered if aquifer characteristics are insufficient to transport replacement water at or above the pumping rate. If the dewatered zone reaches the collector pipe, aerated water will enter the pipe and create a cavitation hazard for the pump. Also, production rates will be sharply reduced. The pumping rate should be regulated to avoid the dewatered condition to the extent possible.

13-6. Bibliography.—

Bowman, B.J. and D.R. Justice, 1992, "Tapping Shallow Ground Water with Horizontal Wells," Proceedings of the Irrigation and Drainage Sessions at Water Forum '92, American Society of Civil Engineering.

Moody, W.T., and R.W. Ribbens, 1965, "Ground Water-Tehama-Colusa Canal Reach No. 3, Sacramento Canals Unit, Central Valley Project, California," memorandum to Chief, Canals Branch, Bureau of Reclamation, December 29, 1965.

Bureau of Reclamation, 1978 "Drainage Manual," revised reprint 1993.

Bureau of Reclamation Special Report, January 1992, "Ground-Water Recharge Plan - O'Neill Unit, Nebraska."

Walton, W.C., 1970, "Groundwater Resource Evaluation," McGraw-Hill, New York, p. 194.

Zangar, C.N., 1948, "Determination of Perforated Pipe Length Embedded in Permeable Soil Necessary to Supply 55 cfs for the N-Bar-N Pumping Plant, Montana Pumping Unit, MRBP," memorandum to R.E. Kruger, Bureau of Reclamation, November 3, 1948.

Chapter XIV

DEWATERING SYSTEMS

14-1. Purposes of Dewatering Systems.—Dewatering is necessary in many construction projects to prevent one or more of the following conditions:

- Unstable natural or excavated slopes

- Unstable, unworkable, or unsuitable subgrade

- Boils, springs, blowouts, or seeps on the slopes or subgrade

- Flooding of excavations or subgrade structures

- Uplift beneath and disturbance of constructed features (example, concrete slabs)

- Dilution, corrosion, or other adverse effects on concrete, metals, or other construction materials

- Threat to the stability of nearby structures

- Threat to the availability and water-levels of surrounding surface and ground water

- Instability of cutoff facilities (example, cofferdams)

- Loss of fines from the foundation

- Threat to life and property

- Construction delays and increased costs

The last two conditions are, of course, a result of the first 10 conditions.

If artesian conditions exist, dewatering may be required even where no water is encountered in the excavation. Water under piezometric head can be pumped to the surface by soil vibrations generated by construction equipment. A properly designed and constructed dewatering system will lower the water-table or piezometric surface adequately to permit safe and dry construction.

The depth to which the piezometric surface must be lowered is influenced by soil texture. Fine texture soils will pump water higher than will coarse textured soils.

In Bureau of Reclamation terminology, dewatering is defined as the removal of ground water or seepage from below the surface of the ground or other constructed surfaces, and the control of such water. Removal of ponded or flowing surface-water is termed unwatering. The two methods are often used conjunctively in construction operations.

14-2. Methods of Dewatering and Soil Stabilization.—In some small excavations, dewatering is done by alternate stage excavation and subsequent gravity drainage to ditches and sumps in the excavation (figure 14-1). In some instances, this process is supplemented by sheet or similar types of piling driven at the edge of the excavation and, if possible, into an impermeable underlying bed (figure 14-2). Similar methods have been used in the past on larger excavations. Over several decades, improvements and developments in well points, pumps, well construction techniques, and increased knowledge of ground-water hydraulics have led to the design of better, more efficient methods using well points, deep well systems, horizontal drains, sand drains, and vacuum and electro-osmosis techniques.

Methods of dewatering include: (1) well-point systems, either standard or educator-type, (2) deep wells, (3) horizontal drains, (4) electro-osmosis, and (5) vertical sand drains or stone columns. Where artesian conditions exist, relief wells to reduce artesian pressures may be used. Depending on conditions, these wells may be pumped, or they may be drained through gravity flow in a collector system. Dewatering of the downstream toe of dams using horizontal drains is frequently used to reduce uplift pressures. Artificial barriers, such as sheet piling, cutoff walls, slurry trenches, freezing soils, or grouting are also used to reduce flow.

In massive hardrock, interception and disposal of water is the principal requirement of a dewatering system. Where the permeability of the rock consists primarily of fractures and other similar openings, grouting may be required in conjunction with the dewatering procedures (Driscoll, 1986).

The choice of dewatering method is determined by local geologic and hydrologic conditions, reason for dewatering, and to some extent, equipment readily available. Where fine-grained soils are

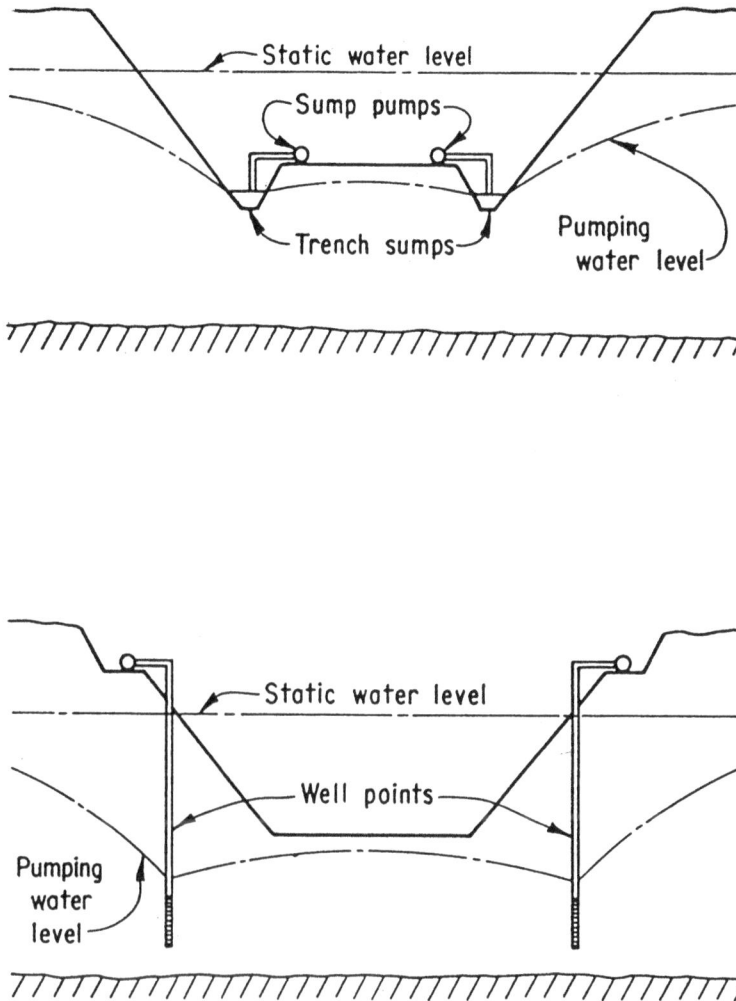

Figure 14-1.—Schematic sections showing dewatering by
pumping from sumps and well-point systems.

involved, closer well spacing and longer time periods will be
required. Layered materials are also likely to be more complicated
to dewater. In such cases, it is very important to have a contractor
experienced in dewatering to design and operate the system. In
addition, where dewatering occurs at the toe of dams or other
critical areas, dewatering methods or pumping rates should be
chosen to minimize removal of fines from the dam or dam
foundation.

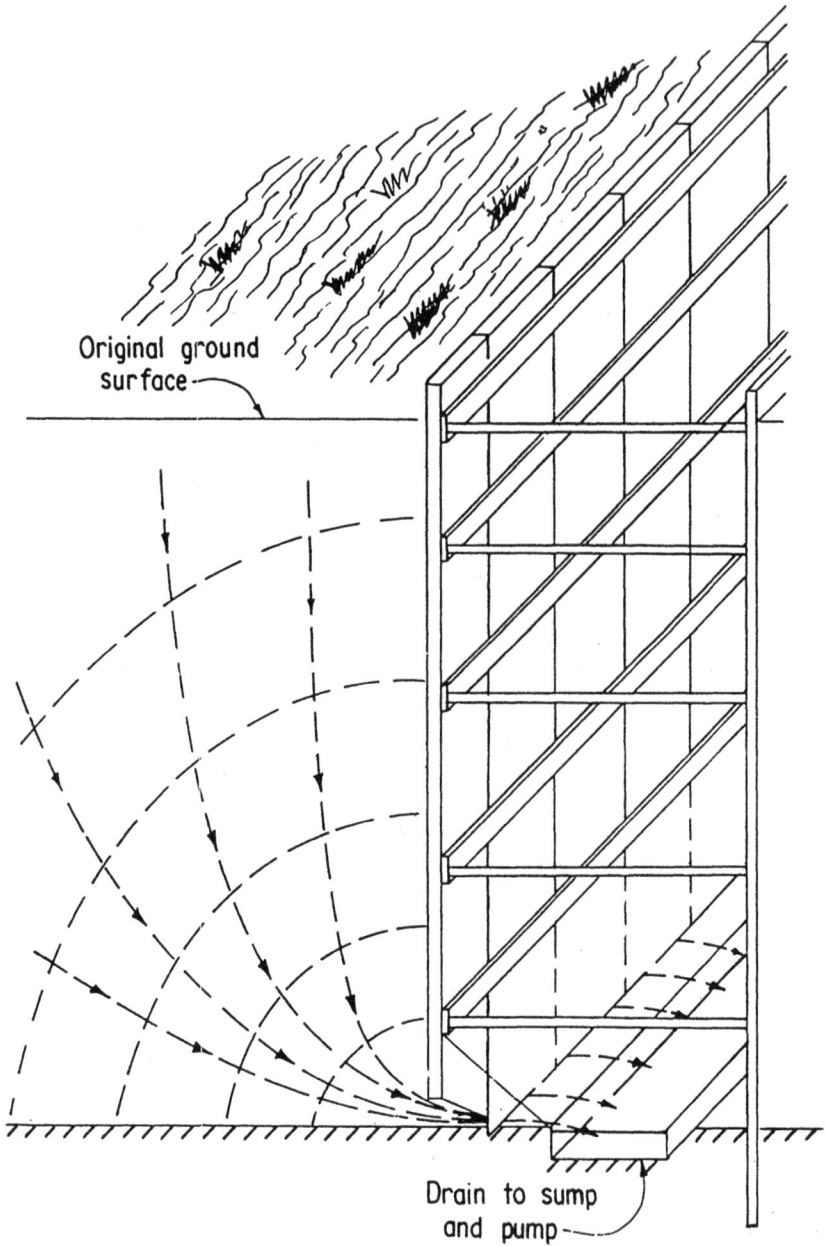

Figure 14-2.—Dewatering by drainage under or
through shoring to a sump.

Driscoll (1986) lists the two most important considerations in the design of a well dewatering system as storativity and transmissivity because these factors control the volume of ground water in the area to be dewatered and the rate at which it can be removed. However, many Bureau of Reclamation projects are in relatively narrow river valleys where boundary conditions rather than transmissivity and storativity may control all but initial stages of inflow to the dewatering system. In such conditions, conducting one or more tests involving simultaneous pumping of several closely spaced wells may be advisable to obtain a clearer picture of interference and boundary effects.

(a) Well-Point Systems.—A well point is a 1-1/2- to 3-1/2-inch-diameter well screen, usually 0.5 to 1 meter (18 to 40 inches) long (figure 14-3). Well points may be jetted (figure 14-4), driven, gravel packed, installed in open holes, or otherwise emplaced. Numerous types of well points are manufactured. Some use wire screen of various types that cover perforated pipe, and others are wire wound, slotted, or perforated. The percentage of open area is variable, and the slots are usually fixed to a relatively small number of sizes. The wire-wound type offers the greatest flexibility in slot size as well as the greatest percentage of open area and most efficient distribution of slots. In some types of installation, regular 102- to 152-millimeter (4- to 6-inch-diameter) well screens and casing may be set. The well point is usually attached to a riser pipe which may be 38 to 76 millimeters (1-1/2 to 3 inches) in diameter. Well-point systems for dewatering consist of a number of well points placed at 0.5 to 2 meters (2- to 6-foot) intervals along a line. The riser pipes are connected to a header usually 150 to 200 millimeters (6 to 8 in) in diameter (figure 14-5) with a swing connection (figure 14-6) which can rotate through 360° horizontally and about 270° vertically.

The swing connector contains a valve that controls water withdrawal from the well point. The yield may vary between well points in any system; consequently, the discharge of each well point must be controlled by adjusting the valve so that drawdown is not large enough to expose the top of the screen and draw air into the system. The header pipe may be up to 183 meters (600 feet) long and placed as straight and level as possible. The pipe is usually connected to a centrifugal pump located in about the middle of the header and which can develop from 5 to 7 meters (15 to 22 feet) of suction lift. Well points operating by suction lift are placed with the top of the screen at depths 5 to 6 meters (15 to 18 feet) below the static water-level.

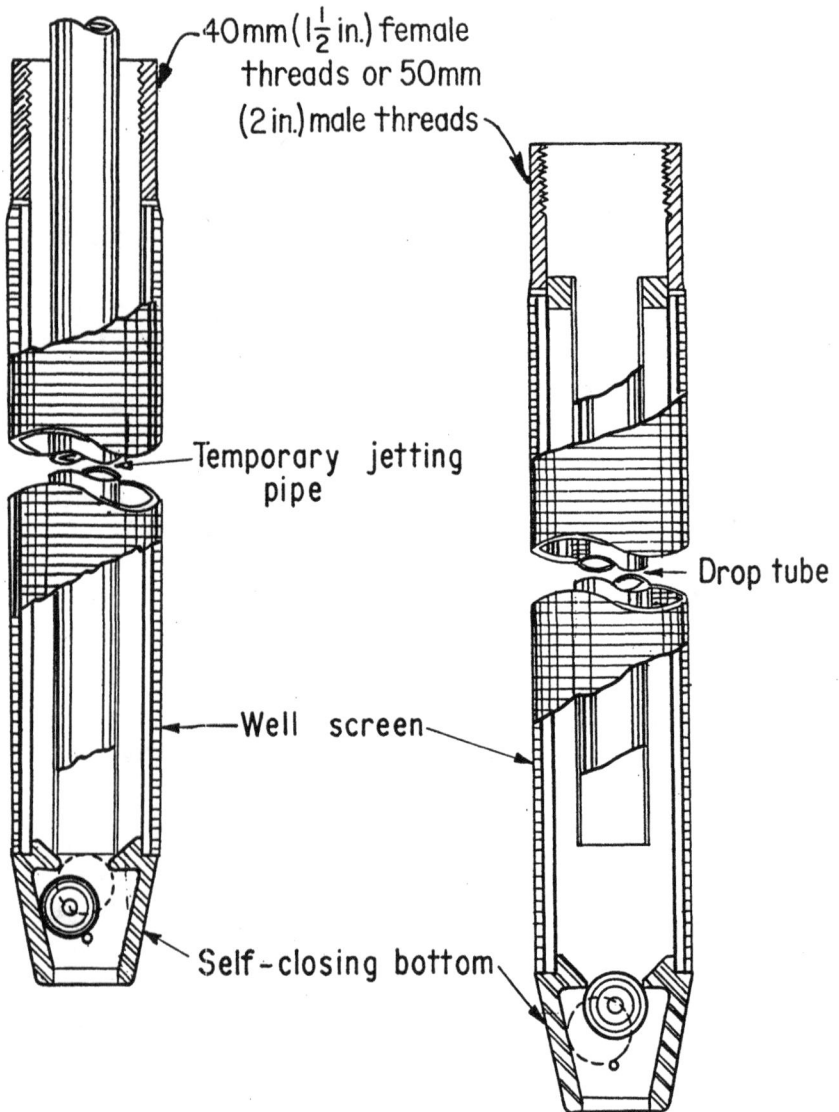

Figure 14-3.—Typical well points equipped with jetting tips.
From (UOP Johnson, 1969).

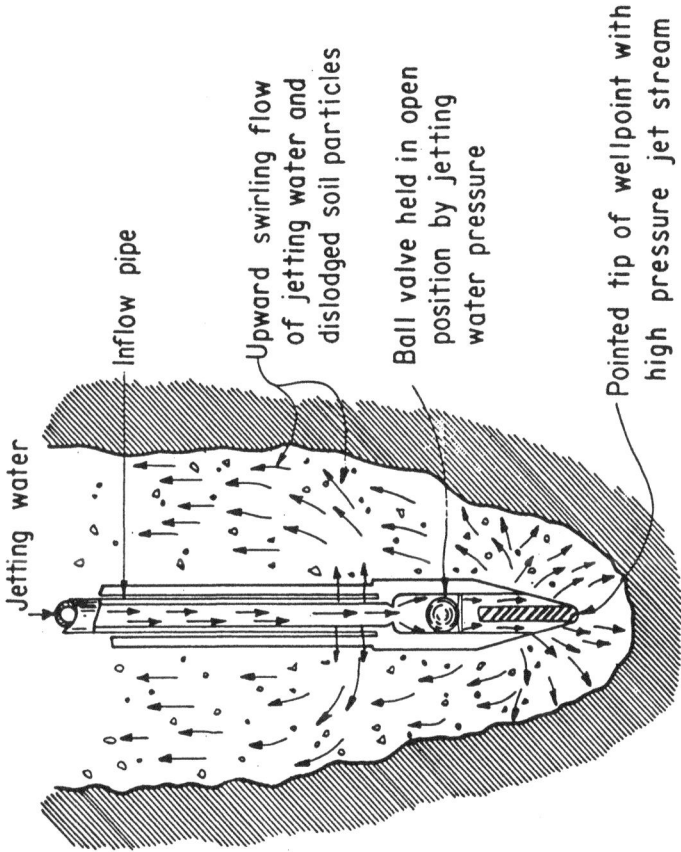

Jetting water

Inflow pipe

Upward swirling flow of jetting water and dislodged soil particles

Ball valve held in open position by jetting water pressure

Pointed tip of wellpoint with high pressure jet stream

(Griffin Wellpoint Corp.)

Figure 14-4.—Installation of a well point by jetting (Griffing Well Point Corp.).

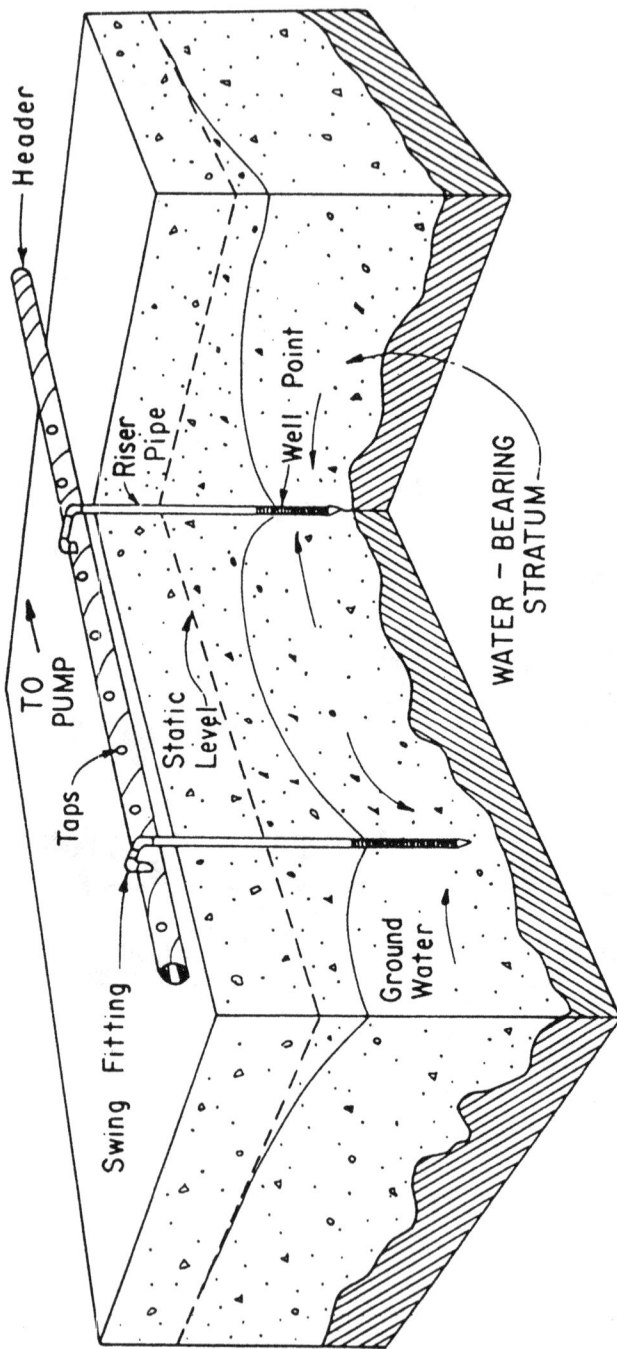

Figure 14-5.—Schematic section of a portion of a well-point dewatering system (UOP Johnson Division, 1969).

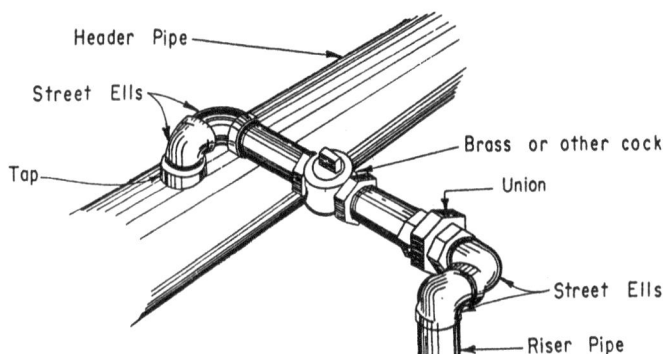

Details of connecting fitting from riser pipe to header pipe. Made up of 25 to 65 mm (1 to 2½ in.) pipe fittings, the connection may be swung horizontally and vertically to meet the top of the riser pipe. This flexibility is important since, in practice, the well point and its riser pipe generally will not match the calculated position.

Figure 14-6.—Swing connection fittings for riser pipe to header pipe connection (UOP Johnson Division, 1969).

An excavation is usually dug to within a few centimeters of the water-table before installing a well-point system so that maximum advantage may be made of the suction lift available (figures 14-1 and 14-7). Standard well-point systems are effective only to a depth of about 5 to 6 meters (15 to 20 feet) because they operate on a suction system. However, the total depth of excavation dewatered by well points can be significantly larger using a stepped or stage system, with each successive stage about 5 meters (15 feet) below the previous step (figures 14-7 and 14-8). When more than two stages are involved, supplemental wells and well points may be required to control certain unstable spots in the excavation (figure 14-9).

(b) Jet Eductor Systems.—Jet eductor systems (sometimes called jet ejector systems) are used primarily where standard well-point systems or wells are likely to be ineffective because of the presence of fine-grained materials. The systems can be either two-pipe or single-pipe systems.

The system operates using a venturi. Supply water at high pressure travels down the well through the supply pipe to the tapered nozzle. The supply water exits the nozzle tip at less than atmospheric pressure, creating a partial vacuum in the suction chamber. This vacuum draws ground water into the suction

FIRST STAGE WELL-POINT SYSTEM

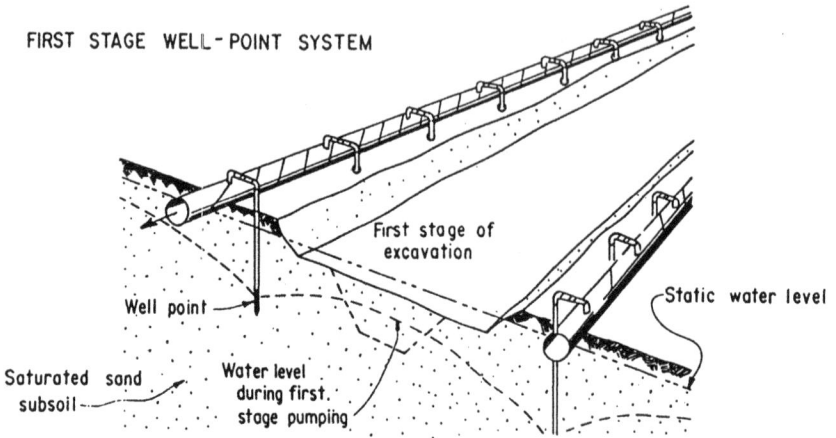

Figure 14-7.—Single stage well-point system
(UOP Johnson Division, 1969).

SECOND STAGE WELL-POINT SYSTEM

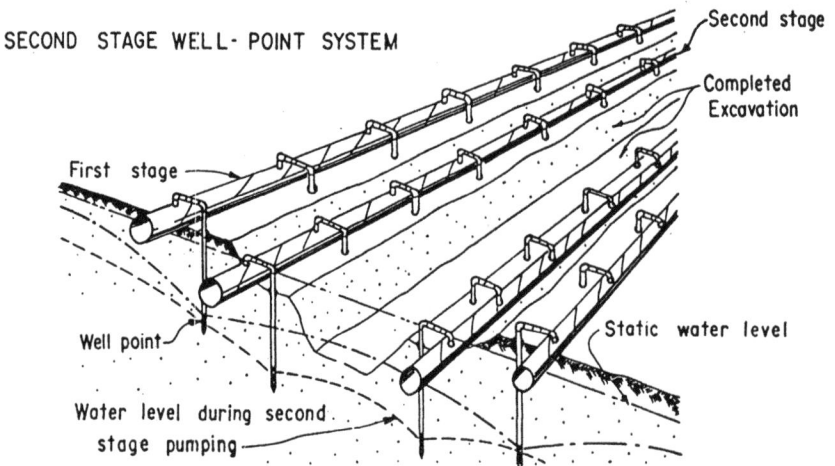

Figure 14-8.—Two stage well-point system
(UOP Johnson Division, 1969).

chamber, where it passes into the venturi. The decrease in velocity in the venturi creates a sufficient increase in pressure to transport the combined flow to the ground surface (Powers, 1981).

Although educator systems are more expensive to install, operate, and maintain than standard well-point systems, in certain cases their advantages may offset the increased costs. They can

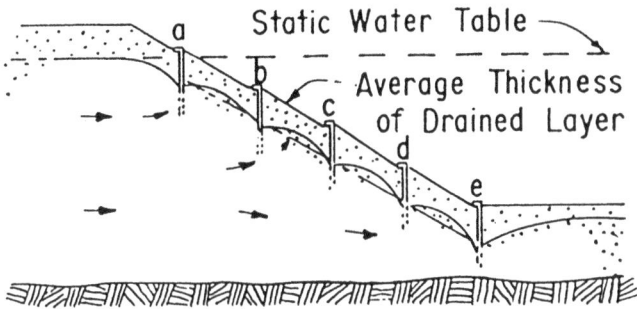

MULTISTAGE WELL POINT SYSTEM
(Stages a through e)

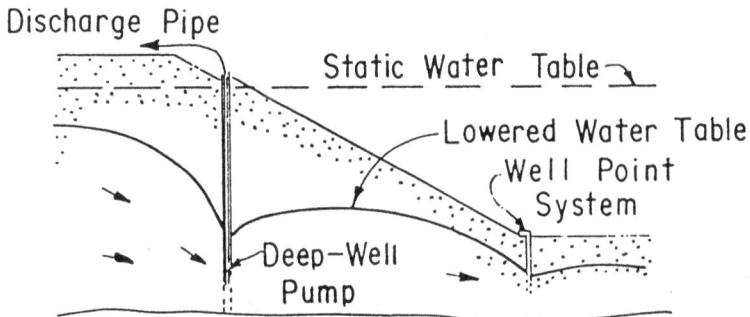

COMBINED DEEP WELL AND
WELL POINT SYSTEMS

Figure 14-9.—Multistage well-point and combined systems. A system combining a deep well pump and a well-point system that may be used as an alternative to a multistage well-point system (Driscoll, 1986).

lift water from depths of 30 meters (100 feet) or more, and thus may obviate the necessity for a multistage system in deep excavations.

Dewatering of one well in the system does not disrupt system operation, as is the case in standard well-point systems. Thus, they have applications where lithologic conditions are variable within the site. Also, in fine-grained material, neither standard well points nor wells are very effective in lowering water-levels over a large area. The vacuum-type educator system creates negative pore pressure or tension in the soil, and thus aids drainage of the fine-grained soils.

Construction details of an educator well are shown on figure 14-10. The well should be gravel packed from the bottom of the hole to within a few meters of the surface of the slowly permeable material. The remainder of the hole, the top of the slowly permeable material, should be sealed with bentonite or other impermeable soil. By maintaining a vacuum in the well screen and pack, flow toward the well or well points is increased, particularly in stratified soils. The system normally requires use of closely spaced well points, and pumping capacity is usually small. Vacuum booster pumps may be required on the headers of individual wells for effective operation.

(c) *Horizontal Drains.*—Where excavations are located on slopes or where landslides have occurred, the excavations may be dewatered and stabilized with horizontal wells or drains. Drains can be installed with drills especially designed for horizontal drilling. The first horizontal drain should be installed near an existing observation well in which the influence and effect of the drawdown can be observed. Using the principle of superposition, approximate drain spacing to achieve an adequate drawdown can be estimated. Gravel packing should be used whenever possible because the pack increases efficiency several fold. Dewatering can usually be accomplished by gravity flow, but pumping can be used if necessary (Moody and Ribbens, 1965). Horizontal drains are also used to dewater the downstream toe of dams. These drains are often polyvinyl chloride (PVC) pipe laid in graded gravel envelope designed to enhance hydraulic flow properties around the pipe while stabilizing the base soils. These drains may consist of perforated pipe on plain pipe laid with open joints (Bureau of Reclamation, 1987). These toe drains are often installed using special trenching machines. Bureau of Reclamation experience indicates that when drains are constructed using specially adapted

Header pipe, vacuum
pump attached

Atmospheric
pressure

Saturated with capillary water

Seal

Static water level

Well point equipped
with suction
lift pump

Sand filter
voids under
vacuum

Water level
in filter

Cock

Discharge header

Cock

Swing fitting

Original ground surface

Pressure header

Static water level

Expanding rubber seal

Sand filter voids
under vacuum

Impermeable bentonite seal

Discharge pipe

Riser pipe
Drop pipe

Jet ejector pump equipped well
or well point. Diagram
illustrates vacuum method of
drainage. A regular installation
is similar but omits the
expanding rubber seal and the
bentonite seal. Submersible
pump installations are similar
but there is only the discharge
pipe and power cable in the
riser pipe.

Jet ejector pumping unit

Well point or screen

Gravel pack

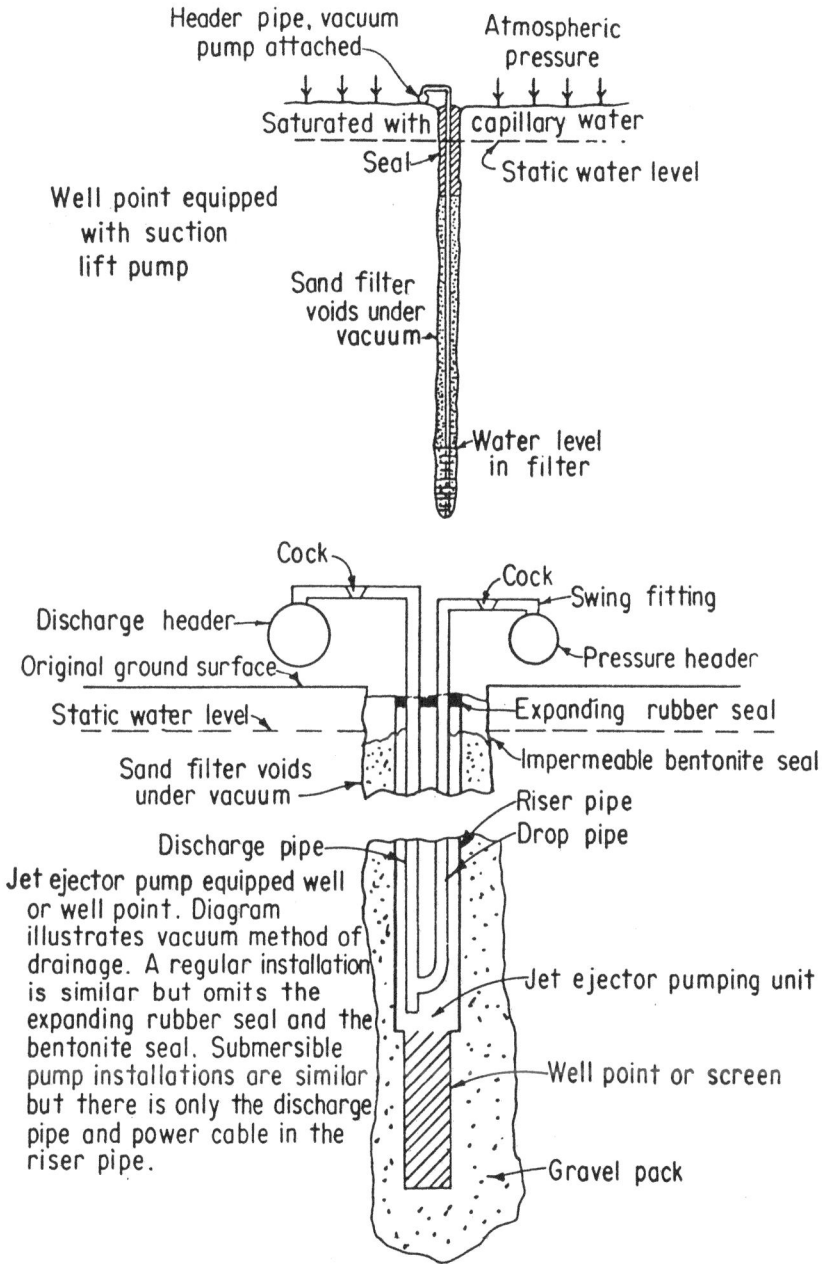

Figure 14-10.—Use of a vacuum to increase well-point efficiency.

trenching machines, fewer problems are encountered when the piezometric level along drain centerline has been lowered in advance of excavation to within 1 meter (4 feet) above drain subgrade or when excavation, drainage tubing, and envelope installation proceeds in a continuous fashion and at a constant rate exceeding 1.5 meters (5 feet) per minute. Continuous flow of envelope material must be maintained within the trenching shield and about the drainage tubing during drain construction to ensure that the constructed drain lies within a continuous envelope. One small gap in the envelope can destroy the entire drain. High piezometric pressures and slow installation rates may cause blockages of the envelope material in the shield. The Bureau of Reclamation *Drainage Manual* (1993) and the Pipe Drains Module of Reclamation's Comprehensive Construction Training Program contain an indepth discussion of this subject.

(d) *Electro-osmosis.*—Electro-osmosis has been used to drain and stabilize saturated fine-grained soils. The installation and operation and maintenance costs are usually higher than with a well-point system, but the electro-osmosis system may accomplish the purpose where a well-point system will not. An electro-osmosis system consists of negative and positive electrodes installed at spacings of 3 to 30 meters (10 to 100 feet). The cathodes (negative electrodes) consist of smaller diameter well points because the rate of discharge is usually small and intermittent. A small suction pump is usually used on the manifold. The anodes (positive electrodes) may be iron or steel pipe, rails, or other available conductors placed midway between the cathodes, or separate lines of cathodes and anodes may be placed 5 to 6 meters (15 to 20 feet) apart (figure 14-11). The current requirement may be relatively high because, even with spacing of about 6 meters (20 feet), 25 or more amperes are required to produce 130 to 150 liters (35 to 40 gallons) per day per well point. Detailed design procedures are not included herein because the system is seldom used (Mansur and Kaufman, 1962; Terzaghi and Peck, 1967).

(e) *Vertical Sand Drains.*—Vertical sand drains have been used in conjunction with well points to facilitate drainage in stratified soils. The usual installation calls for 400- to 500-millimeter (16- to 20-inch) diameter sand drains to be installed on 2- to 3-meter (6- to 10-foot) centers through the materials to be stabilized and extended to the underlying permeable layer where the well points or other wells are placed. Under such conditions, the well points or wells are usually gravel packed and may be designed for

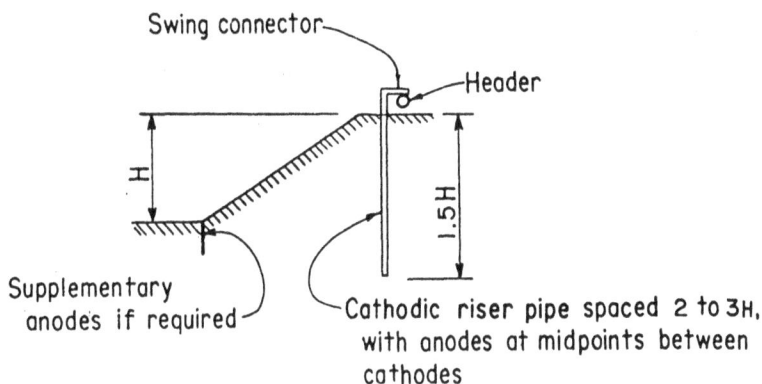

Figure 14-11.—Use of the electro-osmosis technique for dewatering.

vacuum drainage as well. The sand drain backfill is selected according to the same criteria used in selecting a gravel pack (section 14-11) (Driscoll, 1986).

(f) Drainage Wells.—Drainage wells are occasionally used as an alternative to or supplement for deep, open, or buried pipe drains for subsurface drainage. Drainage wells differ little from conventional water-supply wells except that the provision of water is incidental to the lowering of the water-table. However, conditions must be especially favorable to justify consideration of drainage wells as an alternative to conventional drains. Conditions include:

• Presence of a productive aquifer

• Presence of adequate vertical permeability in all materials between the root zone and the aquifer

• Availability of pumping energy (unless artesian flow is adequate to provide the necessary drainage)

In multiwell drainage, the wells are usually spaced to provide mutual interference to maximize the drawdown at the critical midpoint areas between wells.

Investigations necessary to determine the feasibility of using drainage wells generally are extensive and costly. Primary basic data needed include:

- Aquifer transmissivity and storativity

- Areal extent, thickness, and homogeneity of the aquifer

- Mode of occurrence of ground water—whether unconfined, confined, or leaky

- Estimated recharge caused by precipitation, irrigation, and other sources

- Boundary conditions

- Water quality data

The total capacity of wells in a drainage well field must exceed the estimated average recharge to the aquifer to maintain an acceptable depth to the water-table. Capacity up to 1.5 to 2 times greater than the recharge may be necessary in some cases to:

- Establish an initial gradient to the wells

- Permit rapid lowering of the water-table prior to the irrigation season where pumping is not continued year around

- Permit rapid lowering of the water-table in local areas where unforeseen rises in water-levels are caused by above normal precipitation, surface flooding, well failure, or other similar factors

- Permit adequate control within the drained areas

(g) Deep Wells for Pumps.—When artesian pressures are involved or where the soil to be dewatered is part of a relatively uniform thick and permeable aquifer, dewatering can frequently be accomplished with a few deep, high-capacity wells equipped with deep well turbine or other similar pumps. If the transmissivity and storativity of the aquifer can be determined, application of the principle of superposition (section 5-20) can be used to determine approximate well capacity, depth, and spacing required to accomplish the dewatering. The principle is the same as for drainage wells except drainage wells are operated in a free water-table condition. An advantage of deep wells rather than well points is that they often may be located outside the boundaries of the excavation and, hence, offer little or no interference with the excavation operations (figure 14-12).

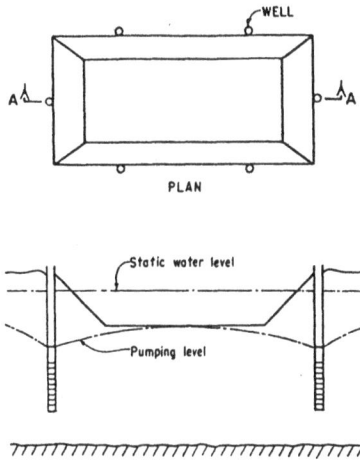

Figure 14-12.—Use of deep wells and high lift pumps for dewatering.

(h) Relief Wells.—Relief wells are used at the toe of dams and levees and other structures such as stilling basins and outlet structures to relieve excess hydrostatic or uplift pressures and stabilize the structure, soils, or strata. They reduce the potential for liquefaction ("sand boils") caused by increased head or seismic activity, and for piping or internal erosion. They are used extensively downstream from dams or on the landward side of levees where pervious materials are overlain by an impervious or semipervious layer which creates confined conditions in the pervious materials. Where they are used to reduce existing hydrostatic head, a collection system is required to lower the water to an acceptable level. In these cases, a gravity drain system is preferred because of the high cost of continuously pumping a number of wells. Relief wells are often used in combination with underseepage control measures such as upstream blankets, downstream seepage berms, and grouting. In planning relief wells, the ground-water hydrologist thus often must work closely with geotechnical and drainage engineers.

A relief well system is a flexible control measure because the number of wells can be increased if hydrostatic levels increase. In addition, wells can be pumped if necessary. Relief wells can also be used to monitor changes in chemical quality which might indicate solutioning or piping. However, maintenance of relief well systems can be high. The conditions inherent in relief well systems are conducive to the growth of iron-related bacteria, and clogging from this or other sources is common. Therefore, special attention should be given to operation, rehabilitation, and maintenance problems at the time the wells are designed.

Where relief wells are installed as a safety precaution in the event of seismic loading, potential pore pressure changes will need to be determined. In layered deposits, considerable difference may exist in pore pressures in various layers.

Careful records of flow rates in relief wells and corresponding reservoir levels need to be kept to evaluate the effectiveness of the wells. Piezometers should be installed at midpoints between wells to provide information on water-levels in the main water-bearing strata. If declines in well discharges are accompanied by declines in water-levels, siltation in the reservoir is probably reducing seepage. However, if water-levels rise while production declines, wells are probably clogging and losing their effectiveness. If so, wells should be treated or redeveloped. Additional wells may need to be installed if treatment and redevelopment do not result in increased flow and reduced piezometric levels.

Installation of relief wells may result in increased seepage discharge. If seepage discharge becomes excessive, alternative solutions to provide stability, such as loading berms, may be necessary.

Stone columns are sometimes used as an alternative to relief wells to provide dissipation of pressures in layered deposits at the toe of dams (U.S. Army Corps of Engineers, 1993 and 1994).

(1) Design.—Design of relief wells must take into account geologic and hydrologic conditions at the site. If layered strata are present, care should be taken not to interconnect separate zones because such a procedure could potentially create higher uplift pressures in previously unaffected zones. Also, chemical constituents of the underseepage should be considered when selecting casing because the water may be corrosive or encrusting.

In general, relief wells should be designed to fully penetrate the permeable zone. However, where impervious material is at great depth, this penetration may not be practical. In these cases, well depths should be determined by the geologic and hydrologic conditions at the site.

Screen length and slot size should be sufficient to minimize head losses in the well. A graded filter should be installed to minimize removal of fines. If the collection system is located below ground level, the casing should still extend to the ground surface for inspection and maintenance. A check valve and rubber gasket should be provided at the top of the well to protect the well from back flooding by silty surface-water (Singh and Sharma, 1976). In general, the criteria used for design are similar to those used for design of a production well except that surface casing for sanitary protection is not always required. Development of the well is important to assure adequate waterflow through the system.

(2) System Design.—The spacing of relief wells is determined by the aquifer parameters and geology and by the required reduction in pore pressure or piezometric head. Where deposits are lenticular, closer spacing of wells may be necessary to increase the interception of local zones of high hydrostatic pressure.

A line system or well field system may be used, although a line system is usually sufficient. The spacing should be sufficiently small that the required drawdown can be easily obtained at the midpoint between wells. Computer programs can be used to estimate drawdowns from a well field.

14-3. Field Investigation for Dewatering Systems.—The amount of investigation necessary to permit design of a satisfactory dewatering system depends upon the geologic and hydrologic complexity of the area and the depth and areal extent of the excavation.

The scope of all activities associated with dewatering projects includes:

- Site review
- Site investigations
- Data collection
- Data interpretation, evaluation, and presentation
- Compiling specifications paragraphs
- Construction considerations
- Supervision or overview of the operations
- Monitoring and evaluation of the operations
- Documentation of results (e.g., final construction report)

Because adequate and appropriate subsurface data are essential to the proper design, installation, and operation of the dewatering facilities, an investigation program cannot be established in detail until some reconnaissance level drilling has been done and general ground-water geologic and hydrologic conditions are determined.

Construction sites that require control of ground water for construction or proper operation of the facility must be identified as early in the planning or design process as possible (Bureau of Reclamation, 1988).

The procedures for field testing and data analysis have been described and discussed in previous chapters. The principles and procedures discussed are directed primarily toward design, maintenance, and operation of permanent production wells. In dewatering an unconfined aquifer, the equations and formulas previously discussed are not always applicable. Drawdowns are usually in excess of 65 percent of the unconfined aquifer thicknesses. Furthermore, well points are closely spaced and estimates of yield, drawdown, interference, and similar factors do not conform to equations usually used. Much depends upon the judgment and experience of the technician, use of empirical equations, and the results of specific investigations conducted prior to design and installation of a dewatering system. If possible, such investigations should be initiated a year or more before construction is undertaken so that data may be obtained on seasonal and annual variation of ground-water levels. The response of ground-water levels to precipitation and to the variation in water stages of adjacent lakes and streams can then be quantified. These data are obtained by periodic measurement of observation holes and piezometers and from water records of surface-water levels and precipitation. Drilling, completion of observation wells, and piezometers are discussed in sections 3-11 and 3-12.

In and adjacent to the area of excavation, exploration holes should be drilled to a depth 1 to 2 times the proposed depth of the excavation. The holes should be carefully logged and, ideally, samples should be taken at 1.5-meter (5-foot) intervals or at each change in formation, whichever is less. Mechanical analyses should be made on samples consisting of silt or coarser materials. When more than one aquifer is encountered in the holes, static water-levels should be measured in each aquifer by means of piezometers. Pumpout aquifer tests, possibly with multiple wells pumping simultaneously, to determine transmissivity, storativity, and boundary conditions should be performed on the aquifers present.

Where use of well points is planned, a test installation may be desirable to determine response to the pumping of one or more well points.

On completion of the exploration program, a three-dimensional diagram of the area of excavation and adjacent land, based on available geologic and hydrologic data, should be prepared and studied as a basis for design of a dewatering program and possible location of additional observation wells. Key exploration holes, both in and adjacent to the proposed excavation, should be completed as observation wells and piezometers (figure 14-13). These locations should be measured periodically prior to construction. As many observation wells as possible should be preserved during construction and measured frequently during construction as a check on the performance of the dewatering system.

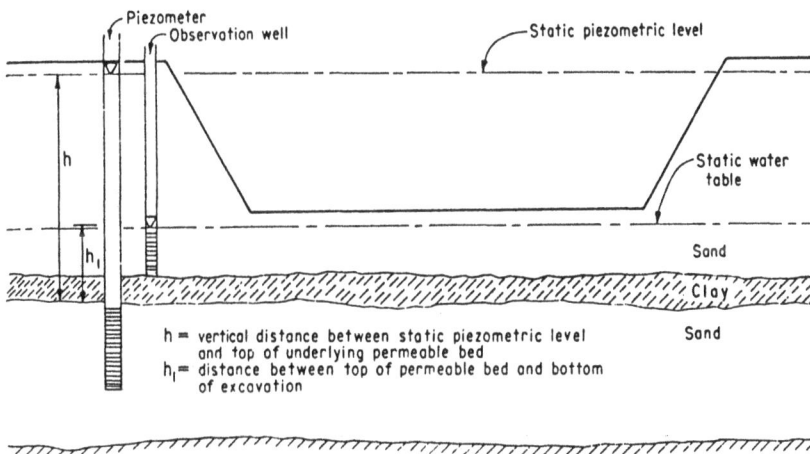

h = vertical distance between static piezometric level and top of underlying permeable bed
h_1 = distance between top of permeable bed and bottom of excavation

Figure 14-13.—Use of piezometers and observation wells to delineate complex ground-water conditions.

14-4. Design of Dewatering System.—Values of many of the aquifer parameters will be estimates only, and the resulting computation of water volume should be taken as an estimate only. It is advisable to use ranges of values of the parameters so that the maximum and minimum potential volume can be estimated.

The volume of water that a dewatering system has to pump to produce a certain drawdown can be determined using the following equations for an unconfined aquifer (Driscoll, 1986):

$$Q = \frac{K(H^2 - h^2)}{0.733 \log R/r} \quad Metric \qquad\qquad 14\text{-}1$$

or:

$$Q = \frac{K(H^2 - h^2)}{1055 \log R/r} \quad English \qquad\qquad 14\text{-}2$$

where:

Q = Discharge, m³/day (gal/min)
K = Hydraulic conductivity, m/day (gpd/ft²)
H = Saturated thickness of aquifer before pumping, m (ft)
h = Depth of water in well while pumping, m (ft)
R = Radius of cone of depression, m (ft)
r = Radius of well, m (ft)

Equation for a confined aquifer:

$$Q = \frac{Kb(H - h)}{0.366 \log R/r} \quad Metric \qquad\qquad 14\text{-}3$$

or:

$$Q = \frac{Kb(H - h)}{528 \log R/r} \quad English \qquad\qquad 14\text{-}4$$

where:

b = Thickness of aquifer, m (ft)
H = Distance from the static water-level to the bottom of the aquifer, m (ft)

The equation for the drawdown at any point within the cone of depression:

$$s = \frac{0.366 \, Q \log R/r}{Kb} \qquad\qquad 14\text{-}5$$

or:

$$s = \frac{528 \ Q \ \log \ R/r}{Kb} \qquad \text{14-6}$$

If the dewatering project is extensive, a computer program may be desirable to design the system. This procedure is particularly advisable in many Bureau of Reclamation projects, where complex hydrologic boundaries may result in a complicated image-well development and little precision in estimates of aquifer parameters.

The following procedure can be used to estimate the number, spacing, and pumping rate required for relief well arrays:

- Select a value for transmissivity, T, in liters (gallons) per day per meter (feet) (use data from pumping tests if available or check literature for T values in that formation or area). Because the calculated value for drawdown is inversely proportional to T, it is important to remember that the value of T in the well array can be significantly different from the value used in the calculation, even when the value comes from a pumping test in the general vicinity. This difference can be caused by variations in lithology, structure, well construction, or other factors. Also, in water-table aquifers, T decreases as the water-table declines because of the decreased aquifer thickness. Calculating drawdowns using a range of T values may therefore be advisable. A doubling or halving of the T value, which is a very real possibility in alluvial aquifers, would approximately halve or double the calculated drawdown.

- Choose a time, t, in days, at which you want to know drawdown (generally should be based on the time in which a certain drawdown needs to occur).

- Choose a pumping rate Q, in liters (gallons) per minute, based on reasonable yields from the aquifer or any other reasonable value.

- Choose a coefficient of storage, S, (dimensionless) based on pumping tests or lithology. General range is 0.00001 (artesian aquifers) to 0.25 (clean gravel in water-table aquifers).

- Compute drawdown, s, for at least two values of distance, r^2, (say 1 meter and 30 meters) using Jacob's approximation, given below. Jacob's formula is valid for small values of radius and large values of time, so some error occurs at large distances. However, considering the possibility of error occurring in choosing T and S, the accuracy is sufficient.

$$s = \left(\frac{0.183Q}{T}\right) Log \left(\frac{2.25Tt}{r^2S}\right) \quad Metric \qquad 14\text{-}7$$

or

$$s = \left(\frac{264Q}{T}\right) Log \left(\frac{0.3Tt}{r^2S}\right) \quad English \qquad 14\text{-}8$$

where:

s = Drawdown at distance r
Q = Pumping rate in liters (gallons) per minute
T = Transmissivity of the aquifer
t = Time in days
r = Radius at drawdown measurement point
S = Coefficient of storage

- Plot s versus r on semilog paper. Because the result is a straight-line plot, only two values of s need to be computed; however, as a check, it is wise to compute s at a third r.

- Choose a well spacing, d, and number of wells such that drawdown in any well, from pumping that well and all other wells in the array at the pumping rate per well, Q, does not exceed the maximum drawdown at which the well can operate effectively (i.e., drawdown cannot exceed depth of well, which depends upon geologic conditions and other factors).

- Using the plot developed earlier, compute the drawdown midway between wells to check that it is sufficient to dewater.

- Adjust Q, d, and number of wells so that drawdown at the well is not excessive, but drawdown at midpoints is sufficient.

It is important to realize that the calculated drawdowns are only approximate. Boundary conditions such as recharge areas, impermeable layers, valley walls, depth to bedrock, etc., will affect the flow to the wells. Errors in coefficient of storage also affect the computation. An order of magnitude difference in S, which is quite possible in artesian aquifers and also possible in water-table aquifers where a pronounced change in lithology exists, could affect the drawdown computation by as much as 50 percent.

14-5. Design of Well Points and Similar Dewatering Systems.—

(a) Suction Lift Well-Point Systems.—The most basic installation is probably that of a single line of well points necessary to dewater a relatively shallow trench for a pipeline or similar installation. The tops of the well points are usually set a minimum of 1 meter (4 feet) below the bottom of the excavation and as close to the edge of the trench as possible without interfering with the work. In some very permeable materials, a line of well points on each side of the excavation may be required to provide adequate dewatering.

An excavation that is large in all directions is usually outlined by a dewatering system.

In relatively low permeability soils such as silty clays, silts, and fine sands, or where such materials overlie a more permeable layer, the well points should be gravel packed except for about 1 to 1.5 meters (4 to 5 feet) at the top of the hole, which is tightly sealed. This construction permits the development of a vacuum in the deeper portions of the hole which encourages horizontal migration of water to the gravel pack and then down the hole to the well point. The amount of water removed may be small, but in many instances it is sufficient to stabilize an otherwise unstable material (figure 14-10). Gravel packing may also be used where clay beds are irregularly distributed through the saturated mass of material to be drained. The selection of screen slot sizes, gravel-pack gradations, etc., is based on the same criteria outlined in section 14-11. Driscoll (1986) gives theoretical and empirical steady-state equations for use in the design of suction lift well-point systems. Calculating the size and shape of the area of influence of this type of well-point system is seldom practical. Drawdowns are usually too great for the usual equations to be applicable, and partial penetration and anisotropy may further complicate the problem.

The depth to which suction lift is effective decreases with increasing altitude. Standard well-point pumps cannot lower the absolute pressure below about 1.6 meters of water (5 inches Hg), although special pumps can lower the absolute pressure to 1.0 meter of water (3 inches Hg) (Powers, 1981).

The vacuum that can be developed is the atmospheric pressure less the absolute pressure in the system. At sea level the atmospheric pressure is about 10.3 meters (33.9 feet) of water, but at an elevation of 1500 meters (5000 feet), normal atmospheric pressure is only about 8.7 meters of water (25 inches Hg). In practice, the suction lift of a well-point system is reduced about 0.3 meter (1 foot) for every 300-meter (1000-foot) increase in elevation. The suction lift is further decreased by system losses caused by factors such as pump efficiency, air entrainment, and possible cavitation in the water pump. In practice, well-point systems at sea level operate at vacuums of 6.2 to 7.6 meters of water (11 to 22 inches Hg), although occasionally higher vacuums are obtained when pumping small quantities of water (Powers, 1981).

Well-point spacing is selected primarily on the basis of judgment and operating experience. In silt and fine sands, 0.5 to 0.75 meters (2- to 2-1/2 feet) is the usual spacing, and as the coarseness of the materials increases, spacing may be increased to about 2 meters (6 feet). Thickness of the aquifer and percent of penetration are also controlling factors. In thin aquifers of less than 5 meters (15 feet), and when the percent of penetration in thicker aquifers is less than about 25 percent, spacing is also small (0.5 to .75 meters [2 to 2-1/2 feet]). Spacing may be increased as the transmissivity and percent of penetration increase.

For silt and other fine-grained materials, well points with diameters of 40 millimeters (1-1/2 inches) generally are satisfactory. The diameter should be increased for more permeable materials. Riser pipes of 25-millimeter (1-inch) diameter are suitable for the smaller diameter well points and should be increased to 50 to 60 millimeters (2- to 2-1/2 inches) for well-point diameters up to 90 millimeters (3-1/2 inches) on suction lift systems.

The well points of the first stage should be set 1 to 1.5 meters (3 to 5 feet) below the bottom of the proposed excavation with maximum drawdown at about 5 meters (15 feet) below the water-table. Following dewatering by the first stage, the excavation is carried to within about 300 millimeters (1 foot) of the water-table, and the second stage is then installed. Theoretically, the procedure could be carried to almost any depth in a thick homogeneous aquifer, but the dewatered thickness is relatively thin on the side of the slope relative to the adjacent saturated thickness (figure 14-8). When three or more stages are required, seepage pressures may cause slope instability. Under such conditions, supplemental deep wells, a deep-well system alone, or a supplemental well-point system should be used.

Where an excavation is underlain by a relatively impermeable bed which limits the drawdown, sufficient dewatering may sometimes be obtained by drilling 300- to 350-millimeter (12- to 14-inch) diameter relatively shallow holes into the impermeable bed, setting the well points into the holes in the impermeable material, and gravel packing around the well points (figure 14-14).

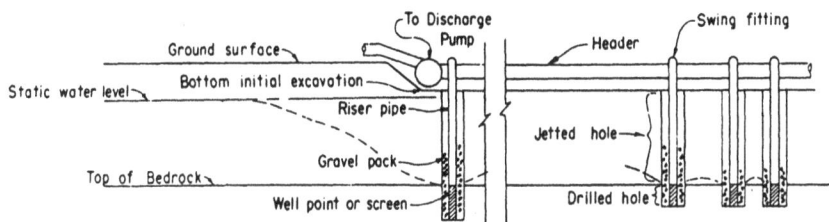

Figure 14-14.—Dewatering a thin aquifer overlying an impermeable material.

An excavation may be underlain by a confining layer of relatively impermeable material which, in turn, is underlain by an aquifer. Water in this aquifer may be under similar or higher head than water above the confining layer. If the material above the layer is dewatered and the bottom of the excavation is lowered, a point may be reached where either the bottom of the excavation will heave or blow out, or boils will occur in areas where the confining layer is thin. Under such conditions, relief wells or well points should be installed in the underlying aquifer to reduce the pressure and stabilize the bottom of the excavation (figure 14-15).

The maximum suction lift which can be obtained with available pumps is about 6 to 7.5 meters (20 to 25 feet); however, friction losses in the system may reduce this lift to about 5 to 5.5 meters (15 to 18 feet). To keep the loss of suction head to a minimum, well points, riser pipes, swing connections, and header pipes should be sized generously. In addition, all joints in the system should be made airtight.

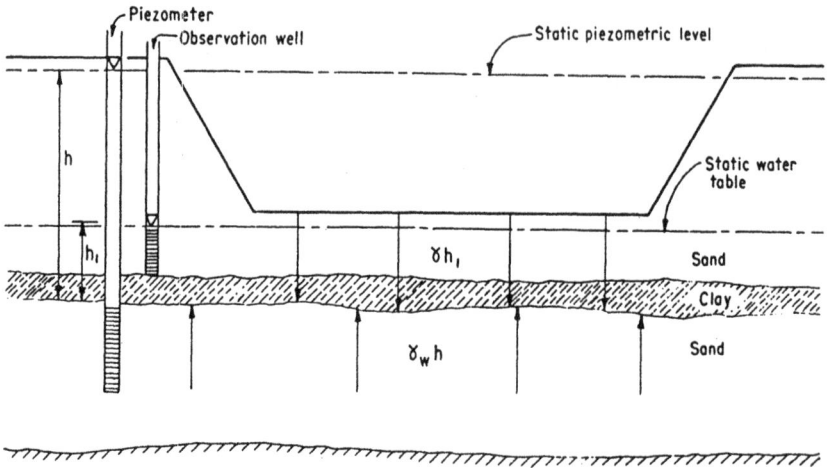

h = vertical distance between static piezometric level and top of underlying permeable bed
h_l = distance between top of permeable bed and bottom of excavation
γ_w = unit weight of water
γ = unit weight of soil and water combined
γh_l = weight of the overlying soil (pressure on confining layer)
$\gamma_w h$ = upward hydrostatic pressure

Figure 14-15.—Factors contributing to blowouts or boils in an excavation.

The pumping and other tests recommended previously in this section should permit an estimate of the probable discharge which will be obtained for well points. From this estimate, required pipe diameters to keep friction losses at a minimum and pump capacities can be estimated. If the thickness, permeability, and storativity of an aquifer are known, a line of closely spaced well points may be considered as a drain, and the equations given in section 14-4 may be used to compute the time required for dewatering (Mansur and Kaufman, 1962). Each line of well points is considered a drain or collector.

(b) Jet Eductor Well-Point Systems.—In some cases, well-point systems are installed using a jet eductor pump (sometimes called jet ejector pump) in each riser pipe rather than pumping by suction lift. Jet eductor pumps operate by an induced suction created by flow of water through a venturi, in addition to normal suction from centrifugal and vacuum pumps. Jet eductor well points have only about a 25- to 35-percent efficiency, but they can lift water from depths of 18 to 30 meters (60 to 100 feet) and develop a vacuum of 5.5 to 6 meters (18- to 20-feet) of head in each hole. The discharge is usually in the range of 45 to 60 liters (12 to 15 gallons) per minute. Well-point spacing is usually 1 to 3 meters (4 to 10 feet).

Two headers are employed. One delivers water under pressure to the venturi in each pump and the other provides discharge.

The riser pipes are usually 90 millimeters (3-1/2 inches) in diameter or larger to permit installation of the jet educator pumps whose intake is usually located a meter above the top of the screen or well point. Well points or screens may be as long as 3 to 4.5 meters (10- to 15-feet) and as much as 150 millimeters (6 inches) in diameter.

At times, particularly in smaller installations, submersible pumps may be used rather than jet educator pumps.

(c) Deep-Well Dewatering Systems.—Deep wells can be used for dewatering thick homogeneous aquifers and to lower heads in underlying artesian aquifers which might give rise to heaving or boils in the floor of an excavation. The design and installation of the wells generally are the same as described in chapter XI except that lower design standards may be adequate because of the temporary nature of the wells. If storativity, transmissivity, boundary conditions, and aquifer thickness are known, wells can usually be located outside the excavation limits.

14-6. Monitoring.—Monitoring to assure that drawdown is occurring as anticipated is an important part of the dewatering program. The monitoring wells should be installed in various locations: (1) at points where drawdown resulting from the system is estimated to be a minimum, (2) at short distances outside the system, and (3) at locations where aquifer parameters may differ significantly from those anticipated. Differences may be caused by individual well or well-point operation or effectiveness, use of values for parameters that do not correspond with actual values, or local geohydrological conditions which differ significantly from those estimated. The cause of the discrepancy should be investigated.

Water-levels should be measured and reviewed daily or, where adequate drawdown is critical from a safety standpoint, twice per shift. If the monitoring wells or other conditions indicate drawdowns are not sufficient, additional wells or well points may need to be installed or pumping rates may need to be increased. In critical situations, a dewatering specialist should be on site at all times to review conditions. In such situations, equipment, supplies, and operating personnel should be readily available to install additional wells or well points, if necessary.

14-7. Installation of Dewatering Wells and Well Points.— Well points are made to be driven in place, jetted down, or installed in open holes. For dewatering purposes, the points are seldom driven. A more common practice is to jet the well point down to the desired depth, flush out the fines leaving the coarser fraction of material to collect in the bottom of the hole, and then drive the point into this coarser material. Additional gravel may be added to fill the hole near or to the ground surface or a seal may be placed over the pack in the upper 600 to 900 millimeters (2 to 3 feet).

A method used in some unstable material consists of jetting down or otherwise sinking temporary casing, into which the well point and riser pipe are installed. As the casing is pulled, gravel may be placed around the well point.

Major dewatering contractors have developed equipment and techniques for mass installation of dewatering wells of all types.

Regardless of the method of installation, developing a well point by pumping, surging, or other means is desirable prior to use.

14-8. Pumps for Dewatering Systems.— Pumps for suction lift well-point systems should have adequate air handling capacity and be capable of producing a high vacuum. Well-point system pumps are usually self-priming centrifugal pumps with an auxiliary vacuum pump which permits developing a vacuum of 6 to 7 meters (20 to 25 feet) of water. The intake of the pump should be as close to the bottom of the excavation as practicable (Mansur and Kaufman, 1962).

Submersible pumps and jet educator pumps are usually standard off-the-shelf items selected for the estimated yields and lifts involved.

Good engineering practice requires excess capacity in the pumps and standby units in the event of pump failure.

14-9. Artificial Ground-Water Barriers.— Natural ground-water dams caused by faulting, dikes, and similar features are frequently encountered in ground-water investigations. Similar barriers may be constructed artificially for the purposes of ground-water control, such as seepage under dams, protection of excavation, and the raising or maintenance of ground-water levels.

(a) Sheet Piling.—Sheet piling is a commonly used method, but of questionable effectiveness in some aspects, for construction of ground-water barriers. If completely interlocking and driven to an impermeable barrier, sheet piling is an effective means of controlling piping and instability but not of stopping subsurface flow. The interlocking of sheet piling does not give a watertight joint and if 10-percent open area is present, about 70 percent of the water will flow through the piling. Although sheet piling barriers may have relatively small effect in decreasing the flow of water, they do cause an increase or dispersal of the exit area with a consequent decrease in the possibility of the formation of boils and development of piping.

Sealed joint pilings that are vibrated into place rather than driven can create an effective vertical barrier. In these installations, a rubber gasket is inserted into the joints between sheets. The sheets, which are flexible, are carried down with a steel driving shoe. The shoe is then extracted, leaving the piling in place.

Where boulders and cobbles are present, sheet piling generally is not feasible, and concrete soldier piles may have to be set in pre-excavated holes.

(b) Cutoff Walls.—Cutoff walls are often placed beneath dams located on permeable foundations. The integrity of such cutoff walls seldom presents a problem because they are commonly placed in open cuts dewatered by wells or well points and protected by cofferdams. To be entirely effective, cutoff walls should provide for 100-percent closure, which may be economically or physically impossible in some instances. Under such conditions, a careful study of soil condition, permeabilities, etc., should be made, usually in combination with flow net analyses, to determine the best design from the standpoint of leakage and stability.

(c) Slurry Trenches.—The recent development of slurry trenches has provided a safer and less expensive method of installing ground-water barriers in unconsolidated materials. Draglines or backhoes, depending upon the depth and size of the excavation, are used to excavate the trench, which is kept full of a bentonite-based fluid. This fluid, which has a weight of 4 to 4.5 kilograms per liter

(9-1/2 to 10 pounds per gallon), develops a filter cake on the side of the trench which reduces seepage of the fluid away from the excavation and exerts hydraulic pressure on the side of the trench which minimizes caving. Excavation by such a process has been carried to depths of up to 30 meters (100 feet). When reaching the top of bedrock, the excavation may be cleaned of pockets of sand by operation of an airlift pump.

The completed excavation can be filled with concrete placed through tremie pipes extending to the bottom. As the concrete is pumped in, the bentonite-based fluid is displaced and flows out at the surface. The trench may also be excavated and backfilled in sections, with the bentonite being diverted to the next section as the completed section is being backfilled.

Another procedure is to displace the bentonite in the trench with the sand and coarser fractions of the excavated material, thus creating a clayey-sand mixture. This procedure results in a watertight and permanent ground-water barrier. In using the slurry trench method, an engineer experienced in the handling of mud should be available to advise on proper use and treatment of the slurry.

(d) Freezing of Soils.—In certain cases, freezing of soils may be effective in reducing ground-water flow. The usual method is to install a row of vertical freeze pipes. The surrounding earth freezes in a vertical cylinder around the pipes. The cylinders then gradually enlarge until a continuous wall is formed. If the freezing process continues, the wall may increase in thickness. In saturated sands, pore water freezes rapidly and stabilization can occur at only a few degrees below freezing. However, in clay soils, some of the water is molecularly bonded to the soil particles, inhibiting rapid freezing. Thus, stabilization may require longer times and lower temperatures (Powers, 1981).

(e) Grouting.—The formation of impermeable barriers to water seepage by grouting is an established practice. Grouting consists of introducing sealing compounds or mixtures, usually under pressure, into rock and soil to fill fractures and voids with stable, insoluble materials. Grouting has usually been successful in fractured hard rocks; however, the results in unconsolidated materials have been variable. Native clays, bentonite, silts, and sand are natural materials which have been used for grouting.

Cement, various chemicals such as combined sodium silicate and calcium chloride, tar, asphalt, and various epoxies have also been used.

The nature of the openings, their size and continuity, the velocity of waterflow through the openings, and other factors influence the type of grouting materials to be used and the method of injection. Injection pressures should be carefully regulated because excessive pressures can cause fracturing that could result in increased, rather than decreased, flow.

Successful grouting is both an art and a science and should be undertaken only under the direction of an experienced and knowledgeable engineer.

14-10. Bibliography.—

American Society of Civil Engineers, 1993, "Design, Construction, and Maintenance of Relief Wells."

Bureau of Reclamation, 1987, *Design of Small Dams*, 3rd edition.

Bureau of Reclamation, 1988, "Dewatering," chapter 17, In: *Engineering Geology Field Manual*, pp. 415-436.

Bureau of Reclamation, 1993, *Drainage Manual*.

Driscoll, F.G., 1986, *Groundwater and Wells*, 2nd edition, UOP Johnson Division, St. Paul, Minnesota.

Mansur, C.I. and R.I. Kaufman, 1962, "Dewatering," chapter 3, In: *Foundation Engineering*, (edited by G.A. Leonards), McGraw-Hill, New York, pp. 241-3507.

Moody, W.T. and R.W. Ribbens, December 29, 1965, "Ground Water-Tehama-Colusa Canal Reach No. 3, Sacramento Canals Limit, Central Valley Project, California," memorandum to Chief, Canals Branch, Bureau of Reclamation.

Powers, J. Patrick, 1981, *Construction Dewatering*, John Wiley & Sons.

Singh, Bharat and H.D. Sharma, 1976, *Earth and Rockfill Dams*.

Terzaghi, K. and R.B. Peck, 1967, *Soil Mechanics in Engineering Practice*, 2nd edition, John Wiley & Sons, New York.

U.S. Army Corp of Engineers, 1993, Technical Engineering and Design Guides As Adapted From the U.S. Army Corps of Engineers, No. 3, "Design, Construction, and Maintenance of Relief Wells," USACE Engineering Manual EM 1110-2-1914.

U.S. Army Corp of Engineers, 1994, Engineering and Design, "Earth and Rock Fill Dams - General Design and Construction Considerations," USACE Engineer Manual EM 1110-2-2300, pp. 6.1-6.6.

UOP Johnson Division, 1986, St. Paul, Minnesota.

UOP Johnson Division, 1969, "Well Point Systems," Bulletin No. 467-D, St. Paul, Minnesota.

Chapter XV

WATER WELL PUMPS

15-1. Introduction.—The function of a pump is to transfer energy from a power source to a fluid thereby creating flow or simply creating greater pressures on a fluid. Pumps are installed in water wells to lift the water in the well to the ground surface and deliver it to the point of use.

A variety of pumps are available to remove water from wells. There are several basic types of pumps:

- Centrifugal pumps.—Contain a rotating impeller mounted on a shaft turned by a power source. They can be vertical turbine or submersible.

- Jet pumps.—Used for shallow water-levels in small-diameter holes. They are actually combined centrifugal and ejector pumps.

- Pneumatic pumps.—Operate on air pressure. They can be bladder-type or displacement-type.

- Airlift pumps.—Use compressed air injected into a discharge line lowered into the well.

- Positive displacement pumps.—They can be piston pumps, often operated by hand or windmill, or rotary peristaltic pumps, used to take water samples from monitoring wells.

- Suction lift pumps.—Generally used in dewatering operations.

15-2. Conventional Vertical Turbine Pumps.—Vertical turbine pumps have the motor installed on the discharge head at ground surface and require a drive shaft extending down the well to the pump located below the water surface.

(a) Turbine Pump Principles.—The vertical turbine pump is often the most suitable pump for ground-water applications, especially for moderate to large discharge rates. Improved materials and design, combined with increased efficiency, have greatly broadened the field of vertical turbine pump application. There are very few ground-water pumping problems that cannot be solved efficiently by using the vertical turbine pump.

Pump selection varies with the type and temperature of the fluid being pumped. The following discussion is based on the pumping of water at temperatures in the range of 4° to 27 °C (40° to 80 °F), which includes most ground-water applications.

The capacity, head (pressure), efficiency, and power requirements of a vertical turbine pump depend on the design of the impeller and bowl assembly, the diameter of the impeller, and the operating speed of the pump.

Choosing a pump with performance characteristics best suited for the application will result in the most efficient application. The basic principles upon which vertical turbine pump characteristics are determined will be discussed briefly to show the effect of different requirements on design.

The pressure developed by a vertical turbine pump is a function of the peripheral velocity of the impeller, which in turn is a function of the impeller diameter and rate of rotation. The pressure is usually expressed in meters (feet) of water. The rotating impeller imparts energy to the water, and the directional vanes in the bowl surrounding the impeller convert this energy to pressure and guide the fluid vertically so that the flow becomes axial with the pump shaft.

(b) Turbine Pump Operating Characteristics.—Pump performance characteristics determined by tests in the manufacturer's laboratory and plotted on a graph furnish an understandable picture of the performance of a particular impeller design. This graph, called a performance curve, is the key to selecting the type of impeller to suit the pumping requirements. The performance curves for two 300-millimeter (12-inch) pumps, shown on figure 15-1, were plotted for a constant speed of 1,760 revolutions per minute (r/min). The difference in the performance curves for the two pumps of the same size and rate of rotation is due to the differences in impeller and bowl design. On figure 15-1, the head-capacity curves vary from the shutoff head on the left of the graph to the maximum capacity to the right. The discharge of the pump can be completely stopped (shut off) by closing the discharge valve. This, however, should not be done for any length of time because the pump bearings are water cooled and will overheat. A performance curve is generally based on the operating characteristics of one stage and shows the relationships between the head capacity, horsepower, and efficiency of a stage. The horsepower curve represents the brake horsepower required at various head capacities.

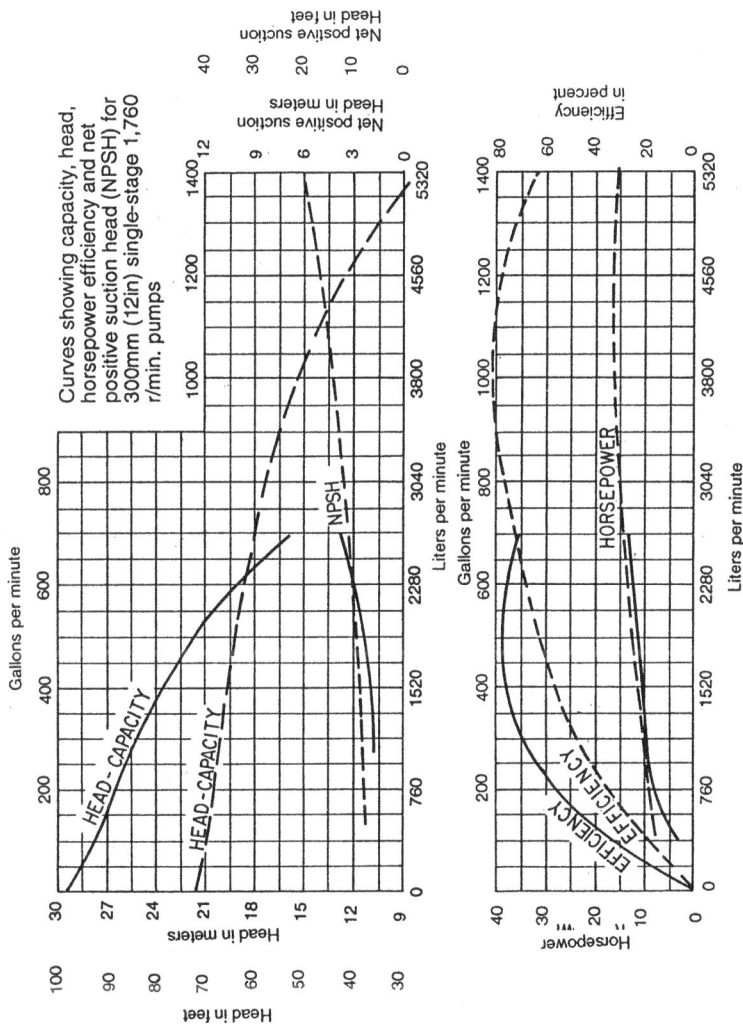

Figure 15-1.—Laboratory performance curves for two 300-millimeter (12-inch) single-stage deep-well turbine pumps.

In the examples that follow, the relationships of capacity, head, and horsepower, as shown on figure 15-1, are valid provided that the rotational speed is constant at 1,760 revolutions per minute. If the speed is changed, the capacity, head, and horsepower will change as follows:

$$N_1/N_2 = Q_1/Q_2 = H_1^2/H_2^2 = bhp_1^3/bhp_2^3$$

where:

Q = capacity (or discharge)
H = total head
bhp = brake horsepower
N = rate of rotation

Theoretically, the efficiency does not change. However, in field service, the efficiency will decrease slightly depending on mechanical losses and other factors. With a variable-speed driver, the rotation speed of a pump may usually be varied as much as 20 percent below the design value without serious loss of efficiency. To illustrate the effect of speed on pump characteristics, calculate the change in head, capacity, and horsepower on the pump shown by the dashed lines on figure 15-1 when the speed is reduced from 1,760 to 1,400 revolutions per minute. From the curves, which are based on the higher speed, at the BEP (point of maximum efficiency.

For operation at 1,760 r/min:

Capacity = 3,800 L/min (1,000 gal/min)
Head = 15.6 meters (52 feet)
Power = 12.7 kw (17 bhp)

For operation at 1,400 r/min:

Capacity varies directly with the speed:

3,800 L/min (1,000 gal/min) x (1,400/1,760) = 3,020 L/min (795 gal/min)

Head varies as the square of the speed:

15.6 meters (52 feet) x (1,400/1,760)^2 = 9.9 meters (32.9 feet)

Power varies as the cube of the speed:

12.7 kw (17 bhp) x (1,400/1,760)^3 = 6.4 kw (8.6 bhp)

The capacity, head, and horsepower relationships for a particular impeller and bowl operating at constant speed can also be varied by changing the diameter of the impeller. This is known as trimming. The impeller should not be trimmed in the field without first consulting the pump manufacturer to determine the effects of trimming.

A steep head-capacity characteristic curve and a flat efficiency curve are desirable for deep-well turbine pumps as the water-level in the well may vary considerably during the life of the pump. As the water-level declines in the well, the head increases, but the capacity decreases only slightly in proportion. Also, with this type impeller, the brake horsepower curve is almost flat, with the power input highest at highest pump efficiency. This characteristic is desirable for protection of the motor from overloading. A steep head-capacity curve is more desirable for service where the capacity is varied slightly by operation of a valve in the discharge line.

(c) Net Positive Suction Head and Submergence.—Net Positive Suction Head (NPSH) is defined as the suction head above vapor pressure at the eye of the impeller. The NPSH available is calculated from the following equation:

$$NPSH = H_p + H_s - H_f - H_v$$

where:

H_p = normal barometric pressure for the elevation of the installation, in meters (feet); see table 15-1

H_s = distance in meters (feet) from the eye of the lowest impeller to the surface of the water in the well while pumping (the water-level must be at least this high); a positive value indicates submergence of the eye of the impeller and a negative value indicates a suction

H_f = head lost through the suction piping, in meters (feet)

H_v = vapor pressure of the water, in meters (feet); see table 15-2

Table 15-1.—Normal barometric pressure at various elevations

Altitude		Barometric pressure, H_2O		Altitude		Barometric pressure, H_2O	
meters	feet	meters	feet	meters	feet	meters	feet
0	0	10.4	34.0	2250	7500	7.8	25.7
150	500	10.2	33.4	2400	8000	7.7	25.2
300	1000	10.0	32.8	2550	8500	7.6	24.8
450	1500	9.8	32.2	2700	9000	7.4	24.3
600	2000	6.9	31.6	2850	9500	7.3	23.8
750	2500	9.4	31.0	3000	10000	7.1	23.4
900	3000	9.3	30.5	3150	10500	6.9	22.4
1050	3500	9.1	29.9	3300	11000	6.8	21.9
1200	4000	9.0	29.4	3450	11500	6.7	21.4
1350	4500	8.8	28.8	3600	12000	6.5	21.0
1500	5000	8.6	28.3	3750	12500	6.4	20.6
1650	5500	8.5	27.8	3900	13000	6.3	20.2
1800	6000	8.3	27.3	4050	13500	6.2	19.8
1950	6500	8.1	26.7	4200	14000	6.0	
2100	7000	8.0	26.2				

Table 15-2.—Vapor pressure of water at various temperatures

°C	°F	Vapor pressure, H_2O		°C	°F	Vapor pressure, H_2O	
		meters	feet			meters	feet
4.4	40	0.085	0.28	16.1	61	0.189	0.62
5.0	41	0.088	0.29	16.7	62	0.195	0.64
5.6	42	0.091	0.30	17.2	63	0.201	0.66
6.1	43	0.097	0.32	17.8	64	0.207	0.68
6.7	44	0.101	0.33	18.3	65	0.216	0.71
7.2	45	0.104	0.34	18.9	66	0.223	0.73
7.8	46	0.107	0.35	19.4	67	0.229	0.75
8.3	47	0.113	0.37	20.0	68	0.241	0.79
8.9	48	0.116	0.38	20.6	69	0.247	0.81
9.4	49	0.122	0.40	21.1	70	0.256	0.84
10.0	50	0.125	0.41	21.7	71	0.262	0.86
10.6	51	0.131	0.43	22.2	72	0.274	0.90
11.1	52	0.134	0.44	22.8	73	0.283	0.93
11.7	53	0.140	0.46	23.3	74	0.293	0.96
12.2	54	0.146	0.48	23.9	75	0.302	0.99
12.8	55	0.152	0.50	24.4	76	0.311	1.02
13.3	56	0.155	0.51	25.0	77	0.323	1.06
13.9	57	0.162	0.53	25.6	78	0.335	1.10
14.4	58	0.168	0.55	26.1	79	0.347	1.14
15.0	59	0.174	0.57	26.7	80	0.357	1.17

A particular pump design requires a certain minimum NPSH to prevent cavitation. The available NPSH at the site must be equal to or greater than the required NPSH. Thus, a requirement for a higher value of NPSH is satisfied by lowering the pump in the well. The NSPH is given on the pump performance curve.

Turbine pumps will operate with a suction lift. The pump suction can be equipped with a length of suction pipe which extends below the waterlevel. The pump bowl assembly must, however, be submerged at startup of the pump. It is advisable for the pump bowls to be submerged when the pump is operating. Submergence avoids corrosion problems resulting from the bowls being alternately submerged and exposed to the atmosphere and eliminates the suction pipe. The pump should be set at a depth such that the top of the bowls is 1.5 meters (5 feet) or more below the estimated lowest water elevation when the well is pumping. Some pumps may require deeper settings because of net positive suction head requirements.

(d) Turbine Pump Construction Features.—Turbine pumps were originally designed for use in drilled wells; thus, the nominal diameter of the bowls are designated by standard well casing sizes. The pump size designation (4, 6, 8 , etc.) indicates the smallest diameter of a well casing of standard size (section 9-3(b)) into which the pump can be installed. For clearance purposes, the outside diameter of the pump bowl is manufactured a few fractions of an inch smaller than the inside diameter of the nominal size casing. In actual design and practice, however, more clearance than this is usually specified. A minimum clearance of 1 inch around the pump bowls (casing diameter of 2 inches larger than the pump diameter) is recommended. More clearance may be required for large pumps or very deep settings.

Where several stages (bowls) are assembled in series on a common shaft, they constitute a multistage pump. The head produced is directly proportional to the number of stages. For example, if the head requirement is 42 meters (140 feet) at 1,900 liters (500 gallons) per minute and one bowl and impeller develop 22 meters (72 feet) of head at 1,900 liters (500 gallons) per minute, then two stages are required to furnish the required performance. As the velocity is converted into pressure in one stage and guided to the next stage, additional pressure energy is added by the second stage and the required horsepower is increased equally by each additional stage.

Deep-well turbine pumps have the line shaft lubricated by oil or water. In oil-lubricated pumps, the line shaft and bearings are enclosed in a tube into which oil is dripped (while the pump is operating) from an oil reservoir mounted on the pump base or discharge head. In a water-lubricated pump, the tube enclosing the shaft and bearings is omitted and the water flowing up the pump column acts as the lubricant. The bearings in an oil-lubricated pump are usually bronze, while those in the water-lubricated pump are made of special types of rubber. Oil-lubricated pumps are generally used when the depth-to-water is 15 meters (50 feet) or more, and water-lubricated pumps are generally used when the depth-to-water is less than 15 meters (50 feet). Water-lubricated pumps can be used where the depth-to-water is more than 15 meters (50 feet), but they must be equipped with a means of prelubricating the bearings before the pump is started. Use of oil-lubricated pumps results in leakage of oil into the well. Thus, oil-lubricated pumps should not be used in situations where contamination of the well water is of concern.

Step drawdown tests are made after completion and development of a well to determine the pumping lift for various discharges. These data must be known to select the correct pump for the well (section 9-15).

(e) *Selection of Pump Bowl and Impeller.*—With the required discharge and head known, the pump can be selected from a manufacturer's pump performance curves. To obtain the lowest initial cost and most economical operation of the pump, the rate of rotation should be kept as high as possible without sacrificing efficiency. The smaller diameter pumps will usually require more stages and will have lower efficiency than the larger pumps. The pump should be selected which will more closely approach the required capacity at the maximum head with the highest efficiencies over the estimated range of heads and the smallest number of bowls.

(f) *Discharge Heads.*—Discharge heads are components which convey the pump discharge from the vertical column pipe to the horizontal discharge. The discharge head is mounted above and attached to the column pipe and below the pump driver. Discharge heads are made of cast iron or fabricated steel, depending on the pump discharge pressure and type of driver. Most are standardized to National Electrical Manufacturers' Association (NEMA) standard dimensions and will permit the use of a matching NEMA standard motor, a right-angle gear drive, a belt drive, or a combination drive. Most heads are designed for a single type of drive, but combination heads are available which permit the use of both a vertical motor and a right-angle gear or belt drive. This permits the use of an engine drive as standby power in the event of an electrical power outage.

Discharge head nomenclature is based on nominal drive base and column and discharge pipe dimensions. For example, a pump may require a motor having a base diameter of either 16 or 20 inches (NEMA motors will have a choice of 2 or 3 base diameters for any size motor) and an 8-inch diameter column pipe and discharge pipe. The discharge head for such an installation would be designated as a 1608 or 2008, or 8 x 8 x 16 or 8 x 8 x 20. The horizontal discharge pipe connection is normally threaded up through 4 inches and flanged for larger pipes. Manufacturers' catalogs for pumping equipment usually contain instructions on the selection of discharge heads for various pump sizes and horsepower requirements.

15-3. Submersible Pumps.—The term "submersible" (also termed submergible) is applied to turbine pumps where the motor is close-coupled beneath the bowl assembly of the pump and both are installed under water. This type of construction eliminates the surface motor, long drive shaft, shaft bearings, and lubrication system of the conventional turbine pump; however, the electrical connections are submerged. Submersible pumps are especially useful for high-head, low-capacity applications such as domestic water-supply. With the exception of the factors discussed below, the selection of submersible pumps is identical to that of conventional deep-well turbine pumps.

The entire weight of the pump, cable, drop pipe, and column of water within the pipe must be supported by the drop pipe. Therefore, the drop pipe and couplings should be composed of a good quality galvanized steel. Cast-iron fittings should not be used where they support pumps and pump columns.

The motors are cooled by the water flowing past the motor to the pump intake. This cooling system permits a different motor design than is possible with air-cooled motors. The submerged motors are designed much longer and of smaller diameter than surface motors of the same horsepower and speed so they can be installed in the well.

To avoid high head losses in flow past the motor and into the pump intake, the pump chamber should be large enough so that velocity of flow does not exceed 1.5 meters (5 feet) per second, and should preferably be nearer 0.3 meter (1 foot) per second. A minimum velocity of 0.3 meter (1 foot) per second is needed to ensure adequate cooling of the motor. Head losses due to high velocities in restricted annular spaces may result in a reduction of the available NPSH at the pump. This may be compensated for by increasing the submergence of the unit below the pumping water-level.

Where the use of large capacity submersible pumps (3,000 gal/min or 400 hp) is contemplated, the manufacturer should be consulted regarding desirable submergence, pump chamber diameter, and length of the pump and motor assembly.

15-4. Jet Pumps.—The jet pump combines two principles of pumping, that of the injector (jet) and that of the centrifugal pump. The jet is actually a pump in itself being powered by the water under pressure from a centrifugal pump (Anderson, 1973). The job of the jet is to create pressure. In shallow wells (depths to 7.5 meters [25 feet]), the jet is built into the pump and raises the pressure to the desirable limits. Thereas in a deep-well (depth from 7.5 meters [25 feet] to 27 meters [90 feet]), the jet is suspended on two pipes and its pressure forces water up to the ground-level pump, which pumps the water into the distribution system (Anderson, 1973).

Jet pumps may be installed directly over the well or alongside it. Since there are no moving parts in the well, straightness and plumpness do not affect the performance of the well. The weight of the equipment in the well is relatively light, being mostly pipe (often plastic), so the load is easily supported by a sanitary well seal (Environmental Protection Agency, 1975). A "pitless adapter" or a "pitless unit" can be designed for the jet pump system.

Jet pumps are inefficient when compared with ordinary centrifugal pumps, but this is not necessarily bad in domestic installations because of other favorable features, such as: (1) adaptability to small wells, down to 50 millimeters (2 inches) inside diameter in deep-lift installations; (2) simple design combined with relatively low equipment and maintenance costs; (3) accessibility at ground surface to all moving parts; (4) and capability of being installed with the moving parts offset from the well (Driscoll, 1986). In some locations, jet pumps may not be completely satisfactory, such as where water-levels are subject to large seasonal variations or where severe corrosion or encrustation causes enlargement or plugging of the nozzle (Driscoll, 1986).

The principal cause of trouble usually occurs during initial installation due to carelessness, permitting foreign material to enter pipes that causes plugged jets (Anderson, 1973). Other common faults are incomplete priming and insufficient operating or discharge pressure.

15-5. Pneumatic pumps.—Pneumatic pumps operate using air pressure and are generally used under special conditions such as contaminant cleanup and monitoring. They are used for purging, sampling, product-only pumping, product pumping with drawdown, gradient control pumping, and low- to moderate-flow pumping. Unlike submersible pumps, they do not require liquid cooling, and thus do not burn out if fluid level drops below pump level. They have no down-hole electrical connections. They are particularly suitable for low flow intermittent pumping.

Pneumatic pumps can be bladder-type or displacement-type. Bladder-type pneumatic pumps have flow rates of only a few liters per minute and are used primarily for sampling, although in some cases, they can be used for purging as well. They can be used for micropurging to minimize the amount of purge water when sampling. Studies have indicated that reproducibility of analytical results is very good when bladder pumps are used. (Muska et al., 1986).

When the bladder pump is operated, water enters the flexible bladder from the bottom and is squeezed up to the surface through a discharge line by gas pressure applied to the outside of the bladder. The separate bladder chamber prevents contact between the pump's air supply and the liquid being pumped, thereby eliminating volatile organic compound (VOC) air emissions when pumping liquids with VOC. Bladder pumps can be set up with timers to control flow.

Cleaning the inner components of bladder pumps may be difficult. Where contaminants are present or suspected, it is advisable to dedicate any bladder pump for usage in a particular well.

The displacement-type pumps are frequently used in purging and in pump-and-treat or other ground-water extraction operations.

They can handle liquids with high solids content, and have higher flow rates (up to 50 L/min or more) than the bladder-type pumps. They have no moving parts except check valves. They do not need surface controls or mechanical timers, but will pump as they are filled. An internal float rises up as the pump is filled, tripping a check valve which opens the air line. The air displaces the water, which then opens up the fluid discharge. The float then falls back down the pump, and the cycle is repeated. The pump requires a compressed air supply and three lines: an air-supply line, a fluid discharge line, and an air-exhaust hose.

Advantages of pneumatic pumps include:

• No shock or explosion hazard

• Lower maintenance—few moving parts

• Ease of installation—are lightweight and can be installed by one person without specialized tools

Disadvantages of pneumatic pumps include:

• Will generally handle only low flows (25 L/min [6.5 gal/min] or less), although certain models can handle up to about 50 L/min (13 gal/min)

15-6. Airlift Pumps.—Water can be pumped from a well by releasing compressed air into a discharge pipe (air line) lowered into a well (Driscoll, 1986). The air will mix with the water and the specific gravity of the water column is lifted to the surface. Because airlift pumping is so inefficient and rather cumbersome and expensive in comparison with the other pumping methods, this method of pumping is rarely used for permanent pumping systems (Driscoll, 1986).

Airlift pumping is generally used only to test well production or take water samples for testing for major constituents. It should not be used where VOC are a concern. A minimum submergence of 30-40 percent of the total tube length is required; however, if submergence is too great, the air pressure may not be sufficient to overcome the head. In such a case, the air tube should be withdrawn somewhat.

15-7. Positive Displacement Pumps.—The positive displacement pump forces or displaces the water through a pumping mechanism. There are several types of positive displacement pumps. In this section, piston pump (reciprocating) and the rotary peristaltic pump will be discussed.

- Piston pumps are used most often in hand-operated wells and windmills. They may be single, double, or triple acting, and are generally small diameter. When the piston is drawn upward, the check valve at the base of the piston is closed by gravity and the water pressure. Pressure is lowered by the moving piston and water flows through the intake valve into the pump cylinder as a result of the pressure differential caused by the stroke of the piston. When the piston moves downward, the valve opens and then closes again when the pressure above it exceeds the pressure below it and the discharge valve opens when the pressure below it exceeds that above it; therefore, the water trapped in the cylinder during the downstroke is forced upward from the piston to the discharge pipe in the next upstroke.

- Rotor peristaltic pumps are actually modifications of the rotor pump. The original rotor pump was designed using gears. The gears fit closely in the housing of the pump and mesh with little clearance. When rotated, the gears squeeze the water from between the teeth of the gears as they mesh together, bringing in a replacement supply of water along the outer surface of the housing at the inlet side of the moving teeth of the gears.

- Lemoineau-Type Pump.—The Lemoineau-type pump is a specially designed positive displacement pump available in both mounted and submersible models. The most widely used pumping element consists of a hard-surfaced, corrosion-resistant, helical-contoured metal rotor which revolves inside a tough, abrasion-resistant, double helical-contoured, flexible rubber stator. At the prescribed speed of rotation, discharge is practically constant regardless of the lift, although the horsepower requirement increases with increased lift. Conversely, since this is a positive displacement pump, discharge varies almost directly with the speed. The power unit and the column pipe and shaft above the pumping unit are similar to those of a water-lubricated pump of the same capacity, and features of the submersible type are similar to a submersible turbine pump. The design of the pump results in high resistance to electrolytic corrosion and damage by sediment-laden water at high heads but low capacities.

15-8. Suction Pumps.—Suction pumps are limited by the suction lift which can be developed. This is dependent upon the atmospheric pressure, and thus is affected by the elevation above sea level. The vacuum that can be developed is the atmospheric pressure less the absolute pressure in the system, which generally is a minimum of about 1.6 meters (5.3 feet) of water. At sea level, the normal barometer reading is about 10.3 meters (34 feet) water, and the theoretical vacuum which can be developed is about 8.7 meters (29 feet) of water. However, the theoretical suction lift is reduced at higher elevations due to the decrease in atmospheric pressure. As a rule of thumb, the theoretical suction lift is reduced by about 0.1 meter (0.3 foot) of water for every 100 meters (330 feet) of elevation increase. In practice, the suction lift is usually limited to where the water-table is about 7 meters (22 feet).

15-9. Estimating Projected Pumping Levels.—Projections of anticipated future ground-water-levels are often difficult and unreliable because of poorly known sources and magnitudes of factors which influence such levels. However, reasonably reliable estimates within the range of pump operation can be made. Seasonal and long-time increases or decreases in static ground-water-levels occur because of seasonal variations in precipitation and long-time changes in the climatic cycle. Long-time declines due to withdrawals imposed on natural fluctuations may reflect normal aquifer development. An extended period of pumping from any well is accompanied by a continuous, slow decline at a constantly diminishing rate unless recharge balances withdrawals. Interference from existing or future wells, installation of recharge facilities, or a change in boundary conditions may cause a change in pumping levels. Deterioration of a well due to corrosion and encrustation may cause a significant decrease in the specific capacity. All these factors should be considered in estimating probable maximum and minimum pumping levels.

Projections of static water-levels in conjunction with analysis of results of pumping tests and the probable pumping schedule are made to determine the required pump characteristics and setting. The drawdown for a given discharge over any period of pumping under conditions prevailing at the time of the test can be approximated by extending the straight-line portion of a semilog time-drawdown plot (sections 9-14 and 9-15). This estimated drawdown is adjusted to compensate for the projected decrease or increase in saturated thickness or the static water-level present at the time of the tests (section 5-1). A judgment estimate of the

possible influence of well interference is added to the adjusted
estimated drawdown to obtain a value of possible maximum and
minimum pumping elevations over the projected future. This
analysis gives an estimate of the minimum and maximum pumping
levels for a given minimum discharge for a specific pumping
schedule.

**15-10. Analysis of Basic Data on Well and Pump
Performance.**—To amplify the previous discussion, a summary of
the procedures and methods in a hypothetical situation is given in
this section.

(a) In a developed area, a review of existing well per-
    formance hydrographs and logs of wells in the vicinity
    prior to drilling the pilot hole indicated that the desired
    yield could be obtained from a well within the following
    range of conditions:

    (1) Depth to static water-level . . . . 245 to 285 feet
    (2) Annual fluctuation in static water-level . . . . . 6 to 8 feet
    (3) Probable drawdown @ 900 gal/min . . . . . . 25 to 40 feet
    (4) Average annual water-level
         decline during a 6-year trend . . . . 5 feet per year
    (5) Depth of existing wells . . . . 400 to 500 feet
    (6) Thickness of saturated aquifer . . . . 200 to 250 feet
    (7) Average age of wells . . . . 15 years
    (8) Problems from encrustation or
         corrosion . . . . little or none reported

(b) The pilot hole showed the following conditions at the
    well site:

    (1) Depth to static water-level . . . . 254 feet
    (2) Thickness of saturated aquifer . . . . 226 feet
    (3) Depth to bottom of aquifer . . . . 480 feet
    (4) Mechanical analyses of the aquifer material and
        study of the log showed an adequate thickness of
        materials was present opposite which a 0.050-inch slot
        (No. 50) screen could be set to furnish the desired
        yield. In an undeveloped area, the pilot hole
        information would probably be supplemented by a
        pumping test to determine aquifer characteristics.

(5) Chemical analysis of the water indicated total
    dissolved solids content of 300 parts per million, pH of
    7.2, and favorable Ryzner and Langlier corrosion and
    encrustation indices.

(c) Basic well design:

   (1) Required minimum yield .... 900 gal/min
   (2) Minimum pump bowl nominal
       diameter .... 10 to 12 inches
   (3) Minimum pump chamber diameter .... 12 to 16 inches
   (4) Pump chamber depth:
       a. Present static water-level (b)(1) .... 254 feet
       b. Possible maximum drawdown (a)(3) .... 40 feet
       c. Decline in static water-level in
          20 years (a)(4) .... 100 feet
       d. Decline in pumping level in 20 years due to
          1-percent annual deterioration of well
          (judgment estimate) .... 2.48 feet
       e. Overlap between screen assembly and pump
          chamber (standard for this type of well) .. 10 feet
       f. Estimated total depth of pump chamber
          casing required: Sum of a. through e. above
          76.2 + 12 + 30 + 2.4 + 3 .... 412 feet
          (254 + 40 + 100 + 8 + 10)
          Since the aquifer thickness is 226 feet
          and the maximum desirable
          drawdown is 65 percent, the
          maximum drawdown is:
          226 x 0.65 = 147 feet.
          The maximum desirable pumping
          level would then be: 254 + 147 =
          401 feet.
          The maximum pump chamber depth is:
          40 + 10 feet overlap = 411 feet
          or 120.2 meters (401 feet)
          of usable pump chamber depth.
   (5) Screen assembly:
       a. 10-inch telescoping screen
          (recommended diameter for
          900 gal/min from table 11-7 in
          section 11-4) with 0.050-inch, No. 50
          slot has about 125 to 130 in^2
          per linear foot, see table 11-9,
          section 11-4. To estimate gal/min

per linear foot of screen at
0.1 ft/s entrance velocity,
multiply square inches of open
area per linear foot of screen by
0.31. This factor is derived by
(0.1 ft/s x 7.48 gal/ft^3 x
60 sec/min)/(144 in^2/ft^2 = 0.31).
For an average area of 127.5 in^2/ft
and an entrance velocity of
0.1 ft/s, the amount of water
entering the well is 127.5 x
0.31 = 39.5 gal/min. Length of
screen = 900 gal/min/39.5 =
22.8 feet (minimum), use 30 feet.

  b. As dictated by aquifer conditions, set
two 15-foot-long screen sections
separated by a 39-foot-long flush
tube section between depths of
411 and 480 feet.

  c. 10-foot section of flush tube overlap.

  d. Sump: 10 feet of blank flush tube extension
on bottom of screen with closed bail
bottom or other seal.

(6) Total well depth:

  a. Pump chamber depth (c)(4)f . . . . 411 feet

  b. Casing and screen assembly below
pump chamber including 10-foot
sump: Sum of (c)(5) a. through d.
30 + 39 + 10 + 10 . . . . 89 feet

    Total depth of well: Sum of a. + b. above less
10-foot overlap . . . . 490 feet

(7) Estimated pump requirements:

  a. Q = 900 gal/min (discharge)

  b. Drawdown (at end of 5 years)
(c)(4)b. + (a)(4) + (c)(4)d.
40 + 25 + 2 . . . . 67 feet

  c. Drawdown (at end of 20 years)
(c)(4)b. + (c)(4)c. + (c)(4)d.
40 + 100 + 8 . . . . 148 feet

  d. Pump lift (at end of 5 years)
(b)(1) + (c)(7)b.
254 + 67 . . . . 321 feet

  e. Initial bowl setting for first 5 years
(c)(7)d. rounded to standard column
lengths . . . . 325 feet

(8) Estimated probable pump head losses:
    a. Length of 8-inch column with
        1½-inch shaft . . . . 325 feet
    b. Column loss at 900 gal/min . . . . 10.4 feet
    c. Discharge head loss . . . . 0.03 foot
    d. Total pump head loss b.+ c. . . . . 10.7 feet
(9) Estimated surface losses:
    a. Elevation of bottom of storage tank . . . . . . 550 feet
    b. Elevation of shutoff elevation of
        storage tank . . . . 56 feet
    c. Effective length 8-inch pipe and
        fittings . . . . 104 feet
    d. Pipe head loss . . . . 0.3 foot
    e. Maximum surface head requirement
        b.+ d . . . . 59 feet
(10) Estimated total head:
    (c)(7)d. + (c)(8)d. + (c)(9)e.
    321 + 10.7 + 59) . . . . 391 feet
(11) Probable pump (from manufacturer's data):
    Bowl diameter . . . . 12-inch nominal
    Head per stage . . . . 80.5 feet
    Number of stages 5
    Horsepower per stage 22
    Bowl efficiency . . . . 82 percent
(12) For 12-inch nominal bowls use 16-inch
    casing. The clearance between 14-inch by
    0.375-inch wall thickness with i.d. of
    13.25 inches and bowls with o.d. of
    11.5 inches would be inadequate.
(13) Final well design:
    a. Casing: 16-inch casing from +1 foot to
        depth of 411 feet.
    b. Screen and casing assembly: 30 feet of 10-inch
        by 0.050-inch slot screen and 59 feet
        of 10-inch casing from 401 feet to
        490 feet.

(d) Results of step and 72-hour production tests on
    completion of well and development:

(1) Elevation of well head . . . . 5011 feet
(2) Static water-level start of test (low water
    period) . . . . 266 feet
(3) Thickness of aquifer . . . . 224 feet
(4) Temperature of water . . . . 54 °F

(5) Step test: 3 steps at 387, 701, and
1,001 gal/min, each step runs 4 hours.
(6) Pump schedule may call for 30-day
continuous pumping. Projection of
drawdown through 900 gal/min parallel to
plot of first step indicates a 22-foot
drawdown in 30 days. Projection of 72-hour
pumping test drawdown to 30 days
indicates 24 feet.

(e) Refinement of pump requirements for first 5 years:

(1) Static water-level at end of 5 years:
(d)(2) + 5 x (a)(4) = 256 + 5 x 5 = 281 feet
(2) Thickness of aquifer in 5 years:
(b)(3) - (e)(1) =   480 - 281 = 199 feet
(3) Drawdown for 900 gal/min for 30 days
at end of 5 years: (d)(6) x = (24) = 27 feet
(4) Pump lift at end of 5 years—no deterioration:
(e)(1) + (e)(3) = 281 + 27 = 308 feet
(5) Pump lift at end of 5 years and 1 percent a
year well deterioration: (e)(1) + (e)(3) x
1.05 = 281 + 27 x 1.05 = 309 feet

(f) Estimate of pump and well performance for first 5 years:

(1) Present water conditions   Minimum pump lift
a. Static water-level (d)(2) . . . . 256 feet
b. Aquifer thickness (d)(3) . . . . 224 feet
c. Drawdown at 900 gal/min (d)(6) . . . . 24 feet
d. Pump lift a.+ c. . . . . 280 feet
(2) Low water conditions in 5 years:
a. Low static water-level (e)(1) . . . . 281 feet
b. Thickness of aquifer (e)(2) . . . . 199 feet
c. Drawdown at 900 gal/min, 30 days:
(e)(3) . . . . 27 feet
d. Pump lift (e)(4) plus 1-percent
deterioration for 5 years . . . . 309 feet
(3) bhp = gal/min times total head in feet divided
by 3,960 times efficiency of the pumping
unit: Efficiency of pumping unit is equal to

the product of the bowl and motor efficiency:
(0.82)(0.90) = 0.74, use 75 percent

$$bhp = \frac{(900)(379)}{(3,960)(0.75)} = 115hp$$

(4) Shaft loss bhp = 4
(5) Total bhp = 119, use 125-bhp (93-kw) motor.

(g) NPSH required:

(1) Net positive suction head required at
      900 gal/min . . . . 15 feet
(2) Vapor pressure of water at 50 °F . . . . 0.4 foot
(3) Barometric pressure at 5000 feet elevation . . . 28.2 feet
(4) Available net positive suction head (3) + (2) . . 28.6 feet
(5) Excessive net positive suction head (4) - (1) . . . 13.6 feet

Theoretically this pump could operate with a 3.9-meter (13-feet)
plus suction lift; but for other reasons, it is preferable that the
bowls be submerged. The 97.5-meter (325-foot) bowl setting
originally estimated will be satisfactory.

The well with the above pump would perform satisfactorily for
5 years or more if the projections regarding aquifer thickness, etc.,
are realized. Eventually, an additional bowl would have to be
added, the bowl setting increased, and a larger motor installed. If
the decrease in aquifer thickness continued, at some still later
date, the yield would have to be reduced and a second well drilled
if the minimum yield requirements were to be met. Before any
such changes were made, a step test of the well would be desirable,
followed by rehabilitation if necessary, and a subsequent step test
before the new pump is specified.

When the basic pump bowls have been selected, the above data
and the charts and tables in the manufacturer's technical manual
permit estimates of additional values for use in the preparation of
designs and specification of components.

**15-11. Additional Factors in Pumping Equipment
Design.**—The hypothetical situation in section 15-10 covered
selection of the basic pumping equipment for the particular
application. Additional data on complete design include diameter
of the column pipe, diameter of the drive shaft, discharge head size

and type, lubrication, power selection, and type of drive. Most of these items have been standardized by the industry, and methods of determining the required components are included in the catalogs of pump curves and equipment issued by the various pump manufacturers.

Most large pump installations use weatherproof electric motors and control equipment. However, a pumphouse may be necessary under certain conditions. To facilitate pulling the pump, a roof hatch located over the well or a removable roof should be provided in the design of a pumphouse.

A pit installation below the ground may be advantageous in some instances; however, such installations are particularly susceptible to flooding and are prohibited by some State regulations.

Pumps operating in corrosive waters may require use of corrosion-resistant metals. The general conditions governing the use of such metals are covered in the discussion on corrosion and encrustation (chapter XVI). Specific solutions for a particular case should be discussed with corrosion specialists and the pump manufacturer.

The type of power unit selected usually depends on the availability and cost of fuel. If electric service is available within a reasonable distance, an electric motor is generally preferred because of lower first cost, lower maintenance cost, and its reliability without regular servicing and periodic attendance. If electric service is not available, an engine fueled with gasoline, diesel fuel, natural gas, or liquid petroleum gas usually must be chosen. Such an engine can either be belted or geared to the pump and can be fitted with many different appurtenances. Small pumps can be powered by windmills where wind conditions are favorable.

15-12. Measuring Pump Performance.—Cost of energy is one of the principal expenses incurred in the operation of pumps. Therefore, pumps should be monitored to ensure that they are operating at or near peak efficiency. Three factors must be measured to check pump efficiency: (1) total head, (2) input horsepower, and (3) quantity of water pumped. When internal combustion engines are used, it is also to be taken simultaneously when the flow, head, and speed are steady. Other flow rates must determine if the whole performance curve of the pump is desired.

15-13. Estimating Total Pumping Head.—The total dynamic head against which the pump is operating includes the vertical distance from the water-level in the well while pumping to the center of the free-flowing discharge, plus all losses in the line between the point of entry of the water and the point of discharge.

If the discharge is maintained under pressure, the pressure required at the pump head to operate the system is added to the lift and line losses to obtain the total head.

Losses in pipe and fittings can be obtained from a hydraulics handbook and pump-column losses from pump manufacturer catalogs or from Standards of the Hydraulic Institute.

15-14. Estimating Horsepower Input.—A convenient method of measuring the power input to electric motors without interrupting their operation is with a hook-on voltameter. Usually there is enough slack in the wires in the motor starter box to permit reading each phase. The method given here pertains to three-phase circuits but can be adapted to others. The power input is obtained by dividing the average current of the three phases by the full load amperes as stamped on the nameplate of the motor. For example, if the average current in the three phases for all 1,800 revolutions per minute, 200-volt, 30-horsepower motor, with a full load current of 75 amperes, is 50 amperes, then 50 divided by 75 equals 67 percent of the full load current, and the power input is 67 percent of the rated horsepower, or 20 horsepower. The voltameter is not only convenient for determining that the motor nameplate voltage is maintained, but it also will reveal any serious unbalance between the three phases. This method of power measurement should usually yield results accurate to within about 3 percent.

Another method of determining power input is with the aid of the watt-hour meter. A watt-hour meter installed on the pump control panel can be used. The procedure is simple to count the number of revolutions of the meter disk for a time interval (3 minutes is usually enough), during which time water discharge measurements are also taken. The electrical input to the motor is given by the formula:

$$hp\ input\ =\ \frac{3,600RK}{746t}$$

where:

R = number of revolutions of the disk in time t
K = meter constant taken from meter nameplate
t = time in seconds for R revolutions

If current and potential transformers are used, the meter constant must be multiplied by the current transformer ratio, the potential transformer ratio, or the product of both, and the computations would then be made as follows:

$$hp \ input = \frac{3,600RKM}{746t}$$

where:

M = transformer ratio

Unless one is experienced in the operation and testing of large, high-voltage motors, a qualified industrial electrician should be consulted prior to testing for pump performance.

Where an internal-combustion engine is used as the prime mover for a pump, the input horsepower can be calculated by methods described in various mechanical engineering handbooks or manufacturers' catalogs.

15-15. Measuring Pump Discharge.—Several means of measuring the discharge of a pump are available, but for freely discharging pumps, a weir or orifice is widely used and each is adaptable to most field situations. Tables and information on weirs, as well as on some other measuring devices, are available in the Bureau of Reclamation *Water Measurement Manual* (1967), and orifice plates have been previously described in section 8-9. Where a closed system is involved, there are several types of flowmeters which can be used.

15-16. Measuring Pump Efficiency.—With measurements of total head, input horsepower, and quantity of water pumped, the efficiency of the installation, expressed as a decimal, may be determined from the following formula:

$$Plant\ efficiency = \frac{Q\ (gal/min) \times total\ head\ (ft)}{3,960 \times input\ horsepower}$$

The pump efficiency may be determined by dividing the plant efficiency by the efficiency of the electric motor or of the engine and drive mechanism:

$$Pump\ efficiency = \frac{plant\ efficiency}{motor\ efficiency}$$

The efficiency of an electric motor is usually between 90 and 95 percent, depending on size and type, but an exact value can be obtained from the information furnished by the manufacturer for the particular motor. The efficiency of an internal-combustion engine is more difficult to obtain because it changes as wear occurs. Plant efficiency (sometimes called wire-to-water efficiency) should be determined, at least annually, as a means of checking wear or changes in pumping conditions. In some areas where power costs are high and pumps are operated a large part of the year, plant efficiency should be checked every 2 months (Fabrin, 1954; Vertical Turbine Pump Association, 1962).

15-17. Selection of Electric Motors.—The designer should consult an electrical specialist for advice and assistance on selecting electric motors. However, the following summary of motor characteristics is included as a guide.

Electric motors are usually selected according to NEMA's standards including its definitions of enclosures and cooling methods.

Dripproof motors are built for a 40 °C ambient temperature rise. These motors are satisfactory where equipment is installed within a shelter.

Splashproof motors are built to tolerate a 50 °C ambient temperature rise. Precipitation coming to the motor at angles less than 100° from the vertical cannot enter the motor. These motors are satisfactory for use in the open where rain, snow, and wind velocities are not excessive.

Weather-protected motors are made with provision for ambient temperature rise of either 40°C type I or 50°C type II. The Type I motor has the ventilation openings so constructed as to minimize the entrance of rain, snow, or airborne particles into the motor. Most are so constructed as to prevent the insertion of a rod 19 millimeters (3/4-inch) in diameter through the ventilation openings. These motors are suitable for installation in the open but screening of the ventilation openings is mandatory. They are used in relatively unprotected locations where extreme adverse weather conditions exist (i.e., areas where hurricanes, repetitious storms, snow, extreme heat, and abundant rain are prevalent).

The number phases, frequencies, and voltage of the motor are usually established in advance by the power service available.

The motor should be selected to deliver the estimated maximum power required by the pump without overloading but with consideration of the service factor and desirable insulation.

Thrust bearings are usually built into the motor and vary in type of construction consistent with the magnitude of the thrust expected. Total thrust consists of the weight of the rotating elements of the pump, the weight of the column of water, and the hydraulic thrust developed by the pump. Most pump manufacturers' catalogs furnish thrust and bearing data.

Pump motors should be equipped with nonreverse protection. This usually consists of a releasing coupling which disengages the motor when the pump is stopped for cause, such as a power failure. The coupling allows the pump drive shaft to spin in reverse as water drains from the column pipe without driving the motor. This eliminates the possibility of the motor turning in reverse or of snapping the drive shaft in the event the power outage is only momentary.

Supply line limitations often limit the amount of inrush power required as a motor is started. If the supply line permits an inrush of 600 percent of the full load current, the most economical control is across-the-line starting. However, if limitations preclude the 600 percent, a reduce-voltage starter should be used.

An electrical specialist should be consulted on all aspects of selection, installation, and operation of electrical pumping equipment

In summary, in selecting an electric motor, the following factors should be considered:

- Power required by the pump and service factor of the motor

- Compatibility of design rotation rates of pump and motor

- Use of shelters or protected motors

- Adequate thrust bearing capacity

- Self-releasing couplings or other nonreverse protection

- Compatibility of pump discharge head, column pipe, and motor dimensions

- Inclusion of thrust horsepower loss in wire-to-water efficiency

- Inrush limitation and need of reduced voltage starting

15-18. Selection of Internal-Combustion Engines.—
Selection of an internal-combustion engine as a source of power for pumps is more complex than selection of an electric motor. Internal-combustion engine horsepower ratings are usually given without consideration for power consumed by accessories and are rated for sea level operation. The developed horsepower decreases with increase in altitude. The maximum developed horsepower is usually rated at a given revolutions per minute and varies with different manufacturers so sheave ratios for belt drives and gear ratios for gear drives must be selected to give compatibility of pump and motor speeds. Most internal-combustion engines undergo up to 25-percent reduction in developed horsepower if used continuously as compared to intermittent use. When an engine is adapted to the use of natural gas or other similar fuel, the BTU rating of the fuel is also a factor in estimating the developed horsepower. Engine manufacturers can furnish data which will permit estimates of the horsepower and speed developed by their engines at various altitudes, BTU content of fuel, and other factors.

15-19. Bibliography.—

American Water Works Association, 1971, "American National Standard for Deep Well Vertical Turbine Pumps, Line Shaft and Submersible Types," AWWA E101-71, No. 45101, American Water Works Association, New York.

Anderson, K.E., 1973, *Water Well Handbook*, 3rd edition, Missouri Water Well & Pump Contractors Assn., Inc.

Bureau of Reclamation, 1967, *Water Measurement Manual*, 2nd edition, Denver, Colorado.

Driscoll, F.G., 1986, *Ground Water and Wells*, 2nd edition, Johnson Division, St. Paul, Minnesota.

Environmental Protection Agency, 1975, *Manual of Individual Water Supply Systems*, Office of Water Program, Water Supply Division.

Fabrin, A.O., 1954, *The Answers to Your Questions About Layne Vertical Turbine Pumps*, Layne Bowler, Inc., Memphis, Tennessee.

Fairbanks Morse and Co., 1959, *Hydraulic Handbook*, 3rd edition, Kansas City, Missouri.

Gibbs, C.W., editor, *Compressed Air and Gas Data*, 2nd edition, Ingersoll-Rand, Woodcliff Lake, New Jersey.

Moore, A.W. and H. Sens, (editors), 1954, *The Vertical Turbine Pump by Johnston*, Johnston Pump Co., Pasadena, California.

Muska, C.F., W.P. Colven, V.D. Jones, J.T. Scogin, B. Looney, and V. Price, Jr., 1986, "Field Evaluation of Ground Water Sampling Devices for Volatile Organic Compounds," Proceedings of the Sixth National Symposium and Exposition of Aquifer Restoration and Ground Water Monitoring, National Water Well Association, Dublin, Ohio, pp. 235-246.

QED GroundWater Specialists, 1994, "Pneumatic Cleanup Pumping Guide," Ann Arbor, Michigan.

"Standards of the Hydraulic Institute," 1955, 10th edition, Hydraulic Institute, New York.

"Turbine Pump Facts," 1962, Vertical Turbine Pump Association, Pasadena, California.

Chapter XVI

WELL AND PUMP COSTS, OPERATION AND MAINTENANCE, AND REHABILITATION

16-1. Well Construction Costs.—Until recently, standards in the water well drilling industry were largely determined by local custom. Accordingly, many design and construction practices were questionable. The situation was further complicated by the geographic concentrations of drilling contractors, reluctance of contractors to move more than 65 to 80 kilometers (40 to 50 miles) from their base of operations, diverse geologic and hydrologic conditions, diverse drilling methods, and seasonal operations. Well construction costs that developed under such conditions tended to be erratic and unpredictable. Although these problems have not been eliminated entirely, the industry has been stabilized by several factors, including:

- Enactment of minimum well construction standards by many States

- Development of more efficient and versatile equipment

- Training of contractors in good engineering and business practices

- The organization of State, regional, and national drillers' associations

Well construction costs show a marked seasonal variation in much of the country. Costs are usually highest in the early spring and lowest from early fall to midwinter. Move-in costs for small wells are usually relatively low, but they can be extremely variable for larger, more complex jobs. This variation on larger jobs may be attributable to the unbalancing of bids to obtain operating funds early in the operation.

The foregoing practices have precluded the establishment of a meaningful well construction cost index similar to those made available to the general construction industry by various engineering publications and reporting services. Consequently, cost estimates must be based on site-specific market surveys. For small jobs, such as a water-supply for a campground, the market survey should be limited to local contractors. Large jobs involving development of entire well fields for municipalities, for instance, may justify a nationwide market survey.

16-2. Pump Costs.—The vertical turbine pump is practically standard equipment for water wells of moderate to large capacity. Manufacturers have essentially standardized all motors, motor controls, pump discharge heads, and column assembly features so that they are generally interchangeable for pumps of a given size, capacity, and rotation rate.

A wide variety of off-the-shelf pumping units of standard construction is available, and some manufacturers offer off-the-shelf units for use in corrosive environments.

Most manufacturers publish manuals and catalogs describing their products which contain performance curves showing head-capacity relationships, bowl efficiencies, horsepower requirements, pump speeds, and net positive suction head requirements. Similar publications are available from motor, electrical control, valve, and flowmeter manufacturers. A review of the available literature will usually permit preparation of specifications which will permit bidding on a competitive basis.

Small wells usually are equipped with jet, lift, or small submersible pumps. Local distributors can usually furnish literature on capacities, costs, etc., for the preparation of specifications and cost estimates.

16-3. Operation and Maintenance Responsibilities.—
Operation and maintenance of wells and pumps are usually the responsibility of ground-water specialists and mechanical design personnel. Primary responsibilities may include pump selection, pump installation, and design of discharge and distribution facilities including controls and housing. Close coordination between these two technical groups assures that certain features will be included in the discharge and distribution facilities to permit proper monitoring and maintenance of the well and pump. The more important of these features include:

- An outlet in the discharge system to permit diversion from the system during future test pumping and water sample collection

- A permanent throttling valve on the discharge

- A permanent air line with valve and gauge for water-level reading

- Access into the pump chamber casing which can also be used to measure water-levels by tape or electric probe

- Ready access to the well to pull the pump and maintain the well

16-4. Operation and Maintenance Basic Records.—All well installations should incorporate basic documentation. This documentation includes well construction, well performance tests, and pump efficiency records. Well construction records include geologic or formation logs, as-built construction diagrams, and mechanical analyses of aquifer and gravel pack (if used) samples. Upon completion, every well should be tested for performance. Wells should be tested for:

- Sand content of the discharge
- Drawdown at various yields (step drawdown)
- Drawdown at design capacity
- Plumbness and alignment
- Water quality analyses

After the permanent pump is installed and adjusted, its efficiency should be tested and results should be recorded. A permanent pump should be tested for wire-to-water efficiency (actual discharge of pump compared to theoretical discharge considering amount of energy used), shut-in head, and conformance to the performance curves furnished by the manufacturer. These data, together with a copy of the well and pump specifications, should be included in the permanent record. These data provide the baseline to which the results of subsequent tests of the entire installation will be compared so that pump and well conditions can be evaluated and the need for rehabilitation or other maintenance can be determined.

The following conditions can generally be assumed for Bureau of Reclamation installations:

- The well was carefully designed, constructed, developed, and tested after completion to permit the determination of specific capacity and related characteristics of the well and the quality of the water.

- The pump was selected to discharge the minimum acceptable volume of water at the estimated maximum probable pump lift and within an acceptable range of efficiencies.

- The pump was installed in the well and tested for conformance to the specifications.

- The pump and motor have received the service and maintenance recommended by the supplier.

16-5. Video Well Surveys. — Loss of well efficiency, development of sand pumping, change in quality of water, or well failure are all causes of concern and usually require well rehabilitation or replacement. A video survey of the well is one of the most economical and helpful tools for determining the nature of the problem and possible method of rectification.

Closed-circuit video equipment gives excellent black and white or color views down the well (longitudinal) and, with some equipment, horizontal (radial) views of the side of the hole. The horizontal views are particularly valuable in close-up viewing of suspected corrosion or encrustation of screens or perforated zones.

Some equipment will operate in holes as small as 75 millimeters (3 inches), but generally, a 150- to 200-millimeter (6- to 8-inch) hole is required. Video surveying services are available from several Federal agencies and commercial operators.

Prior to making any video survey of a well, an effort should be made to clarify and reduce the turbidity of water in the well. Many procedures have been used to address this problem, but none have been markedly successful. However, the following procedure is recommended:

- If the water has a pH below 7, about 1 kilogram (2 pounds) of slaked lime, $Ca(OH)_2$, per 4,000 liters (1,000 gallons) of water in the well should be added and thoroughly dispersed through the total well depth by surging before adding the coagulant as described in the following paragraph.

- If the pH of the water is above 7, the alkalizing agent is not necessary and about 0.25 kilogram (0.5 pound) of alum ($Al_2(SO_4)_3$) or ferric sulfate ($Fe(SO_4)_4$) per 4,000 liters (1,000 gallons) of water in the casing and

screen should be added to the well. The well should
then be strongly agitated with a surge block or similar
tool through the entire depth of water for at least
30 minutes for each 30 meters (100 feet) of water in
the well. The surging and agitation will loosen any
existing biomass or even some light mineral deposits
from the well screen. If the intent of the video survey
is to view these deposits, then the video should be
attempted without surging and bailing the well.

- The coagulant should be added to the well at least
  3 days, and preferably a week, before inserting the
  camera into the hole.

- If a layer of oil is present on top of the water in the well,
  an effort should be made to bail the oil out before adding
  the coagulant. This procedure is usually not wholly
  successful, so the camera lens should be wetted with a
  strong solution of detergent as it is placed in the hole.
  This procedure will prevent the lens from being coated
  with oil, which would considerably reduce the definition
  of the image.

- If repairs or rehabilitation are performed on the well, the
  video operation should be repeated to provide a record
  for comparison on future inspections.

**16-6. Routine Observations and Measurements on Large
Capacity Wells.**—Irrigation wells and other large capacity wells
are often operated seasonally in conjunction with agricultural
requirements. Proper monitoring and preventive maintenance can
eliminate or substantially reduce operating costs of these wells. A
general monitoring program should include the following
measurements and observations:

• Static ground-water-level measurements a week or two
  before the pumping season begins.

• Measurements for drawdown, discharge, and power usage
  shortly after the start of the pumping season. These
  measurements should be made after at least 8 hours of
  continuous well operation.

- Sand content of the discharge should be measured 5 minutes and 30 minutes after initial startup. Most wells will produce a small amount of sand when pumping is initiated after a long idle period. The sand pumpage should not increase from year to year and should return to near normal condition after 30 minutes of pumping.

- In multiwell fields, each well should be tested individually and drawdowns in the adjacent wells should be measured during the test.

- Static water-level measurements at intervals of several months during the pumping season (a minimum of 12 hours of nonoperation should be allowed before measuring static levels).

- Total seasonal amount of water pumped and power used for determining wire-to-water efficiency.

- End-of-season water sample taken for water quality analysis.

- At the end of the pumping season each well should be measured, if possible, to determine the total depth.

- Static water-level measurements each year about midway between the end of the pumping season and the beginning of the following pumping season.

- Continuous hydrographs should be plotted of the static levels, pumping levels, and specific capacities of each well.

If sand or other material has accumulated in the bottom of a well to a level where it has encroached on the screen, or may encroach on the screen during the next pumping season, the pump should be pulled and inspected, and the well should be bailed clean before any other tests or measurements are made. Where possible, the following measurements and tests should be made:

- Static water-level

- A step test at about the same rates and for the same period of time as was made when the well was initially completed

- Closed-in head

• Wire-to-water efficiency

• Sand content of the discharge 5 minutes and 30 minutes after pumping is started

• Water samples taken for quality analyses

The tests should be analyzed and the results should be compared with those of the initial tests made when the well was completed.

Preventive maintenance includes routine lubrication and servicing of each well installation. During routine lubrication and servicing, the following should be observed and recorded:

• Any increase in sand content of the discharge
• Decrease in discharge
• Excessive heating of the motor
• Excessive oil consumption
• Excessive vibration
• Sounds possibly attributable to cavitation
• Cracking or uneven settlement of the pump pad or foundation
• Settlement or cracking of the ground
• Change in ground surface gradient around the well

16-7. Interpretation of Observed or Measured Changes in Well Performance or Conditions.—As discussed above, observations made during preventive maintenance and servicing can reveal changes in well performance or conditions. The following paragraphs offer some guidance in interpreting changes in well performance or conditions.

• A decrease in specific capacity without a proportional decline in the static water-level may indicate blockage of the screen or gravel pack by encrustation, or collapse of casing or screen.

Should the specific capacity during a step test show a decline of 10 percent or more from the original step test at a given discharge, the well should be surveyed with a dolly or bailer (section 12-8) to determine the location and extent of possible contributing conditions. If collapse appears to be the problem, the well should be inspected with video equipment (section 16-5) to determine the location and nature of the collapse. If collapse is not the problem,

the inside of the well should be scraped and the sediment that was
subsequently bailed from the bottom should be examined to
determine the chemical composition, nature, and extent of the
encrustation material as a basis for a plan or rehabilitation.

- An increase in sand content of the discharge, particularly if
  it is associated with a measurable accumulation of sand in
  the bottom of the well, may indicate enlargement of slot sizes
  by corrosion; settlement of gravel pack beneath a bridge
  leaving an unpacked zone opposite a screened section; a
  break in the casing or screen, usually at a joint; or failure of
  a packer seal. Mechanical and mineralogical examination of
  a sample bailed from the bottom of the hole and comparison
  with the original description of the aquifer and gravel pack
  materials made during construction of the well may give
  some indication of the nature of the difficulty. If the
  material is noticeably smaller in grain size than the grain
  size of any aquifer screened in the well, or if the material
  contains the full range of sizes of the gravel pack, either the
  casing or the screen are probably broken. If all the material
  is smaller than the screen slot sizes, a bridge in the gravel
  pack has probably developed. If the above interpretations of
  grain size and distribution are not applicable, the problem
  may be caused by enlargement of a slot size by corrosion. A
  problem apparently caused by bridging can frequently be
  corrected by redevelopment while pouring water down the
  gravel refill tremies and the addition of gravel pack material
  (section 11-12). The other problems usually require a video
  survey to be made of the well to more clearly assess the
  problem. Decisions can then be made concerning the
  practicability of rehabilitation and the procedures to be
  followed.

- Settlement or cratering of the land surface around a well, the
  development on the ground surface of small drainage
  channels toward the well, and cracking and settlement of
  pump pads and foundations are all indicative of settlement of
  the well structure. In some areas, the problem may be
  associated with land subsidence caused by excessive pumping
  of the aquifers. Usually, the problem is related to poor well
  design, construction, or development, and results from
  excessive pumping of sand. In many instances, the sand
  pumping is complicated by collapse of casing or screen,
  bridging of gravel packs, and similar deterioration. When
  such conditions are encountered and as a basis for

rehabilitation, the well should be taken out of service,
sounded for depth, and surveyed photographically to
determine whether any structural damage has occurred. If
the well cannot be shut down because of the need for water,
the casing should be temporarily supported by welding heavy
I-beams to it (section 11-2).

The foregoing problems are related primarily to the well and are
the most commonly encountered. Many of them may occur because
of conditions that were not considered in the original well design;
others are caused by inadequate investigation prior to construction
or the attempt to standardize on a particular well design. In any
event, any failure or deterioration should be thoroughly
investigated and recorded along with the rehabilitation program
used and the success of the program. These data should be made a
permanent part of the well records and used as a guide in the
design and construction of wells drilled in the area in the future.

- Decline in pump discharge and head may be caused by
  deterioration of the pump or simultaneous deterioration of
  both the well and the pump. A decrease in shutin head and
  significant decrease in discharge without a corresponding
  decline in static water-level and specific capacity is a
  common occurrence. The condition is usually caused by
  (1) improper adjustment of the impeller because of wear or
  other causes, (2) a hole in the column pipe, or (3) erosion or
  corrosion of the impeller or bowls. The latter condition is
  usually associated with considerable vibration when the
  pump is running.

- Excessive vibration of the pump may result from imbalance
  of the impeller or from the pump being installed in a crooked
  well. If the condition cannot be corrected by adjusting the
  impellers, the pump should be pulled and repaired or
  replaced. The cause of the problem should be thoroughly
  investigated and made a part of the permanent well and
  pump record. If the vibration is caused by a crooked well,
  routine maintenance will not cure the problem, and a new
  well may be needed.

A pump which makes a crackling noise similar to gravel being
thrown on a tin roof is probably experiencing cavitation, a form of
erosion, of the impellers. This condition is particularly true if
the discharge is surging and irregular and contains considerable

air. The condition usually results from a decline in the static water-level, encrustation, or accumulation of sand in the screen. Any of these conditions results in excessive drawdown for the pump and a decline below that required in the available net positive suction head. If the condition is caused by a decline in the static water-level, it can usually be corrected by lowering the bowls. Severe cases may require additional stages and a larger motor in addition to lowering the bowls. The well should also be checked for possible encrustation of the screen or other causes of reduced efficiency.

- Excessive heating of the motor is occasionally encountered and is usually associated with an overload condition and the consumption of excessive electrical energy. Excessive heating may be caused by:

  - A poorly adjusted impeller which is dragging on the bowls
  - A packing gland that is too tight
  - Improper or unbalanced voltage
  - Poor electrical connection
  - Improper sizing of the motor

Occasionally, excessive heating is caused by trash that has lodged in the bowls or blockage of the impellers or bowl channels byproducts of corrosion and encrustation. Correction entails pulling the pump for repair. These conditions may also be associated with inadequate discharge. Where overheating is encountered, the installation should be checked by an electrician, as a first step, to determine whether the trouble is in the power system or in the pump rather than in the well.

- Occasionally, a noticeable increase in oil consumption is encountered in oil-lubricated pumps. The excessive consumption may be caused by a hole in the wall of the oil tubing or excessive wear on a packing gland in the tubing. These conditions can result in a decrease in differential pressure in the oil tubing and loss of oil into the well. The first condition can result in inflow of water into the tubing and formation of an emulsion of water and oil. The emulsion lacks adequate lubricating qualities and can result in excessive wear or bearing burnout. The escape of oil into the well can result in the accumulation of oil floating on the water surface in the casing. With adequate pump submergence, this latter condition may not cause serious trouble. However, if drawdown increases for any reason, oil

may be drawn into the pump, or oil may leak through the
screen to the aquifer causing impairment of water quality.
In addition, the presence of oil may preclude accurate
measurement of static water and pumping water-levels.

* Small capacity wells of less than 500 liters per minute
  (125 gallons per minute) commonly have 150 millimeters
  (6-inch) or smaller casing and screen, and materials used in
  their construction are relatively light in weight. Although
  the observations and measurements outlined above for large
  capacity pumps and wells are equally applicable, they are
  usually difficult to justify economically.

  Pumps may be shaft-driven vertical or submersible turbines,
  ejector, cylinder, or suction types of various kinds. The cost
  of the well construction is minor compared to a large
  capacity well. In many instances, when a small capacity well
  fails, replacement is less expensive than rehabilitation.
  Although the continued observation and periodic inspection
  and testing of small capacity wells can seldom be justified
  economically, the wells should be checked at least once a
  year for discharge, drawdown, specific capacity, sand content
  of discharge, effective depth, and static and pumping water-
  levels. Many manufacturers' handbooks give methods of
  testing and evaluating the condition of their pumps. The
  literature should be consulted and, where practical, the
  recommendations should be applied.

16-8. Well Rehabilitation Planning.—Most Reclamation
water storage, conveyance, and control structures are subject to
periodic inspection and testing. However, wells and pumps are
often neglected. Wells deteriorate over time, and the deterioration
may not be readily discernible during operation, thus eluding
recognition until the well fails.

Well deterioration is difficult to monitor because the greater
portion of both well and pump are located beneath the ground
surface. Problems usually develop slowly to a critical point and
then accelerate rapidly to failure. Rehabilitation may be possible if
the deterioration can be recognized before reaching the critical
point. If the deterioration continues too long, the potential of
successful rehabilitation is substantially decreased. A loss of
50 percent or more in well efficiency usually means the well cannot
be successfully rehabilitated by routine redevelopment methods.

Efficiency may be reduced because of encrustation, corrosion, or other factors which tend to reduce the intake area of the screen or permeability of the adjacent aquifer.

Water well rehabilitation includes the repair of wells which:

- Have experienced failure of the screen or casing,
- Have begun to pump sand,
- Have experienced a change in water quality, or
- Have shown a marked decrease in efficiency

Normally, well rehabilitation does not include deepening or other major changes in the well structure. Construction of a replacement well may be necessary if rehabilitation of a well is impracticable.

A major problem in well rehabilitation may be in determining the exact nature of the deterioration because the screen and other components most likely to deteriorate are not subject to direct visual inspection or testing. Accordingly, well rehabilitation usually involves the risk of further damaging a well or destroying its usefulness. However, the element of risk can be reduced by data collection and planning prior to undertaking the work. Data should include:

(a) Original design and construction (as-built conditions)
    (1) When drilled
    (2) Method of drilling
    (3) Materials log
    (4) Geophysical logs
    (5) Casing log
        a. Length and diameter
        b. Wall thickness
        c. Type and location of joints
    (6) Screen or perforated casing description
        a. Type and material
        b. Length and diameter
        c. Slot size
        d. Depth of settings
        e. Type and location of joints
    (7) Grout or seals
        a. Type and composition
        b. How placed

(b) Mechanical analysis of aquifer material and gravel pack

(c) Relationship of aquifer material or gravel pack to screen slot
     opening

(d) Method and completeness of development

(e) Original pump test results: step and constant yield tests,
     sand content of discharge

(f) Ground-water hydrographs in the area

(g) Water quality test results

(h) Summary of historical performance and operation

(i) Summary of needed maintenance and rehabilitation

A number of proprietary methods have been developed to
rehabilitate wells and improve well yields. These methods
generally use specialized mixtures of chemicals, such as hot water
and acid mixtures, liquid carbon dioxide, or specific combinations
of acids, disinfectants, or surfactants. Often, these proprietary
methods also employ specially designed tools, injection nozzles, or
some other physical method to enhance the chemical action.

16-9. Sand Pumping.—Most wells pump sand to some degree.
However, proper design and adequate development usually can
limit sand pumping to an acceptable concentration. Excessive sand
pumping is accompanied by numerous undesirable side effects.
Pump bowls and impellers may be eroded by sand and necessitate
frequent replacement. Not all sand entering a well is pumped out
with the discharge. Some of the larger grained portion of the sand
settles to the bottom of the well where it may encroach on the
screen and reduce well efficiency. This reduction results in
increased drawdowns, increased entrance velocity, and perhaps
accelerated corrosion and encrustation.

The sand in the discharge may collect in pipelines and channels,
thereby reducing their carrying capacity and necessitating periodic
cleaning. Furthermore, sufficient sand may enter a well to create
fairly large cavities in the aquifer around the screen. As the
cavities collapse, the casing or screen may be broken or deformed.
In severe cases, subsidence at the surface may damage the entire
installation. Where sprinkler systems are directly supplied from a
well, excessive sand may block the pipelines and erode the orifices
in the sprinkler heads, which changes the distribution pattern.

If sand pumping is caused by a broken screen or casing or faulty packer, the location and nature of the break can usually be determined by sounding and can be verified by a video survey of the well. In a few instances, the break may consist only of a parting of the casing or screen at a point with no offset of the axis. This break is the easiest type to repair. In most cases, the break will be associated with displacement of the axis and possibly deformation of the casing and screen on one or both sides of the break. The best procedure is to run a hydraulic or mechanical casing swage into the well to round out and, if possible, realign the casing or screen. In some instances, the casing and screen may be so far out of line that realignment may be impossible without causing buckling of the casing elsewhere. No fully satisfactory solution exists in this circumstance, although the well possibly can be modified by inserting a liner to produce sand-free water at a lower discharge and specific capacity. Where approximate or complete realignment is possible, the well can be repaired by inserting a liner through the break and anchoring the liner in place either by using a hydraulic swage or by cementing it in place. This procedure usually will reduce the yield or specific capacity and may require the use of a different pump because of the reduced inside diameter.

If a broken or defective seal is involved, several possibilities for repair of nontelescoped assemblies may exist. Removing a swaged lead seal is almost impossible without pulling the casing or screen. One solution is to telescope 3 meters (10 feet) or more of liner with neoprene rubber seals sized to both the smaller and larger casings into the smaller casing. Another solution is to swage a liner into the smaller casing with the end extending about 0.9 meter (3 feet) above the original lead packer and then fill the annular space with a neat cement grout.

If the problem is caused by either localized enlargement of screen slots or a hole in the casing or screen as a result of corrosion, a liner can sometimes be swaged in place in the corroded section. The liner in all cases should be made of the same material as the casing or screen within which it is placed.

In some instances, because of poor initial slot selection or enlargement of slot sizes over the greater length of a screen because of corrosion, insertion of a liner is impractical. If telescoping construction has been used, the screen may be pulled and a new screen with smaller slots or casing made of more corrosion-resistant materials may be installed. This procedure is

impossible, however, where single string design has been used.
Nevertheless, rehabilitation has been made on single string designs
by ripping the original screen to increase the open area and then
telescoping a smaller diameter screen inside it. This procedure
may be used as a temporary expedient but is not recommended as
a permanent repair. Well efficiency is reduced and will deteriorate
rapidly, although several years' service may be obtained from the
well. In some instances, stainless steel screen has been installed
inside low carbon steel. This practice is not recommended. The
same material should be used as in the original screen; otherwise,
aggravated corrosion of the original screen and blockage of the
inserted screen by incrusting corrosion products are almost certain
to result.

Where sand pumping has resulted from settlement and bridging
of a gravel pack, the best procedure is to vigorously redevelop the
well while injecting large quantities of water into the pack from
the surface. This procedure will usually cause the bridge to
collapse, thereby reestablishing the integrity of the pack.

Additional material should then be added to replenish the pack.
In older wells, the bridging may be a result of local cementation of
the pack and the foregoing procedure may be ineffective. In such a
case, the well should be acidized, and then another attempt should
be made to cause collapse of the bridge.

Where casing settlement has occurred, the structure should first
be supported by welding parallel I-beams of adequate strength and
length to opposite sides of the casing collar at the ground surface
(section 11-4).

The well should then be surged vigorously while water is injected
into the gravel pack or, if no pack is present, water should be
applied to the caved area around the top of the well. Selected
gravel pack material should be added to the pack or to the caved
area around the well as required. This procedure will fill existing
caverns and cause bridging to collapse and ensure a stable
condition before further rehabilitation is undertaken. Excessive
subsurface movement during this work may aggravate the
situation or cause additional casing failure, which may make
further work impractical. During work of this type, care should be
taken to ensure that rapid settlement does not endanger workers
or equipment.

Collapse from excessive hydraulic differential head may result from overpumping a well which has inadequate entrance area in the perforations or screen or from loss of intake area because of encrustation. This condition is seldom a problem where casing with an adequate diameter to wall thickness ratio has been used. Necessary corrective measures may include swaging the affected component to full diameter or reducing pumping.

16-10. Decline in Discharge.—A decline in discharge and an increase in drawdown are usually caused by:

- A decline in the static water-level

- The installation of additional nearby wells that have over-lapping areas of influence

- An accumulation of sediment on the bottom of a well sufficient to cover a significant part of the screen

- Collapse of the screen

- Encrustation of the screen and gravel pack

Where the decline in yield is the result of a decline in the static water-level or interference from other wells, the situation may be repairable by merely lowering the pump bowl and, if necessary, adding additional bowls and a larger motor. If regular measurements are made of static and pumping levels in the well, the cause is usually apparent.

The possibility of decline in yield caused by accumulation of sand over part of the screen can be determined by sounding the well. The solution is to bail the well clean. However, such an accumulation is usually indicative of other problems. The discharge should be tested for sand content and, if too high, the investigations outlined in section 16-7(b) should be made.

If collapse of the casing or screen is suspected, lowering a bailer or dolly down the hole on a cable will usually show the approximate location of the problem (section 12-7). If collapse is indicated, a television or photographic survey should be made to determine the nature of the damage and the possibility of repair. If the casing or screen is not broken, the problem may be corrected

by using a hydraulic or mechanical casing swage. If a break has
occurred, however, corrective measures should be taken as
described in section 16-9.

16-11. Corrosion.—Corrosion problems develop in many forms
and from many causes as described in section 11-2. Corrosion that
has caused the screen or casing to fail will result in sand pumping
and accumulations of sand in the bottom of the well. The
corrective measures described in section 16-9 should be taken.
However, before proceeding with repairs, one should ensure that
materials used will be compatible with the original materials
(section 11-2).

16-12. Encrustation.—Section 11-3 deals with encrustation
considerations at the design stage. This information is equally
valuable in maintenance activities but will not be repeated here.

An aggravated mineral encrustation problem in an existing well
can often be overcome by reducing the rate of pumping and,
consequently, the entrance velocity of the water flowing into the
well. This procedure may be accomplished by installing a smaller
pump and operating the well longer to obtain a required quantity.
Where this practice is not possible, the rate of deposition may be
decreased by installing several wells and pumping them at a
relatively low rate to obtain the needed volume of water.

Because encrustation cannot be entirely avoided, rehabilitation of
permanent wells should be anticipated in many cases. Good
practice involves carrying out such rehabilitation at stated
intervals before the problem becomes too acute at any one well.

Mineral encrustation caused by bacteria often is a factor in the
decline of yield of a well. *Crenothrix* or similar organisms form a
slimy, gelatinous mass which accumulates on the screen or other
metal parts in the well. This substance may appear in the well
water as fine, short, reddish-brown filaments in a gelatinous
matrix. The mass not only blocks the screen but gives a
disagreeable taste and odor to the water and fosters aggravated
corrosion of the ferrous metal parts. Where experience has shown
that such organisms infect wells in an area, newly drilled and
serviced wells and pump installations should be thoroughly
sterilized as a preventive measure. Similarly, sterilization of
infected wells may, in some cases, destroy the organisms and
alleviate the problem.

16-13. Use of Explosives and Acidizing.—If all other possibilities are eliminated, the problem of reduced yield is usually one of encrustation of the screen or pack. The first step in rehabilitation is to scrape the inside of the screen with a steel disk on a drill stem or rod to break loose some of the incrusting materials which will settle to the bottom of the well. These scrapings should be examined to determine the nature and chemical composition of the encrustation. If the encrustation consists primarily of calcium, magnesium and iron carbonates, or iron hydroxides, rehabilitation using sulfamic or hydrochloric acid may be possible.

If iron and manganese compounds constitute over 20 percent of the material, other than included sands, corrosion should be suspected as a contributing factor. If the molecular ratio of $Fe(OH)_3$ (ferric hydroxide) to FeS (ferrous sulfide) is 3 to 1, sulfate-reducing bacteria are probably a contributing factor.

After the nature of the encrustation has been determined, re-examination of the video survey of the well is recommended to assess the extent and location of concentrated zones of encrustation.

If the encrustation does not appear to be heavy and the condition of the screen is believed to be good, a single string of 150 grains per meter (50 grains per foot) of Prima Cord, cut to the length of each screen section, can be fired within each screen section. In some instances, two shots may be advisable, but in no case should more than one string be fired at a time. Shooting will crack and break the encrustation and cause it to be more readily attacked by the subsequent acid treatment. The practice usually results in some encrustation being broken out of the screen and settling to the bottom of the well. This debris should be removed by bailing before acidizing.

Shooting should not be attempted in any well that has pumped appreciable sand or where evidence exists that the casing or screen might not be fully supported by earth materials.

Sonar Jet cleaning, a patented process, consists of shooting a series of small explosive charges in a well with a slight delay between the detonation of each successive charge. This process has not been extensively used by the Bureau of Reclamation, and reports of its effectiveness are variable. Sonar Jet cleaning should

be considered where other procedures have been poor or ineffective.
Whether the well is shot with Prima Cord or by Sonar Jet, it may
require acidization subsequent to shooting.

When complete rehabilitation is impractical, shooting alone will
often temporarily improve well performance. This procedure may
be done by lifting the pump head from the base and moving it to
one side. The Prima Cord is lowered alongside the column pipe
into the well and detonated within the screen or screens, and the
well is developed by placing the pump back in position and
rawhiding. The above procedure is a temporary expedient at best
and should be followed by a more complete program as soon as
conditions permit.

For successful well acidizing, the acids must be strong and the
products of the reaction must be soluble. The more commonly used
acids are muriatic or hydrochloric (HC1), sulfuric (H_2SO_4), and
sulfamic (amino sulfamic) (H_2NSO_3H).

Where iron or manganese constitute a significant part of the
encrustation and the pH of the acid solution reaches about 3, the
iron and manganese compounds form insoluble precipitates which
settle out. Under these conditions, chelating agents should be used
to keep the iron and manganese compounds in solution so that
they may be readily pumped from the well. Commonly used
chelating agents are:

- Citric Acid-(COOH)CH_2C(OH) (COOH) CH_2COOH
- Phosphoric Acid-H_3PO_4
- Tartaric Acid-HOOC (CHOH) COOH
- Rochelle Salt-$KN_aC_4H_4O_6$
- Glycolic Acid-(HOCH)$_2$COOH

Usual amounts of chelating agents used are:

0.45 kilogram (1 pound) of agent to 7 kilograms (15 pounds) of
sulfamic acid powder

0.9 kilogram (2 pounds) to each 3.8 liters (gallons) of 15-percent
HCl

1.8 kilograms (4 pounds) to each 3.8 liters (gallons) of H_2SO_4

Sulfuric acid is seldom used in acidizing wells because the reaction of sulfuric acid with calcium carbonate forms calcium sulfate (gypsum), which is relatively insoluble and difficult to remove from the well.

In addition, even when inhibited, sulfuric acid is aggressive to most metals, particularly copper alloys. It should be used only as a last resort when two or more treatments with less active acids have been unsuccessful, and the only other alternative is the construction of a new well.

Muriatic or hydrochloric acid was for years the most commonly used acidizing agent and is still popular. However, hydrochloric acid should not be used even in inhibited form on wells equipped with type 304 or 308 stainless steel screens, casing, or other components because it causes stress corrosion cracking of these alloys. The damage caused by the acid may not show up for some time after treatment of the well. However, use of hydrochloric acid would probably be safe with type 316 or 321 stainless steel.

Muriatic acid is available commercially in three strengths, but that most commonly used for well acidizing is 18° Baume or 27.92 percent hydrochloric acid. The acid is usually used full strength. The volume of water within each screen section is estimated, and 2 to 2-1/2 times as much acid as water is placed in the well through a plastic or black iron pipe within each screened section. Diethylthiourea or similar inhibitor in the amount of 0.2 kilogram per 380 liters (0.5 pound per 100 gallons) of acid is used as well as chelating agents if required. The acid is normally left in the well for 4 to 6 hours; the well is then surged with a surge block for 15 to 20 minutes at about 1-hour intervals, after which the solution is pumped or bailed from the well and properly disposed.

Work involving hydrochloric acid is hazardous and should be carried out by specialized well-servicing firms employing experienced personnel and special equipment.

Sulfamic acid is being used increasingly for well treatment. It is more costly than hydrochloric acid but is much more convenient, safer to use, and more easily shipped and stored. It is not as aggressive as hydrochloric acid and more time is required for an equivalent treatment. If the work is done by the regular maintenance crew or a local contractor, sulfamic acid is usually less expensive than treatment with hydrochloric acid.

The product of the reaction of sulfamic acid with calcium carbonate is calcium sulfamate, which is highly soluble and readily pumped from the well. If iron compounds make up a considerable part of the encrustation, a chelating agent should be used. Although not highly aggressive, sulfamic acid should not be used on copper alloy screens and other components without an inhibitor.

The solubility of sulfamic acid is as follows:

Water temperature (C)	Solubility (kg/L)
0	0.62
5	0.65
10	0.70
15	0.75
20	0.81

Sulfamic acid in the amount of 0.65 kilogram per liter of water is equivalent in reaction to 3.8 liters of 18° Baume (27.97 percent) hydrochloric acid and 1.4 kilograms of 15 percent hydrochloric acid per 3.8 liters of water.

In English units, sulfamic acid in the amount of 5.5 lb/gal of water is equivalent in reaction to 1 gallon of 18° Baume (27.97 percent) hydrochloric acid and 3 pounds of 15 percent hydrochloric acid per gallon of water.

Such sulfamic acid concentrations cannot be obtained as true solutions; however, a slurry can be mixed and pumped into the well. The mix proportions in 380 liters (100 gallons) of water include 135 kilograms (300 pounds) of sulfamic acid, 9 kilograms (20 pounds) of citric acid, 7.7 kilograms (17 pounds) of diethylthiourea, 1.5 kilograms (3.5 pounds) of pluronic F 68 or L 62, and 68 kilograms (150 pounds) of sodium chloride. The chemicals are dissolved and suspended in water at the surface in a volume equivalent to the volume of water within the well casing and screen. The slurry is dumped or poured into the well through a black iron or plastic pipe which initially extends to the bottom of

the well. The pipe is raised in 1.5- or 3-meter (5- or 10-foot) stages
as sufficient solution is added to displace an equivalent volume of
water in the well.

The solution is left in the well for 12 to 24 hours, during which
time it is surged for 15 to 20 minutes at hourly intervals. When
the solution in the well, when tested with litmus or similar paper,
shows a pH of between 6 and 7, the acid may be considered
exhausted. The solution should then be pumped from the well and
the well should be tested for yield and drawdown. If a marked
improvement is apparent, the well should then be redeveloped,
sterilized, tested as a new well, and put back into production. If
the improvement on initial testing after acidizing is relatively
slight, another treatment should be made.

When acidizing a well, a tank of concentrated sodium
bicarbonate solution should be available to permit neutralizing the
acid in event of an accident. In addition, workers should wear
protective rubber shoes, clothing, gloves, hood, and goggles. Until
all components are mixed in water, a filter respirator should also
be worn. Special equipment may be required (i.e., mixing tanks
and piping fabricated of black iron, plastic, or wood).

Encrustation may consist primarily of silica, clay particles, and
other materials resistant to normal acid treatment. Under such
circumstances, successful acidizing usually entails the use of
hydrofluoric and similar strong acid. In many instances, because
of the specialized nature of the treatment, the cost of such
treatment may approach or exceed that of a new well. In some
instances, where telescoping screen construction has been used,
removal and cleaning of the screen above ground may be preferable
to acidization.

16-14. Chlorine Treatment.—Where screen blockage is caused
by slime-forming organisms, chlorine gas may be an effective
treatment agent. Chlorine gas is dangerous to use without
experienced personnel and adequate equipment. Usually a 45- to
68-kilogram (100- to 150-pound) pressure cylinder of chlorine gas is
used. The cylinder is mounted on a scale to permit checking on the
rate of feed. The gas is fed through a plastic or black iron pipe
which extends to approximately the bottom of the well. The
bottom of the pipe should be centered with an appropriate device
so that the released gas does not impinge directly on the casing or
screen, and an approved feeder should be employed to avoid back

sucking. The cylinder is opened slowly, one full counter-clockwise
turn of the valve. Rate of discharge of the cylinder should not
exceed 18 kilograms per 24 hours.

When the cylinder is exhausted, the chlorine in the well can be
neutralized by adding sodium hydroxide or calcium hydroxide to
the water prior to pumping it to waste. Hypochlorite solutions are
cheaper, more convenient, and safer to use than gas but generally
are less effective.

Sufficient hypochlorite should be added to the water in the well
to give an estimated 1,000-mg/L chlorine content. The hypochlorite
is poured or pumped into the well, then thoroughly mixed and
diffused by surging for about 30 minutes. The solution is left in
the well for about 6 hours, during which it is surged for 15 to
20 minutes at hourly intervals, then pumped or bailed to waste.
Following these steps, the well can usually be adequately
redeveloped by rawhiding. This treatment can usually be carried
out without pulling the pump.

16-15. Rehabilitation of Rock Wells.—The previous
discussion has been primarily applicable to cased and screened
wells in unconsolidated materials. The fractures and other voids in
uncased rock wells may become clogged and sealed by deposition
products similar to encrustation of a well screen.

Hydrochloric acid of 18° Baume strength is commonly used full
strength to treat rock wells. A volume of acid equal to about
2.5 times the volume of water in the well is pumped or poured into
the well through a plastic or black iron pipe extending to the
bottom of the well. The pipe is raised as acid displaces the water
in the well. If the water-level in the well is within the casing, an
inhibitor should be used. The acid is permitted to remain in the
well for at least 6 hours, surged for 15 or 20 minutes at hourly
intervals, and then pumped to waste.

In some open hole rock wells, use of explosives has been more
effective than acid treatment, and at times a combination
treatment has been used in which acidization follows shooting.

Within the open hole, 4-1/2-kilogram (10-pound) shots of 50- to
60-percent dynamite are used at 1.5-meter (5-foot) intervals. Shots
should not be fired within 3 meters (10 feet) of a shale formation or
within 15 meters (50 feet) of the bottom of the casing. Shots are

fired separately beginning at the bottom of the open hole. After shooting is completed, the well should be bailed clean and developed.

In very hard rock, up to 45 kilograms (100 pounds) of explosive have been used in shots 3 to 3.5 meters (10 to 12 feet) apart. The amount of powder to use and the spacing is a matter of judgment based on experience. One or more test shots are advisable when operating in unfamiliar rocks.

Nitramine has also been used instead of dynamite. Although more expensive than dynamite, it is much safer and easier to handle. One can of nitramine is equivalent to 0.7 kilogram (1.6 pounds) of 50-percent or 0.45 kilogram (1 pound) of 60-percent dynamite.

On completion of acidizing or shooting work, the well should be thoroughly developed in the same manner as a new well.

16-16. Hydrofracturing.—In fractured rocks, yields can sometimes be increased by hydrofracturing. Hydrofracturing has been used in oil production since the late 1940's to increase yield and is also used in methane gas production. Its use in the water well industry began in the 1950's, and in recent years has seen expanded application. Hydrofracturing is replacing some other methods of increasing fracturing, such as using explosives or dry ice. In hydrofracturing, water and often some propping material (usually sand or very small plastic spheres) is injected into wells at high pressures. The high injection pressures clean out fines or increase spacing in existing fractures and often will separate previously closed fractures. The propping material holds the fractures open after the injection is stopped. Injection must occur in an uncased well, or the casing must be removed during the operation. Packers can be used to isolate zones for the hydrofracturing operations. Because of the specialized nature of much of the equipment and techniques required in hydrofracturing, it is generally advisable to contract hydrofracturing work to a contractor experienced in the method.

Hydrofracturing can also be used to increase capacity of rock media used for disposal of hazardous wastes or saline water; however, such applications require extreme care to avoid contaminating freshwater aquifers in the vicinity. Although generally used in igneous rocks, hydrofracturing can also be used in sandstones, limestone, or other rocks where secondary

permeability is present. It generally is not suitable in
unconsolidated or soft rocks. Hydrofracturing is used for
individual wells where initial yield is insufficient to supply daily
household use. In such cases, an increase in yield of even 2 liters
per minute (0.5 gallon per minute) may be sufficient to provide an
adequate supply. Hydrofracturing is also used in larger capacity
wells, in cases where yield has declined since initial operation. In
these cases, silting-in of fractures or dewatering of initial supply
fractures may have caused the decline in yield. Because other
factors may affect the decline in yield, however, possible
alternative causes such as biofouling (section 16-12) should be
thoroughly investigated before hydrofracturing is used.

Although contractors report a success ratio of 90 to 97 percent
in increasing well yield through hydrofracturing (Smith, 1989),
hydrofracturing may actually decrease well yield or even collapse
the borehole wall. In addition, damage to nearby wells, through a
decrease in water-levels or yields or production of fines, could
occur. Contamination may occur if fractures are opened to or near
the ground surface.

Determination of the causes of fracture orientation, which
depends on rock properties such as elasticity as well as tectonic
stresses, can aid in planning the hydrofracturing program. If the
three mutually perpendicular principal stresses are unequal, the
fracture is most likely to part along a plane perpendicular to the
least principal axis (Hubbert and Willis, 1972). Where local
normal faults exist, fractures tend to run parallel to the fault
strike. The least principal stress is generally horizontal, and the
fractures tend to be vertical. However, where folding and thrust
faulting are present, the least principal stress tends to be vertical,
and fractures tend to be horizontal (Smith, 1989).

Preliminary testing in the well can also provide useful
information. Packer tests at varying depths can aid in
determining local zones of water inflow. Knowledge of the fracture
orientation and spacing, which can be obtained from borehole
camera and some borehole geophysical logging surveys (chapter IV), can aid in determining location of packers to isolate the
zones of maximum potential fracturing.

Pressures used in hydrofracturing generally range from
7,000 to 20,000 kPa (1,000 to 3,000 lb/in^2). This pressure is
generally not enough to fracture solid rock, which may require

140,000 to 700,000 kPa (20,000 to 100,000 lb/in^2) (Smith, 1989). However, the lower pressure may be enough to lift blocks along bedding plane or open previously tight fractures.

16-17. Bibliography.—

Baski, Hank, 1987, "Hydrofracturing of Water Wells," *Water Well Journal*, June 1987, pp. 34-35.

Grange, J.W., and E. Lund, May 1969, "Quick Culturing and Control of Iron Bacteria," *Journal of the American Water Works Association*, vol. 6, No. 5, pp. 242-245.

Hubbert, M.K., and D.G. Willis, 1972, "Natural and Induced Fracture Orientation, Underground Waste Management and Environmental Implications," AAPG Memoir 18, American Association of Petroleum Geologists, Tulsa, Oklahoma, pp. 239-257.

Kuhlman, F.W., November 1959, "Corrosion of Iron in Aqueous Media," *Canadian Mining and Metallurgical Bulletin*, No. 52, pp. 713-729.

Moehrl, K.E., February 1961, "Corrosion Attack in Water Wells," *Corrosion*, vol. 17, No. 2, pp. 26-27.

Pallotta, Christopher, 1989, "Hydrofracturing: Realistic Solution to Low-Yield Wells." *Water and Wastewater International*, vol. 4, issue 3, June 1989, pp. 27-31.

Pennington, W.A., March 1965, "Corrosion of Some Ferrous Metals in Soil with Emphasis on Mild Steel and On Gray and Ductile Cast Iron," Bureau of Reclamation Chemical Engineering Branch Report No. ChE-26.

Ryznar, J.W., April 1944, "A New Index for Determining the Amount of Calcium Carbonate Scale Formed by Water," *Journal of the American Water Works Association*.

Smith, Stuart A., 1989, *Manual of Hydraulic Fracturing for Well Stimulation and Geologic Studies*.

Williamson, W.H., and D.R. Wooley, 1980, Hydraulic Fracturing to Improve the Yield of Bores in Fractured Rock, Dept. of National Development and Energy, Australian Water Resources Council, Research Project 78/98.

APPENDIX

International System (SI metric)/
U.S. Customary Conversion Tables

LENGTH

To convert from	To	Multiply by
angstrom units	nanometers (nm)	0.1
	micrometers (μm)	1.0×10^{-4}
	millimeters (mm)	1.0×10^{-7}
	meters (m)	1.0×10^{-10}
	mils	$3.937 \ 01 \times 10^{-6}$
	inches (in)	$3.937 \ 01 \times 10^{-9}$
micrometers	millimeters	1.0×10^{-3}
	meters	1.0×10^{-6}
	angstrom units (A)	1.0×10^{4}
	mils	0.039 37
	inches	$3.937 \ 01 \times 10^{-5}$
millimeters	micrometers	1.0×10^{3}
	centimeters (cm)	0.1
	meters	1.0×10^{-3}
	mils	39.370 08
	inches	0.039 37
	feet (ft)	$3.280 \ 84 \times 10^{-3}$
centimeters	millimeters	10.0
	meters	0.01
	mils	0.3937×10^{3}
	inches	0.3937
	feet	0.032 81
inches	millimeters	25.40
	meters	0.0254
	mils	1.0×10^{3}
	feet	0.083 33
feet	millimeters	304.8
	meters	0.3048
	inches	12.0
	yards (yd)	0.333 33

To convert from	*To*	*Multiply by*
yards	meters	0.9144
	inches	36.0
	feet	3.0
meters	millimeters	1.0×10^3
	kilometers (km)	1.0×10^{-3}
	inches	39.370 08
	yards	1.093 61
	miles	$6.213\ 71 \times 10^{-4}$
kilometers	meters	1.0×10^3
	feet	$3.280\ 84 \times 10^3$
	miles	0.621 37
miles	meters	$1.609\ 34 \times 10^3$
	kilometers	1.609 34
	feet	5,280.0
	yards	1,760.0
nautical miles (nmi)	kilometers	1.8520
	miles	1.1508

AREA

To convert from	*To*	*Multiply by*
square millimeters	square centimeters (cm^2)	0.01
	square inches (in^2)	1.550×10^{-3}
square centimeters	square millimeters (mm^2)	100.0
	square meters (m^2)	1.0×10^{-4}
	square inches	0.1550
	square feet (ft^2)	$1.076\ 39 \times 10^{-3}$
square inches	square millimeters	645.16
	square centimeters	6.4516
	square meters	6.4516×10^{-4}
	square feet	69.444×10^{-4}

To convert from	*To*	*Multiply by*
square feet	square meters	0.0929
	hectares (ha)	9.2903×10^{-6}
	square inches	144.0
	acres	$2.295\ 68 \times 10^{-5}$
square yards	square meters	0.836 13
	hectares	8.3613×10^{-5}
	square feet	9.0
	acres	$2.066\ 12 \times 10^{-4}$
square meters	hectares	1.0×10^{-4}
	square feet	10.763 91
	acres	2.471×10^{-4}
	square yards (yd^2)	1.195 99
acres	square meters	4046.8564
	hectares	0.404 69
	square feet	4.356×10^{4}
hectares	square meters	1.0×10^{4}
	acres	2.471
square kilometers	square meters	1.0×10^{6}
	hectares	100.0
	square feet	107.6391×10^{5}
	acres	247.105 38
	square miles (mi^2)	0.3861
square miles	square meters	$258.998\ 81 \times 10^{4}$
	hectares	258.998 81
	square kilometers (km^2)	2.589 99
	square feet	$2.787\ 84 \times 10^{7}$
	acres	640.0

VOLUME-CAPACITY

To convert from	*To*	*Multiply by*
cubic millimeters	cubic centimeters (cm^3)	1.0×10^{-3}
	liters (l)	1.0×10^{-6}
	cubic inches (in^3)	$61.023\ 74 \times 10^{-6}$

To convert from	*To*	*Multiply by*
cubic centimeters	liters	1.0×10^3
	milliliters (ml)	1.0
	cubic inches	$61.023\ 74 \times 10^{-3}$
	fluid ounces (fl oz)	33.814×10^{-3}
milliliters	liters	1.0×10^{-3}
	cubic centimeters	1.0
cubic inches	milliliters	16.387 06
	cubic feet (ft^3)	$57.870\ 37 \times 10^{-5}$
liters	cubic meters	1.0×10^{-3}
	cubic feet	0.035 31
	gallons	0.264 17
	fluid ounces	33.814
gallons	liters	3.785 41
	cubic meters	$3.785\ 41 \times 10^{-3}$
	fluid ounces	128.0
	cubic feet	0.133 68
cubic feet	liters	28.316 85
	cubic meters (m^3)	$28.316\ 85 \times 10^{-3}$
	cubic dekameters (dam^3)	$28.316\ 85 \times 10^{-6}$
	cubic inches	1,728.0
	cubic yards (yd^3)	$37.037\ 04 \times 10^{-3}$
	gallons (gal)	7.480 52
	acre-feet (acre-ft)	$22.956\ 84 \times 10^{-6}$
cubic miles	cubic dekameters	$4.168\ 189 \times 10^6$
	cubic kilometers (km^3)	4.168 18
	acre-feet	3.3792×10^6
cubic yards	cubic meters	0.764 55
	cubic feet	27.0
cubic meters	liters	1.0×10^3
	cubic dekameters	1.0×10^{-3}
	gallons	264.1721
	cubic feet	35.314 67
	cubic yards	1.307 95
	acre-feet	8.107×10^{-4}

To convert from	*To*	*Multiply by*
acre-feet	cubic meters	1233.482
	cubic dekameters	1.233 48
	cubic feet	43.560×10^3
	gallons	325.8514×10^3
cubic dekameters	cubic meters	1.0×10^3
	cubic feet	$35.314\ 67 \times 10^3$
	acre-feet	0.810 71
	gallons	$16.417\ 21 \times 10^4$
cubic kilometers	cubic dekameters	1.0×10^6
	acre-feet	$0.810\ 71 \times 10^6$
	cubic miles (mi^3)	0.239 91

TEMPERATURE

To convert	*To*	*Solve*
degrees Celsius (C)	degrees Kelvin (K)	$K = C - 273.15$
	degrees Fahrenheit (F)	$F = (C \times 1.8) + 32$
	degrees rankine (R)	$R = C \times 1.8 + 491.69$
degrees Kelvin (K)	degrees Fahrenheit (F)	$F = (K - 255.91) \times 1.8$
	degrees Celsius (C)	$C = K + 273.15$
	degrees rankine (R)	$R = K \times 1.8$
degrees Fahrenheit (F)	degrees Celsius (C)	$C = (F - 32)/1.8$
	degrees rankine (R)	$R = F - 459.69$
	degrees Kelvin (K)	$K = (F + 459.69)/1.8$
degrees rankine (R)	degrees Kelvin (K)	$K = R/1.8$
	degrees Celsius (C)	$C = (R/1.8) - 273.69$
	degrees Fahrenheit (F)	$F = R - 459.69$

ACCELERATION

To convert from	*To*	*Multiply by*
feet per second squared	meters per second squared (m/s²)	0.3048
	G's	0.031 08
meters per second squared	feet per second squared (ft/s²)	3.280 84
	G's	0.101 97
G's (standard gravitational acceleration)	meters per second squared	9.806 65
	feet per second squared	32.174 05

VELOCITY

To convert from	*To*	*Multiply by*
feet per second	meters per second (m/s)	0.3048
	kilometers per hour (km/h)	1.097 28
	miles per hour (mi/h)	0.681 82
meters per second	kilometers per hour	3.60
	feet per second (ft/s)	3.280 84
	miles per hour	2.236 94
kilometers per hour	meters per second	0.277 78
	feet per second	0.911 34
	miles per hour	0.621 47
miles per hour	kilometers per hour	1.609 34
	meters per second	0.447 04
	feet per second	1.466 67
feet per year (ft/yr)	millimeters per second (mm/s)	$9.665\ 14 \times 10^{-6}$

FORCE

To convert from	To	Multiply by
pounds	newtons (N)	4.4482
kilograms	newtons	9.806 65
	pounds (lb)	2.2046
newtons	pounds	0.224 81
dynes	newtons	1.0×10^{-5}

MASS

To convert from	To	Multiply by
grams	kilograms (kg)	1.0×10^{-3}
	ounces (avdp)	0.035 27
ounces (avdp)	grams (g)	28.349 52
	kilograms	0.028 35
	pounds (avdp)	0.0625
pounds (avdp)	kilograms	0.453 59
	ounces (avdp)	16.00
kilograms	kilograms (force)-second squared per meter ($kgf \cdot s^2/m$)	0.101 97
	pounds (avdp)	2.204 62
	slugs	0.068 52
slugs	kilograms	14.5939
short tons	kilograms	907.1847
	metric tons (t)	0.907 18
	pounds (avdp)	2000.0

To convert from	To	Multiply by
metric tons (tonne or megagram)	kilograms	1.0×10^3
	pounds (avdp)	$2.204\ 62 \times 10^3$
	short tons	1.102 31
long tons	kilograms	1016.047
	metric tons	1.016 05
	pounds (avdp)	2240.0
	short tons	1.120

VOLUME PER UNIT TIME FLOW

To convert from	To	Multiply by
cubic feet per second	liters per second (l/s)	28.316 85
	cubic meter per second second (m^3/s)	0.028 32
	cubic dekameters per day (dam^3/d)	2.446 57
	gallons per minute (gal/min)	448.831 17
	acre-feet per day (acre-ft/day)	1.983 47
	cubic feet per minute (ft^3/min)	60.0
gallons per minute	cubic meters per second	0.631×10^{-4}
	liters per second	0.0631
	cubic dekameters per day	5.451×10^{-3}
	cubic feet per second (ft^3/s)	2.228×10^{-3}
	acre-feet per day	4.4192×10^{-3}
acre-feet per day	cubic meter per second	0.014 28
	cubic dekameters per day	1.233 48
	cubic feet per second	0.504 17
cubic dekameters per day	cubic meters per second	0.011 57
	cubic feet per second	0.408 74
	acre-feet per day	0.810 71

VISCOSITY

To convert from	To	Multiply by
centipoise	pascal-second (Pa · s)	1.0×10^{-3}
	poise	0.01
	pound per foot-hour (lb/ft · h)	2.419 09
	pound per foot-second (lb/ft · s)	$6.719 \ 69 \times 10^{-4}$
	slug per foot-second (slug/ft · s)	$2.088 \ 54 \times 10^{-5}$
pascal-second	centipoise	1000.0
	pound per foot-hour	$2.419 \ 09 \times 10^{3}$
	pound per foot-second	0.671 97
	slug per foot-second	20.8854×10^{-3}
pound per foot-hour	pascal-second	$4.133 \ 79 \times 10^{-4}$
	pound per foot-second	$2.777 \ 78 \times 10^{-4}$
	centipoise	0.413 38
pound per foot-second	pascal-second	1.488 16
	slug per foot-second	31.0809×10^{-3}
	centipoise	$1.488 \ 16 \times 10^{3}$
centistokes	square meters per second (m²/s)	1.0×10^{-6}
	square feet per second (ft²/s)	$10.763 \ 91 \times 10^{-6}$
	stokes	0.01
square feet per second	square meters per second	9.2903×10^{-2}
	centistokes	9.2903×10^{4}
stokes	square meters per second	1.0×10^{-4}
rhe	1 per pascal-second (1/PA · s)	10.0

FORCE PER UNIT AREA
PRESSURE—STRESS

To convert from	To	Multiply by
pounds per square inch	kilopascals (kPA)	6.894 76
	[1]meters-head	0.703 09
	[2]mm of Hg	51.7151
	[1]feet of water	2.3067
	pounds per square foot (lb/ft^2)	144.0
	std. atmospheres	68.046x10^{-3}
pounds per square foot	kilopascals	0.047 88
	[1]meters-head	4.8826x10^{-3}
	[2]mm of Hg	0.359 13
	[1]feet of water	16.0189x10^{-3}
	pounds per square inch	6.9444x10^{-3}
	std. atmospheres	0.472 54x10^{-3}
short tons per square foot	kilopascals	95.760 52
	pounds per square inch (lb/in$_2$)	13.888 89
[1]meters-head	kilopascals	9.806 36
	[2]mm of Hg	73.554
	[1]feet of water	3.280 84
	pounds per square inch	1.422 29
	pounds per square foot	204.81
[1]feet of water	kilopascals	2.988 98
	[1]meters-head	0.3048
	[2]mm of Hg	22.4193
	[2]inches of Hg	0.882 65
	pounds per square inch	0.433 51
	pounds per square foot	62.4261

---

[1] Column of H_2O (water) measured at 4 °C.

[2] Column of Hg (mercury) measured at 0 °C.

To convert from	*To*	*Multiply by*
kilopascals	newtons per square meter (N/m²)	1.0x10³
	³mm of Hg	7.500 64
	⁴meters-head	0.101 97
	³inches of Hg	0.2953
	pounds per square foot	20.8854
	pounds per square inch	0.145 04
	std. atmospheres	9.8692x10⁻³
kilograms (f) per square meter	kilopascals	9.806 65x10⁻³
	³mm of Hg	73.556x10⁻³
	pounds per square inch	1.4223x10⁻³
millibars (mbar)	kilopascals	0.10
bars	kilopascals	100.0
std. atmospheres	kilopascals	101.325
	³mm of Hg	760.0
	pounds per square inch	14.70
	⁴feet of water	33.90

MASS PER UNIT VOLUME
DENSITY AND MASS CAPACITY

To convert from	*To*	*Multiply by*
pounds per cubic foot	kilogram per cubic meter (kg/m³)	16.018 46
	slugs per cubic foot (slug/ft³)	0.031 08
	pounds per gallon (lb/gal)	0.133 68
pounds per gallon	kilograms per cubic meter (kg/m³)	119.8264
	slugs per cubic foot	0.2325

---

³ Column of Hg (mercury) measured at 0 °C.

⁴ Column of H_2O (water) measured at 4 °C.

To convert from	To	Multiply by
pounds per cubic yard	kilograms per cubic meter	0.593 28
	pounds per cubic foot (lb/ft^3)	0.037 04
grams per cubic centimeter	kilograms per cubic meter	$1.0\text{x}10^3$
	pounds per cubic yard	$1.6856\text{x}10^3$
ounces per gallon (oz/gal)	grams per liter (g/l)	7.489 15
	kilograms per cubic meter	7.489 15
kilograms per cubic meter	grams per cubic centimeter (g/cm^3)	$1.0\text{x}10^{-3}$
	metric tons per cubic meter (t/m^3)	$1.0\text{x}10^{-3}$
	pounds per cubic foot (lb/ft^3)	$62.4297\text{x}10^{-3}$
	pounds per gallon	$8.3454\text{x}10^{-3}$
	pounds per cubic yard	1.685 56
long tons per cubic yard	kilograms per cubic meter	1328.939
ounces per cubic inch (oz/in^3)	kilograms per cubic meter	1729.994
slugs per cubic foot	kilograms per cubic meter	515.3788

VOLUME PER UNIT AREA PER UNIT TIME
[5]HYDRAULIC CONDUCTIVITY (PERMEABILITY)

To convert from	To	Multiply by
cubic feet per square foot per day	cubic meters per square meter per day $(\text{m}^3/(\text{m}^2 \cdot \text{d}))$	0.3048
	cubic feet per square foot per minute $(\text{ft}^3/(\text{ft}^2 \cdot \text{min}))$	$0.6944\text{x}10^{-3}$
	liters per square meter per day $(\text{l/m}^2 \cdot \text{d})$	304.8

---

[5] Many of these units can be dimensionally simplified. For example, $\text{m}^3/(\text{m} \cdot \text{d})$ can als be written m^2/d.

To convert from	*To*	*Multiply by*
	gallons per square foot per day (gal/(ft^2· d))	7.480 52
	cubic millimeters per square millimeter per day (mm^3/(mm^2· d))	304.8
	cubic millimeters per square millimeter per hour (mm^3/mm^2· h))	25.4
	cubic inches per square inch per hour (in^3/(in^2· h))	0.5
	cubic centimeters per square centimeters per second (cm^3/cm^2/s)	3.52x10^{-4}
gallons per square foot per day	cubic meters per square meter per day (m^3/(m^2· d))	40.7458x10^{-3}
	liters per square meter per day (l/(m^2· d))	40.7458
	cubic feet per square foot per day (ft^3/(ft^2· d))	0.133 68

VOLUME PER CROSS SECTIONAL AREA PER UNIT TIME [6]TRANSMISSIVITY

To convert from	*To*	*Multiply by*
cubic feet per foot per day (ft^3/(ft · d))	cubic meters per meter per day (m^3/(m · d))	0.0929
	gallons per foot per day (gal/(ft · d))	7.480 52
	liters per meter per day (l/(m · d))	92.903
gallons per foot per day	cubic meters per meter per day (m^3/(m · d))	0.012 42
	cubic feet per foot per day (ft^3/(ft · d))	0.133 68

---

[6] Many of these units can be dimensionally simplified. For example, m^3/(m· d) can als be written m^2/d.

INDEX

☆U.S. GOVERNMENT PRINTING OFFICE: 1995- 676-690